INTERNATIONAL SERIES OF MONOGRAPHS ON PHYSICS

International Series of Monographs on Physics

151. R. Blinc: *Advanced ferroelectricity*
150. L. Berthier, G. Biroli, J.-P. Bouchaud, W. van Saarloos, L. Cipelletti: *Dynamical heterogeneities in glasses, colloids and granular media*
149. J. Wesson: *Tokamaks, Fourth edition*
148. H. Asada, T. Futamase, P. Hogan: *Equations of motion in general relativity*
147. A. Yaouanc, P. Dalmas de Rotier: *Muon spin rotation, relaxation, and resonance*
146. B. McCoy: *Advanced statistical mechanics*
145. M. Bordag, G.L. Klimchitskaya, U. Mohideen, V.M. Mostepanenko: *Advances in the Casimir effect*
144. T.R. Field: *Electromagnetic scattering from random media*
143. W. Götze: *Complex dynamics of glass-forming liquids - a mode-coupling theory*
142. V.M. Agranovich: *Excitations in organic solids*
141. W.T. Grandy: *Entropy and the time evolution of macroscopic systems*
140. M. Alcubierre: *Introduction to 3+1 numerical relativity*
139. A. L. Ivanov, S. G. Tikhodeev: *Problems of condensed matter physics - quantum coherence phenomena in electron-hole and coupled matter-light systems*
138. I. M. Vardavas, F. W. Taylor: *Radiation and climate*
137. A. F. Borghesani: *Ions and electrons in liquid helium*
136. C. Kiefer: *Quantum gravity, Second edition*
135. V. Fortov, I. Iakubov, A. Khrapak: *Physics of strongly coupled plasma*
134. G. Fredrickson: *The equilibrium theory of inhomogeneous polymers*
133. H. Suhl: *Relaxation processes in micromagnetics*
132. J. Terning: *Modern supersymmetry*
131. M. Mariño: *Chern-Simons theory, matrix models, and topological strings*
130. V. Gantmakher: *Electrons and disorder in solids*
129. W. Barford: *Electronic and optical properties of conjugated polymers*
128. R. E. Raab, O. L. de Lange: *Multipole theory in electromagnetism*
127. A. Larkin, A. Varlamov: *Theory of fluctuations in superconductors*
126. P. Goldbart, N. Goldenfeld, D. Sherrington: *Stealing the gold*
125. S. Atzeni, J. Meyer-ter-Vehn: *The physics of inertial fusion*
123. T. Fujimoto: *Plasma spectroscopy*
122. K. Fujikawa, H. Suzuki: *Path integrals and quantum anomalies*
121. T. Giamarchi: *Quantum physics in one dimension*
120. M. Warner, E. Terentjev: *Liquid crystal elastomers*
119. L. Jacak, P. Sitko, K. Wieczorek, A. Wojs: *Quantum Hall systems*
117. G. Volovik: *The Universe in a helium droplet*
116. L. Pitaevskii, S. Stringari: *Bose-Einstein condensation*
115. G. Dissertori, I.G. Knowles, M. Schmelling: *Quantum chromodynamics*
114. B. DeWitt: *The global approach to quantum field theory*
113. J. Zinn-Justin: *Quantum field theory and critical phenomena, Fourth edition*
112. R.M. Mazo: *Brownian motion - fluctuations, dynamics, and applications*
111. H. Nishimori: *Statistical physics of spin glasses and information processing - an introduction*
110. N.B. Kopnin: *Theory of nonequilibrium superconductivity*
109. A. Aharoni: *Introduction to the theory of ferromagnetism, Second edition*
108. R. Dobbs: *Helium three*
107. R. Wigmans: *Calorimetry*
106. J. Kübler: *Theory of itinerant electron magnetism*
105. Y. Kuramoto, Y. Kitaoka: *Dynamics of heavy electrons*
104. D. Bardin, G. Passarino: *The Standard Model in the making*
103. G.C. Branco, L. Lavoura, J.P. Silva: *CP Violation*
102. T.C. Choy: *Effective medium theory*
101. H. Araki: *Mathematical theory of quantum fields*
100. L. M. Pismen: *Vortices in nonlinear fields*
99. L. Mestel: *Stellar magnetism*
98. K. H. Bennemann: *Nonlinear optics in metals*
94. S. Chikazumi: *Physics of ferromagnetism*
91. R. A. Bertlmann: *Anomalies in quantum field theory*
90. P. K. Gosh: *Ion traps*
87. P. S. Joshi: *Global aspects in gravitation and cosmology*
86. E. R. Pike, S. Sarkar: *The quantum theory of radiation*
83. P. G. de Gennes, J. Prost: *The physics of liquid crystals*
73. M. Doi, S. F. Edwards: *The theory of polymer dynamics*
69. S. Chandrasekhar: *The mathematical theory of black holes*
51. C. Møller: *The theory of relativity*
49. J. Wesson: Tokamaks, *Fourth edition*
46. H. E. Stanley: *Introduction to phase transitions and critical phenomena*
32. A. Abragam: *Principles of nuclear magnetism*
27. P. A. M. Dirac: *Principles of quantum mechanics*
23. R. E. Peierls: *Quantum theory of solids*

Dynamical Heterogeneities in Glasses, Colloids and Granular Media

Ludovic Berthier
Laboratoire des Colloides, Verres et Nanomateriaux, Université Montpellier II, France
Giulio Biroli
Institut de Physique Théorique, CEA Saclay, France
Jean-Philippe Bouchaud
Science & Finance, Capital Fund Management, Paris, France
Luca Cipelletti
LCVN, UMR 5587 Université Montpellier 2 and CNRS, France
Wim van Saarloos
Instituut-Lorentz, LION, Leiden University, The Netherlands

OXFORD
UNIVERSITY PRESS

OXFORD
UNIVERSITY PRESS

Great Clarendon Street, Oxford OX2 6DP

Oxford University Press is a department of the University of Oxford.
It furthers the University's objective of excellence in research, scholarship,
and education by publishing worldwide in

Oxford New York

Auckland Cape Town Dar es Salaam Hong Kong Karachi
Kuala Lumpur Madrid Melbourne Mexico City Nairobi
New Delhi Shanghai Taipei Toronto

With offices in

Argentina Austria Brazil Chile Czech Republic France Greece
Guatemala Hungary Italy Japan Poland Portugal Singapore
South Korea Switzerland Thailand Turkey Ukraine Vietnam

Oxford is a registered trade mark of Oxford University Press
in the UK and in certain other countries

Published in the United States
by Oxford University Press Inc., New York

First published 2011

British Library Cataloguing in Publication Data
Data available

Library of Congress Cataloging in Publication Data
Data available

Typeset by SPI Publisher Services, Pondicherry, India
Printed in Great Britain
on acid-free paper by
CPI Antony Rowe, Chippenham, Wiltshire

ISBN 978–0–19–969147–0

1 3 5 7 9 10 8 6 4 2

Preface

Many materials of industrial importance are amorphous, from window glasses and plastic bottles to emulsions, foams, and dense granular assemblies. Understanding the formation of these different types of disordered solids via the so-called glass and jamming transitions is a challenge that resisted a large research effort in condensed-matter physics over the last decades, and that is of interest to several fields from statistical mechanics and soft matter to material sciences and biophysics. From a fundamental point of view, the question is to know whether the sudden, but continuous freezing is due to a genuine underlying phase transition, or if it is a mere crossover with little universality in the driving mechanisms.

In recent years, evidence has mounted that the dynamical slowing down of supercooled liquids, colloids and granular media might indeed be related to the existence of genuine phase transitions, but of very peculiar nature. Contrasting with usual phase transitions, the dynamics of these materials dramatically slows down with nearly no changes in their conventional structural properties. One of the most interesting consequences of these ideas is the existence of dynamical heterogeneities, which have been discovered to be, in the space-time domain, the analog of critical fluctuations in standard phase transitions. Dynamic heterogeneity is now recognized to be a key feature of glassy dynamics, after a major research effort from a large community of researchers originating from a broad spectrum across condensed-matter physics, from liquid-state theorists to experimentalists working on foams or grains. Yet, there is no textbook specifically dedicated to this large body of work and research community, and we felt that this gap needed to be filled.

The present book provides a broad and up-to-date overview of the current understanding of dynamic heterogeneity in glasses, colloids and granular media. It is multidisciplinary in nature. It contains chapters dedicated to theory, numerical simulations and experiments. Its content spans from the physics of molecular glass formers to soft glassy materials, foams and grains and covers both equilibrium aspects related to the glass transition, and non-equilibrium features such as aging phenomena. The book contains formal chapters about recent theoretical developments and very phenomenological ones concerned with more practical and experimental aspects.

Such a broad scope would have been difficult to cover by a single author. We have therefore mustered different scientists, who have all made important contributions to the field. The book in fact originated from a workshop organized by three of us and held in the Lorentz Center in Leiden in September, 2008. The workshop had an unusual format: relatively long, in-depth oral contributions were systematically followed by a summary of the presented work and a critical discussion by a "discussant". Although standard in economics for example, this format is not very common in physics and should be encouraged. It has indeed fostered a very animated and insightful meeting.

We have tried to have the book reflect the special atmosphere of the conference by insisting that most chapters should be jointly written by researchers working on closely related topics, but usually with different points of view, backgrounds, or interests, and sometimes from competing groups! Out of this original construction, where some consensus had to be reached within each chapter, has emerged a series of contributions that has no equivalent in the published literature. The resulting book has, we believe, a much broader, equilibrated and lively perspective than conventional review articles.

A particular effort was made in each chapter towards explaining the topic in very pedagogical terms, such that a broad range of readers—graduate students and more experienced researchers, experimentalists and theoreticians—can benefit from the book. The book is hopefully well suited to people who want to learn and discover what dynamic heterogeneities are and why they are important and useful. It is also meant to be a valuable piece for researchers who already know the field and want to understand or delve into a more detailed aspect they know less about, or bring forth some experiments or simulations they are unaware of. Each chapter contains a large number of references, making the book useful also for a literature search. Finally, we made sure that each chapter can be read independently from the rest of the book, while making many cross-references between chapters to guide the readers.

The book opens with yet another unusual item: the parallel interviews of four condensed-matter luminaires: Jorge Kurchan, Jim Langer, Tom Witten and Peter Wolynes. We have asked them to answer, in an informal way, eleven questions about the glass transition and dynamics heterogeneities, the last one being: *If you met an omniscient God and were allowed one single question on glasses, what would it be?* The result is both delightful and insightful, and we hope the reader will enjoy these lively snippets of science as much as we did.

We hope that this book will fulfill a useful function in bringing together, within a single cover, most of the recent developments on dynamic heterogeneity studies. We would like to thank all the authors for their efforts and care in preparing their chapters, all the referees who accepted to review the chapters and helped the authors to produce a text of high scientific quality and rigor. We also thank the speakers, discussants and participants of the Lorentz Center workshop in Leiden in September 2008 who contributed to a successful meeting and eventually gave birth to this book.

L. Berthier
G. Biroli
J.-P. Bouchaud
L. Cipelletti
W. van Saarloos
Montpellier, Paris, Leiden
October 2010

Contents

List of contributors

Christiane ALBA-SIMIONESCO
Laboratoire Léon Brillouin, UMR 12, CEA-CNRS, 91191- Saclay, France

Jean-Louis BARRAT
Laboratoire Interdisciplinaire de Physique, Université Joseph Fourier Grenoble 1 et CNRS, 140 rue de la physique – Domaine Universitaire, BP 87–38402 Saint Martin d'Hères cedex, France

Ludovic BERTHIER
Laboratoire Charles Coulomb, Université Montpellier 2 and CNRS, UMR 5221, 34095 Montpellier, France

Giulio BIROLI
Institut de Physique Théorique, CEA, IPhT, 91191 Gif sur Yvette, France and CNRS URA 2306, France

Jean-Philippe BOUCHAUD
Science & Finance, Capital Fund Management 6-8 Bd Haussmann, 75009 Paris, France

Luca CIPELLETTI
Laboratoire Charles Coulomb, Université Montpellier 2 and CNRS, UMR 5221, 34095 Montpellier, France

Federico CORBERI
Dipartimento di Fisica E. R. Caianiello, Italy

Leticia F. CUGLIANDOLO
Université Pierre et Marie Curie - Paris VI, Laboratoire de Physique Théorique et Hautes Energies, 4 Place Jussieu, Tour 13, 5émeétage, 75252 Paris Cedex 05, France

Olivier DAUCHOT
SPEC, CEA-Saclay, URA 2464 CNRS, 91 191 Gif-sur-Yvette, France

Douglas J. DURIAN
University of Pennsylvania, Department of Physics and Astronomy, Philadelphia, PA 19104-6396, USA

Silvio FRANZ
Université Paris Sud, CNRS, LPTMS, UMR8626, Orsay F-91405, France

Juan P. GARRAHAN
School of Physics and Astronomy, University of Nottingham, Nottingham NG7 2RD, UK

Peter HARROWELL
School of Chemistry, University of Sydney, Sydney NSW 2006, Australia

Nathan ISRAELOFF
Department of Physics, Northeastern University, Boston, Massachusetts 02115, USA

Robert L. JACK
Department of Physics, University of Bath, Bath BA2 7AY, UK

Jorge KURCHAN
PMMH, ESPCI, 10 rue Vauquelin, CNRS UMR 7636 , Paris, France 75005

François LADIEU
Service de Physique de l'Etat Condensé (CNRS URA 2464), DSM/IRAMIS CEA Saclay, Bat. 772, Gif-sur-Yvette, F-91191 Cedex, France

James S. LANGER
Physics Department, University of California, Santa Barbara, California, 93106, USA

Anaël LEMAÎTRE
Université Paris-Est {Navier {CNRS UMR 8205 {ENPC {LCPC 2 allée Képler, 77420 Champs-sur-Marne, France

Denis L'HÔTE
Service de Physique de l'Etat Condensé (CNRS URA 2464), DSM/IRAMIS CEA Saclay, Bat. 772, Gif-sur-Yvette, F-91191 Cedex, France

Andrea J. LIU
Department of Physics and Astronomy, University of Pennsylvania, Philadelphia, PA 19104, USA

Sidney R. NAGEL
James Franck Institute, The University of Chicago, Chicago, Illinois, 60637, USA

Ranko RICHERT
Department of Chemistry and Biochemistry, Arizona State University, Tempe, Arizona 85287, USA

Guilhem SEMERJIAN
LPTENS, Unité Mixte de Recherche (UMR 8549) du CNRS et de l'ENS associéëa l'université Pierre et Marie Curie, 24 Rue Lhomond, 75231 Paris Cedex 05, France

Peter SOLLICH
King's College London, Department of Mathematics, Strand, London WC2R 2LS, UK

Gilles TARJUS
LPTMC, CNRS-UMR 7600, Université Pierre et Marie Curie, 4 Place Jussieu, 75252 Paris cedex 05, France

Cristina TONINELLI
LPMA Univ.Paris VI-VII, CNRS UMR 7599, Case courrier 188, 4 Place Jussieu, 75252 Paris Cedex 05, France

Martin van HECKE
Kamerlingh Onnes Laboratory, Universiteit Leiden, Postbus 9504, 2300 RA Leiden, The Netherlands

Wim van SAARLOOS
Instituut-Lorentz, LION, Leiden University, P.O. Box 9506, 2300 RA Leiden, The

Netherlands. Present address: FOM Foundation, P.O. Box 3021, 3502 GA Utrecht, The Netherlands

Eric R. WEEKS
Department of Physics, Emory University; Mail stop 1131/002/1AB 400 Dowman Dr., Atlanta GA 30322-2430, USA

Thomas A. WITTEN
The James Franck Institute and The Department of Physics, The University of Chicago, 929 East 57th Street, Chicago, Illinois 60637, USA

Peter G. WOLYNES
Department of Chemistry and Biochemistry, University of California at San Diego, La Jolla, California 92093-0371, USA

Matthieu WYART
Physics Department, New York University, 4 Washington Place, New York, NY 10003, USA

Hajime YOSHINO
Department of Earth and Space Science, Faculty of Science, Osaka University, Toyonaka 560–0043, Japan

1
Scientific interview

Jorge KURCHAN, James S. LANGER,
Thomas A. WITTEN and Peter G. WOLYNES

Abstract

Four leading scientists in the field of the glass transition answer to a series of questions formulated by the Editors. No specific format has been imposed to their answers: the scientists were free to answer or not to any given question, using how much space they felt appropriate. After a first round of answers, each author was given the opportunity to read the answers by his colleagues and to adjust his own answers accordingly.

The questions were:

- Q1) In your view, what are the most important aspects of the experimental data on the glass transition that any consistent theory explain? Is dynamical heterogeneity one of these core aspects?
- Q2) Why should we expect anything universal in the behavior of glass-forming liquids? Is the glass-transition problem well defined? Or is any glassy liquid glassy in its own way?
- Q3) In spin-glasses, the existence of a true spin-glass phase transition has been well established by simulations and experiments. Do you believe that a similar result will ever be demonstrated for molecular glasses?
- Q4) Why are there so many different theories of glasses? What kind of decisive experiments do you suggest to perform to rule out at least some of them?
- Q5) Can you briefly explain, and justify, why you believe your pet theory fares better than others? What, deep inside, are you worried about, that could jeopardize your theoretical construction?
- Q6) In the hypothesis that RFOT forms a correct skeleton of the theory of glasses, what is missing in the theoretical construction that would convince the community?
- Q7) Exactly solvable mean-field glass models exhibit an extraordinary complexity requiring impressive mathematical tools to solve them. Are you worried that for

both the spin and structural glass problems it proves so hard to establish the validity of mean-field concepts in finite dimensions? Why are finite-dimensional versions of mean-field glasses always behaving in a very non-mean-field manner?

- Q8) In your view, do the recent ideas and experimental developments concerning jamming in granular media and colloids contribute to our understanding of molecular glasses, or are they essentially complementary?
- Q9) If a young physicist asked you whether he or she should work on the glass problem in the next few years, would you encourage him or her and if so, which aspect of the glass problem would you recommend him or her to tackle? If no, what problem in condensed-matter theory should he or she tackle instead? If yes, what particular aspect of the glass problem seems the most exciting at present?
- Q10) In twenty years from now, what concepts, ideas or results obtained on the glass transition in the last twenty years will be remembered?
- Q11) If you met an omniscient God and were allowed one single question on glasses, what would it be?

1.1 Jorge Kurchan answers

Note: only references to papers that are difficult to identify are given here. For a general presentation, see the review article (Cavagna, 2009).

Q1) In your view, what are the most important aspects of the experimental data on the glass transition that any consistent theory explain? Is dynamical heterogeneity one of these core aspects?

The mere existence of heterogeneities is not in itself an additional fact, but a mathematical necessity: there have to be heterogeneities if there is a time scale that grows faster than Arrhenius in a system of soft particles. The only way a system can construct zero modes or higher barriers is through cooperativity. Of course, if one has access to the detailed space-time form of the heterogeneities, then this is extra information.

A model of fragile glasses should capture the relation between growing time scales and vanishing entropy. I would add that fluctuation–dissipation relations should reproduce the Lennard-Jones results of Barrat and Berthier Barrat and Berthier (2002), a useful discriminant since the precise form of these relations is model dependent. All these aspects are subject to the uncertainties associated with short times.

If we make a multidimensional scatter plot of the parameters describing existing glass formers, we shall find that there are correlations. Those between fragility, specific heat and stretching exponents have been discussed by Wolynes and coworkers, and there must be others that I do not know. To the extent that they are significant, correlations must be explained. If they are imperfect, this has to be explained too.

Having said this, I think that all in all it will be the internal logic and the tractability of the theory—*the fact that the objects it invokes and relates have a clear definition*—that will make it satisfactory.

Q2) Why should we expect anything universal in the behavior of glass-forming liquids? Is the glass-transition problem well defined? Or is any glassy liquid glassy in its own way?

Here one should separate two levels. The word "glassy" is used (a) in the general sense of slow (or *slowing*) dynamics. (b) for "fragile" glasses with their rather specific features of entropy and time scales.

The general sense (a) includes systems such as the two- and three-dimensional spin-glasses (including those having Potts or p-spin interactions), "dirty" ferromagnets, crystal ripening and defect annealing, quasicrystal annealing, strong and fragile glasses, and also kinetically constrained models, frustration-limited domains, "Backgammon", "car-parking", Tower of Hanoi, and a myriad other models. The class of fragile glasses (b) is much more specific: it includes a group of systems having slowing down of dynamics in a characteristic (non-power-law) super-Arrhenius way, with what seems to be a concomitant decrease in entropy. Polymers and some molecular glasses are of this kind.

In the general class (a) there might be many ways of being glassy. One might still expect some universal features that are just a consequence of the dynamics being slow, or entropy production small. There are relations between time scale increase and dynamics correlations, bounds for the violations of the fluctuation–dissipation relation, generic features of response of the system to driving, etc., that apply to every slow system. We would like to know much more, as this might be a front where out-of-equilibrium thermodynamics might develop—it being less ambitious than the generic non-equilibrium problem.

Class (b) is much more restrictive, to the point that some consider it empty. Super-Arrhenius behavior—the logarithm of the time scale growing faster than linearly in the inverse temperature—implies growing dynamic *and static* length scales (see Q1), and this leads to the search of some form of hidden order, with perhaps complete ordering eventually avoided. Here, the expectation is that things are much more universal in the sense of applying to fewer systems but for those, essentially in the same way. Nobody will be happy with a mechanism, an order parameter or a length scale, that is defined differently for polymers and a monoatomic glasses.

Q3) In spin-glasses, the existence of a true spin-glass phase transition has been well established by simulations and experiments. Do you believe that a similar result will ever be demonstrated for molecular glasses?

The question whether there is a spin-glass transition in a magnetic field is still very much unresolved, and is the object of an impassioned and contradictory literature. The spin-glass example is interesting, as it illustrates a new development that we are witnessing: simulation time scales are for the first time overlapping experimental ones. What in fact we discover is that there is quite a good continuity between experiments and simulations, *and that there are issues that neither will resolve.*

To illustrate how simulations may have a paradoxical effect, consider the example of the JANUS aging simulations The JANUS (2008). They measure two-time correlation functions of a 3D Edwards–Anderson model, up to times corresponding 0.1 s of experiments. They propose a form for the long-time correlations:

$$C(t, t_{\rm w}) = A(t_{\rm w}) \left(1 + \frac{t}{t_{\rm w}}\right)^{-1/\alpha(t_{\rm w})} \tag{1.1}$$

The factor $A(t_{\rm w})$ is familiar from experimental fits: it is important—and unknown—whether it goes to zero or not as $t_{\rm w} \to \infty$. If $A(t_{\rm w})$ goes to zero with $t_{\rm w}$ it means that the Edwards–Anderson parameter is zero (in a ferromagnet $A(t_{\rm w}) \to 0$ would be interpreted as evidence for Kosterlitz–Thouless-like order). If this were the case for spin-glasses, *neither* the droplet *nor* the replica symmetry-breaking scenario would apply in three dimensions, so it is quite an issue. Up to now it was thought that the time dependence of $A(t_{\rm w})$ in experiments might be a consequence of the spins being not quite Ising-like, but the JANUS work has proven that similar results are observed in simulations, where the nature of the spins is obviously under control. These simulations cannot resolve conclusively the value of $\lim_{t_{\rm w}\to\infty} A(t_{\rm w})$. This might sound like a paradox, but what they have done, in "validating" the experimental results, is to prove that even experimental times are too short to resolve the issue.

In fact, it has been argued on the basis of extrapolating simulation results that the longest correlation lengths achieved in spin-glass experiments are about twenty "lattice sites" (see Fig. 1 of Berthier and Bouchaud (2002)). In conclusion: simulations have convinced us that experimental times are short.

(Somebody may ask: why should we care, then, about longer times and length scales? Is it legitimate to take more seriously a length because we think it would ideally continue to grow in times beyond the experimental ones?)

Q4) Why are there so many different theories of glasses? What kind of decisive experiments do you suggest to perform to rule out at least some of them?

As I mentioned above, in many systems (e.g. colloids, spin glasses) simulations have quite literally caught up with experiments. Note how different the situation is in the field of strongly interacting electrons, where nature thermalizes easily what computers cannot simulate—a consequence of the fact that nature often solves the "sign problem" much better than computers, while it is as bad as computers in implementing glassy dynamics.

A new, decisive experiment, will probably be a thought experiment, somewhat like question Q11. The one we have suggested with D. Levine is to consider a large snapshot of a glass configuration, as equilibrated as possible. Take patches of size ℓ and look for places where they repeat to some significant precision. Is there a length $\ell_{\rm o}$ such that there is a crossover from a high frequency of repetition for patch sizes $\ell < \ell_{\rm o}$ to an exponentially low, essentially random frequency of repetition for $\ell > \ell_{\rm o}$? Does $\ell_{\rm o}$ grow as the glass ages? Is it infinite in an equilibrium ideal glass (as it is in all other forms of order we know: crystalline, quasicrystalline, amorphous tilings...)? Note that this definition of the length $\ell_{\rm o}$ is model independent, cfr. the end of Q2.

Q5) and Q6) Can you briefly explain, and justify, why you believe your pet theory fares better than others? What, deep inside, are you worried about, that could jeopardize your theoretical construction? In the hypothesis that RFOT forms a correct skeleton of the theory of glasses, what is missing in the theoretical construction that would convince the community?

In the glass literature there are plenty of phenomenological arguments, of useful representations, of metaphors—sometimes accompanied by a model that illustrates them. RFOT is unique in that it has some microscopic grounds, it is quite constrained (we cannot add or remove pieces at will), and, above all, has time and again given results that were not expected. Consider the following (the "you" here is a rhetoric device, it refers to no-one in particular):

- You read Derrida's paper on the random energy model (REM) and you realize that the transition is precisely the one proposed by Kauzmann for an ideal glass (at T_{K}), with the random energy levels playing the role of states.
- You are unsatisfied with the REM because it is somewhat *ad hoc*. So, recognizing that the REM is an example of the *one-step replica symmetry-breaking* kind, you look in the literature for other more microscopic models of this kind. You find the mean-field Potts glass, spin glasses with multispin interactions, etc.
- These models being microscopic, you may as well study their dynamics. Remarkably, in the high-temperature phase, the dynamics turns out to be described by the mode-coupling (MCT) equations, which are currently studied for glasses. There is a mode-coupling transition at a certain $T_{\mathrm{d}} > T_{\mathrm{K}}$, known already to be an artifact of the approximation. Within your perspective you understand perfectly well why the MCT transition has to go away in finite dimensions, and only one at the Kauzmann transition may remain—thus forcing you to identify the point at which viscosity diverges T_{o} with the thermodynamic temperature T_{K}.
- You are, however, uncomfortable with having used models with quenched disorder, so you try mean-field "glass" models without quenched disorder. You find that almost every model you can construct has a static and dynamic behavior just like the ones you have studied, something you can prove within mean-field. A new surprise is that often an isolated deep state appears that is given by number theoretical properties (see Marinari, *et al.* (1994) for a number-theoretic solution and Marinari, *et al.* (1995) for the mean-field version of the fully frustrated model): you have unexpectedly encountered what looks like a mean-field version of the crystalline state.
- Now that you have a microscopic model, you are not forced to restrict the dynamics to the high-temperature phase. So, you quench the models to low temperatures, and you find that the system does not equilibrate: it "ages", just like true glasses. When you look at the properties of observables, you discover that the slow fluctuations behave as if they were "thermalized" in an effective temperature of the order of that of the glass transition. You discover that "fictive temperatures" have been around since the 1940s, and it is likely that what you have discovered is precisely that—minus the exhilarating freedom of

phenomenological discourse. In particular, you see explicitly that the effective temperatures are *not* just a consequence of the system staying in a pre-quench configuration: *"hot" backbones are reconstructed time and again during aging.* You also study how the system responds to forces that do work on it: you find the generic phenomenology of "shear thinning" of supercooled liquids, and in some cases you can explain the much more rare "shear thickening" Sellitto and Kurchan (2005) of certain glasses.

- Now you go back to the energy and free-energy landscape, something that you could not do in a pure mode-coupling context. You find that the main features (importance and location of saddles, marginality, etc.) are extremely close to those discussed many years ago by Goldstein, to the point that you wonder whether he secretly knew the solution you have.

So far, you are encouraged by what seems an amazing unification, such as happen in other, more fortunate fields of physics. In fact, you begin to think that a substantial part of the twentieth-century phenomenological thinking was a non-mathematical construction of a mean-field picture. This also implies that the mean-field picture inherits the criticisms to these ideas. At this point, however, you begin to encounter difficulties. You know that the dynamic (MCT) transition you have obtained is a mean-field illusion, one that has pre-empted (and thus hidden from you) the ideal glass one. Furthermore, stable metastable states of free-energy density higher than the fundamental (as you invoked to define configurational entropy) are somewhat arbitrarily defined beyond mean-field. You conclude that an important part of the theory of glassy dynamics will be, unfortunately, *even qualitatively* beyond mean-field.

- Technically speaking, this means that you have to calculate corrections to the dynamics that are *non-perturbative* in the mean-field parameter: these are the "activated processes". A true calculation of these is beyond everyone's capabilities, but an estimate for a *finite-size* mean-field system gives you a time scale that diverges at T_K in a way that does not quite conform to a Vogel–Fulcher law. You need to move more resolutely into finite dimensions.
- More trouble: when you attempt to simulate finite-dimensional versions of the very models that inspired you (random Potts, p-spin) nothing like a Kauzmann transition appears. Worse: some models without quenched disorder that worked well at the mean-field level (e.g. the fully frustrated Marinari, *et al.* (1994, 1995) model) becomes quite different from a glass in finite dimensions.

  Your problem is not that you do not have three-dimensional models of fragile glasses without quenched disorder (there are plenty, hard spheres for example), but that these do not interpolate easily with a mean-field version that you can control (see last question in Q7). An example that works, studied by Dotsenko Dotsenko (2004) is that of particles in d dimensions interacting with an oscillatory (cosine) potential, with, say, a finite range cutoff. As the range tends to infinity, one obtains exactly a case of mean-field RFOT.
- If you are interested in thermodynamics (as opposed to dynamics) you may still use a mean-field approach with spatial degrees of freedom, somewhat in the

same spirit of the local mean-field approximations for ferromagnetic systems. The numbers obtained for the glass transition and the configurational entropy are quite encouraging. The whole picture for the thermodynamics of hard-sphere glasses is also quite satisfactory (see, however, the caveat in Q7(b)). In the same spirit, you can work with variable-range interactions, with the long-range limit being the mean-field theory.

- But for true dynamics with "activated processes" you need to plunge into true finite dimensions. Because you cannot do anything analytically, you work phenomenologically, with the mean-field theory in mind. (Your phenomenologist friends, whom you so criticized before, are gloating.) Above the Kauzmann temperature the metastable (now unstable) states of mean-field break into pieces, the "mosaics". One invokes a "library" of mosaics (with, it turns out, extremely brief "books"—the size of a Haiku) that can occupy a given volume, and argues on the basis of an interplay between the entropy associated to their multiplicity and their free energy. You obtain in this way a form for the relaxation time, and suggestions of relations between parameters (cfr. Q1) that turn out to be rather well respected, and, much more importantly, sometimes unexpected.
- One last worry. For temperatures at which the dynamics has virtually ground to a halt, you interpret this as a consequence of a particle cooperation that extends to a few (six at best, three in the computer) interparticle distances. It seems that many models will be able to achieve this, without a true coherence length that is building up—and how is one to distinguish? Are you not defending the mosaics on the basis of what they would become, if given unrealistically large times?

This should not be the end of the story. The mosaic phenomenology may be considered as a promising start, but a lot will depend on to what extent finite-dimensional RFOT can rise above phenomenology in practice. RFOT is a remarkable building, with the upper floor made of wood for lack of budget. The fact that the top floor stands out as a different, temporary construction is just a proof of the ambitious character of the whole enterprise.

Q7a) Exactly solvable mean-field glass models exhibit an extraordinary complexity requiring impressive mathematical tools to solve them. Are you worried that for both the spin and structural glass problems it proves so hard to establish the validity of mean-field concepts in finite dimensions?

This is worrying. Mean-field theory for fully connected systems is quite hard, both statically and dynamically. Extended to randomly dilute (Viana–Bray-like) systems, it becomes daunting. Even analytic results need considerable numerical effort to extract numbers from them. What will become of this in finite dimensions? Who is going to do this, and if somebody can, who will be able to read it? This is all the more worrying because the force of RFOT in the mean-field version is that its pieces are quantitatively defined, so it is disappointing if in finite dimensions it loses this edge.

Q7b) Why are finite-dimensional versions of mean-field glasses always behaving in a very non-mean-field manner?

It would seem that finite-dimensional models either keep you dynamically away from their Kauzmann transition (if you want to approach it in equilibrium), or they allow you to get close and then you see that it is not as you expected it: either there is no transition or it is spin-glass-like (there are many articles on this, by S. Franz, Binder and coworkers, Katzgraber and Young, on variants of disordered p-spin and Potts models). Does this mean that the Kauzmann transition is avoided in finite dimensions, mirroring the situation of the MCT dynamic transition? In other words, is it possible that non-perturbative corrections to the equilibrium calculation destroy the Kauzmann transition? If this is so, we do not know through which mechanism this would happen. (Years ago, Stillinger proposed one, but it relies on the identification of "state" with""inherent structure".)

Q8) In your view, do the recent ideas and experimental developments concerning jamming in granular media and colloids contribute to our understanding of molecular glasses, or are they essentially complementary?

Systems of hard, frictionless particles in mechanical equilibrium are hypo- or isostatic, the static equations are either underdetermined or just determined—like a table with two or three legs touching the floor. From this, one argues that they are extremely sensitive to perturbations, the breaking of a particle contact has a far-reaching effect.

The sensitivity mentioned above is manifested in the appearance of many soft vibration modes (which become manifest if the pressure is slightly released or the potential is softened), and divergent length scales in the hard/jammed limit. In conclusion, the infinite-pressure point for a system of *hard* spheres is a critical point. The question is: is this critical point, and its divergent lengths, responsible for the "glass order", or is it something that *happens* to a system that is amorphous for other reasons?

The example of a FCC crystal of spheres with very small polydispersity is eloquent. As soon as polydispersity is turned on, there is a proliferation of soft modes beyond the usual acoustic modes of the crystal. At infinite pressure we have a critical point just as in an amorphous system. And yet, the nature of order that makes the system solid is that of a crystal, as soon as the pressure is decreased slightly. This suggests that the same is true for amorphous systems: the nature of the order that makes the system a solid at finite pressures is not related to the "jamming" criticality as found at infinite pressure. This is not to say that this criticality may not have important consequences for a system at finite pressures, and that it may be especially relevant for granular and even colloidal matter.

Q9) If a young physicist asked you whether he or she should work on the glass problem in the next few years, would you encourage him or her and if so, which aspect of the glass problem would you recommend him or her to tackle? If no, what problem in condensed-matter theory should he or she tackle instead? If yes, what particular aspect of the glass problem seems the most exciting at present?

"If the nature of the present impasse you fully understand, and a contribution you still feel you can make, then go for it you must."

Q10) In twenty years from now, what concepts, ideas or results obtained on the glass transition in the last twenty years will be remembered?

It is difficult to say, as in the glass field an analytical success often leads to a relapse into even more vigorous phenomenological discourse, the case of effective temperatures is a case in point. There is a reason for this: one does not accept easily that something as familiar as glasses, which are not even quantal, should require complicated mathematics.

Is it possible that there will be new developments starting from the theory of low-dimensional amorphous order, as in non-periodic tilings? It is amazing how little attention the glass community has paid to lessons that could be learned from quasicrystals and in general non-periodic systems. Perhaps rightly so, I do not know.

In any case, my feeling is that in twenty years time many perceived contradictions between different pictures may be seen as only apparent. If you say *"the glass transition is only dynamic"*, this might stir controversy. Instead: *"I understand that a slowing time scale requires a growing equilibrium correlation length, we know that in real systems the dynamics has slowed down considerably by the time this correlation length is three, so I wish to understand how this comes about at such modest equilibrium correlation lengths"*—will be accepted by everyone. Similarly, I would expect that mode-coupling theory will be eventually accepted as part of RFOT, something that has not yet happened, perhaps for sociological reasons.

Q11) If you met an omniscient God and were allowed one single question on glasses, what would it be?

Are there general, non-trivial principles in non-equilibrium thermodynamics (perhaps for the restricted case of slow dynamics)? What I have in mind is something like Prigogine's Minimum Dissipation Principle, or Jaynes' Maximal Entropy Prescription—*but true.*

1.2 James S. Langer answers

Q1) In your view, what are the most important aspects of the experimental data on the glass transition that any consistent theory explain? Is dynamical heterogeneity one of these core aspects?

I am not sure that dynamical heterogeneity is an absolutely essential aspect of glassy systems; but I hope it is, because it seems to emerge naturally from the physical pictures that make most sense to me. I think the most important aspects that we need to understand are the dramatic slowing of molecular rearrangements that occurs with decreasing temperature, and the relation between this dynamic behavior and the thermodynamic behavior seen at about the same temperatures.

The difficulty in drawing definite conclusions about heterogeneity—from either experiments or numerical simulations—is that too many different mechanisms can be occurring and competing with each other at the same time. These different phenomena are essentially impossible to disentangle because everything is happening so slowly in the most interesting circumstances. For example, we saw numerical evidence at the meeting in Leiden that crystallites form during the cooling of some (all?) binary glass-forming liquids. Does the system fail to complete a first-order phase transition because the driving force for coarsening is too small at high temperatures? Or because the kinetics becomes anomalously slow at lower temperatures? Or are we seeing a slowly fluctuating equilibrium phase of crystallites in coexistence with eutectic fluids? If so, is this a glass?

Another example—Harrowell (Widmer-Cooper and Harrowell, 2006, 2007) and coworkers have provided evidence that regions in which molecules are most likely to be mobile may be regions in which the underlying solid-like non-crystalline material is elastically soft, i.e. where there are localized low-frequency vibrational modes. Is this "dynamic heterogeneity?" Or is it just an intrinsic property of any disordered material?

Q2) Why should we expect anything universal in the behavior of glass-forming liquids? Is the glass-transition problem well defined? Or is any glassy liquid glassy in its own way?

I think we should keep open minds about these questions. Our job as theorists is to look for general concepts, and we can't let ourselves be completely discouraged by the possibility that our "universalities" may not always be so universal, or that one supposedly "universal" behavior turns into another one as the temperature or other parameters are changed.

We've found some promising candidates for general concepts in recent decades. For example, there is the "frustration" picture of a glass-forming liquid, in which the energetically favored local correlations are inconsistent with ordered, space-filling states. This picture implies some measure of universality in the intrinsic slowing-down process, because the frustration length scales ought to become irrelevant in comparison with some dynamic length scale (what dynamic length scale?) as the temperature decreases. But this picture can be spoiled by competing equilibration processes such as growth of crystallites, or by a crossover from one dominant relaxation mechanism to another as a function of temperature. The ways in which universality can be broken are likely to be strongly system dependent and, therefore, inconveniently complex.

Another potentially universal concept that I like increasingly these days is the idea that the degree of disorder in a glassy material can be characterized by an effective temperature. Jorge Kurchan has been one of the pioneers in this area. (For example, see (Cugliandolo *et al.*, 1997). Some of my own recent work on this topic is described in (Langer and Manning, 2007; Bouchbinder and Langer, 2009).) This idea is turning out to be especially useful for understanding non-equilibrium phenomena such as shear-banding instabilities and even possibly shear fracture. A temperature-like measure of disorder that plays a central role in glass dynamics (sometimes equivalent to the role played by free volume) might be a definitively glassy feature, although it might also be useful for describing other disordered systems such as polycrystalline solids with high densities of dislocations and other defects.

Q3) In spin-glasses, the existence of a true spin-glass phase transition has been well established by simulations and experiments. Do you believe that a similar result will ever be demonstrated for molecular glasses?

I suspect that-spin glasses, even those with short-ranged interactions, belong to a different universality class from molecular glasses. In the latter case, I take seriously the simple argument that there can be no true glass transition. In an indefinitely large, amorphous system with slightly soft interactions, there is always some way for two molecules to move around each other. All that is needed is a thermally activated fluctuation in which the neighboring molecules squeeze out of the way and thereby open enough room for the molecular rearrangement to occur. This process, or some energetically more efficient version of it, may require an increasingly large amount of activation energy as the temperature decreases, because we must insist that the rearrangement results in a stable new configuration that doesn't just jump back to the original one during the next thermal fluctuation. Thus, the spontaneous rearrangement rate may rapidly become unobservably slow at low temperatures, giving the appearance of a sharp transition to infinite relaxation time. But mathematically there is no transition unless some kind of long-range order intervenes. I don't think that that argument applies to spin-glasses where spins occupy fixed positions and the random interactions are frozen.

In accord with my remarks in (Q2), however, I don't think that this argument—that there is no glass transition—should stop us from trying to make theories of glasses. In fact, I think that this argument may be too academic to be useful. We know that glassy slowing down sets in and becomes increasingly pronounced at characteristic "glass temperatures," which must correspond to characteristic energies and therefore characteristic dynamic mechanisms. The proper goal of a theory is to identify these mechanisms and their ranges of validity. The likelihood that they become irrelevant, or are dominated by other mechanisms at higher and lower temperatures, doesn't make it any less important to understand them

Q4) Why are there so many different theories of glasses? What kind of decisive experiments you suggest to perform to rule out at least some of them?

I'm not sure that there are really "so many" theories of glasses, at least not so many that try to be comprehensive, and that are sharply enough formulated to make unambiguous predictions. In fact, I'm not sure that there are *any* such theories.

There are, indeed, a great many *models* of glassy behavior. Most of these focus only on some but not all of the features that we think are characteristics of glasses. By doing this, they have contributed a great deal to the conceptual tool kit that we use for interpreting data. The kinetically constrained models are certainly in this category. They have proven to be very useful phenomenologically; but by definition they focus primarily on kinetics as opposed to dynamics, and they haven't yet made a clear enough connection to molecular motions to satisfy me. The mean-field models are the opposite. They produce fascinating insights about thermodynamics and some kinds of glass-like transitions. But, for reasons that I will say more about, I worry that they may be intrinsically incapable of addressing the most important dynamic questions. There are also various special-purpose models, designed to describe specific phenomena. Here, I am thinking of the trapping models, or the anomalous-diffusion model that several of us have used to interpret scattering experiments as well as diffusion and viscosity data. These are all complementary and mostly non-contradictory approaches toward understanding glassy behavior. Few of them pretend to be comprehensive theories.

There is also, of course, the wonderful world of glassy models that we can study by numerical simulation these days—hard spheres, soft spheres, molecules with all kinds of structures and local coordinations, even polymers. Some of these simulations have played the roles of well-controlled, benchmark experiments. Simulations become essential parts of the theoretical picture when they are carried out in direct connection with the development of theoretical models.

I am optimistic about the prospects for sorting out these different ideas by means of the increasingly powerful experimental tools that we have available these days. My recommended projects, i.e. those that appeal to me as a theorist, are the ones that bring a wide range of instrumental techniques to bear on a few well-characterized glass-forming systems, and that go as close as possible to where we think there might be a glass transition. As I said in Leiden, I'd like to see neutron-scattering data for systems at temperatures close to the glass temperature, so that we can compare the scattering measurements with the data for viscosity and diffusion that are available there.

Q5) Can you briefly explain, and justify, why you believe your pet theory fares better than others? What, deep inside, are you worried about, that could jeopardize your theoretical construction?

The closest I come to having a "pet theory" is what I've called the "excitation-chain mechanism" (Langer, 2006*a*,*b*, 2008). I've tried to insist that this is not yet really a "theory;" it is at most a "framework" for a theory. It's my best attempt so far to implement my bias about where the crux of the problem lies.

In my opinion, the glass problem is primarily a dynamics problem that needs to be addressed at the molecular level. We should look at a non-crystalline array of molecules interacting via short-ranged forces, and ask why, and under what circumstances, it becomes anomalously difficult for these molecules to undergo rearrangements. More specifically, we should choose some class of rearrangements, and compute the rates at which they are induced spontaneously by thermal fluctuations. The typical rearrangement that I have in mind occurs when a molecule moves out of an energetically favored position, forming the glassy equivalent of a fully dissociated vacancy–interstitial pair, leaving a lower density at one place and producing a higher density somewhere else. This is an important class of fluctuations because it determines diffusion and viscosity coefficients, and also is relevant to non-linear irreversible behavior. I hope that, once I understand this process, other pieces of the puzzle—including glassy thermodynamics—will fall together more easily.

Accordingly, I have asked what thermally activated molecular motions might most efficiently produce such irreversible fluctuations. My philosophical model for this was the Fisher–Huse (Fisher and Huse, 1988*a*,*b*) estimate of relaxation rates in spin-glasses, which makes a specific guess about the intermediate states that enable a correlated cluster of spins to change from one stable microstate to another. I was guided more specifically by numerical simulations of molecular glass dynamics by Glotzer and coworkers (Glotzer, 2000), in which chain-like motions appeared during thermally activated transitions between inherent structures.

The excitation-chain picture has allowed me to use little more than a back-of-the-envelope calculation to derive a Vogel–Fulcher formula for the transition rates. I also have been able, with not much more than dimensional analysis, to connect this dynamic result to thermodynamic behavior near the glass temperature. Most importantly for my recent research, the excitation chains have given me a way to understand how the rate factor for chain-enabled molecular rearrangements fits into theories of plastic deformation and failure. These results emerge naturally from the basic picture, with few *ad hoc* assumptions; but a systematic derivation of them, starting from molecular models, has eluded me so far. That's why I call this just the framework for a theory.

Do I worry that this could be completely wrong? Of course I do. I have tried to find a more systematic derivation, using function-space methods to compute probabilities for chains of molecular displacements in random environments. I learned enough from the function-space effort to be sure that my estimates break down if I try to push them too far into the low-temperature regime. I hope to come back to this effort some day. At the moment, I can only guess that this is qualitatively the kind of theory we'll need if we are ever going to have a real "theory of glasses," i.e. a theory that starts with first principles and makes predictions.

Q6) In the hypothesis that RFOT forms a correct skeleton of the theory of glasses, what is missing in the theoretical construction that would convince the community?

A large part of my motivation for thinking about excitation chains was my discomfort with the random first-order transition (RFOT) theory (Xia and Wolynes, 2000, 2001; Lubchenko and Wolynes, 2007*a*). In my last evening in Leiden, Peter Wolynes and I spent several hours talking about these issues, and I hoped that we had found a point of view that would reduce our differences of opinion to well-defined technical issues. Maybe we made some progress toward that goal. Here's how I see the present state of the discussion.

I have been dubious about the RFOT theory, partly because it didn't seem to address the issues at what I thought was the right level dynamically, and partly because it seemed to be internally inconsistent. These difficulties were most apparent in the earlier statements of the theory, where Wolynes and colleagues postulated that the transition state leading from one glassy configuration to another is an entropically favored droplet. They called this droplet an "instanton," and clearly were thinking about droplet models of first-order phase transitions. Such models are well understood in conventional nucleation problems, where the initial state is a thermodynamically metastable phase, and the transition state contains a coexisting, marginally stable droplet of the nucleating phase. The transition takes place when that droplet fluctuates over its activation barrier and starts to grow, converting the metastable system to a thermodynamically stable one.

The problem with the nucleation version of the RFOT theory is that the initial state is not a thermodynamic phase at all, but is one of an extensive number of microstates; and yet the droplet free energy is computed by summing over all of the microstates as if one were computing the thermodynamic free energy of the glass itself—which, in principle, is what is being done. This is a manifestly non-orthodox theoretical procedure. It is not meaningful to talk about coexistence or first-order phase transitions between microstates, on the one hand, and states of thermodynamic equilibrium on the other. Moreover, the entropic contribution to the droplet free energy is computed by making the Gibbsian assumption that a statistically significant fraction of the microstates is sampled in times shorter than any other time of interest. But the time of interest here is precisely that sampling rate—the rate at which the system moves from one microstate to another—and the problem is to understand why that rate is so small. Thus, there seems to be an internal contradiction.

Recent presentations of the RFOT theory resemble an interesting reinterpretation by Bouchaud and Biroli (Bouchaud and Biroli, 2004). Their idea is to understand the sum over microstates in the droplet free energy as a sum over system trajectories leading from the initial state to all possible final microstates within the spatial region occupied by the droplet. Each of these trajectories has a small *a priori* probability of occurring. But, if the region is big enough, there are enough of these trajectories that the probability of one of them being realized becomes appreciable. Accordingly, the transition state is the smallest region of a microstate that allows enough such trajectories that the region becomes unstable against a spontaneous molecular rearrangement. The mathematics is unchanged from the original RFOT theory, but the language is quite different.

In rough outline, the Bouchaud–Biroli interpretation looks a little like droplet nucleation. In fact, the interpretation of nucleation in terms of system trajectories is what I used in my "instanton" papers over forty years ago (Langer, 1967, 1969). There, however, I knew that the individual molecular motions occur very rapidly compared to the rate at which a thermodynamically (almost) stable droplet changes its size and shape in response to thermal fluctuations. Therefore, I could use coarse-grained quantities such as surface energies and diffusion constants to model the motion of the droplet as it makes its way over the nucleation barrier. In other words, I had a complete dynamic description of this transition state, including the relevant rate factors, and I knew that this description was appropriate for the time scales and length scales at which it was being used.

The reinterpreted RFOT theory has few such reliable ingredients. No attempt is made to understand what fraction of the microstates within a region is actually accessible to the initial state. The assumption is simply that they are all equally accessible and carry equal weight. This assumption may be related to the surface-energy problem, which is the least well understood ingredient of the RFOT theory in any of its versions. In order for the theory to make sense, the surface energy must scale like the square root of the number of molecules in the droplet instead of being proportional to the geometric surface area. The arguments presented in support of this result are based on statistical analyses that I find hard to justify, especially in view of the fact that this transition state apparently contains no more than a few tens of molecules near the glass temperature, and considerably fewer at higher temperatures. In contrast, the excitation-chain mechanism does not require that the transition state be a compact entity at all. The square root of the number of molecules emerges naturally in the same place as in the RFOT theory, and the two analyses share many features beyond this point in their developments.

It must be said in its favor that the the RFOT theory identifies the right groups of parameters and puts them in the right places for comparison with experiment. (The excitation-chain picture does this almost as well.) My sense is that we are all on roughly the right track here, but that we are missing some theoretical tools. Perhaps we just need a more powerful mathematical starting point. A truly new idea might be better.

Q7) Exactly solvable mean-field glass models exhibit an extraordinary complexity requiring impressive mathematical tools to solve them. Are you worried that for both the spin and structural glass problems it proves so hard to establish the validity of mean-field concepts in finite dimensions? Why are finite-dimensional versions of mean-field glasses always behaving in a very non-mean-field manner?

My worry about the mean-field models goes back to my experiences with the droplet calculations that I mentioned in (Q6). I was using simple field theories; but it was clear that no perturbation expansion, no matter how many graphs it summed or how non-linearly self-consistent it might be, could produce the interesting results, which always had essential singularities at the important places. This is well known. Such difficulties arise whenever the crucial physics involves statistically improbable, localized entities such as critical droplets, which don't naturally emerge in mean-field approximations, and especially not in the exactly solvable models. The localized features are too easily lost when theories start with very long-ranged interactions or very high dimensionalities, and try to perturb away from those limits.

Therefore, I suspect that the exactly solvable mean-field theories are not going to come to grips with the dynamics of real molecular glasses near their glass temperatures. Nor are the mode-coupling theories likely to be successful in this regard. They do a good job of expanding away from liquid-like states, but break down at lower temperatures, just when the dynamics becomes most interesting. It seems possible that we might figure out how to use such theories non-perturbatively and induce them to tell us about dynamic heterogeneities. (Schweizer has made an important effort along these lines (Saltzman *et al.*, 2008).) Even then, however, I am afraid we would be starting at the wrong place. If the transition state relevant to anomalously slow relaxation is, in fact, a delicately balanced, non-compact entity that involves only tens and not millions of molecules, then conventional field-theoretic approaches seem doomed to failure.

Q8) In your view, do the recent ideas and experimental developments concerning jamming in granular media and colloids contribute to our understanding of molecular glasses, or are they essentially complementary?

I think that the jamming ideas and the work on hard spheres and athermal granular systems have been enormously valuable, and am almost sure that there are deep connections between these systems and molecular glasses. For purposes of studying glass physics, we should exclude the complications of force chains and bridging configurations and the like that occur in static granular materials. Even without these complications, however, the connection between jamming transitions and glass transitions seems to be complex and subtle. Understanding that connection could shed light on both kinds of phenomena.

Q9) If a young physicist asked you whether he or she should work on the glass problem in the next few years, would you encourage him or her and if so, which aspect of the glass problem would you recommend him or her to tackle? If no, what problem in condensed-matter theory should he or she tackle instead? If yes, what particular aspect of the glass problem seems the most exciting at present?

The first part of this question—whether I encourage young scientists to go into any area of materials theory, especially in the United States these days—requires an essay by itself. Let me say just that I am very grumpy about the state of funding in these areas, and even grumpier about the decline in fundamental research in areas such as materials engineering and materials chemistry. Those fields once were homes for research in amorphous materials. They are now seriously in need of new ideas, but basic research has mostly disappeared in them.

At the same time, however, I am enormously enthusiastic about the prospects for major progress in the non-equilibrium aspects of glass physics. I started thinking about the glass problem only a few years ago, because I needed to understand some aspects of it in connection with my attempts to make theories of deformation and failure of non-crystalline materials. Among other projects at the moment, I am working with two graduate students on applications of these theories to shear failure in earthquake faults. We've found some fascinating results; and in this case the seismologists are gratifyingly receptive.

I admit that I am leery about encouraging young people to tackle the core problems in glass physics, where both the scientific and sociological barriers are uncomfortably high. But there are great opportunities in related areas, especially including biological phenomena. In short, I think that this field has outstanding intellectual and practical potential, which is why I've chosen to work in it.

Q10) In twenty years from now, what concepts, ideas or results obtained on the glass transition in the last twenty years will be remembered?

Primarily, I hope to be surprised by the answer to this question. However, I find it hard to believe that we'll have forgotten some of the basic phenomenologies and concepts that have emerged in recent decades. For example, I think that the Kauzmann paradox, the Adam–Gibbs connection between dynamics and some kind of entropy, the definition of inherent structures, the concept of fragility, and the like will have lasting value. But I hope that we'll have found new ways to think about these phenomena in another decade or so.

Q11) If you met an omniscient God and were allowed one single question on glasses, what would it be?

After having written my answers to some of these questions, especially (5) and (6), I think I'd ask this God to tell me what mathematical tools provide a natural language for talking about glassy dynamics problems. I'm afraid, however, that His/Her reply would be something like this:

"&*%$* you Langer! I've already given you a computer. Anyway, you should know by now that not every problem has a tractable mathematical solution. You've already seen lots of examples of no-solution problems in nuclear-structure physics, astrophysics, turbulence, etc. where the best you can do is compute special cases."

Then I'd meekly ask, if I were allowed to do so—"But haven't you made it seem as if there is something universal about glassy dynamics? And, if there is something universal, doesn't that mean that there is something fundamental to be discovered, and some natural way to describe it?"

To this, I'm afraid that She/He would answer: "Maybe".

1.3 Thomas A. Witten answers

Q1) In your view, what are the most important aspects of the experimental data on the glass transition that any consistent theory explain? Is dynamical heterogeneity one of these core aspects?

Form of divergent dynamic modulus with temperature or density. Dynamical heterogeneity is one of these core aspects insofar as their behavior has common features of many systems. Molecular mobility near walls or other constrained regions.

Q2) Why should we expect anything universal in the behaviour of glass-forming liquids? Is the glass-transition problem well defined? Or is any glassy liquid glassy in its own way?

This point of view says that the local motions that allow flow in a system where motion is highly constrained are dictated by the local molecular packing and bonding geometry. This point of view might well be sufficient to explain all practical measurements. Likewise, the world might have been such that we could never approach a critical temperature more closely than ten per cent. In that case we could never make an experimentally decisive demonstration of critical universality. For features of critical phenomena like the operator product expansion, such decisive demonstrations are in fact lacking, I believe. Nevertheless, one cannot describe the passage from fluid to glassy state fully in principle without invoking a diverging number of degrees of freedom. In such cases, more is necessarily required beyond an accounting of local motional constraints.

Q3) In spin-glasses, the existence of a true spin-glass phase transition has been well established by simulations and experiments. Do you believe that a similar result will ever be demonstrated for molecular glasses?

It is not in our control whether the evidence will become as convincing as it is for a spin glass. It is in our control to devise new realizations of glassiness and new ways to probe glassy systems in the problematic region. I think there is lots of scope for these realizations and probes to give insight

Q4) Why are there so many different theories of glasses? What kind of decisive experiments you suggest to perform to rule out at least some of them?

To me the most plausible approach is to test the logical basis of the theories, e.g. mode coupling, with abstract models, chosen to fulfill the hypotheses of the theories, but not to represent molecules.

Q5) Can you briefly explain, and justify, why you believe your pet theory fares better than others? What, deep inside, are you worried about, that could jeopardize your theoretical construction?

My pet theory is the notion that the enabling motions that permit fluidity are spatially extended vibrational modes with the properties of weakly localized wave-functions. That is all I want for Christmas. The justification is the form of the normal modes in a marginally jammed granular pack, reported by Nagel *et al.* and explained by Matthieu Wyart. Encouragement for this view comes from the extended nature of the relaxation events in, e.g. Dauchaut's grain-pack experiments. The worry is that a satisfactory account of low-temperature properties has to embrace two-level systems and ultrasound echos. These features seem to entail very local modes with a finite set of energy states.

Q6) In the hypothesis that RFOT forms a correct skeleton of the theory of glasses, what is missing in the theoretical construction that would convince the community?

[Tom Witten did not wish to answer to this question]

Q7) (a) Exactly solvable mean-field glass models exhibit an extraordinary complexity requiring impressive mathematical tools to solve them. Are you worried that for both the spin and structural glass problems it proves so hard to establish the validity of mean-field concepts in finite dimensions? Why are finite-dimensional versions of mean-field glasses always behaving in a very non mean-field manner?

Yes I am worried. But is this so different from the development of critical phenomena? An advance in mathematical understanding was needed, and at first the new things were not clearly grasped and so they were hard to explain and distill into a simple form.

Q8) In your view, do the recent ideas and experimental developments concerning jamming in granular media and colloids contribute to our understanding of molecular glasses, or are they essentially complementary?

I think the jamming phenomena serve as a useful extension of molecular glasses. Such extensions are surely needed in order to come to a simple understanding of how mobility turns itself off.

Q9) If a young physicist asked you whether he or she should work on the glass problem in the next few years, would you encourage him or her and if so, which aspect of

the glass problem would you recommend him or her to tackle? If no, what problem in condensed-matter theory should he or she tackle instead? If yes, what particular aspect of the glass problem seems the most exciting at present?

Since the glass field is very mature, this young physicist should be very smart and confident. The essence of the problem is the proliferation of local geometrical constraints. I think there is a lot of scope to devise artificial geometrical systems that embody is idea and reveal how it works asymptotically.

Q10) In twenty years from now, what concepts, ideas or results obtained on the glass transition in the last twenty years will be remembered?

The simplest forms of mode-coupling theory, marginal jamming of athermal systems, quench memory phenomena.

Q11) If you met an omniscient God and were allowed one single question on glasses, what would it be?

Are we on the right track at all (like early SU(3) leading to modern quantum chromodynamics) or are we wandering in the wilderness (like S matrix theory)? (God would know what I meant.)

1.4 Peter G. Wolynes answers

Q1) In your view, what are the most important aspects of the experimental data on the glass transition that any consistent theory explain? Is dynamical heterogeneity one of these core aspects?

At the very start, I think the most commonplace observation about structural molecular glasses requires explanation from a satisfying theory: the existence of sensibly rigid but amorphous objects! This rigidity is not an illusion but reflects a separation in time scales of individual molecular motions from collective motions—a separation spanning many orders of magnitude. Theories must explain how you get this rigidity without observing order, or show us how there is, in fact, broken spatial symmetry of some specific kind but how it is that we have so far failed to see it after very diligent efforts. An example of the first sort of explanation is the random first-order transition theory based on the existence of truly aperiodic minima. In a sense, the RFOT theory is the minimal generalization of the theory of ordinary first-order transitions like crystallization to the situation in which there is a diverse set of structures each without periodicity (Wolynes, 1989).

The RFOT explanation for the initial origin of rigidity coincides with the explanation provided by the dynamic transition in mode-coupling theory (MCT) (Jäckle, 1986). Both MCT and the density-functional-based form of RFOT theory are semi-quantitatively accurate in predicting the plateau of the scattering function, i.e. the Debye–Waller factor that quantitatively reflects the separation between vibrational and glassy motions. An example of the latter type of explanation based on a hidden broken symmetry that we generally don't observe is the icosahedratic theory of Nelson (Nelson *et al.*, 1983) (icosahedral order is seen in some metallic melts). In my view postulating *a priori* models with activated events or defects in a rigid background from the get-go, fails to answer satisfactorily this primary question of the origin of rigidity. For this reason, I do not find merely postulating lattice models "coarse grained" in some manner at all intellectually satisfying on this most important conceptual point—how the coarse graining is done is the key to answering this fundamental puzzle.

The experimental investigation of structural glasses has revealed far more detailed facts that need explanation beyond the mere fact of rigidity, and since I believe in a unified theory I think it is important, ultimately, that no observation be left unexplained.

For example, I would require that an acceptable theory give a recipe, at least, for how to calculate any of its characteristic temperatures or energy scales from the intermolecular forces. This "chemical" attitude of mine is not universally shared in the physics community. I would not insist, however, on quantitative accuracy of such a calculational scheme at present since there are no small parameters apparent in the first formulation of the problem. Beyond this, a theory must explain three qualitative behaviors: non-Arrhenius temperature dependence, non-exponential time dependence and aging.

The non-Arrhenius temperature dependence of transport must be explained in a universal way. Those cases where Arrhenius behavior has been found can be continuously connected to non-Arrhenius systems by doping (e.g. adding a little Na_2O to SiO_2 changes silica from having seemingly Arrhenius behavior to having non-Arrhenius behavior). Owing to this fact, I again think that merely postulating simple activated behavior at some elementary excitation level is a non-starter. Clearly, separate theories for "fragile" and "strong" systems should be ruled out.

The non-exponential time-dependent behavior of glassy molecular liquids must be explained, again in a universal way, although special sources of heterogeneity may exist in some systems, e.g. phase separation. Twenty-five years ago a hotly debated question was whether this non-exponential behavior implied heterogeneous patterns of motion in space or only in time. Experiments have finally answered this question quite clearly. This dynamical heterogeneity, is then, indeed, according to my taste, a core aspect of the glass phenomenon—in a sense it is the clearest manifestation of there being a true diversity of states. Elementary defect models are very hard pressed to explain the details of the spatial dynamical heterogeneity (they require kinetic constraints), even when such models can fit low-order correlation functions.

A theory of glasses must give an explanation of the phenomena seen in aging systems, by definition, since the glass is a system no longer in equilibrium. The near-Arrhenius temperature dependence of aging in a non-equilibrium glass clearly points out the necessity of having activated transitions of some type emerging from the theory.

I would hold that, in addition to the above absolute requirements for any theory of glasses, a satisfying theory should provide some answer to why the thermodynamic properties of supercooled liquids correlate as well as they do with their dynamical properties. This personal view again, is not held universally. Many theorists argue that the thermodynamic–kinetic connections noticed for decades, starting with Simon and Kauzmann more than half a century ago and emphasized for quite a while by Angell are just coincidences or are the result of some "law of corresponding states" pointing to a common model of some unspecified sort. I do not see how practical-minded people aware of the range of data already known can be satisfied with an explanation based on multiple coincidences. Also, why do the correspondences continue to pile up as more systems are studied? I would also put the burden on those advocating the mere corresponding states viewpoint to take a stance on the common model and show how to calculate numbers for a few systems. All evidence (e.g. pressure dependence, crosslink dependence of characteristic temperatures) points to the configurational entropy being the key quantity on which any law of corresponding states is based.

Along with this view of the importance of kinetic thermodynamic correlations, I believe a theory should tell us either how the Kauzmann entropy crisis (i.e. the impending vanishing of the configurational entropy at low temperature) is realized or precisely how it is avoided—without something happening, even the third law of thermodynamics comes into jeopardy with the total entropy for some systems extrapolating to zero only a bit below the conventional Kauzmann temperature.

Q2) Why should we expect anything universal in the behaviour of glass-forming liquids? Is the glass-transition problem well defined? Or is any glassy liquid glassy in its own way?

I was surprised when I first came across the fact that quantitative universal behavior was experimentally found for glass-forming liquids. One could imagine that after explaining the general origin of rigidity a satisfactory theory could find numerous classes of glassy behavior. For example, if we took the icosahedratic picture as correct as the origin of rigidity in metallic glasses, we might imagine borosilicate glasses to find their rigidity to be based on some other platonic or neo-platonic (?) form, van der Waals liquids another, etc. This multiplicity of mechanisms would be like the existence of the finite but large number of crystallographic group symmetries. Ordinary crystallization is quite sensitive to molecular details (e.g. the even–odd effect for alkane crystal melting points) but glass formation is much more robust and changes smoothly with composition. The glassy behavior of molecular liquids, however, seems to be quite universally expressed in terms of their configurational entropy S_{c}, and the fluctuations of configurational entropy as measured by their configurational heat capacity ΔC_{v}. These measures are sufficient to predict properties and otherwise one can be indifferent to the precise chemical and structural origin of the aperiodicity.

According to our understanding from the random first-order transition theory, the observed quantitative universality for molecular liquids comes not from an asymptotically close approach to a critical point but arises from the near universality of the Lindemann parameter measuring the scale of vibrational motions in any amorphous state. This Lindemann parameter universality is connected with the clear separation of vibrational motions from the slow motions in dense collections of particles and is predicted by density-functional and microscopic mode-coupling theories that are the starting points of the RFOT theory. It may be connected to geometrical features of space-filling structures in general. Still, there are hints of an even stronger more universal behavior reflecting something like an approach to a real phase transition. The dependence of the complexity (i.e. the total configurational entropy) of a dynamically rearranging region on the time scale alone, for example follows from random first-order transition theory in a general way without assuming a universal Lindemann parameter. Such a data collapse has been observed by Capaccioli *et al.* (Capaccioli *et al.*, 2008). Unfortunately, it seems that those observations are not yet in a truly asymptotic regime of large correlation lengths, so a good but imperfect data collapse was found. I believe the observed mesoscopic scale of correlations confines the accuracy of such data collapse to about what we see in those experiments. A correlation range of five particle spacings as predicted by RFOT theory means accuracy in the 10% range is about all that can be expected.

While molecular glasses have common patterns of behavior, different liquids do have individual characteristics because molecular liquids may manifest other phase transitions or near phase transitions in the deeply supercooled regime. The clearest examples are phase-separating glasses (like pyrex!!) but probably also include water, silicon and pure silica (SiO_2), for which there is strong circumstantial evidence of quite distinct amorphous forms. Some important substances, such as "amorphous" silicon may not be amorphous at all, but probably possess poorly developed periodic crystalline order (they are anisotropic often!). Despite the occurrence of these phase transitions deep in the supercooled state, I don't think merely saying everything is a result of an avoided phase transition quantitatively explains the universal patterns since one is clearly not in the asymptotic regime of any known avoided transition.

So, although I do not think each liquid will be glassy in its own way I believe molecular liquids will differ from each other quantitatively. Also, if we go outside the class of structural glasses, more complex and distinct glassy behaviors will be found in systems such as colloids, granular assemblies and microemulsions. These need not be described by the same properties as the molecular glasses.

Q3) In spin-glasses, the existence of a true spin-glass phase transition has been well established by simulations and experiments. Do you believe that a similar result will ever be demonstrated for molecular glasses?

"Ever" is a long time, so I hate to answer this question. Straightforward paths to a demonstration of a true phase transition for molecular glasses seem to be blocked by the near-universal relation between time scale and correlation length. Explicitly said, if we accept the predictions of random first-order transition theory for molecular glasses or simply take the empirical results of Capaccioli *et al.* at face value it will be very difficult to achieve by cooling on human time scales an amorphous state with a large enough correlation length to be considered unambiguously in the asymptotic regime. The correlation length at a laboratory glass time scale of one hour reaches only five molecular diameters and waiting even years does not lengthen the scale much.

Are there routes other than cooling that could be used to prepare low configurational entropy or long correlation length glassy systems from molecules? Perhaps. . . . , but nothing just on the horizon appears certain to go far. One promising laboratory route might be the preparation of glasses by vapor deposition. Ediger has prepared amorphous o-terphenyl with an apparent configurational entropy of about half that of a typically slow-cooled glass (Kearns *et al.*, 2008). Jake Stevenson and I have argued using RFOT theory that this is about all the further one can go in decreased configurational entropy (Stevenson and Wolynes, 2008). Indeed our arguments suggest the correlation length may not be much longer than in an ordinarily cooled glass. Would a correlation length of 10 satisfy people as being in the asymptotic limit? I don't know, but I am sure more careful investigation of such hyperstable glasses merits the attention of theorists and experimenters alike.

As in the laboratory, computational paths to slower cooling rates are unpromising with classical computers. Moore's law is trumped by the Vogel–Fulcher law. True quantum computers may fare better but I suspect mere quantum simulator devices will also seriously suffer from glassy slowing effects despite some suggestions to the contrary in the literature (Brooke *et al.*, 1999).

A major mathematical advance in constructing computationally low-energy aperiodic structures could also lead to a breakthrough on this score: rather than cooling maybe someone can find tricks for creating amorphous packings on the computer in a constructive way—a kind of super vapor deposition algorithm, a souped-up version of Bennett's assembly algorithms. Another possible route on the computer, perhaps, could start by replicating finite examples of low-energy structures, then doing some sort of detailed renormalization group-like construction, who knows? There are certainly practical motives arising from coding theory that suggest that exploring this option could lead to better error-correcting codes; for this reason smart people are surely looking at this. The mathematicians did finally put to rest the centuries old Kepler conjecture on periodic fcc sphere packing in recent years, so optimism is not entirely silly.

Another approach would be to show the existence of true phase transitions in less direct ways. In contrast to the magnetic lattice models, series analysis tools have been underutilized for molecular fluids in my opinion. Does the Gaussian core model have a glass transition? Convincing use of such tools may be blocked by essential singularities connected with droplet excitations, however. Also, searching for physical systems or mathematical models that interpolate between spin-glasses and molecular glasses may help resolve the issue of whether there is a true phase transition.

I would like to take the liberty, however, to suggest that oversimplifying the question to one of the existence of a "true phase transition" misses the point of understanding glassy phenomena as we encounter them in the real world. Like flames, glasses are very often examples of "intermediate asymptotics". The ultimate state of most simple molecular liquids will be a periodic crystal a la the Kepler conjecture. We call configurations with few or small numbers of crystallites, glasses. Nevertheless, in the thermodynamic, long-time limit, strictly such crystalline configurations will dominate for simple liquids. We believe that a restriction to "non-crystalline" configurations somehow allows us to forget about this problem of delimiting the amorphous state but this may not be true. In any case, suppose we could not partition off the crystalline part of the phase space sharply. Would the "glass problem" go away and be unworthy of study if the correlation length could only grow to 25 particle diameters before crystallization hit? I don't think so. Still, it would be nice to know if there were such a spinodal limit for crystallization, I suppose. It might help a lot in materials synthesis.

Q4) Why are there so many different theories of glasses? What kind of decisive experiments do you suggest to perform to rule out at least some of them?

Diversity and unity are the yin and yang of theoretical physics. We must always strive to achieve a balance. Multiple ways of looking at the same problem are essential both for present understanding and future theoretical progress—it is not better to look at the diffusion equation as describing a random walk than it is to view the equation as a continuum hydrodynamic theory, the two viewpoints illuminate each other. To reach balance, we must strive to see how diverse theories relate to each other, ultimately forming a united viewpoint. During the development of theories, imbalances between diversity and unity can often arise.

Creativity without discipline and laziness on the part of theoreticians are the main culprits that account for any unbalanced diversity of glass theories. Theorists are perhaps too good at coming up with theories and modifying them. That is a crucial part of our job, after all, but many theorists choose not to be disciplined either by conceptual simplicity or by experimental facts—if one believes that most experimental observations, for example the thermodynamic–kinetic correlations, are just coincidences because one is not studying the asymptotic regime, one is tempted to feel free to formulate theories that can fit data on a case by case basis using adjustable parameters. A very incomplete theory can seem plausible if one uses enough adjustable parameters. Also, if the main conceptual problems are ignored by the starting point of the theory, for example by ignoring the amazing fact of amorphous rigidity by postulating rigidity from the beginning, numerous detailed models can be constructed to account, again on a case by case basis, for each specific piece of quantitative data.

Laziness presents two more serious problems. First, laziness makes it seem there are more theories than there really are. It is hard work to show how theories are related to each other, so seeking relations between theories is avoided by many people. Secondly, I sometimes feel new elements are added to existing theories, thus multiplying their diversity only to avoid work. I think professionals should deplore this kind of laziness more than we do. I feel you really should avoid throwing in a new exponent or adjustable constant if you do it just because you can't be bothered to calculate something. In this respect we have all been spoiled by the history of conventional critical phenomena. The remarkable scaling laws found in magnets, the liquid–gas transition, liquid helium, etc. still tempt us to disdain the calculation of "non-universal" quantities—like the characteristic temperatures at which phenomena occur (that is supposed to be the business of engineers, I guess). But a claim of complete understanding really requires such calculations. In conventional critical phenomena the asymptotic regime was accessible and determination of the exponents seemed to be the most essential feature needed for understanding the basics—but for molecular glasses in the laboratory pre-asymptotic effects enter too. These pre-asymptotic effects are painful to calculate so we again try to avoid calculating them as long as we possibly can.

Computer simulations of ordinary phase transitions could be pushed into the regime of asymptopia more readily for conventional phase transitions than it has been possible to do for glassy ones. It is perhaps worth recalling, however, there was even confusion in the "good old days" of the early 1970s about whether simulations could give true phase transitions or not.

Not getting to asymptopia means we have a "strong-coupling" problem. Other strong-coupling problems, like the Mott transition in hard condensed-matter physics and confinement in QCD, have suffered from unbalanced diversity and unity during their intellectual history too.

Vagueness of a theory when combined with laziness of theoreticians is a really big problem in ruling a theory out. If a model is vague it cannot be tested. Some of the diverse set of glass theories seem to me to be quite vague—that is another reason I recommend requiring their advocates to show a path back to the molecular forces! But also when the model is finally made definite you still must work out its consequences and live with them. I have explained my aesthetic reasons against being satisfied with facilitated Ising models with trivial thermodynamics—they assume a rigid background to start with, have many adjustable parameters, etc. But it is only recently that such models have been firmly ruled out by calculation: through hard work, Biroli, Bouchaud and Tarjus showed that these theories cannot explain the heat capacity of liquids and glassy kinetics with the same parameters (Biroli *et al.*, 2005). Nevertheless, some of the salesmen of such theories just went ahead adding more adjustable parameters, "renormalizing" the difficulty away! This would be merely sad, except for the fact that so few people seem to have noticed what happened, judging from citations to the incorrect work.

You can see that I believe it is important to recognize that many specific theories being bandied about, have *already* been falsified by experiment. These include pure mode-coupling theory (without activated events) and facilitated lattice models with trivial thermodynamics. At the same time, while shown to be incorrect or incomplete many elements of those theories as well as those of several other theories are appropriately left standing by experiment. Some elements of these theories are part of the RFOT construction, like the perturbative parts of mode-coupling theory. Likewise, Nussinov has put forward arguments that connect the somewhat less well developed uniformly frustrated "avoided" phase transition models to random first-order transitions (Nussinov, 2004). In this sense, distinguishing sharply between avoided transition models and RFOT theory is premature in my view.

As an example of another set of connections I note that rigidity percolation and even constraint counting a la Phillips also emerge when RFOT theory is used to treat network glasses (Hall and Wolynes, 2003). Thus, again some parts of those theories are basically correct and consistent with my own thinking based on RFOT ideas.

Likewise the "effects" central to some models, such as "facilitation"—the way in which a mobile region allows its neighbors to move more readily, are indeed real effects. A theory failing to account for these effects is incomplete. RFOT theory does give a way of describing facilitation effects as emergent phenomena. I have used that connection recently in treating rejuvenation of glasses (Bhattacharyya *et al.*, 2008). Nevertheless, I again emphasize that models that merely postulate facilitation from the start miss the main point of explaining the origin of rigidity.

Some attempts to treat activated events can be seen as variations on the mainline of thinking about RFOT theory. More than twenty years ago Randy Hall and I assumed a fixed size of activated event but treated the mismatch energy as growing as the system is compressed (Hall and Wolynes, 1987). I think this view doesn't explain the kinetic–thermodynamic correlation although it continues to resurface as "shoving" models to use Jeppe Dyre's term (Dyre, 2006). Ken Schweizer adds activated events to an RFOT-MCT framework but treats activated motions as largely being single-particle events (Schweizer, 2007). I think this can only be correct near the crossover regime, and therefore is important for colloids but I don't think the approximation is good in the deeply supercooled regime.

Overall, I believe the RFOT theory does well in balancing theoretical unity with diversity. I continue to spend my time trying to think up new experimental approaches that will illuminate the key features of molecular glasses. Most of these involve coming up with a more information-rich experimental picture on the appropriate long time scales.

The biggest surprise from the RFOT theory is the size of the rearranging regions and how the CRR size scales with thermodynamics. While these aspects have been confirmed in a few cases by direct observation still more direct measurements of the size of the CRRs in a larger range of systems are needed to establish universality. Likewise, understanding the shape of rearranging regions in supercooled liquids and glasses remains interesting. I believe directly visualizing amorphous systems on long-time laboratory scales will eventually confirm the RFOT picture of relatively compact activated events in the deep glassy region that become more ramified near the crossover. Better imaging will rule out even more clearly point-defect models than the thermodynamic argument has already done.

Q5) Can you briefly explain, and justify, why you believe your pet theory fares better than others? What, deep inside, are you worried about, that could jeopardize your theoretical construction?

Einstein is said to have remarked that it was curious that typically the only person who doubts the results of an experimentalist is the experimentalist himself and the only one who believes the results of a theorist is the theorist himself. In this spirit, I must say I have become quite convinced of the basic validity of the ideas in RFOT theory, despite being very much aware of gaps in its rigorous construction—it is made up of a set of approximations that work well for molecular liquids but that may break down for other systems. It has seemed to me that once you embark on constructing a theory of long-lived aperiodic structures most of the key features of RFOT theory follow very naturally: A diversity of structures (i.e. configurational entropy) is almost a tautology. That configurational entropy must destabilize any single one of them seems self-evident to me and provides a natural driving force for activated events, etc. Accepting these basic features still leaves many open questions but I think uniquely sets you on developing the set of approximations made in RFOT theory as those of greatest simplicity. Other approaches such as frustrated phase-transition scenarios, to the extent they differ from RFOT raise all sorts of questions of specificity that need answering before we can use them to illuminate our understanding of molecular liquids—please tell us what transition is being avoided! (Again, the icosahedratic theory does this.) Likewise, I would ask excitation chain theorists: please tell us what you are being excited from, etc.

In any event, I think it is fair to say the RFOT theory is, at present, the most completely developed theory of structural glasses in a mathematical sense. The RFOT theory describes a path from the fundamental forces to experimental observables. The mode-coupling theory also starts from the forces and has predicted interesting phenomena—the re-entrant glass transition of adhesive spheres, etc. But, since the MCT dynamic transition is part of the RFOT framework this success from starting microscopically would be shared by an RFOT static density-functional approach. Most other theories of the glass transition don't yet say how to start from the intermolecular forces—leaving the problem of how to "coarse grain" liquid-glass reality for future work.

The mathematical framework of the RFOT theory starts out in a very conservative fashion. Like the van der Waals theory of the liquid-gas transition, RFOT theory is acknowledged to be exact in the mean-field limit of long-range forces. Admittedly, the mean-field phenomena with replica symmetry breaking are richer than ordinary phase transitions but an extraordinary amount of beautiful work has illuminated this limit (Parisi, 2006). This mean-field theory provides a good basis for such concepts as effective temperature when systems are out of equilibrium.

In contrast to the liquid–gas transition where dynamics is a secondary feature, though, dynamics essentially connected to the finite range of the forces is important for the glass problem. The problem is to reconcile diversity with locality. Again, as anticipated by Kirkpatrick and myself more than twenty years ago, the Kac limit of long-but not infinite-range interactions can be analyzed rather completely (Kirkpatrick and Wolynes, 1987*b*). Silvio Franz has done beautiful work showing how the first corrections to the infinite-range limit correspond to the simplest RFOT theory of activated events as instantons (Franz, 2007), paralleling but with much greater rigor recent work with Dzero and Schmalian (Dzero *et al.*, 2005).

The least-developed part of the mathematical framework for the RFOT theory is the role of critical fluctuations that build on each other and thus go beyond what would be found perturbing around the infinite range. In the Kac limit, there are two correlation lengths, one of which diverges at the dynamic transition, which without instantons can be made arbitrarily sharp by tuning the input range of the interactions. The other correlation length involves the instantons. This mosaic length scale diverges when the configurational entropy vanishes. In real molecular systems with short-range forces these scales mix somehow, since the dynamical transition most definitely disappears as a strict divergence. Kirkpatrick, Thirumalai and I explored the consequences of assuming that there is only one divergent length (Kirkpatrick *et al.*, 1989). (The title of our paper was "Scaling Concepts for the Dynamics of Viscous Liquids near *an* Ideal Glass Transition".) This single-length hypothesis is perhaps an overly strong hypothesis but its specificity leads to simple predictions, e.g. the Vogel–Fulcher scaling results as the only one consistent with a strict heat-capacity discontinuity, etc. There is evidence for multiple correlation lengths from the recent simulations of Biroli, *et al.* showing a thermodynamic signature of growing amorphous order (Biroli *et al.*, 2008). Also, strictly speaking the explicit magnetic analogy proposed by KTW that was also explored recently by my research group allows multiple correlation lengths (Stevenson *et al.*, 2008). The magnetic analogy suggests that an ideal glass phase transition occurs strictly near a point not a line in the phase diagram.

I think this question of the multiplicity of length scales is the origin of the mild controversy presently found within the RFOT-friendly community over the proper exponents to employ for the barrier scaling: the original ones of Kirkpatrick and mine that do not involve "wetting" or the ones that satisfy simple hyperscaling. The wetting-free exponents are also those obtained from the literal single instanton calculation in the Kac limit.

Finally, but in my view, the most important reason I favor my "pet theory" is that the RFOT theory already works when it comes to the comparison with experiment! The strongest version of the theory (assuming only a single length and extrapolating mismatch energies to the molecular scale) makes numerous predictions about the correlation of thermodynamic data with kinetics that are well satisfied. These correlations are predicted without the use of adjustable parameters. Many of these predicted correlations are discussed in detail in the review article I wrote with Vas Lubchenko (Lubchenko and Wolynes, 2007*a*; Böhmer, 1998*b*). They include in the classical regime:

- the correlation of activation energy at the glass transition ("the fragility") with the heat-capacity discontinuity;
- the prediction of the stretching exponent from the fragility;
- the prediction of how the "non-linearity" of the aging behavior in the glass correlates with fragility;
- the prediction of how the crossover temperature correlates with configurational entropy.

All of these predictions do not depend on there ever actually being a complete Kauzman entropy crisis—they depend on the local values of the configurational entropy and its derivatives. I am also surprised so many people object to using the Kauzmann temperature as at least a fiducial temperature since there is overwhelming evidence for at least a crossover at the Kauzmann point!

A very natural but admittedly bolder quantization of the RFOT framework made by Lubchenko and me also explains a large number of observations about the low-temperature properties of glasses like two-level systems and the boson peak. Although those calculations are on a much less secure footing than the purely classical arguments of the RFOT theory the confluence of those results with experiment also gives me confidence that the basic notions of the semiclassical RFOT theory are correct. I was very surprised that RFOT explained the weird universality of low-temperature properties that had been emphasized by Tony Leggett. Indeed Lubchenko and I had been pursuing Leggett's interacting defect scenario with little success before we tried to use RFOT ideas. This history is at least a good example of why theorists believe their published theories when others do not—they have usually tried other alternatives and felt the pain of failure!

These numerous comparisons of the theory with experiment either are telling us the RFOT theory approach's approximations are reasonably sound for phenomena in the observed time and temperature range *or* that there is some conspiracy of cancelling errors that is misleading us.

Except for such a possible conspiracy, then, I have no specific worries about what would jeopardize the RFOT theory construction in the observed experimental range for molecular liquids. I would share with many in the community, however, worries about how RFOT theory will fare in "asymptopia"—perhaps there is always some simple ordering that intervenes to prevent the ideal glass transition at low enough temperature. (How this would happen for an atactic polymer, I don't know!)–perhaps, the entropic droplet excitations themselves cut off the thermodynamic replica symmetry breaking transition, etc. as Michael Eastwood and I discussed (Eastwood and Wolynes, 2002) etc.

Even with these worries, I would ask, is it then preferable to start the analysis of real data by an expansion about this ultimate asymptotic state rather than use measured thermodynamic quantities in the appropriate regime as input? In any event my own feeling is that any such approach would still lead to an RFOT-like theory as an approximation when the entropy density is not too low.

Q6) In the hypothesis that RFOT forms a correct skeleton of the theory of glasses, what is missing in the theoretical construction that would convince the community?

I don't know. The psychology of the community eludes me. Twenty-five years ago people told me they would be happy if there was a theory of structural glasses that would reproduce the macroscopic experimental trends, the more remarkable strange behaviors, and be exact in some limit. Those desires have been satisfied. Others, at that time, said they would like to see evidence for a growing length scale, which was predicted by RFOT theory in 1987. Evidence for the growing length has been directly found through non-linear NMR experiments (Tracht *et al.*, 1998) and through imaging approaches (Russell and Israeloff, 2000). Less direct determinations from rigorous inequalities related to macroscopic observables showing a growing correlation length agree even numerically with predictions of the strong form RFOT theory. Those predictions in the year 2000, were made years before the observations were analyzed (Berthier *et al.*, 2005). Still reviewers of papers using RFOT ideas demand another "smoking gun" from experiments.

I suspect some uneasiness about RFOT theory comes from the lack of comfort many people have with activated dynamics in general. With a handful of exceptions physical chemists really misunderstood transition-state theory, even when they used it, until thirty years ago. Ordinary nucleation theory for ordinary first-order transitions still confuses people, I have noticed. I believe all the results obtained using quasi-equilibrium Kramers theory like arguments in the existing RFOT theory can be obtained purely dynamically using Fokker–Planck equations or path integrals just as has been done for isomerizations in chemical physics. I don't know if achieving such comfort with reaction theory will be sufficient to convince the entire statistical-physics community.

So you can see I am mostly at a loss as to what the community needs to be convinced. Perhaps the other interviewees can clue me in on what needs to be done and I can work on it!

Q7) Exactly solvable mean-field glass models exhibit an extraordinary complexity requiring impressive mathematical tools to solve them. Are you worried that for both the spin and structural glass problems it proves so hard to establish the validity of mean-field concepts in finite dimensions? Why are finite-dimensional versions of mean-field glasses always behaving in a very non mean-field manner?

Let me first make some remarks about your first statement that the exactly solvable mean-field glass models "exhibit an extraordinary complexity requiring impressive mathematical tools to solve them". This statement certainly characterizes how most people feel today. Yet at the very same time that I am impressed with the mathematical tools that have been invented to solve these problems, notably replica theory, I think our feeling of complexity about them is transitory. Any new formalism in theoretical physics seems complex to those who put it together and their contemporaries. Remember, again, it was Einstein who characterized one theory as "a witches brew of determinants sufficiently complex so that it can never be falsified." Many would be tempted to say the same today about replica approaches to glassy phenomena. The theory Einstein was talking about was Heisenberg's quantum mechanics!

Replica methodology and its interpretation do indeed seem very intricate—to me too. But I can see younger workers have little trouble with them. I suspect with still more pedagogical effort coming generations will find the ideas and manipulations of replica theories like operator quantum mechanics to be quite natural. After getting that off my chest, let me answer your specific questions.

I am aware of the continuing problems of establishing which aspects of the replica symmetry-breaking mean-field theory apply to ordinary spin-glasses in finite dimension. Although people seem agreed on the usefulness of mean-field theory for estimating critical temperatures, at the next level of detail, everything seems to be discussed rather turbulently. Again, for ordinary spin-glasses everyone agrees something a bit beyond a simple phase transition enters both experiments and simulations: spin-glasses exhibit significant aging, sensitivity to perturbations, etc. But there the consensus stops. Deciding whether these complex features reflect "broken replica symmetry" or "chaos" in a droplet picture or some hybrid of the two has taken much work. It has required developing precise definitions, doing big simulations, etc. I personally have not gotten too worked up about the controversy—primarily because I have not seen how the needed precision of argumentation helps us with the practicalities of understanding the problem of real random magnets or tells us how to use these ideas in other contexts like proteins, structural glasses, neural networks, etc. Perhaps I am a bit too American in my pragmatism, but it seems like the length scale at which the controversy rages keeps getting longer and longer with increasing years of study, making me feel that the controversy is more and more distant from being relevant to other problems where we might use the notions of glass theory.

Beyond this big controversy about replica symmetry breaking, still going back and forth between the Anglo-Saxons and the Continental Europeans focused on ordinary Ising spin glasses, however, results more worrisome to me have appeared in the study of Potts glasses, whose lack of up-down symmetry, recommends them to us at the mean-field level as analogs of structural glasses. Several peculiarities seem to be at work conspiring to cloud the simulations of the Potts systems. First, it was surprising how hard it was to even check the strict mean-field theory out in the infinite-range limit! Binder's calculations (Branigan *et al.*, 2003) and simulations carried out by others show a particularly slow convergence to the thermodynamic limit with system size even for the infinite-range Potts-glass model. This convergence issue also seems to have surfaced in the recent work by Parisi on long-range models that mimic the Kac limit (Franz and Parisi, 1999). Next, even in the strict infinite-range Potts-glass sample-to-sample fluctuations seem to play a big role in the obvious estimators signalling the transition. This difficulty, which occurs even for the infinite-range limit, must reflect itself even more starkly in tests of the RFOT mosaic scenario for finite-range Potts glasses.

These worries coming from studying finite-dimensional spin glasses do not, however, overwhelm my confidence in using RFOT ideas for structural molecular glasses for two technical reasons—both having to do with the "hardness" of the structural glass transition. Let me explain my terms.

Even in mean-field random first-order transitions are quite different from Ising spin-glass transitions—the latter are "soft", while RFOTs are "hard", but to varying degrees depending on the system. My characterization of the Ising spin-glass transitions as "soft" is my way of saying the Edwards–Anderson order parameters, q, varies continuously in magnitude at the transition (just as does the magnetization of a simple ferromagnet) at the transition value of q leads to a small restoring force for q fluctuations. This means the Ising spin glass is "soft" and yields all kinds of additional instabilities. Coping with these instabilities even at mean-field puts us on the long and winding road from the first mean field theories of Ising spin glasses through the radically brilliant one step replica symmetry-breaking solution proposed by Bray and Moore for the Sherrington–Kirkpatrick model to the triumphant Parisi hierarchical ansatz! At finite dimension the same sort of instabilities (tamed globally by the Parisi ansatz) lead to a zoo of critical replicon modes according to de Dominicis and to the annoyingly high upper critical dimension $d = 6$ or $d = 8$ according to Fisher. These soft modes are also at the core of the difficulty of testing the mean-field constructs in less than 8 dimensions for the Ising system.

The situation should be better for Potts glasses since their mean-field transition is "hard"—the Edwards–Anderson order parameter is discontinuous. This means no unstable modes appear at the mean-field transition—Bray and Moore's single step of RSB, inadequate for Ising systems, is enough—and everything should be wonderful! But, alas, there are droplets!!

The problem is that the discontinuity of q for Potts glasses, while strictly present, is small when the number of components is not huge! The transition is, while technically "hard", a bit squishy. The small discontinuity in q does formal miracles for those who love to use Landau expansions but it means the transition is effectively nearly soft: the interface energy cost for making a droplet is very small, so entropic droplets proliferate and interact with each other. Michael Eastwood and I made estimates suggesting these droplet excitations should have a big influence on the configurational entropy made for Potts glasses—enough perhaps to endanger the thermodynamic random first-order transition (Eastwood and Wolynes, 2002). This view would also be consistent with what we find by constructing the explicit magnetic analogy a la Stevenson *et al.* (Stevenson *et al.*, 2008). In the magnetic analogy, lowering the average surface tension as happens for a not very hard transition, while keeping its fluctuations and the field fluctuations (analogous to configurational entropy fluctuations) fixed wipes out the replica symmetry-breaking transition. When the transition is wiped out an analysis along the lines of Tarzia and Moore (Tarzia and Moore, 2007) would seem to be a good starting point.

Because of the small Lindemann parameter, structural glasses are much "harder" than are the Potts glasses with modest numbers of components that have been studied. The analog of the Edwards–Anderson order parameter, the plateau, in the first peak of the structure factor as revealed by neutron scattering is nearly about 0.9. This means the transition is quite hard, moving the surface costs of entropic droplets way up, so they cannot significantly renormalize the impending Kauzmann entropy crisis. The large activation free energy of entropic droplets, however, slows motions down as it must, making simulations of structural glasses difficult to do in the deep glassy regime.

I think the unusual "hardness" of molecular structural glasses is at the root of the quantitative success of the simplest RFOT theory as it currently stands. The existing RFOT theory just adds a non-perturbative decoration of droplets to the mean-field theory. This level of calculation is OK for structural molecular glasses but many other "glassy" systems will not be so easily analyzed since they are softer. We must learn to cope with all the issues of critical fluctuations, replica Goldstone modes, etc. if we want to completely understand such systems as glassy microemulsions, foams, etc.

Q8) In your view, do the recent ideas and experimental developments concerning jamming in granular media and colloids contribute to our understanding of molecular glasses, or are they essentially complementary?

The study of jamming is important in its own right. The experimental results showing very strong divergences near jamming also seem crisp and therefore are especially intriguing. None of them made sense to me in the context of molecular glasses until the recent work of Mari, *et al.* (Mari *et al.*, 2009), which was a real revelation. The connection they found between jamming with an extreme sort of non-equilibrium aging in which the pressure becomes infinite in the mean-field limit opens up the connection to the RFOT viewpoint on molecular glasses.

I think the replica approach makes clear, however, that the origin of the rigidity of near-equilibrium thermal glasses is not precisely the same as that for far from equilibrium jamming systems, however. The single J-point formalism that has been discussed so widely therefore seems way too oversimplified. The replica theory has a continuum of jammed structures. Much more work needs to be done, however, to go beyond mean-field theory in uniting jamming rigidity that reflects far from equilibrium aging with the rigidity of glasses, which is a near to equilibrium phenomenon. Clearly, additional theoretical innovations will be needed since the analog of fictive temperature is not set up so simply for a jammed granular or colloidal system as it is in a cooling protocol of a molecular liquid. Probably, there is some sort of turbulence cascade-like viewpoint (a la Kolmogorov) that will tell us how fictive temperature is generated and transported in these driven systems from large scales to short or vice versa. Perhaps more radical ideas involving intermittency or avalanches are needed. Progress on understanding driven jammed glasses will be essential to understanding structure formation in biology at the supramolecular level—the inside of cells is constantly being built and rebuilt alongside the usual diffusive motions of the individual macromolecular inanimate colloids that make up the cytoplasm. I consider exploring the connection between molecular glasses and jammed colloidal systems therefore a high priority.

Q9) If a young physicist asked you whether he or she should work on the glass problem in the next few years, would you encourage him or her and if so, which aspect of the glass problem would you recommend him or her to tackle? If no, what problem in condensed-matter theory should he or she tackle instead? If yes, what particular aspect of the glass problem seems the most exciting at present?

Yes, absolutely. In my view coping with diversity is the key problem of glassiness and is the key issue in many areas of theoretical physics. The "single-state" viewpoint still dominates throughout theoretical physics. I view this single-state tyranny as a road-block in many areas of physical, biological, and social science. We need young people to join the revolution. Examples where better coping with diversity would help include:

- "Hard" condensed-matter physics. The paradigms of Sommerfeld, Peierls and Landau still dominate—nearly everything is an attempted perturbation about the single-state Fermi liquid or some single-ordered structure. Those a bit more daring add two states or a quantum critical point to their thinking. Joerg Schmalian impressed me many years ago that we must take seriously the glassy behavior seen in the phase diagram of cuprates (Schmalian and Wolynes, 2000). Dobrosavljevic has gone further in suggesting a glassy system might provide the infamous "quantum critical point" so-desired by many as an explanation of the pseudo-gap phase (Panogopoulos and Dobrasavljevic, 2005).

  When I recall the ancient controversy in chemistry between whether molecules exhibit tautomerism (where different structures isomerize classically) or exhibit mesomerism (where different structures quantum mechanically superimpose via resonance), I come to feel that all the resonating valence-bond work starting with Anderson, might make more sense when viewed via the energy-landscape language of glassy systems.
- Particle physics and cosmology—Landscapes have entered these subjects leading to much controversy, even in the popular press (Susskind, 2005). When I read research papers in these areas I feel that this community is still groping with how to deal with diversity, and could learn something from the efforts of the condensed-matter community in studying glasses. In any case I am grateful to David Gross and his collaborators, already, for their early work on mean-field spin-glasses that helped RFOT theory get started. Why such a distinguished particle physicist worried about such a question is unclear. I will have to ask him!

- Biological physics—The study of protein dynamics and particularly protein folding continues to be influenced by glass physics. I have found this to be a two-way street for information flow.

  But diversity is a key issue in all of biological physics not just at the single-molecule level of proteins. Problems where diversity matters range from gene regulation (are cell types minima on a landscape? (Sasai and Wolynes, 2003)), the structure and dynamics of individual cells (what determines the shape of a cell?) and, of course, the problem of memory, where already a vigorous interchange with glass physics was developed a quarter century ago through the pioneering work of Hopfield.
- Econophysics—Didn't Keynes make it clear there were multiple states in the economy? His statement that "In the long run, we are all dead!" is a classic appreciation of the importance of long-lived, i.e. glassy states. Multiple solutions to satisfying individuals in their social and economic interchanges are obviously possible. Bouchaud's writings on this type of issue are very inspiring (Borghesi and Bouchaud, 2007).

In all of the above areas of theoretical science the battle even to acknowledge diversity as a factor has just begun. People really want to fit systems into the Procrustean beds of the single-state techniques they learned in school years ago. At the same time they feel no progress in understanding diversity has been made in any problem that clearly exhibits diversity. Thus, one of the psychological hold-ups to progress is really settling the issue of what's going on in molecular glasses. I think getting complete control of the molecular-glass problem is therefore widely important to the progress of physical theory as a whole.

I believe a young physicist who wants to work on any challenging problem in physics will eventually have to learn about glasses. She or he may also be able to directly contribute to the study of glasses and supercooled liquids, proper, while doing so.

In the study of molecular glasses over the next years four areas of enquiry stand out in my mind. The first is how to exploit the existing universal ideas from the RFOT theory to address the chemical and molecular details of liquids and amorphous materials: Water, that most biologically relevant and reactive solvent, has already been explored keeping in mind its amorphous state(s). I think the ideas from RFOT theory can help clarify many problems here. Amorphous semiconductors have exciting electronic properties for which local bonding "defects" are crucial. How can we look at such defects in a local energy-landscape picture? Vas Lubchenko has exciting ideas on this question.

Secondly, the properties of aging and rejuvenating systems need much more study both in the laboratory and from a formal theoretical point of view. Is time reparametrization symmetry an actual but emergent property away from the mean-field limit?

Thirdly, the truly intermediate regime around the crossover temperature may have a deeper representation than in the current RFOT mean-field droplet approach. Within RFOT theory this transition comes about from the ramification of the entropic droplets and resembles the Hagedorn transition of string theory (Stevenson *et al.*, 2006*a*). Is there a way to see strings formally as real dynamic entities with microscopically calculated interactions? I am not satisfied with all the approximations made in Langer's theory. Is there a way to picture some sort of duality between the situation above and below the crossover—mobile strings below morphing into partially rigid structures above?

Fourthly, the origin of amorphousness still needs clarification. RFOT theory takes aperiodic minimal energy structures as given. Simulations show aperiodic minimia exist so that is enough. Are these minima just superpositions of defects a la the icosahedratic theory for simple fluids? Or is such a view something like Landau's construction of turbulent states via multiple periodic instabilities. He thought a system becoming turbulent just continued bifurcating but instead strange attractors were found at the onset of turbulence. Like turbulence, is spatial chaos intrinsic and more subtle?

There is plenty to do and time is a-wastin'!

Q10) In twenty years from now, what concepts, ideas or results obtained on the glass transition in the last twenty years will be remembered?

Results from the last 20 years or so should already be part of the curriculum. The mere idea of an existence of an energy landscape with great diversity is non-controversial and is a powerful way even for undergraduates to picture many systems ranging from liquids to proteins. There is no reason not to teach about some formal aspects of glass-transition theory: strict mode-coupling theory (as a resummed perturbation theory), density functional treatments of aperiodic minima, basic replica methods and how the entropy crisis occurs in simple models. No one really disputes these are mathematically correct any longer and these ideas are the starting point for future advances. I think the entropic droplet and mosaic notions may be modified but will survive further examination. The future of current issues including the critical phenomena aspects and effects of interactions between activated events ("facilitation") will I hope be soon enough settled to be remembered. Simple facilitated models will survive, not because of their direct applicability to molecular liquids, but because of their pedagogic simplicity and applicability to other systems.

Q11) If you met an omniscient God and were allowed one single question on glasses, what would it be?

The *New York Times* has already quoted one scientist as derisively referring to RFOT theory as "metaphysical", so I really don't want to go further by entering into theology. My own thinking parallels a joke that made the rounds in America a few years ago after the truly tragic mismanagement of the Hurricane Katrina relief:

A devout believer was living in the predicted path of the Hurricane. TV reports urged all those in the path to evacuate to safer ground. The believer did not go saying "I trust in the Lord. He will save me." The flooding began and the water came up to the level of the porch. At this time the sheriff came by in a boat, asking the believer to come with him to safety. The believer did not go, saying, "I trust in the Lord. He will save me." The water continued to rise, finally reaching the roof of the house.

The believer had already climbed out onto the roof. A helicopter appeared and the National Guard lowered a ladder telling the believer to come on up to be whisked away. The believer stayed on his roof citing his trust that the Lord would save him.

The believer drowned.

Being a true believer he did go to Heaven. There he asked God, "Why didn't you save me?" The Lord answered "I allowed meteorologists to predict the storm's path. You didn't listen. I sent the sheriff with his boat to get you. You didn't listen. Finally I sent a helicopter. I really don't think it is fair to ask me that question!"

1.4.1 Reflections on the interview responses, by Peter G. Wolynes

Despite differences of tone and emphasis between the respondents I think there are many aspects of the glass transition on which there is agreement. Many of the issues raised by Langer, Witten and Kurchan were already discussed by me in my first response. I limit myself primarily here to directing the reader to specific papers that discuss a few more of the issues raised by the other responses.

Langer spells out some specifics about his discomforts with RFOT theory. Many of his worries have already been mentioned in existing papers on RFOT theory. Doubtless these issues haven't been completely resolved to everyone's satisfaction but should provide an agenda for future work or reformulations.

It is worth pointing out that the early RFOT theory papers contrasted the nucleation-like arguments of RFOT theory and those for ordinary first-order transitions and emphasized that there were differences as to growth of unstable phases, etc. The reformulation of the nucleation-style arguments by Lubchenko and Wolynes was aimed at aging (Lubchenko and Wolynes, 2004) but is indeed quite consistent with the contemporary Biroli–Bouchaud arguments for equilibrated samples and makes clear why the problem is distinct from the traditional nucleation of an unstable phase. The Lubchenko–Wolynes paper discusses in a critical way the quasi-equilibrium assumption of the RFOT theory: RFOT theory does indeed postulate that the only constraints on motion across the transition state for escape from a minimum are those that are in fact generated by the direct interactions. Quantitatively, these constraints are accounted for in the density—functional starting point. Indeed, other additional topological constraints may enter for some systems—entangled polymers were noted as an example by LW. "Short cuts" to escape from a present local minimum would also not satisfy the quasi-equilibrium assumption: such short cuts were discussed by the LW paper as well in the language of defect models. Specifically, the role of string-like excitations that can give short cuts for minimum escape has been discussed by Stevenson, Schmalian and myself (Stevenson *et al.*, 2006*b*) and their importance to secondary relaxation has been explored by us recently (Stevenson and Wolynes, 2009).

Witten raises the issues of soft modes and two-level systems that are apparent at cryogenic temperatures. Although my remarks were aimed at the high-temperature, nearly classical phenomena, the RFOT theory does make some predictions about the relevant quantum low-temperature properties that depend on dynamic modes, see Lubchenko and Wolynes, Phys. Rev. Lett. (Lubchenko and Wolynes, 2001), PNAS

(Lubchenko and Wolynes, 2003) and a long article in Advances in Chemical Physics (Lubchenko and Wolynes, 2007*c*).

I found Kurchan's responses very much on the mark and agree with the gestalt of his description of the RFOT approach. I particularly enjoyed his metaphor of the RFOT theory as a building with the "upper floor made of wood for lack of budget." (May I quote this next time I apply for research funds?) In this regard, however, I would remind seekers of the Holy Grail of glass physics of the lesson taught by Henry Jones, Jr. in the film "Indiana Jones and the Last Crusade": the power of the Grail is not entirely determined by its materials of construction. Choose wisely!

References

Barrat, J.L. and Berthier, L. (2002). *J. Chem. Phys.*, **116**, 6228.

Berthier, L., Biroli, G., Bouchaud, J.P., *et al.* (2005). *Science*, **310**, 1797–1800.

Berthier, L. and Bouchaud, J.-P. (2002). *Phys. Rev. B*, **66**, 054404.

Bhattacharyya, S. M., Bagchi, B., and Wolynes, P.G. (2008). *PNAS*, **105**, 16077–16082.

Biroli, G., Bouchaud, J.P., and Tarjus, G. (2005). *J. Chem. Phys.*, **123**, 044510.

Biroli, G., Bouchaud, J.P., Cavagna, A., *et al.* (2008). *Nature Phys.*, **4**, 771–775.

Borghesi, C. and Bouchaud, J.P. (2007). *Quality & Quantity*, **441**, 557–568.

Bouchaud, J.-P. and Biroli, G. (2004). *J. Chem. Phys.*, **121**, 7347.

Bouchbinder, E. and Langer, J.S. (2009). *Phys. Rev. E*, **80**, 031132.

Branigan, C., Kob, W., and Binder, K. (2003). *J. Phys. A*, **36**, 10847–10866.

Brooke, J., Bitko, D., Rosenbaum, T., and Aeppli, G. (1999). *Science*, **284**, 779–781.

Capaccioli, S., Ruocco, G., and Zamponi, F. (2008). *J. Phys. Chem B*, **112**, 10652–10658.

Cavagna, A. (2009). *Phys. Rep.*, **476**, 51.

Cugliandolo, L., Kurchan, J., and Peliti, L. (1997). *Phys. Rev. E*, **55**, 3898.

Dotsenko, V. (2004). *J. Stat. Phys.*, **115**, 823.

Dyre, J. (2006). *Rev. Mod. Phys.*, **78**, 953–972.

Dzero, M., Schmalian, J., and Wolynes, P.G. (2005). *Phys. Rev. B*, **72**, 100201.

Eastwood, M.P. and Wolynes, P.G. (2002). *Europhys. Lett.*, **60**, 587–593.

Fisher, D.S. and Huse, D.A. (1988*a*). *Phys. Rev. B*, **38**, 373.

Fisher, D.S. and Huse, D.A. (1988*b*). *Phys. Rev. B*, **38**, 386.

Franz, S. (2007). *J. Stat. Phys.*, **126**, 765–780.

Franz, S. and Parisi, G. (1999). *Europhys. J. B.*, **8**, 417–422.

Glotzer, S.C. (2000). *J. Non-Cryst. Solids*, **274**, 342.

Hall, R. W. and Wolynes, P. G. (1987). *J. Chem. Phys.*, **86**, 2943–2948.

Hall, R. W. and Wolynes, P. G. (2003). *Phys. Rev. Lett.*, **90**, 085505.

Kearns, K.L., Swallen, S.F., M. Ediger *et al.* (2008). *J. Phys. Chem.*, **112**, 4934–4942.

Kirkpatrick, R., Thirumalai, D., and Wolynes, P. G. (1989). *Phys. Rev. A*, **1045–1054**(40).

Kirkpatrick, R. and Wolynes, P. G. (1987*a*). *Phys. Rev. A*, **35**, 3072–3080.

Kirkpatrick, R. and Wolynes, P. G. (1987*b*). *Phys. Rev. B*, **36**, 8552–8564.

Langer, J.S. (1967). *Ann. Phys.*, **41**, 108.

Langer, J.S. (1969). *Ann. Phys.*, **54**, 258.
Langer, J.S. (2006*a*). *Phys. Rev. E*, **73**, 041504.
Langer, J.S. (2006*b*). *Phys. Rev. Lett.*, **97**, 115704.
Langer, J.S. (2008). *Phys. Rev. E*, **78**, 051115.
Langer, J.S. and Manning, M. L. (2007). *Phys. Rev. E*, **76**, 056107.
Lubchenko, V. and Wolynes, P.G. (2001). *Phys. Rev. Lett.*, **87**, 195901.
Lubchenko, V. and Wolynes, P.G. (2004). *J. Chem. Phys*, **121**, 2852–2865.
Lubchenko, V. and Wolynes, P.G. (2007*a*). *Annu. Rev. Phys. Chem.*, **58**, 235–266.
Lubchenko, V. and Wolynes, P.G. (2007*c*). *Adv. Chem. Phys.*, **136**, 95–206.
Lubchenko, V. and Wolynes, P. G. (2003). *Proc. Natl. Acad. Sci. USA*, **100**, 1515–1518.
Mari, R., Krzakala, F., and Kurchan, J. (2009). *Phys. Rev. Lett.*, **103**, 025701.
Marinari, E., Parisi, G., and Ritort, F. (1994). *J. Phys A*, **27**, 7647.
Marinari, E., Parisi, G., and Ritort, F. (1995). *J. Phys A*, **28**, 327.
Nelson, D.R. (1983). *Phys. Rev. Lett.*, **50**, 982–985.
Nussinov, Z. (2004). *Phys. Rev. B*, **69**, 014208.
Panogopoulos, C. and Dobrasavljevic, V. (2005). *Phys. Rev. B*, **72**, 014536.
Parisi, G. (2006). *Proc. Natl Acad. Sci. USA*, **103**, 7948–7955.
Russell, E.V. and Israeloff, N.E. (2000). *Nature*, **408**, 695–698.
Saltzman, E.J., Yatsenko, G., and Schweizer, K.S. (2008). *JPCM*, **20**, 244129.
Sasai, M. and Wolynes, P. G. (2003). *Proc. Natl. Acad. Sci. USA*, **100**, 2374–2379.
Schmalian, J. and Wolynes, P. (2000). *Phys. Rev. Lett.*, **85**, 836–839.
Schweizer, K. (2007). *J. Chem. Phys.*, **127**, 164506.
Sellitto, M. and Kurchan, J. (2005). *Phys. Rev. Lett.*, **95**, 236001.
Stevenson, J.D., Schmalian, J., and Wolynes, P.G. (2006*a*). *Nature Phys.*, **21**, 268–274.
Stevenson, J.D., Schmalian, J., and Wolynes, P.G. (2006*b*). *Nature Phys.*, **21**, 268–274.
Stevenson, J.D., Walczak, A.M., Hall, R.W., and Wolynes, P.G. (2008). *J. Chem. Phys.*, **129**, 194505.
Stevenson, J.D. and Wolynes, P.G. (2008). *J. Chem. Phys.*, **129**, 234514.
Stevenson, J.D. and Wolynes, P. G. (2009). *Nature Phys.*, **6**, 62–68.
Susskind, L. (2005). *The Cosmic Landscape*. Little, Brown & Company.
Tarzia, M. and Moore, M.A. (2007). *Phys. Rev. E*, **75**, 031502.
The JANUS, collaboration (2008). *Phys. Rev. Lett.*, **101**, 157201.
Tracht, U., Wilhelm, M., Heuer, A., Feng, H., Schmidt-Rohr, K., and Spiess, H.W. (1998). *Phys. Rev. Lett.*, **81**, 2727–2730.
Widmer-Cooper, A. and Harrowell, P. (2006). *Phys. Rev. Lett.*, **96**, 185701.
Widmer-Cooper, A. and Harrowell, P. (2007). *J. Chem. Phys.*, **126**, 154503.
Wolynes, P. G. (1989). In *Proceedings of the International Symposium on Frontiers in Science* (ed. H. Frauenfelder Festschrift, S. Chan, and P. G. DeBrunner), pp. 39–65. Am. Inst. Physics.
Xia, X. and Wolynes, P.G. (2000). *Proc. Natl. Acad. Sci. USA*, **97**, 2990.
Xia, X. and Wolynes, P.G. (2001). *Phys. Rev. Lett.*, **86**, 5526.

2
An overview of the theories of the glass transition

Gilles TARJUS

Abstract

The topic of the glass transition gives rise to a a wide diversity of views. It is, accordingly, characterized by a lack of agreement on which would be the most profitable theoretical perspective. In this chapter, I provide some elements that can help sorting out the many theoretical approaches, understanding their foundations, as well as discussing their validity and mutual compatibility. Along the way, I describe the progress made in the last twenty years, including new insights concerning the spatial heterogeneity of the dynamics and the characteristic length scales associated with the glass transition. An emphasis is put on those theories that associate glass formation with growing collective behavior and emerging universality.

2.1 Introduction

Anyone who takes a fresh look at the literature on the glass transition cannot fail to be struck by the wide diversity of views and approaches. There is a broad consensus on the fact that understanding supercooled liquids, glasses and glass formation represents a deep, interesting, mysterious, and fundamental problem... yet, there is no general agreement concerning what actually makes the problem deep, interesting, mysterious, and fundamental, nor on the paths to be explored to solve it. Even worse, there does not seem to exist among the scientific community a shared view of what kind of theoretical achievement would be needed to declare the problem solved. As a result, claims that the problem has already been solved are not uncommon, but encounter a high level of skepticism.

The main purpose of the present review is neither to detail the main features of the phenomenology of glass formation nor to thoroughly assess the validity of the proposed theoretical descriptions, a necessarily very subjective exercise anyhow. Many such reviews have already been published (Angell, 1995; Ediger *et al.*, 1996; Debenedetti,

1996; Ngai, 2000; Tarjus and Kivelson, 2001; Debenedetti and Stillinger, 2001; Alba-Simionesco, 2001; Cavagna, 2009; Berthier and Biroli, 2009)[1] and there is no point in adding one more opus of the sort. I would rather like to give the reader a few indications to get oriented in the apparent jumble formed by the various approaches. Along the way I will also try to identify intrinsic features of the phenomenology that may represent obstacles to a full resolution of the problem and important questions that remain unsettled.

Before moving on to a more detailed presentation, it is worth stressing that some form of consensus does exist on at least a few points. For instance, it is by now established that the observed glass "transition" is not a *bona fide* phase transition, but rather a dynamical crossover through which a viscous liquid falls out of equilibrium and appears solid on the experimental time scale. (I will nonetheless use the term glass transition without quotation marks in what follows as this is common practice, but the reader should keep in mind the present warning.) The glass-transition temperature T_g at which this, indeed quite sharp, crossover occurs depends on cooling rate or observation time; it is fixed by some operational convention and, typically, corresponds to a viscosity that reaches 10^{13} Poise or a relaxation time that is of the order of 10^2 s. Another widely accepted starting point is that equilibrium statistical mechanics is the relevant framework to describe glass-forming systems, including supercooled liquids that are in a metastable state compared to the crystal; the latter is therefore excluded from theoretical descriptions of glass formation. [This seems reasonable but may sometimes require some caution: see (Cavagna, 2009) for a pedagogical discussion.]

Considered with some historical perspective, the field seems to be only very slowly evolving. Most of the theories and concepts that are at the forefront of nowadays discussions about the glass transition were around by the mid-1980s. Besides the defect-diffusion (Glarum, 1960), free-volume (Cohen and Turnbull, 1959; Turnbull and Cohen, 1961) and configurational-entropy (Gibbs and Di Marzio, 1958; Gibbs, 1960; Adam and Gibbs, 1965) approaches that date back to the 1960s, this is for instance the case for the energy-landscape picture (Stillinger and Weber, 1983), the mode-coupling (Bengtzelius *et al.*, 1984; Leuthesser, 1984) and the random first-order transition (Singh *et al.*, 1985; Kirkpatrick and Wolynes, 1987*a*,*b*) theories, the concepts of kinetic constraints (Palmer *et al.*, 1984; Fredrickson and Andersen, 1984) and of geometric frustration (Kléman and Sadoc, 1979; Sethna, 1983; Nelson, 1983*a*,*b*; Nelson and Widom, 1984; Sachdev and Nelson, 1985; Sadoc and Mosseri, 1984), all approaches that will be discussed in more detail below. Similarly, many of the experimental advances were operational twenty years ago. This observation certainly generates a distressing feeling that the glass-transition problem, one of the oldest puzzles in physics, does not get any closer to a resolution. However, there is room for optimism. First, new ideas and standpoints do emerge. This is illustrated by the relatively recent surge of experimental, numerical and theoretical work on a previously overlooked aspect of the relaxation in glass-forming systems: the increasing heterogeneous character of the dynamics and the associated

[1] Topical reviews on a variety of theoretical approaches will be mentioned further down in the text.

growth of space-time correlations as one approaches the glass transition, a property that represents the central topic of this book. In addition, progress has been made along several lines of research, with the development of models and theoretical tools, and sometimes of crisper and possibly testable predictions.

2.2 A diversity of views and approaches

2.2.1 What is there to be explained?

What is there to be explained about the glass transition? What is the scope of the description, i.e. what is the range of phenomena and of systems to be included? Answering those questions is already a perilous task that requires *a priori* choices. The word "glass" and the qualifier "glassy" are used in a great variety of contexts to describe systems with unusually sluggish dynamics, in which the degrees of freedom of interest remain (apparently) disordered. Such systems can get stuck in an arrested state when the dynamics becomes too slow to be detectable. Some generic phenomena are then observed in a slew of apparently unrelated materials. As an illustration, consider what goes under the term of "aging" (Struik, 1978; Bouchaud *et al.*, 1998). Aging describes the fact that the properties of a system depends on its "age", i.e. on the elapsed time since it has been prepared. This is most clearly seen in two-time response and correlation functions that, in the aging regime, lose the time–translation invariance found in equilibrium states and depend on the waiting time between preparation of the system and beginning of the measurement. This aging phenomenon, in which an apparent relaxation time is set by the waiting time, is encountered in very different materials; to list a few: polymer glasses (plastics), molecular glasses, colloidal gels, foams, spin glasses, vortex glasses, electron glasses. Aging therefore appears as a "universal" property of all out-of-equilibrium systems whose relevant degrees of freedom look frozen on the experimental time scale, while still exhibiting some residual motion. In the above examples, the "glassy" degrees of freedom are associated with molecules, monomers, colloidal particles, bubbles, spin magnetic dipoles, vortices or electrons, and "glassiness" can either be self-generated or result from the presence of quenched disorder due to impurities and defects. At this level of generality, it is unclear whether one should look for a basic principle of out-of-equilibrium dynamics or first subdivide the systems into classes characterized, say, by different aging exponents.

While it can certainly be a worthy strategy to try solving the "glassy" problem in its full blown but rather ill-defined generality, I will take here a more restrictive view that focuses on the glass transition of liquids and polymers. In any case, this can be taken as a starting point. Other systems and more phenomena could be fruitfully brought into the picture, but, in my opinion, only to the extent that their behavior is argued to closely resemble that of glass-forming liquids and polymers. Needless to say that this approach may not be shared by the whole "glass community".

Viewed from the "liquid side", i.e. upon cooling a liquid (or, possibly, compressing it, see below), the phenomenon of glass formation stands out by the spectacular slowing down of relaxation and the related increase of viscosity that both take place in a continuous manner as temperature is decreased: one observes changes of up to 14 orders of magnitude in the main (α) relaxation time and the viscosity for a 30% variation in temperature. In the same temperature range, the average static correlations between pairs of molecules in the liquid, as probed through static structure factors, barely change. Provided of course crystallization is bypassed, this unique feature, which combines a dramatic evolution of the dynamics and an apparently very modest structural change, is seen in virtually all kinds of liquids (inorganic and organic, ionic salts and metallic alloys, polymer melts), liquids with a variety of molecular shapes and a wide range of intermolecular forces.

The phenomenology of glass-forming liquids and polymers is characterized by salient generic trends as temperature decreases. Besides the already-mentioned dramatic increase of α relaxation time and viscosity, whose temperature dependence is typically described by a faster-than-Arrhenius form, and the concomitant bland behavior of the static structure factors, the main qualitative features can be summarized as follows: a marked non-exponential time dependence of the relaxation functions (or, equivalently, a marked non-Debye behavior of the frequency-dependent susceptibilitites), the appearance and the development of several relaxation regimes, an increasingly heterogeneous character of the dynamics, with the coexistence over an extended period of time of fast and slow regions, a significant decrease of the entropy, with the difference between the entropy of the supercooled liquid and that of the corresponding crystal dropping by up to a factor of 3 between the melting and the glass-transition temperatures. (Again, because the way to present the phenomenology always comes with a pre-conceived picture, some researchers would probably want to add a few items to the list of central features.)

The above trends and dependences on temperature can be fitted, with reasonable accuracy, by means of various functional forms. This fact supports the generic and "universal" character of the glass-forming behavior. The fitting formulas on the other hand include temperature-independent but material-dependent parameters that contain the specific and non-universal aspects of the problem. For instance, the degree of super-Arrhenius behavior, i.e. of departure from Arrhenius temperature dependence of the α relaxation time or the viscosity, is material specific: it is embodied in several (alternative) indices that quantify how "fragile"(Angell, 1985) a glass former is. One may of course dream of being ultimately able to predict these material-specific quantities from a first-principle, microscopic theory. In the mean time and more modestly, it is at least possible to look for empirical correlations among them. Numerous attempts of this kind have been made over the years, with varying degrees of robustness. They can be roughly classified into three groups: correlations among characteristics of the slow dynamics, correlations between slow dynamics and thermodynamics, and correlations between properties of the slow and of the fast dynamics. In the first group, one finds for instance correlations between the fragility (see above and (Angell, 1985)) and the degree of non-exponential behavior of the relaxation functions (the "stretching" parameter) (Böhmer *et al.*, 1993) or between the stretching of the relaxation and the amount of decoupling in the temperature dependences of the translational and the rotational diffusion motions (Ediger, 2000). In the second group are the correlation between the heat-capacity jump at the glass transition T_g and the fragility (Angell, 1985) or else between the decrease of the "configurational" entropy (excess entropy of the supercooled liquid over that of the crystal) and the increase of the relaxation time (Richert and Angell, 1998; Martinez and Angell, 2001). Finally, in the third group, one can list many proposed correlations between the fragility and various characteristics of the fast dynamics, either in the liquid or the glass (relative amplitude of the Boson peak (Sokolov *et al.*, 1993), Poisson ratio of elastic moduli (Novikov and Sokolov, 2004), non-ergodicity parameter (Scopigno *et al.*, 2003), etc.). Whether one discards as fortuitous and wobbly, or else takes seriously some of these correlations is strongly connected to the type of description of the glass transition one is aiming at.

2.2.2 Sorting out the theoretical approaches

Even with the restriction that one focuses on what governs glass formation in liquids and polymers, there is plenty of room for diversity and controversy. A central question, to which different answers are given and that is therefore far from purely rhetorical, is whether a general theory of glass formation is possible? By "general theory" is meant here a theory that explains the main features associated with the glass transition, without having to rely on a full description of the microscopic details that vary from one substance to another (Kivelson and Tarjus, 2008). A corollary to the existence of such a theory is that the phenomenon displays some form of "universality". Although challengeable, a number of elements point toward a positive answer, which most certainly explains the attraction that the field has exerted on theorists over the years.

As briefly summarized above, many pieces of the phenomenology of glass-forming liquids appear general enough to be described by master curves (with admittedly a number of material-specific adjustable parameters) and empirical correlations are (more or less successfully) established between various properties. All of this suggests some common underlying explanation. Beyond this, the dramatic super-Arrhenius rise of the α relaxation time with decreasing temperature, which for a given glass former is very similar in a wide variety of dynamical measurements (with a few well-identified and interesting exceptions corresponding to a decoupling of the motions either of different entities, as the decoupling between conductivity and viscosity in ionic mixtures (Ediger *et al.*, 1996), or of the same entities probed on very different length scales, as the decoupling between local rotational relaxation and translational diffusion in molecular liquids and polymers (Ediger, 2000; Sillescu, 1999)), seems indicative of growing collective behavior. Such a growth is of course what partly washes out the influence of molecular details, which can then be incorporated in effective parameters, and can lend itself to a general theory.

Attributing the viscous slowing down in supercooled liquids and polymers to a collective phenomenon is at the core of most theories. Collective behavior implies the existence of at least one supermolecular length scale that can be associated with growing correlations or growing coherence in the system as temperature decreases. This issue of length scales has indeed become central in most recent discussions and developments concerning the glass transition. More will be said below on this, but for now it is worth recalling a few points. First, the nature of the putative supermolecular lengths characterizing the slowdown of relaxation and flow is quite elusive. Nothing much happens in the static pair correlations, which are the only measures of the structure that are easily accessible experimentally. As a consequence, if spatial correlations of one sort or another grow, they must be of rather subtle and unusual nature. The experimental observation that has recently put length scales at the forefront is that of the heterogeneous character of the dynamics. It has been shown through both numerical and experimental work that the dynamics becomes spatially correlated as one approaches the glass transition, with an emerging length scale related to the typical size of the dynamical heterogeneities (see (Ediger, 2000; Sillescu, 1999) and the other chapters of this book). It is not clear, however, if this "dynamical" length is the one controlling the viscous slowing down. Another keypoint, and a most annoying fact, is that, irrespective of their precise definition, supermolecular lengths in glass-forming systems never seem to grow bigger than a few nanometers, i.e. 5 to 10 molecular diameters at most (Ediger, 2000; Berthier *et al.*, 2005; Dalle-Ferrier *et al.*, 2007; Richert *et al.*, 2011). Collective behavior is thus present, but it may never fully dominate local, molecular effects in the experimentally accessible range. The absence of a singularity with diverging time- and length-scales that could be seen or closely approached in experiments can be viewed as an intrinsic difficulty of the glass-transition problem.

Looking for a general theory does not exhaust the possible lines of research on glass formation. For instance, one may want to know how the processes by which molecules move change between the high-temperature liquid and the glass transition or to describe the viscous slowing down in terms of specific mechanisms for relaxation and flow. Universality, considered in a weaker sense than when associated with diverging length scales at critical points, could still be observed if a well-identified mechanism prevails. This would then be more akin to the dislocation description of plastic motion for crystals (Orowan, 1934; Polanyi, 1934). Finally, it could also well be, but this would be disappointing to many, that understanding glass formation lies in chemistry with no way out of microscopic and substance-specific computations or measurements, the generic character of the phenomenology summarized above being then only superficial.

To contrast the various theoretical approaches of the glass transition, it is helpful to sort out their starting points and perspectives according to pairs of related opposites:

(i) *Liquid side versus amorphous-solid side.*

Above, I have emphasized the approach of the glass transition from the liquid side, as it shows the most remarkable pieces of phenomenology. However, viewed from the "solid side", glass formation may be quite fascinating too. What makes the rigidity of an amorphous phase devoid of long-range order?, what is the specificity of the associated vibrational modes and elastic response?, how does the glass "melt"? are all challenging questions (see e.g. (Wyart, 2005)). The glass is for sure a frozen liquid, but it has also been suggested to envisage the viscous liquid as a "solid that flows" (Dyre, 2006).

(ii) *Coarse-graining, scaling and underlying critical point versus atomic-motion level and relaxation mechanisms.*

Growing collective behavior is a well-established cause of emerging universality, and it is then tempting to apply the recipes that have been very successful in statistical physics: coarse-graining procedures and use of effective theories, search for critical points around which to organize scaling analysis. Being biased toward the possibility of such a general theory of the collective behavior associated with glass formation, I may somewhat overlook in the following the numerous "microscopic" approaches that focus on describing the actual motion of the molecules and the way it changes as temperature is decreased. Such detailed information is hardly obtainable from experiments, with the possible exception of colloidal suspensions and driven granular media, but is provided by computer simulations (Kob, 2005). In addition, as alluded to above, it may also well be that the apparent universality of the slowdown of dynamics leading to glass formation originates from the predominance of a specific relaxation mechanism rather than from an underlying critical point.

(iii) *Real three-dimensional space versus configurational space.*

That relaxation in a viscous liquid seems well described in terms of thermally activated events suggests that the system may be temporarily trapped in "metastable states" and, accordingly, that a fruitful framework is provided by a description of the "energy landscape" and its topographic properties (Goldstein, 1969; Stillinger and Weber, 1983; Debenedetti and Stillinger, 2001). This configurational-space picture, which focuses on the hypersurface formed by the potential energy or some coarse-grained free energy as a function of all configurational variables, is one of the paths followed to describe glassy systems. Alternatively, one may attempt a real-space description of the phenomena. For instance, activated processes correspond to rare, localized events that are more easily describable from a real-space than from a configurational-space perspective. A similar conclusion applies to the dynamical heterogeneities.

(iv) *Kinetics versus statics.*

The observed glass transition is undoubtedly a kinetic crossover and the most spectacular phenomena, slowing down and dynamical heterogeneities, pertain to the dynamics. Yet, on general grounds (see below), one may expect some thermodynamic and structural underpinning. Whether the physics of glass formation can be understood on the basis of a purely dynamical origin with no thermodynamic signature or requires a thermodynamic or structural explanation represents a central issue. The case for the latter option is *a priori* difficult to pin down due to the absence of clear evidence showing growing thermodynamic or structural correlations as one approaches the glass transition. (It can of course be argued that such correlations are subtle, and consequently hard to detect, and that their growth is limited, with an impact on the dynamics that is enormously amplified.) On the other hand, a purely kinetic explanation overlooks the thermodynamic aspects of the phenomenology, such as the rapid decrease of the "configurational" entropy.

Before closing this section, I would like to come back to the theories that relate the apparently universal features of glass formation to some collective behavior controlled by underlying phase transitions. As no such transitions are observed in experiments, the putative singularities must be located outside of the physically accessible range of parameters: the postulated critical points and phase transitions may then be either *avoided* or *unreachable*. "Avoided" implies that the singularity only appears as a crossover in real life, at a temperature that is above the experimental T_g; "unreachable" means that the transition occurs at a temperature below T_g, being for that reason inaccessible in experiments. In line with point (iv) made above, the nature of the hypothesized critical points and transitions can moreover be either dynamic or static. A more detailed description along these elements of classification will be provided in Sections 2.4 and 2.5.

2.3 Elements of theoretical strategies

As briefly sketched in the preceding sections, the pursuit of a theoretical description of the glass transition is multiform. An overview of theories should then try to do justice to the diversity of concepts, models, tools, ideas, and frameworks that have been put forward in the context of glass-forming systems. The motivation for such an exercise is not only oecumenical. First, at the present time, no theoretical approach has been accepted as the definite answer to the glass problem: different paths may ultimately prove fruitful and complementary. Secondly, some of the constructs devised in the course of studies on glass formers may find a life of their own and be transposable to other fields. The list given below does not pretend to be exhaustive.

2.3.1 Minimal models and simplifying concepts

An obvious strategy in physics to address a seemingly general phenomenon is trying to identify the main ingredients and get rid of all superfluous microscopic details. Through some level of coarse graining, choice of relevant degrees of freedom and use of simplifying concepts, minimal models and effective theories can then be proposed as starting points for further studies. To pick up a well-known example in a closely related field: the study of spin glasses has been put on a firm basis with the proposal by Edwards and Anderson of their lattice hamiltonian model. In this case, two physical ingredients, frustration and quenched disorder, have been incorporated in a simple classical Ising hamiltonian (Edwards and Anderson, 1975). The order parameter and the associated phase transition are then adequately described (at least at a minimum level), and the detailed nature of the spin degrees of freedom (except for symmetry), of the interactions among the latter, of the distribution of the quenched disorder, and of the underlying lattice is essentially irrelevant to an understanding of the spin-glass phenomenology. This, by far, does not mean that the spin-glass problem is solved. Central questions concerning the Edwards–Anderson model, such as the nature of the spin-glass phase and the existence of a transition in non-zero magnetic field, are still awaiting a definite answer. But the starting point is well accepted.

The situation is unfortunately not as clear for glass-forming liquids. There are models of liquids, but they can hardly be considered as minimal models for a theory of glass formation. This is for instance the case of the Lennard-Jones models for atomic liquid mixtures and of the hard-sphere models for repulsive colloidal suspensions, which are at the heart of most computer simulation studies (Kob, 2005). Such "realistic" atomistic models still incorporate a wealth of microscopic details without allowing one to tune in any significant way the salient features of glass formation such as the fragility.

To build minimal models or effective theories of glass-forming systems, several routes have been proposed. They rely on various ingredients and concepts that are hypothesized to account for the origin of the increased sluggishness and of the heterogeneous character of the relaxation as temperature decreases, be they cooperativity, free volume, jamming, kinetic constraints, facilitation, local packing constraints, geometric frustration, etc. They may also involve analogies with spin-glass and other disordered models (Kirkpatrick and Wolynes, 1987*a*,*b*; Kirkpatrick and Thirumalai, 1987, 1989; Sethna *et al.*, 1991; Moore and Drossel, 2002), uniformly frustrated models (Nelson, 2002; Kim and Lee, 1997; Grousson *et al.*, 2002*a*,*b*), etc. Some of these attempts will be further discussed below.

2.3.2 Looking for a localized relaxation mechanism in real space

That a full understanding of glass formation will ultimately require knowledge of the principal relaxation mechanism(s) is rather undisputed, as is the fact that relaxation in glass-forming liquids proceeds via localized events in space. On the other hand, looking for such mechanisms may not necessarily be considered as the highest priority for building a theory. Here, however, I more specifically refer to those approaches that put emphasis on mechanisms without invoking underlying phase transitions.

A line of research indeed posits that understanding glass formation only requires identifying a dominant relaxation mechanism that is supposed to take over as the liquid becomes more viscous. Examples taking this perspective include the excitation-chain (Langer, 2006*a*,*b*) and single-particle barrier-hopping (Saltzman and Schweizer, 2006) approaches, the description in terms of quasiparticles (Boué *et al.*, 2009), and the elastic models (Dyre, 2006). In the "shoving model" (Dyre, 2006) for instance, localized flow events are assumed to take place in compact regions whose rearrangement involves an increase of volume. The activation energy then comes from the elastic cost associated with "shoving" the rest of the system, which is taken as a solid on the short time scale of the event. In this work, the super-Arrhenius increase of the activation energy is related to the temperature dependence of the infinite-frequency shear modulus. Slowing down occurs with no growing length scale (hence with no underlying singularity), since the typical size of the rearranging regions stays constant with decreasing temperature.

It is also worth noting that computer simulations of atomistic models offer an opportunity to look for the main mechanisms for relaxation and flow in real space. In this vein, most simulation studies on the heterogeneity of the dynamics in glass-forming liquids have focused on dynamical clustering, providing some evidence for strings and microstrings (Donati *et al.*, 1998; Vogel *et al.*, 2004), or other types of clusters (Appignanesi *et al.*, 2006). More recently, Harrowell and coworkers (Widmer-Cooper *et al.*, 2008; Widmer-Cooper and Harrowell, 2009) have investigated the collective moves of the atoms, in the form of strain events, localized reorganizations, soft modes, etc., that could be responsible for the irreversible nature of structural relaxation and flow in supercooled liquids. As already discussed, these attempts could provide a description of glass-forming liquids at a level that is comparable to that of plastic flow in crystalline materials by the dislocation-based theory.

2.3.3 Developing a specific statistical-mechanical framework

As stressed in the Introduction, there is a general agreement concerning the fact that supercooled liquids, and more generally glass-forming systems above their glass transition, can be treated by equilibrium statistical mechanics (provided that one excludes the crystal phase). In this general setting, specific frameworks have been devised to study glass formers. Interestingly, these theoretical constructs developed in the context of glasses may prove useful, or even have already proven so, in other scientific fields. I sketch some of these developments below.

One such framework rests on a "topographic" view of the phenomena associated with glass formation. The core object is the "landscape" formed by the potential energy surface plotted as a function of the coordinates of all the atoms in the liquid. This leads to a reformulation of statistical mechanics that has been put forward by Stillinger and coworkers (Stillinger and Weber, 1983; Stillinger, 1995; Debenedetti and Stillinger, 2001). Focus is shifted to a study of the properties of the (temperature-independent) landscape, i.e. the statistics of the minima and the saddle points, the characteristics of their neighborhoods, the connectivity graph between minima, etc. (Wales, 2003). The potential descriptive power of the approach comes from the remark, nicely articulated in (Goldstein, 1969), that the viscous liquid at low enough temperature spends most of the time vibrating around typical energy minima (later called "inherent structures" by Stillinger (Stillinger and Weber, 1983)) with only rare, localized reorganization events that are associated with thermally activated barrier hopping between minima. The potential-energy landscape formalism then provides a vista for rationalizing, in a qualitative way, collective phenomena occurring in supercooled liquids (Stillinger, 1995). It also represents a framework in which to compute properties of specific systems by means of numerical simulations (Keyes, 1992; Sastry *et al.*, 1998; Sciortino, 2005; Heuer, 2008). The daunting difficulty of course comes from the dimensionality of configurational space. With its $3N$ dimensions, N being the number of atoms, it makes the landscape technically hard to characterize and, in effect, conceptually hard to visualize.

With some (usually unspecified) coarse graining in mind, it may be fruitful to go from an energy landscape to a free-energy one, in which energy minima separated by small barriers are grouped into a free-energy state. Such a construction is well defined at a mean-field level, and classes of complex free-energy landscapes with multiple metastable states have been found and thoroughly characterized in theoretical studies of systems with quenched disorder, mostly mean-field spin-glass models (Castellini and Cavagna, 2005). Supercooled liquids and the associated glasses, however, have no quenched disorder. Remarkably, some powerful tools introduced in the context of spin glasses (Mézard *et al.*, 1988), such as the replica method and generating functional approaches, have been generalized to investigate, at a mean-field level at least, the complex free-energy landscape of glass-forming liquids (Monasson, 1995; Castellini and Cavagna, 2005; Mézard and Parisi, 2009; Franz and Parisi, 1998). This point will be further discussed in Section 2.5.2.

The two other formalisms that I would like to briefly mention also represent, if not reformulations, at least new twists in statistical mechanics. The first one is the "iso-configurational ensemble" (Widmer-Cooper *et al.*, 2004; Widmer-Cooper and Harrowell, 2006) that is formed by all trajectories that start from an identical configuration of particles with random initial velocities sampled from an equilibrium Maxwell distribution. Together with such notions as "dynamic propensity", it has been introduced to study whether the static structure of a glass-forming liquid, without *a priori* knowledge of which of its features might be relevant, influences the heterogeneous character of the dynamics (Widmer-Cooper *et al.*, 2004; Widmer-Cooper and Harrowell, 2006). The second formalism goes somewhat beyond the realm of equilibrium statistical mechanics: it considers the space of trajectories and introduces an (*a priori* unphysical) external field that couples to the mobility of the system in such a way that it can drive the latter out of equilibrium in an immobile, non-ergodic phase (Garrahan *et al.*, 2007; Hedges *et al.*, 2009). The presence of a non-equilibrium transition between ergodic and non-ergodic phases can then be investigated either in idealized models or in atomistic ones (see also Section 2.4.2).

2.3.4 In search of a growing length scale

For those theories that associate glass formation with some sort of universality and emerging collective behavior, a keypoint is the characterization of a length that grows as temperature is decreased and that is directly connected to the slowdown of relaxation. A related, but not quite coincident, issue is that of the existence of a length associated with the increasing spatial heterogeneity of the dynamics.

Recently, there has been a major effort to go beyond heuristic or *ad hoc* definitions of length scales in glass-forming liquids, such as the cooperativity length introduced in the Adam–Gibbs postulate (Adam and Gibbs, 1965) of cooperatively rearranging regions (see below) or various measures of dynamical clustering. This has led to introducing appropriate correlation functions that are in principle computable.

Much progress has been made concerning the growing spatial correlations in the dynamics. It has been realized that the latter, which are associated with fluctuations around the averaged dynamics and with the phenomenon of dynamical heterogeneities, are describable through multipoint space-time correlation functions. Information on the corresponding "dynamical" correlation length can be extracted from a 4-point correlation function that describes how far the dynamics at a given point in space affects the dynamics at another point and, with some plausible assumptions, from the associated dynamic susceptibility (Dasgupta *et al.*, 1991; Franz *et al.*, 1999*b*; Franz and Parisi, 2000; Berthier *et al.*, 2007*a*,*b*). Direct or indirect evidence for a "dynamical" length that grows as temperature decreases has been obtained in this general framework. The topic being at the heart of the present book and developed in the other chapters, I will not dwell more on it, except to mention an interesting result obtained in this context: it has been shown that the dynamical singularity predicted by the mode-coupling theory of glass-forming liquids comes with a divergence of the above described "dynamical" length (Franz and Parisi, 2000; Biroli and Bouchaud, 2004), thereby quashing a previously widespread belief that the relaxation time diverges with no accompanying diverging length scale.

A separate issue, which has deep consequences on the theoretical picture of glass-forming systems, is whether there also exists a growing static length as temperature decreases. If there is a true divergence of the α relaxation time at a non-zero temperature (a hypothesis that of course cannot be verified experimentally), heuristic and rigorous arguments (Montanari and Semerjian, 2006; Semerjian and Franz, 2011) prove that it must come with the divergence of a static correlation length, albeit a complicated one involving point-to-set correlations. If, on the other hand, the relaxation time only diverges at zero temperature, the situation is not as clear-cut. One could envisage a relaxation process whose temperature behavior is characterized by an Arrhenius behavior with a purely local energy barrier, so that no static correlations grow as the relaxation time increases (and diverges at zero temperature). It would therefore seem more relevant to study the behavior of the relaxation time normalized by the local, "bare" relaxation time: $\tau(T)/\tau_0(T)$ with e.g. $\tau_0(T) \sim \exp(\frac{E_\infty}{T})$. With the assumption that the rigorous bounds between length scales and time scales derived by Montanari and Semerjian (Montanari and Semerjian, 2006) are of general validity, one then concludes that the growth of this scaled relaxation time must come with a growing static correlation length. This suggests that the super-Arrhenius temperature dependence of the α relaxation time of fragile glass-forming liquids, for which, even after normalization by a "local" Arrhenius-like relaxation time, $\tau(T)/\tau_0(T)$ behaves as $\exp(\frac{\Delta E(T)}{T})$ with $\Delta E(T)$ increasing as temperature decreases, is indicative of collective behavior with a concomitantly growing static correlation length.

There are several possible, and not mutually excluding, strategies to search for static correlations associated with glass formation. Since no interesting correlations show up in simple 2-body structural measures such as the static structure factor, one may then think of: (i) studying how far amorphous boundary conditions influence the system, thereby looking at static point-to-set correlations such as those discussed above (Bouchaud and Biroli, 2004; Montanari and Semerjian, 2006; Franz and Montanari, 2007; Cavagna *et al.*, 2007; Biroli *et al.*, 2008; Cavagna, 2009), (ii) using finite-size analysis for chosen thermodynamic quantities (Doliwa and Heuer, 2003; Fernandez *et al.*, 2006; Karmakar *et al.*, 2009), (iii) looking at a crossover size in pattern repetition within a configuration (Kurchan and Levine, 2011), or else (iii), provided that the locally preferred arrangement of the molecules in the liquid has been properly identified, investigating the static pair correlations of the associated local order parameter, which amounts to considering multiparticle correlations as in bond-orientational order parameters (Steinhardt *et al.*, 1981; Shintani and Tanaka, 2006; Kawasaki *et al.*, 2007; Sausset and Tarjus, 2010). Up to the present at least, these procedures are not experimentally realizable in liquids. However, they have been numerically tested on liquid models with encouraging preliminary results concerning the increase of a static length with decreasing temperature.

2.4 Theories based on an underlying dynamical transition

As discussed above, glass formation takes place in liquids and polymers without any observed singularity. Universality and detail independence, if indeed a genuine property of the glass-transition phenomenon, are explainable on the basis of underlying critical points that control the physics of the viscous slowing down but are either avoided or unreachable. In this section, I will discuss theories that involve purely dynamical transitions with no thermodynamic signature.

2.4.1 Mode-coupling theory: an avoided dynamical transition at $T_c > T_g$

The *mode-coupling* theory of glass-forming liquids predicts a dynamical arrest without any significant change in the static properties (Götze, 1991; Götze, 2008; Götze and Sjögren, 1992). The latter are assumed to behave smoothly, and the viscous slowing down results from a non-linear feedback mechanism affecting the relaxation of the density fluctuations. Formally, the theory involves a set of non-linear integro-differential equations describing the evolution of the dynamic structure factor $S(q,t)$, which is the wavevector- and time-dependent pair correlation function of the density fluctuations. These equations have been originally derived by using the Zwanzig–Mori projection-operator formalism (Götze, 1991; Götze, 2008; Götze and Sjögren, 1992), but they can also be obtained within a dynamical field-theoretical framework (Das and Mazenko, 1986; Kim and Mazenko, 1980; Miyazaki and Reichman, 2005; Kim and Kawasaki, 2008; Andreanov *et al.*, 2006). The crux of the approach consists in formulating an approximation, the "mode-coupling approximation", that allows one to close formally exact dynamical equations and write down a tractable set of self-consistent equations for $S(q,t)$. The key input that comes into the approximate theory is the static structure factor, $S(q) \equiv S(q,t=0)$.

The solution of the self-consistent equations predicts a slowdown of the relaxation of $S(q,t)$ with decreasing temperature that is physically attributed to a "cage effect" and to the feedback mechanism mentioned above. This solution exhibits a dynamical freezing at a critical point T_c that represents a transition from an ergodic to a non-ergodic state with no concomitant singularity in the thermodynamics of the system (Bengtzelius *et al.*, 1984; Leuthesser, 1984; Götze, 1991; Götze, 2008). The α relaxation time diverges in a power-law fashion for $T > T_c$,

$$\tau(T) \sim (T - T_c)^{-\gamma}, \tag{2.1}$$

and several specific predictions are made concerning the scaling behavior near to T_c.

Early on, it was realized that the dynamical arrest at T_c could not describe the observed glass transition at T_g, nor a transition to a putative ideal glass at a temperature below T_g, and that T_c should rather be located above T_g. The singularity at T_c must then be interpreted as "avoided" and manifesting itself as a crossover in the phenomenology of glass-forming liquids (Götze, 1991; Götze, 2008). The mode-coupling approach can thus at best describe the dynamics of moderately supercooled liquids, for which its main achievement is the predicted appearance of a two-step relaxation process as temperature decreases, as indeed observed in experiments and in simulations. The mode-coupling theory has also proven a versatile scheme to study additional systems, such as repulsive and attractive colloids (Zaccarelli *et al.*, 2004), and phenomena. It has been, for instance, generalized to investigate aging dynamics in the non-ergodic phase (Cugliandolo and Kurchan, 1993; Bouchaud *et al.*, 1998), non-linear rheology of a variety of glassy systems (Fuchs and Ballauff, 2005; Fuchs and Cates, 2009), or more recently dynamical heterogeneities and multipoint space-time correlations (Berthier *et al.*, 2007*a*,*b*; Biroli *et al.*, 2006; Tarzia *et al.*, 2010), all cases for which it provides non-trivial predictions.

What is the nature of the mode-coupling approximation and why is the singularity avoided? The traditional answers (Götze, 1991; Götze, 2008; Götze and Sjögren, 1992) invoking freezing due to a "local cage effect" (see above) and avoidance due to "hopping mechanisms" are not satisfactory. A clearer picture has emerged from work on *a priori* unrelated systems, mean-field "generalized" spin-glasses (Kirkpatrick and Thirumalai, 1987, 1989; Cavagna, 2009; Biroli and Bouchaud, 2009), and further developments. The analogy with mean-field spin-glasses will be discussed in more detail in the section on the random first-order transition theory. To make a long and elaborate story short, let me summarize the findings as follows:

(i) The mode-coupling approximation and the associated self-consistent equations have a mean-field character (Kirkpatrick *et al.*, 1989; Andreanov *et al.*, 2009; Biroli and Bouchaud, 2009).
(ii) In a free-energy landscape picture, the ergodicity-breaking transition corresponds to the disappearance of unstable directions for escape and to the emergence of infinite barriers between the dominant states (Cavagna, 2009).
(iii) The divergence of the relaxation time at T_c is not due to a purely local effect, but involves the divergence of a length characterizing spatial correlations in the dynamics (Franz and Parisi, 2000; Biroli and Bouchaud, 2004).
(iv) Being of mean-field character, the mode-coupling singularity is affected in real, finite-dimensional, systems by fluctuations whose influence is two-fold (Andreanov *et al.*, 2009). First, there is a standard effect that can often be handled through perturbative expansions: the exponents describing the scaling behavior are modified below an upper critical dimension, which in the present case is found equal to $d = 8$ (Biroli and Bouchaud, 2007; Berthier *et al.*, 2007*a*,*b*; Franz *et al.*, 2010). In addition, and with more severe consequences, rare localized events corresponding to thermally activated processes destroy the singularity; these intrinsically non-perturbative phenomena operate in all finite dimensions.

This puts the mode-coupling approach on a much firmer basis. The down side is that a major theoretical breakthrough is needed to incorporate non-perturbative effects beyond the mean-field picture (see Section 2.5.2).

2.4.2 Dynamical facilitation and kinetically constrained models: unreachable dynamical critical point at $T = 0$ and avoided dynamical first-order transition

As for the mode-coupling theory, the radical perspective taken by the *dynamical facilitation* approach (Garrahan and Chandler, 2010) is that the main characteristics of glass-forming liquids can be described by purely dynamical arguments. Building on the observation that the static pair correlations change only weakly with temperature and that the associated correlation length always stays of the order of the molecular diameter, this approach indeed assumes that, after some suitable coarse graining, static correlations become negligible and that, accordingly, thermodynamics is trivial. In this picture, glassiness, cooperativity and heterogeneity in the dynamics result from effective kinetic constraints that emerge at low enough temperature when mobility in a supercooled liquid is concentrated in rare localized regions, the rest of the system being essentially frozen. Such "mobility defects" are taken as the effective degrees of freedom that, together with the postulated kinetic rules that constrain their motion, form the basis of the theoretical description. "Facilitation" in this context means that mobility defects trigger mobility in neighboring regions.

This general scenario has led to the formulation of families of models generically referred to as "kinetically constrained models" (Ritort and Sollich, 2003). These models rest on a hamiltonian for non-interacting variables (spins or particles on a lattice) combined with specific constraints on the allowed moves of any such variable. These constraints involve a dependence on the local neighborhood, e.g. a particle can hop to a different site only if the number of nearest neighbors at the original and the target sites is less than a fixed threshold value. These models are sufficiently tractable to allow detailed analytical and numerical calculations. It has been shown that cooperativity of the relaxation, with a super-Arrhenius temperature dependence of the relaxation time, is a property of many of the models and that heterogeneity of the dynamics naturally emerges as a central feature of all models. The kinetically constrained models qualitatively reproduce many aspects of the slow dynamics of glass-forming liquids, but, more specifically, they provide a consistent picture of the dynamical heterogeneities in a space-time setting (Garrahan and Chandler, 2002; Ritort and Sollich, 2003; Garrahan and Chandler, 2010).

Beyond the variety of models and associated behaviors, it has been suggested that the apparent universality of the dynamical properties in glass-forming systems could come from the existence of a dynamical critical point at zero temperature (Whitelam *et al.*, 2004, 2005; Jack and Mayer, 2006; Garrahan *et al.*, 2007; Hedges *et al.*, 2009). Dynamical scaling analysis can then be organized about this zero-temperature singularity. For instance, inspired by the low-T behavior of some "hierarchical" kinetically constrained models (Jäckle and Eisinger, 1991), Garrahan and Chandler (Garrahan and Chandler, 2003, 2010) have proposed to describe the super-Arrhenius temperature dependence of glass-forming liquids by a Bässler-type expression (Bässler, 1987),

$$\tau(T) \sim \exp\left(\frac{A}{T^2}\right) \tag{2.2}$$

for temperatures much below an "onset" that marks the beginning of facilitated dynamics with rare mobile regions.

The facilitation approach puts the emphasis on a space-time picture of glassy dynamics. At a descriptive level, dynamical heterogeneities in glass-forming liquids can be for instance associated to the proximity to a first-order non-equilibrium transition in trajectory space that is characterized by the coexistence of a mobile and an immobile phase (Garrahan *et al.*, 2007; Hedges *et al.*, 2009) (see Section 2.3.4).

As with most theories (with the possible exception of the mode-coupling theory discussed above), the present approach comes up against the difficulty of deriving rather idealized models from realistic systems. This is by no means benign. For instance, the simple assumption that mobility is conserved, i.e. that there is no spontaneous appearance or disappearance of mobility defects, is questionable (Candelier *et al.*, 2009). More importantly, the facilitation approach cannot address the aspects of the glass-forming phenomenology that involve thermodynamics (e.g., the behavior of the entropy and of the heat capacity (Biroli *et al.*, 2005)) nor the non-trivial static correlations that are argued to accompany the increase of relaxation time in fragile glass formers (see above).

2.5 Theories based on an underlying thermodynamic or static transition

In this section will be considered theoretical approaches that relate the (hypothesized) collective behavior of glass-forming systems and associated universality to underlying thermodynamic or static critical points. I will stress those theories that are amenable to a genuine statistical-mechanical treatment and involve effective hamiltonian models. However, before doing so, it is worth mentioning two phenomenological pictures, the free-volume and the configurational-entropy models, as they have been influential for the thinking on the glass transition, and are still commonly used to rationalize data on liquids and polymers at a semiempirical level.

2.5.1 Free-volume and configurational-entropy models: unreachable transition points at $T_0 < T_g$

Free-volume models rest on the assumption that molecular transport in viscous fluids only occurs when voids having a volume large enough to accommodate a molecule form by the redistribution of some "free volume" (Cohen and Turnbull, 1959; Turnbull and Cohen, 1961; Grest and Cohen, 1981). The latter is loosely defined as some surplus volume that is not taken up by the molecules. In the standard presentation, a molecule in a dense fluid is mostly confined to a cage formed by its nearest neighbors and the local free volume v_f is that part of a cage space that exceeds the volume taken by a molecule. It is then assumed that between two events contributing to molecular transport, a reshuffling of free volume among the cages occurs at no cost of energy and that the local free volumes are statistically uncorrelated. This leads to an expression for the viscosity,

$$\eta(T) = \eta_0 \exp\left(\frac{K}{v_f(T)}\right), \tag{2.3}$$

with K essentially constant, which is also known as the Doolittle equation (Doolittle, 1951).

The free-volume mechanism fundamentally relies on a hard-sphere picture in which thermal activation plays no role. For application to real liquids and polymers, temperature enters through the fact that molecules and monomers are not truly hard and that the constant-pressure volume is temperature dependent. An underlying transition comes into play when further assuming that all free volume is consumed at a non-zero temperature $T_0 < T_g$, which, when inserted in Eq. (2.3), gives the Vogel–Fulcher–Tammann (VFT) expression,

$$\eta(T) = \eta_0 \exp\left(\frac{DT_0}{T - T_0}\right), \tag{2.4}$$

also called the Williams–Landel–Ferry (WLF) formula in the context of polymers and widely used to fit experimental data.

The *configurational-entropy* picture on the other hand rests on the idea, popularized by Goldstein (Goldstein, 1969), that relaxation in a deeply supercooled liquid approaching the glass transition is best described by invoking motion of the representative state point of the system on the potential-energy hypersurface (Gibbs, 1960) (see Section 2.3.3). In this view, the slowing down of relaxation and flow with decreasing temperature is related to a decrease of the number of available minima and of the associated "configurational entropy". The Adam–Gibbs approach (Adam and Gibbs, 1965) represents a phenomenological attempt to make this relation more precise. Structural relaxation is assumed to take place through increasingly cooperative rearrangements of groups of molecules. Any such group, called a cooperatively rearranging region, is assumed to relax independently of the others. The effective activation energy for relaxation is then equal to the typical energy barrier per molecule, which is taken as independent of temperature, multiplied by the number of molecules that are necessary to form the smallest cooperatively rearranging region. This latter number goes as the inverse of the configurational entropy per molecule $s_c(T)$, which leads to the following expression for the α relaxation time:

$$\tau(T) = \tau_0 \exp\left(\frac{C}{Ts_c(T)}\right), \tag{2.5}$$

with C a constant.

If the configurational entropy vanishes at a non-zero temperature T_0, an assumption somewhat analogous to that made in the free-volume model (see above),[2] the relaxation time diverges at this same non-zero temperature. In particular, if the configurational entropy is identified with the entropy difference between the supercooled liquid and the crystal, the Adam–Gibbs theory correlates the extrapolated divergence of the relaxation time at T_0 with the extrapolated vanishing of the excess entropy at the so-called Kauzmann temperature T_K (Kauzmann, 1948). Equation (2.5) also gives

[2] The vanishing of the configurational entropy at a non-zero temperature is, however, found in the Gibbs–di Marzio approximate Flory–Huggins mean-field treatment of a lattice model of linear polymeric chains (Gibbs and Di Marzio, 1958).

back the VFT/WLF formula, Eq. (2.4), by assuming that the configurational entropy vanishes linearly at T_0.

Note that in both the free-volume and the Adam–Gibbs configurational-entropy approaches, the unreachable thermodynamic transition temperature T_0 does not appear as a necessary ingredient of the theoretical description, but rather as a convenient input to derive the empirical VFT/WLF expression.

2.5.2 Random first-order transition theory: an unreachable thermodynamic critical point at $T_K < T_g$ coupled with an avoided dynamical transition at $T_c > T_g$

The *random first-order transition* theory (Kirkpatrick *et al.*, 1989; Lubchenko and Wolynes, 2007) can be seen as a three-stage construction. The foundations are formed by an intricate mean-field theory of glass formation. In the 1980s, Wolynes and coworkers (Kirkpatrick and Wolynes, 1987*a*,*b*; Kirkpatrick and Thirumalai, 1987, 1989; Kirkpatrick *et al.*, 1989) realized that many (postulated) elements of the description of glass-forming liquids, the Kauzmann-like thermodynamic "catastrophe" associated with the extrapolated vanishing of the configurational entropy (Kauzmann, 1948), the mode-coupling dynamical singularity and the emergence of a complex (free) energy landscape with a multitude of trapping minima, could all be tied together in a coherent scenario for which explicit realizations were provided by mean-field "generalized" spin glasses (such as the random p-spin and Potts-glass models with infinite-range interactions).

It has been by now established that the scenario is not only realized in spin models with quenched disorder. It is more generally characteristic of a whole class of glass-forming systems described at a mean-field level, with a transition to an ideal glass phase that is second order in the usual thermodynamic sense (with, e.g., no latent heat) but is accompanied by a discontinuous jump in the order parameter (Lubchenko and Wolynes, 2007; Cavagna, 2009; Biroli and Bouchaud, 2009; Mézard and Parisi, 2009). This "random first-order transition" (also called "one-step replica symmetry breaking") phenomenology, with a high-temperature ergodicity-breaking transition at T_c and a low-temperature thermodynamic glass transition at T_K that are separated by a regime in which an exponentially large (in system size) number of metastable free-energy states dominates the thermodynamics while trapping the dynamics, has been found in several standard liquid models when treated within mean-field-like approximations (Singh *et al.*, 1985; Mézard and Parisi, 1997; Franz *et al.*, 1999*a*; Parisi and Zamponi, 2010; Chaudhuri *et al.*, 2008). Specific methods, involving generating functionals and replica formalism (see also Section 2.3.3), have been developed for this purpose.

The mean-field character of the above-mentioned results is manifest in the existence of metastable states whose lifetime is infinite (in the thermodynamic limit). In realistic models (called "finite-range", "finite-dimensional" in the standard terminology used in this context), metastability is destroyed by nucleation events. Ergodicity is then restored by thermally activated processes (in the absence of a genuine thermodynamic phase transition) and the dynamical transition at T_c is smeared out, as already alluded to in the section on the mode-coupling theory.

The activated relaxation mechanisms that take over must be described in a nonperturbative way, and this represents the second stage of the theory. Kirkpatrick, Thirumalai, and Wolynes (Kirkpatrick *et al.*, 1989) have proposed a description of the liquid below the crossover at T_c as a "mosaic state" and a dynamical scaling theory close to T_K based on "entropic droplets", the driving force for nucleation being provided by the non-zero "configurational entropy" (associated with the number of free-energy minima, as found at the mean-field level). Super-Arrhenius temperature dependence of the α relaxation time follows, with an effective activation barrier given in terms of the length ξ_* characterizing the mosaic cells and the entropic droplets by

$$E(T) \sim \Delta_0 \, \xi_*(T)^{\psi}, \tag{2.6}$$

with

$$\xi_*(T) \sim \left(\frac{\sigma_0}{T s_c(T)} \right)^{\frac{1}{d-\theta}}, \tag{2.7}$$

where Δ_0 and σ_0 are two "bare" energy scales (in appropriate units), d the dimension of space, and ψ and θ are two critical exponents. The latter are predicted to be both equal to 3/2 in $d = 3$ by Kirkpatrick *et al.* (Kirkpatrick *et al.*, 1989), which leads to an Adam–Gibbs-type of formula for the relaxation time (see Eq. (2.5)) with the configurational entropy per particle $s_c(T)$ vanishing at the ideal glass transition T_K. More recently, the mosaic scenario has been reformulated in a way that makes a direct connection between the mosaic length ξ_* and a point-to-set correlation length (Bouchaud and Biroli, 2004; Biroli and Bouchaud, 2009), which therefore allows for possible testing (Franz and Montanari, 2007; Cavagna *et al.*, 2007; Biroli *et al.*, 2008; Cavagna, 2009).

The last stage of the approach is formed by phenomenological input and additional modeling that are used to make contact with a broad range of experimental data in glass-forming liquids and polymers, and in glasses as well (Xia and Wolynes, 2000; Lubchenko and Wolynes, 2007).

The random first-order transition approach starts with a sophisticated mean-field theory that is both robust and appealing. Going beyond this and addressing non-perturbative effects is a formidable task. However, what makes a firm derivation crucial in the present case is that the mean-field scenario of a complex free-energy landscape could be very fragile to the introduction of fluctuations arising in finite-range, finite-dimensional systems. It has been, for instance, argued that the whole scenario is then destroyed (Moore, 2006), and there is so far little evidence that it indeed persists in finite dimensions.

2.5.3 Frustration-based approach: an avoided thermodynamic critical point at $T^* > T_g$

The concept of *frustration* quite generally describes situations in which one cannot minimize the energy function of a system by merely minimizing all local interactions

(Toulouse, 1977). In the context of liquids and glasses, frustration is attributed to a competition between a short-range tendency for the extension of a locally preferred order and global constraints that forbid the periodic tiling of the whole space with the local structure (Sadoc and Mosseri, 1999; Nelson, 2002; Tarjus *et al.*, 2005). A prototypical example is that of local tetrahedral order in three-dimensional one-component liquids in which the atoms interact through spherically symmetric pair potentials: despite being more favorable locally, extended tetrahedral or icosahedral order is precluded at large distances and cannot give rise to long-range crystalline order.

Frustration has first been invoked by Frank (Frank, 1952) to explain at a geometric, structural level the resistance to crystallization and the degree of supercooling of a liquid, an explanation that has since received direct experimental (Shen *et al.*, 2009) and numerical (van Meel *et al.*, 2009*a*,*b*) confirmation. It has subsequently been used to describe the structure of glasses, more specifically metallic glasses and network-forming systems (Kléman and Sadoc, 1979; Nelson, 1983*a*,*b*; Nelson and Widom, 1984; Sachdev and Nelson, 1985; Sadoc and Mosseri, 1984). More recently, a step toward the formulation of a frustration-based theory of the glass transition has been to realize that, under rather generic conditions, frustration gives rise to the phenomenon of "avoided criticality" (Kivelson *et al.*, 1995; Chayes *et al.*, 1996; Nussinov *et al.*, 1999; Nussinov, 2004; Tarjus *et al.*, 2005). The latter expresses the fact that the ordering transition that may exist in the absence of frustration disappears as soon as an infinitesimal amount of frustration is introduced.

A frustration-based theory of the glass transition has been put forward, based on three plausible, but not fully established, propositions (Kivelson *et al.*, 1995; Kivelson and Tarjus, 1998; Viot *et al.*, 2000; Tarjus *et al.*, 2005): (i) the existence in a liquid of a locally preferred structure, an arrangement of molecules that minimizes some local free energy, (ii) the impossibility for this local order to tile the whole space, which expresses ubiquitous frustration, and (iii) the possibility to construct an abstract system in which the effect of frustration can be turned off, e.g. by tinkering with the metric of space. The resulting phenomenon of frustration-induced avoided criticality then naturally leads to collective behavior on a mesoscopic scale. In physical terms, the spatial extension of the locally preferred structure generates superextensive strain that prevents long-range ordering; below a crossover temperature T^* corresponding to the transition in the absence of frustration, this results in the breaking up of the liquid into domains, whose size and further growth are limited by frustration. This occurs provided the ordering transition in the unfrustrated space is accompanied by a diverging, or at least large, correlation length and provided frustration in a given liquid is weak enough that the transition is only narrowly avoided in physical space.

In this frustration-limited domain picture (Kivelson *et al.*, 1995; Kivelson and Tarjus, 1998; Viot *et al.*, 2000; Tarjus *et al.*, 2005), T^* marks the onset of anomalous, supermolecular behavior and it can be used to organize a scaling description of the viscous slowing down and other collective properties of glass-forming liquids. For instance, the α relaxation time and the viscosity are predicted to follow a super-Arrhenius activated temperature dependence below the crossover T^*, the effective activation energy being expressed as $E(T) = E_\infty + \Delta E(T)$ with

$$\Delta E(T) = BT^* \left(\frac{T^* - T}{T^*}\right)^\psi \tag{2.8}$$

for $T < T^*$ and the exponent ψ argued from statistical-mechanical and phenomenological arguments to be close to $8/3$ in $d = 3$. The "fragility" of a glass former, i.e. the departure from Arrhenius behavior, is quantified by the parameter B that is inversely proportional to the degree of frustration. In this approach, a large fragility is therefore associated with a small frustration, which implies a closer proximity to the avoided transition and a larger extent of collective behavior. Moreover, as in the random first-order transition theory, the heterogeneity of the dynamics primarily stems from the "patchwork" or "mosaic" character of the configurations.

The frustration-based theory leads to the formulation of effective minimal models of glass-forming systems for which genuine statistical-mechanical treatments and numerical computations are possible (Grousson *et al.*, 2002*a*,*b*; Sausset *et al.*, 2008; Sausset and Tarjus, 2010). It faces, however, two main difficulties. The first one is technical, and it is shared by all other theories of relaxation in deeply supercooled liquids: the phenomena associated with avoided criticality are intrinsically non-perturbative and a full-blown resolution going beyond phenomenological modeling and limited computations is *a priori* very hard. The second one concerns the very physical basis of the approach: the ubiquitousness of frustration in glass-forming liquids is a reasonable postulate but it requires confirmation. The icosahedral example has presumably no value in molecular liquids and polymers, and one still awaits proper identification of a locally preferred structure for molecules of non-spherical shapes that form most real fragile glass formers.

As a final comment, I would like to point out that an appealing property of the frustration approach, whether or nor it represents the "general theory" of the glass transition, is that it produces microscopic models in which the fragility of a glass former, and more generally the degree to which collective behavior can develop, can be varied at will by tuning the amount of frustration, while keeping the other parameters such as the interaction potentials fixed (Sausset *et al.*, 2008).

2.5.4 Jamming scenario: an unreachable static critical point at $T = 0$

The *jamming* scenario and the associated phase diagram may be viewed as a grand unification scheme in which the glass transition of liquids and polymers is taken as one example of a more general phenomenon in which the sluggish response of a condensed-matter system leads to an amorphous arrested state with no observable macroscopic flow (Liu and Nagel, 1998; Liu *et al.*, 2011). At the heart of this approach is the realization that temperature, packing fraction and stress act similarly on a disordered system on the verge of rigidity. This allows one to draw analogies between granular materials, emulsions and foams under the effect of external stress or forcing, repulsive colloidal suspensions as a function of concentration, and glass-forming liquids and polymers with temperature as the control variable (Liu and Nagel, 2001; Liu *et al.*, 2011).

The jamming paradigm may prove useful in suggesting systematic experimental investigations of a given material as a function of several parameters potentially controlling its jamming or unjammming behavior: for instance glass formation and the associated phenomena could be studied in liquids and polymers not only through temperature or pressure changes but also by varying the applied stress. However, interesting as they may be, a heuristic phase diagram and broad-based comparisons do not make a theory. A step forward in establishing a jamming-based theory has been the evidence for the existence of a well-identified, genuine, critical point located at zero temperature (O'Hern *et al.*, 2002, 2003; Schwarz *et al.*, 2006).

"Point J", the jamming critical point, is observed in model systems of spherical particles interacting through finite-range repulsive potentials as one compresses the system via a non-equilibrium protocol. Scaling laws and a diverging (static) correlation length are found around point J (O'Hern *et al.*, 2002, 2003; Schwarz *et al.*, 2006), and the marginally rigid jammed solid close to point J shows an anomalous elastic response characterized by the presence of soft (zero or low-frequency) vibrational modes (Wyart, 2005; Xu *et al.*, 2007; Liu *et al.*, 2011). Point J is hypothesized to control the jamming behavior of soft-condensed systems such as foams and emulsions, of hard objects such as solid colloidal particles and grains, as well as glass-forming liquids and polymers.

Several issues have been raised concerning the jamming scenario and its associated zero-temperature critical point:

(i) The robustness of point J with respect to physically relevant factors not included in the original formulation: friction for grains, asphericity of particle shape, thermal fluctuations, longer-ranged interaction potentials and attractive forces; some of these factors can be accounted for by an appropriate generalization (Liu *et al.*, 2011; van Hecke, 2010), but the latter seems more problematic (Berthier and Tarjus, 2009).
(ii) The uniqueness of point J. The precise location of point J is protocol dependent and actually a whole range of J points seems to exist at zero temperature (Krzakala and Kurchan, 2007; Chaudhuri *et al.*, 2010). In itself, this is not a fatal blow to the jamming picture, but it requires some caution.
(iii) The relevance of point J to the glass transition of liquids and polymers, which is the main focus of this overview. The interactions among particles in liquids and polymers involve long-range attractive interactions. As already mentioned, treating the latter as a mere perturbation is highly questionable (Berthier and Tarjus, 2009).

Whereas some specific phenomena associated with slow dynamics in liquids and anomalous elastic response in glasses may profitably be envisaged within the jamming context, it is at present unclear if the physics of glass formation can be reduced to a jamming behavior controlled by a zero-temperature point J.

2.6 Concluding remarks

How to assess the validity of the proposed theories? This seemingly trivial question is more complex than it may sound at first. Indeed, theories mostly rationalize existing data. In addition to providing a narrative to explain the qualitative trends that characterize glass formation, they reproduce quantitative features of the phenomenology, but at the expense of adjustable parameters. New predictions that would not involve additional assumptions and could be crisply checked in experiments are rare if not non-existent.

Experimental observations do put constraints on the theories. Phenomena, trends, orders of magnitude and correlations found in glass-forming systems should of course be reproduced. However, there is always some leeway. First, some of the observations may be discarded by the proponents of a theory as irrelevant or out of the scope of their approach. Secondly, many of the theories, especially those attempting to describe the viscous regime close to the glass transition, have phenomenological input and cannot just be taken as "first-principle" approaches. (Even the mode-coupling theory, which is often considered as the archetype of a "microscopic" theory and certainly makes detailed predictions based on liquid structure factors, cannot be tested without bias because of the loosely defined, and therefore adjustable, location of the crossover replacing the dynamical singularity T_c in real systems.) The resulting unavoidable use of fitting parameters allows for flexibility in comparing theory with experiment. Not surprisingly, several theories can claim success in reproducing observed properties, for instance the temperature dependence of the relaxation time and the viscosity, in spite of the quite different functional forms and underlying physics they entail.

The above considerations also apply to comparisons with data obtained through computer simulation of realistic models of liquids. Simulations may nonetheless offer a possibility of testing the theoretical premises of the proposed approaches in some depth. Such studies have already been undertaken on basic issues such as the existence of growing point-to-set correlations and of a non-zero surface tension between amorphous glassy metastable states, evidence for locally preferred structures in atomic and molecular liquids and relevance of frustration, robustness of the critical jamming point J to the introduction of friction or of long-ranged attractive forces, the appearance of slow anomalous modes associated with marginal rigidity, etc.

Constraints on theories also lie in their internal consistency. However, it may be hard to maintain full rigor when tackling "non-perturbative" effects that, in my opinion, are at the core of the glass-transition phenomenon (see above). More likely, one will at best be able to study limiting cases and possibly to make asymptotic predictions when the postulated collective behavior is dominant. In this endeavor, as in computer simulations, progress seems to be limited by the absence of a widely accepted and sufficiently tractable minimal model. A reasonable strategy to assess the validity of the candidates for a "general theory" of the glass transition would then be to devise models in which the collective behavior is so exaggerated that asymptotic predictions can be cleanly proved or disproved, without having to worry about subdominant effects (Kivelson and Tarjus, 2008). Such models should of course allow one to go continuously, with quantitative but no qualitative changes, from this extreme, but testable, behavior to that of real glass-forming systems.

To conclude this overview of theories of the glass transition, I would like to raise an apparently incongruous question: to what extent could various theoretical approaches be compatible or even complementary? The scenarios and narratives of glass formation produced by most theories do seem completely at odds. However, this may be less so for the developed frameworks, the working concepts and the proposed relaxation mechanisms that underly these theories. To illustrate this with a few examples: frustration is undoubtedly compatible with the emergence of a complex free-energy landscape and with a description of supercooled liquids as mosaic states, and none of these theoretical elements are in contradiction with the development of kinetic constraints on the motion of some effective degrees of freedom and with the presence of facilitation; in a somewhat different vein, elastic models could be extended and combined with the effect of a growing length scale as the glass transition is approached. Going further in a fruitful way of course requires a serious elaboration to avoid the risk of diluting even more the assessment of the theoretical constructions. However, at this stage where, despite the progress made in the last 25 years, full resolution of the glass-transition problem does not seem to be around the corner, it may be wise to take an open view on the possible complementarity of different theoretical perspectives.

References

Adam, G. and Gibbs, J. H. (1965). *J. Chem. Phys.*, **43**, 139.

Alba-Simionesco, C. (2001). *C. R. Acad. Sci. Paris IV*, **2**, 203.

Andreanov, A., Biroli, G., and Bouchaud, J.-P. (2009). *Europhys. Lett.*, **88**, 16001.

Andreanov, A., Biroli, G., and Lefèvre, A. (2006). *J. Stat. Mech.*, P07008.

Angell, C. A. (1985). In *Relaxation in Complex Systems* (ed. K. L. Ngai and G. B. Wright), Arlington. Office of Naval Research.

Angell, C. A. (1995). *Science*, **267**, 1924.

Appignanesi, G. A., Rodriguez Fris, J. A., Montani, R. A., and Kob, W. (2006). *Phys. Rev. Lett.*, **96**, 057801.

Bässler, H. (1987). *Phys. Rev. Lett.*, **58**, 767.

Bengtzelius, U., Götze, W., and Sjölander, A. (1984). *J Phys. C*, **17**, 5915.

Berthier, L. and Biroli, G. (2009). In *Encyclopedia of Complexity and Systems Science*, Berlin. Springer.

Berthier, L., Biroli, G., Bouchaud, J. P., Cipelleti, L., El-Masri, D., L'Hôte, D., Ladieu, F., and Pierno, M. (2005). *Science*, **310**, 1797.

Berthier, L., Biroli, G., Bouchaud, J.-P., Kob, W., Miyazaki, K., and Reichman, D. R. (2007*a*). *J. Chem. Phys.*, **126**, 184503.

Berthier, L., Biroli, G., Bouchaud, J.-P., Kob, W., Miyazaki, K., and Reichman, D. R. (2007*b*). *J. Chem. Phys.*, **126**, 184504.

Berthier, L. and Tarjus, G. (2009). *Phys. Rev. Lett.*, **103**, 170601.

Biroli, G. and Bouchaud, J. P. (2004). *Europhys. Lett.*, **67**, 21.

Biroli, G. and Bouchaud, J. P. (2007). *J. Phys.: Condens. Matter*, **19**, 205101.

Biroli, G. and Bouchaud, J.-P. (2009). *preprint arXiv:0912.2542*.

Biroli, G., Bouchaud, J.-P., Cavagna, A., Grigera, T. S., and Verrocchio, P. (2008). *Nature Phys.*, **4**, 771.

Biroli, G., Bouchaud, J.-P., Miyazaki, K., and Reichman, D. R. (2006). *Phys. Rev. Lett.*, **97**, 195701.

Biroli, G., Bouchaud, J.-P., and Tarjus, G. (2005). *J. Chem. Phys.*, **123**, 044510.

Böhmer, R., Ngai, K. L., Angell, C. A., and Plazek, D. J. (1993). *J. Chem. Phys.*, **99**, 4201.

Bouchaud, J.-P. and Biroli, G. (2004). *J. Chem. Phys.*, **121**, 7347.

Bouchaud, J.-P., Cugliandolo, L. F., Kurchan, J., and Mézard, M. (1998). In *Spin Glasses and Random Fields* (ed. A. P. Young), Singapore. World Scientific.

Boué, L., Lerner, E., and Procaccia, I. (2009). *J. Stat. Mech.*, P11010.

Candelier, R., Dauchot, O., and Biroli, G. (2009). *Phys. Rev. Lett.*, **102**, 088001.

Castellini, T. and Cavagna, A. (2005). *J. Stat. Mech.*, P05012.

Cavagna, A. (2009). *Phys. Rep.*, **476**, 51.

Cavagna, A., Grigera, T. S., and Verrocchio, P. (2007). *Phys. Rev. lett.*, **98**, 187801.

Chaudhuri, P., Berthier, L., and Sastry, S. (2010). *Phys. Rev. Lett.*, **104**, 165701.

Chaudhuri, P., Karmakar, S., and Dasgupta, C. (2008). *Phys. Rev. Lett.*, **100**, 125701.

Chayes, L., Emery, V. J., Kivelson, S.A., Nussinov, Z., and Tarjus, G. (1996). *Physica A*, **225**, 129.

Cohen, M. H. and Turnbull, D. J. (1959). *J. Chem. Phys.*, **31**, 1164.

Cugliandolo, L. F. and Kurchan, J. (1993). *Phys. Rev. Lett.*, **71**, 173.

Dalle-Ferrier, C., Thibierge, C., Alba-Simionesco, C., Berthier, L., Biroli, G., Bouchaud, J.-P., Ladieu, F., L'Hôte, D., and Tarjus, G. (2007). *Phys. Rev. E*, **76**, 041510.

Das, S. P. and Mazenko, G. F. (1986). *Phys. Rev. A*, **34**, 2265.

Dasgupta, C. (1991). *Europhys. Lett.*, **15**, 307.

Debenedetti, P. G (1996). *Metastable Liquids.* Princeton, Princeton University Press.

Debenedetti, P. G. and Stillinger, F. H. (2001). *Nature*, **410**, 259.

Doliwa, B. and Heuer, A. (2003). *J. Phys.: Condens. Matter*, **15**, S849.

Donati, C., Douglas, J., Kob, W., Plimpton, S.J., Poole, P.H., and Glotzer, S.C. (1998). *Phys. Rev. Lett.*, **80**, 2338.

Doolittle, A. K. (1951). *J. Appl. Phys.*, **22**, 1471.

Dyre, J. (2006). *Rev. Mod. Phys.*, **78**, 953.

Ediger, M. D. (2000). *Annu. Rev. Phys. Chem.*, **51**, 99.

Ediger, M. D., Angell, C. A., and Nagel, S. R. (1996). *J. Phys. Chem.*, **100**, 13200.

Edwards, S. F. and Anderson, P. W. (1975). *J. Phys. F: Met. Phys.*, **5**, 965.

Fernandez, L. A., Martin-Mayor, V., and Verrocchio, P. (2006). *Phys. Rev. E*, **73**, 020501(R).

Frank, F. C. (1952). *Proc. Roy. Soc. London*, **215A**, 43.

Franz, S., Cardenas, M., and Parisi, G. (1999*a*). *J. Chem. Phys.*, **110**, 1726.

Franz, S., Donati, C., Parisi, G., Glotzer, S. C., and B, *Philos. Mag.* (1999*b*). **79**, 1827.

Franz, S. and Montanari, A. (2007). *J. Phys. A: Math. Gen.*, **40**, F251.

Franz, S. and Parisi, G. (1998). *Physica A*, **261**, 317.

Franz, S. and Parisi, G. (2000). *J. Phys.: Condens. Matter*, **12**, 6335.

Franz, S., Parisi, G., Ricci-Tersenghi, F., and Rizzo, T. (2010). *preprint arXiv: 1008.0996*.

Fredrickson, G. H. and Andersen, H. C. (1984). *Phys. Rev. Lett.*, **53**, 1244.

Fuchs, M. and Ballauff, M. (2005). *Colloids Surf. A*, **270–271**, 232.
Fuchs, M. and Cates, M. E. (2009). *J. Rheol.*, **53**, 957.
Garrahan, J. P. and Chandler, D. (2002). *Phys. Rev. Lett.*, **89**, 035704.
Garrahan, J. P. and Chandler, D. (2003). *Proc. Natl. Acad. Sci. USA*, **100**, 9710.
Garrahan, J. P. and Chandler, D. (2010). *Annu. Rev. Phys. Chem.*, **61**, 191.
Garrahan, J. P., Jack, R. L., Lecomte, V., Pitard, E., van Duijvendijk, K., and van Wijland, F. (2007). *Phys. Rev. Lett.*, **98**, 195702.
Gibbs, J. H. (1960). In *Modern Aspects of the Vitreous State* (ed. J. D. Mackenzie), London. Butterworths.
Gibbs, J. H. and Di Marzio, E. A. (1958). *J. Chem. Phys.*, **28**, 373.
Glarum, S. H. (1960). *J. Chem. Phys.*, **33**, 639.
Goldstein, M. (1969). *J. Chem. Phys.*, **51**, 3728.
Götze, W. (1991). In *Liquids, Freezing, and the Glass Transition* (ed. J. P. Hansen, D. Levesque, and J. Zinn-Justin), Amsterdam. North Holland.
Götze, W. (2008). *Complex Dynamics of Glass-forming Liquids: A Mode-coupling Theory*. Oxford, Oxford University Press.
Götze, W. and Sjögren, L. (1992). *Rep. Prog. Phys.*, **55**, 241.
Grest, G. S. and Cohen, M. H. (1981). *Adv. Chem. Phys.*, **48**, 455.
Grousson, M., Tarjus, G., and Viot, P. (2002*a*). *Phys. Rev. E*, **65**, 065103(R).
Grousson, M., Tarjus, G., and Viot, P. (2002*b*). *J. Phys.: Condens. Matter*, **14**, 1617.
Hedges, L. O., Jack, R. L., Garrahan, J. P., and Chandler, D. (2009). *Science*, **323**, 1309.
Heuer, A. (2008). *J. Phys.: Condens. Matter*, **20**, 373101.
Jack, R. L., Mayer, P., and Sollich, P. (2006). *J. Stat. Mech.* P03006.
Jäckle, J. and Eisinger, S. (1991). *Z. Phys. B*, **84**, 115.
Karmakar, S., Dasgupta, C., and Sastry, S. (2009). *Proc. Natl. Acad. Sci. USA*, **106**, 3675.
Kauzmann, W. (1948). *Chem. Rev.*, **43**, 219.
Kawasaki, T., Araki, T., and Tanaka, H. (2007). *Phys. Rev. Lett.*, **99**, 215701.
Keyes, T. (1992). *J. Chem. Phys.*, **101**, 5081.
Kim, B. and Kawasaki, K. (2008). *J. Stat. Mech.*, P02004.
Kim, B. and Lee, S. J. (1997). *Phys. Rev. Lett.*, **78**, 3709.
Kim, B. and Mazenko, G. F. (1980). *Adv. Chem. Phys.*, **78**, 129.
Kirkpatrick, T. R. and Thirumalai, D. (1987). *Phys. Rev. Lett.*, **58**, 2091.
Kirkpatrick, T. R. and Thirumalai, D. (1989). *J. Phys. A*, **22**, L149.
Kirkpatrick, T. R., Thirumalai, D., and Wolynes, P. G. (1989). *Phys. Rev. A*, **40**, 1045.
Kirkpatrick, T. R. and Wolynes, P. G. (1987*a*). *Phys. Rev. A*, **35**, 3072.
Kirkpatrick, T. R. and Wolynes, P. G. (1987*b*). *Phys. Rev. B*, **36**, 8552.
Kivelson, D., Kivelson, S. A., Zhao, X.-L., Nussinov, Z., and Tarjus, G. (1995). *Physica A*, **219**, 27.
Kivelson, D. and Tarjus, G. (1998). *Philos. Mag. B*, **77**, 245.
Kivelson, S. A. and Tarjus, G. (2008). *Nature Mater.*, **7**, 831.
Kléman, M. and Sadoc, J. F. (1979). *J. Phys. Lett. (Paris)*, **40**, L569.
Kob, W. (2005). In *Handbook of Material Modelling* (ed. S. Yip), Berlin. Springer.

Krzakala, F. and Kurchan, J. (2007). *Phys. Rev. E*, **76**, 021122.
Kurchan, J. and Levine, D. (2011). *J. Phys. A: Math. and Theor.*, **44**, 035001.
Langer, J. S. (2006*a*). *Phys. Rev. Lett.*, **97**, 115104.
Langer, J. S. (2006*b*). *Phys. Rev. E*, **73**, 041504.
Leuthesser, E. (1984). *Phys. Rev. A*, **29**, 2765.
Liu, A. J. and Nagel, S. R. (1998). *Nature*, **396**, 21.
Liu, A. J. and Nagel, S. R. (ed.) (2001). *Jamming and Rheology: Constrained Dynamics on Microscopic and Macroscopic Scales.* London, Taylor and Francis.
Liu, A. J., Nagel, S. R., van Saarloos, W., and Wyart, M. (2011). *This book.*
Lubchenko, V. and Wolynes, P. G. (2007). *Ann. Rev. Phys. Chem.*, **58**, 235.
Martinez, L.-M. and Angell, C. A. (2001). *Nature*, **410**, 663.
Mézard, M. and Parisi, G. (1997). *Phys. Rev. Lett.*, **79**, 2486.
Mézard, M. and Parisi, G. (2009). *preprint arXiv: 0910.2838.*
Mézard, M., Parisi, G., and Virasoro, M. (1988). *Spin Glass Theory, and Beyond.* Singapore, World Scientific.
Miyazaki, K. and Reichman, D. R. (2005). *J. Phys. A: Math. Gen.*, **38**, 20.
Monasson, R. (1995). *Phys. Rev. Lett.*, **75**, 2847.
Montanari, A. and Semerjian, G. (2006). *J. Stat. Phys.*, **125**, 22.
Moore, M. A. (2006). *Phys. Rev. Lett.*, **96**, 137202.
Moore, M. A. and Drossel, B. (2002). *Phys. Rev. Lett.*, **89**, 217202.
Nelson, D. R. (1983*a*). *Phys. Rev. Lett.*, **50**, 982.
Nelson, D. R. (1983*b*). *Phys. Rev. B*, **28**, 5515.
Nelson, D. R. (2002). *Defects and Geometry in Condensed Matter Physics.* Cambridge, Cambridge University Press.
Nelson, D. R. and Widom, M. (1984). *Nucl. Phys. B*, **240**, 113.
Ngai, K. L. (2000). *J. Non-Cryst. Solids*, **275**, 7.
Novikov, V. N. and Sokolov, A. P. (2004). *Nature*, **431**, 961.
Nussinov, Z. (2004). *Phys. Rev. B*, **69**, 014208.
Nussinov, Z., Chayes, L. N., Rudnick, J., and Kivelson, S. A. (1999). *Phys. Rev. Lett.*, **83**, 472.
O'Hern, C. S., Langer, S. A., Liu, A. J., and Nagel, S. R. (2002). *Phys. Rev. Lett.*, **88**, 075507.
O'Hern, C. S., Silbert, L. E., Liu, A. J., and Nagel, S. R. (2003). *Phys. Rev. E*, **68**, 011306.
Orowan, E. (1934). *Z. Phys.*, **89**, 614.
Palmer, R. G., Stein, D. L., Abrahams, E., and Anderson, P. W. (1984). *Phys. Rev. Lett.*, **53**, 958.
Parisi, G. and Zamponi, F. (2010). *Rev. Mod. Phys.*, **82**, 789.
Polanyi, M. M. (1934). *Z. Phys.*, **89**, 660.
Richert, R. and Angell, C. A. (1998). *J. Chem. Phys.*, **108**, 9016.
Richert, R., Israeloff, N., Alba-Simionesco, C., Ladieu, F., and L'Hôte, D. (2011). *This book.*
Ritort, F. and Sollich, P. (2003). *Adv. Phys.*, **52**, 219.
Sachdev, S. and Nelson, D. R. (1985). *Phys. Rev. B*, **32**, 1480.
Sadoc, J.-F. and Mosseri, R. (1984). *J. Physique*, **45**, 1025.

Sadoc, J.-F. and Mosseri, R. (1999). *Geometrical Frustration*. Cambridge, Cambridge University Press.
Saltzman, E. J. and Schweizer, K. S. (2006). *J. Chem. Phys.*, **125**, 044509.
Sastry, S., Debenedetti, P. G., and Stillinger, F. H. (1998). *Nature*, **393**, 554.
Sausset, F. and Tarjus, G. (2010). *Phys. Rev. Lett.*, **104**, 065701.
Sausset, F., Tarjus, G., and Viot, P. (2008). *Phys. Rev. Lett.*, **101**, 155701.
Schwarz, J. M., Liu, A. J., and Chayes, L. (2006). *Europhys. Lett.*, **73**, 560–566.
Sciortino, F. (2005). *J. Stat. Mech.*, P05015.
Scopigno, T., Ruocco, G., Sette, F., and Monaco, G. (2003). *Science*, **302**, 849.
Semerjian, G. and Franz, S. (2011). *This book*.
Sethna, J. P. (1983). *Phys. Rev. Lett.*, **51**, 2198.
Sethna, J. P., Shore, J. D., and Huang, M. (1991). *Phys. Rev. B*, **44**, 4943.
Shen, Y. T., Kim, T. H., Gangopadhyay, A. K., and Kelton, K. F. (2009). *Phys. Rev. Lett.*, **102**, 057801.
Shintani, H. and Tanaka, H. (2006). *Nature Phys.*, **2**, 200.
Sillescu, H. (1999). *J. Non-Cryst. Solids*, **243**, 81.
Singh, Y., Stoessel, J. P., and Wolynes, P. G. (1985). *Phys. Rev. Lett.*, **54**, 1059.
Sokolov, A. P., Rössler, E., Kisliuk, A., and Quitmann, D. (1993). *Phys. Rev. Lett.*, **71**, 2062.
Steinhardt, P. J., Nelson, D. R., and Ronchetti, M. (1981). *Phys. Rev. Lett.*, **47**, 1297.
Stillinger, F. H. (1995). *Science*, **267**, 1935.
Stillinger, F. H. and Weber, T. A. (1983). *Science*, **225**, 983.
Struik, L. C. E. (1978). *Aging in Amorphous Polymers and Other Materials*. Amsterdam, Elsevier.
Tarjus, G. and Kivelson, D. (2001). In *Jamming and Rheology: Constrained Dynamics on Microscopic and Macroscopic Scales* (ed. A. J. Liu and S. R. Nagel), London. Taylor and Francis.
Tarjus, G., Kivelson, S. A., Nussinov, Z., and Viot, P. (2005). *J. Phys.: Condens. Matter*, **17**, R1143.
Tarzia, M., Biroli, G., Lefèvre, A., and Bouchaud, J.-P. (2010). *J. Chem. Phys.*, **132**, 054501.
Toulouse, G. (1977). *Commun. Phys.*, **2**, 115.
Turnbull, D.J. and Cohen, M. H. (1961). *J. Chem. Phys.*, **34**, 120.
van Hecke, M. (2010). *J. Phys.: Condens. Matter*, **22**, 033101.
van Meel, J. A., Charbonneau, B., Fortini, A., and Charbonneau, P. (2009*a*). *Phys. Rev. E*, **80**, 061110.
van Meel, J. A., Frenkel, D., and Charbonneau, P. (2009*b*). *Phys. Rev. E*, **79**, 030201(R).
Viot, P., Tarjus, G., and Kivelson, D. (2000). *J. Chem. Phys.*, **112**, 10368.
Vogel, M., Doliwa, B., Heuer, A., and Glotzer, S.C. (2004). *J. Chem. Phys.*, **120**, 4404.
Wales, D. J. (2003). *Energy Landscapes: Applications to Clusters, Biomolecules and Glasses*. Cambridge, Cambridge University Press.
Whitelam, S., Berthier, L., and Garrahan, J. P. (2004). *Phys. Rev. Lett.*, **92**, 185705.
Whitelam, S., Berthier, L., and Garrahan, J. P. (2005). *Phys. Rev. E*, **71**, 026128.

Widmer-Cooper, A. and Harrowell, P. (2006). *Phys. Rev. Lett.*, **96**, 185701.
Widmer-Cooper, A. and Harrowell, P. (2009). *Phys. Rev. E*, **80**, 061501.
Widmer-Cooper, A., Harrowell, P., and Fynawever, H. (2004). *Phys. Rev. Lett.*, **93**, 135701.
Widmer-Cooper, A., Perry, H., Harrowell, P., and Reichman, D. R. (2008). *Nature Phys.*, **4**, 711.
Wyart, M. (2005). *Ann. Phys. (Paris)*, **30**, 1.
Xia, X. Y. and Wolynes, P. G. (2000). *Proc. Natl. Acad. Sci. USA*, **97**, 2990.
Xu, N., Wyart, M., Liu, A. J., and Nagel, S. R. (2007). *Phys. Rev. Lett.*, **98**, 175502.
Zaccarelli, E., Sciortino, F., and Tartaglia, P. (2004). *J. Phys.: Condens. Matter*, **16**, S4849.

3

Overview of different characterizations of dynamic heterogeneity

Ludovic BERTHIER, Giulio BIROLI,
Jean-Philippe BOUCHAUD and Robert L. JACK

Abstract

Dynamic heterogeneity is now recognized as a central aspect of structural relaxation in disordered materials with slow dynamics, and was the focus of intense research in the last decade. Here, we describe how initial, indirect observations of dynamic heterogeneity have recently evolved into well-defined, quantitative, statistical characterizations, in particular through the use of high-order correlation and response functions. We highlight both recent progress and open questions about the characterization of dynamic heterogeneity in glassy materials. We also discuss the limits of available tools and describe a few candidates for future research in order to gain a deeper understanding of the origin and nature of glassiness in disordered systems.

3.1 Introduction

3.1.1 Dynamical heterogeneity in glassy materials

The glass transition is often cited as a profound outstanding problem in condensed-matter physics. This field may be contrasted with that of simple liquids, for which the broad picture is now well established, and appropriate theoretical methods are well developed (Hansen and McDonald, 1986). Why, then, is the glass problem so difficult?

From a theoretical perspective, a central difficulty arises from the importance of fluctuations in glassy systems. Both the liquid and the glass have disordered structure, so even if all molecules in the system are identical, they experience different local environments. In the liquid, these differences can be neglected: one may infer the behavior of the system from that of a typical particle in a typical environment. Thus, for example, microscopic properties, such as the rate at which particles diffuse in the liquid, are directly related to bulk properties, such as the viscosity. However, as the glass transition is approached, it becomes increasingly difficult to characterize "typical" particles and "typical" environments because a variety of different behaviors emerges. Within a given interval of time, some particles may move distances comparable to their size, while others remain localized near their original positions. Thus, on these time scales, we can refer to "mobile" and "immobile" particles. Of course, on long enough time scales, ergodicity ensures that particles become statistically identical.

In case this discussion seems rather abstract, we refer the reader to Fig. 3.1. Here, we show several systems in which mobile and immobile particles can be identified in particular trajectories using different methods. Strikingly, these images reveal that particles with different mobilities do not appear randomly in space but are clustered. This observation suggests that structural relaxation in disordered systems is a non-trivial dynamical process. In its narrow sense, the term "dynamical heterogeneity" encapsulates the spatial correlations observed in Fig. 3.1. However, the term is frequently used in a broader sense, referring to a range of fluctuation phenomena that arise from deviations from the "typical" behavior (Ediger, 2000).

Over the last decade, it has become clear from experiments and computer simulations that a variety of glassy systems display the kind of clusters shown in Fig. 3.1. Experimentally, their existence can be inferred from experiments in molecular liquids (Ediger, 2000; Sillescu, 1999; Richert, 2002; Richert *et al.*, 2010), while direct observation of single-particle motion makes them vivid in colloids and granular media (Weeks *et al.*, 2000; Keys *et al.*, 2007; Kegel and van Blaaderen, 2000; Dauchot

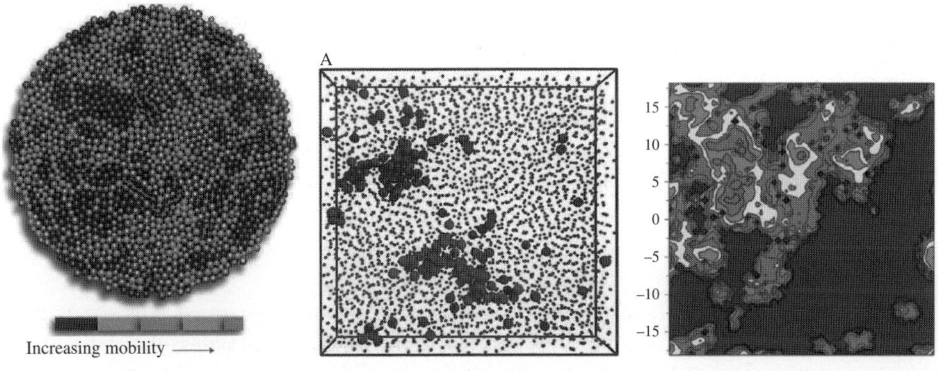

Fig. 3.1 Three examples of dynamical heterogeneity. In all cases, the figures highlight the clustering of particles with similar mobility. (Left) Granular fluid of ball bearings, with a gray scale showing a range of mobility increasing from dark to light (Keys *et al.*, 2007). (Center) Colloidal hard-sphere suspension, with most mobile particles highlighted (Weeks *et al.*, 2000). (Right) Computer simulation of a two-dimensional system of repulsive disks. The color scheme indicates the presence of particles for which motion is reproducibly immobile or mobile, respectively from dark to light (Widmer-Cooper *et al.*, 2008).

et al., 2005). In computer simulations, spherical particles with simple pair potentials have been used as models for both colloidal and molecular systems, with dynamically heterogeneous behavior clearly present in a variety of models. Dynamical heterogeneity has also been investigated in a large number of more schematic lattice models, such as kinetically constrained or lattice glass models (Ritort and Sollich, 2003). This important body of experimental and computational observations has also stimulated important theoretical developments since they represent a new set of observations against which existing theories can be confronted.

So far, the reader may be unconvinced of the difficulty of the problem. After all, theoretical methods for analyzing systems with large fluctuations and correlated domains already exist: the methods developed to describe critical phenomena, such as field theory and the renormalization group. These ideas are indeed central to this chapter, but their application to glassy liquids has required substantial new insight into the nature of the relevant fluctuations and observables. The reason is that the distinction between mobile and immobile particles is in essence dynamical. Therefore, if one analyses static snapshots of viscous liquids, there is little evidence of increasing fluctuations or heterogeneity, at least when analyzed using standard liquid-state correlation functions. Instead of an ensemble of snapshots, one must apply the methods of critical phenomena to an ensemble of "movies" (i.e. trajectories, or dynamical histories) of the system. This will be the approach that we will follow in later sections.

3.1.2 Suitable probes for the emergence of glassiness

Returning to Fig. 3.1, the observation of clusters of mobile particles (or at least of regions with correlated mobility) raises many questions. What is the nature of the clusters? Indeed, are they of the same nature in each case? What determines their size? How is their size distribution related to the relaxation time of the system, if at all? Do the particles in fast or slow clusters have different local environments that can be characterized by any simple structural measure? From a theoretical point of view, it is natural to ask whether the correlated regions represent the cooperatively rearranging regions imagined long ago by Adam–Gibbs (Adam and Gibbs, 1965); whether they mirror some soft elastic modes in the system; how they might connect to locally ordered domains; or whether they reveal the presence of localized defects that facilitate structural rearrangement.

However, before addressing these ambitious questions, we must answer a more prosaic one. How can the "clusters of mobile and immobile particles" be defined and characterized? In Fig. 3.1, the mobile particles were identified by a threshold on their displacement, over a particular time scale. Indeed, many early papers have suggested different ways to define "mobility" and "clusters" that are similar in spirit but different in practice, as we detail in the following sections. However, one must certainly evaluate how strongly the cluster properties depend on thresholds or time scales, which might hinder firm conclusions and may prevent fair comparison between different systems.

In more recent years, it has become possible to define and measure observables that can be determined without arbitrariness in a range of systems, are amenable to analytic theory and scaling arguments, and may sometimes be inferred from experimental data even in molecular liquids. These are known as "four-point correlation functions" and are now broadly accepted as standard tools for analyzing dynamical heterogeneity.

Within this toolbox, a central role is played by the four-point dynamical susceptibility $\chi_4(t)$. Loosely speaking, it measures the number of particles involved in correlated motion on timesscales of the order of t. To interpret the dynamical susceptibility, it is useful to invoke an analogy with critical phenomena, in which a (static) correlation length ξ diverges at a critical temperature T_c, accompanied by the spontaneous appearance of an order parameter. This divergence is associated with a diverging susceptibility χ, which may be measured either through the fluctuations of the order parameter or through the response of the order parameter to its conjugate field. When considering such phenomena in glasses, a problem arises, in that a static order parameter and its conjugate field are not known. Instead, a fairly good dynamical order parameter is given by any generic dynamical two-point correlator, e.g. density–density, displaying the slowing down of the dynamics. This motivated the use of a four-point dynamical susceptibility associated to spontaneous fluctuations of the dynamical order parameter. This allows the identification of a dynamic length scale, $\xi_4(t)$, and a susceptibility, $\chi_4(t)$, by analogy with conventional critical phenomena.

Returning to Fig. 3.1, simulation studies and recent experiments indicate that the clustering of mobile particles is directly linked with an increasing susceptibility χ_4 and an increasing correlation length scale ξ_4. The central part of this chapter will be devoted to a discussion of these four-point functions. We will discuss some of the insights that they have revealed into the nature of glassy behavior in liquids, colloids, and granular media. However, these studies also revealed that interpretation of four-point functions may be somewhat ambiguous, while direct measurements of correlation length scales remain difficult. Additionally, the averaging procedure inherent in the four-point functions means that they may obscure important features of dynamic heterogeneity such as the cluster shape and the nature of interfaces between clusters of mobile and immobile particles. Towards the end of the chapter, we will discuss a range of alternative observables that complement the information available from four-point functions.

3.2 Observables for characterizing dynamical heterogeneity

3.2.1 Two-point observables and their inadequacy

We begin with a review of some two-point functions that are used to characterize simple liquids. We will show that these functions are largely blind to the dynamically heterogeneous behavior shown in Fig. 3.1, motivating the discussion of more discriminating observables.[1]

[1]There are other examples of systems for which standard two-point correlators are blind to interesting intermittent (or heterogeneous) effects, such as turbulence or financial markets, and for which higher-order correlations are informative.

For any fluid of particles, a natural quantity to measure is the structure factor,

$$S(q) = \frac{1}{N} \langle \rho_{\mathbf{q}}(t) \rho_{-\mathbf{q}}(t) \rangle, \tag{3.1}$$

where brackets indicate an ensemble average, and the Fourier component of the density is

$$\rho_{\mathbf{q}}(t) = \sum_{i=1}^{N} e^{i\mathbf{q}\cdot\mathbf{r}_i(t)}, \tag{3.2}$$

with N being the number of particles and $\mathbf{r}_i(t)$ being the position of particle i at time t. The structure factor gives information about the strength of density fluctuations on a length scale $2\pi/|\mathbf{q}|$. However, its behavior in the vicinity of the glass transition is unremarkable, with no hint of the dynamic clustering of Fig. 3.1. Although static heterogeneities in the density would directly imply the existence of dynamic heterogeneity, the reverse is not true. Thus, dynamic heterogeneity related to the motion of particles has a much more subtle origin. Note that more complicated static correlation functions have been studied (Debenedetti, 1996), especially in numerical work, all attempting to identify fluctuations of some prescribed sort of local order (translational, orientational, etc.). Until now, there are no strong indications of a diverging, or even substantially growing, length scale (Menon and Nagel, 1995; Fernandez *et al.*, 2006). In order to see the growth of a static amorphous correlation, one possibility is to introduce the so-called "point-to-set" correlation function that we discuss later, see Section 3.4.4.

We therefore turn to dynamical observables. A quantity relevant for light- and neutron-scattering experiments is the intermediate scattering function,

$$F(\mathbf{q}, t) = \frac{1}{N} \langle \rho_{\mathbf{q}}(t) \rho_{-\mathbf{q}}(0) \rangle . \tag{3.3}$$

Measurements of this function by neutron scattering in supercooled glycerol (Wuttke *et al.*, 1996) are shown for different temperatures in Fig. 3.2. These curves suggest a first, rather fast, relaxation to a plateau followed by a second, much slower, relaxation. The plateau is due to the fraction of density fluctuations that are frozen on intermediate time scales, but eventually relax during the second relaxation. The latter is called "alpha-relaxation", and corresponds to the structural relaxation of the liquid. The plateau is akin to the Edwards–Anderson order parameter, defined for spin-glasses that measures the fraction of frozen spin fluctuations (Binder and Kob, 2005). Note that the Edwards–Anderson parameter continuously increases from zero below the critical temperature in the conventional spin-glass transition (Mézard *et al.*, 1988), while for structural glasses, a finite plateau value seems to emerge above any putative transition.

The full decay of the intermediate scattering function can be measured only within a relatively small range of temperatures. In order to track the dynamic slowing down from microscopic to macroscopic time scales, other correlators have been studied. A very popular one is the measurement of the dielectric linear susceptibility, which can be followed over up to 18 decades of frequency (Lunkenheimer and Loidl, 2002). It is

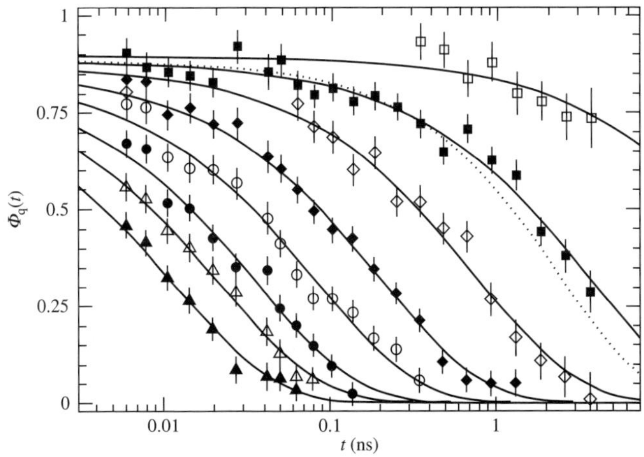

Fig. 3.2 Temperature evolution of the normalized intermediate scattering function, $\phi_q(t) = S(\mathbf{q}, t)/S(\mathbf{q}, 0)$, for supercooled glycerol (Wuttke *et al.*, 1996). Temperatures decrease from 413 K to 270 K from left to right. The lines are fits with a stretched exponential form.

generally accepted that different dynamic probes reveal similar temperature dependences for the relaxation time, at least as long as the probes measure local motion. In broad terms, the essential features on supercooling are a dramatic increase in the correlation time, and a broad distribution of time scales in the system characterized in the time domain by non-exponential relaxation functions.

It is increasingly accepted that the presence of such a broad distribution of time scales in glassy systems is associated with the presence of mobile and immobile domains. However, the size and shape of these domains, or even their very existence, cannot be deduced directly from $F(\mathbf{q}, t)$. We are therefore motivated to consider more advanced correlation functions.

3.2.2 Indirect evidence: Intermittency and decoupling phenomena

We now return to the presence of mobile and immobile particles in the supercooled phase. Simulation studies are ideal for studying fluctuation properties, since accurate trajectories for all particles are accessible, over long time scales. These features are displayed in Fig. 3.3, which shows that while the averaged mean-squared displacements are smooth functions of time, time signals for individual particles clearly exhibit specific features that are not observed unless dynamics is resolved both in space and time. In this figure, we observe that particle trajectories are very intermittent, being composed of a succession of long periods of time where particles simply vibrate around well-defined locations, separated by rapid "jumps". Vibrations were previously inferred from the plateaux observed at intermediate times in the mean-squared displacements or intermediate scattering functions, but the existence of jumps that are statistically widely distributed in time cannot be revealed from averaged quantities only. The fluctuations in Fig. 3.3 suggest, and direct measurements confirm, the importance played

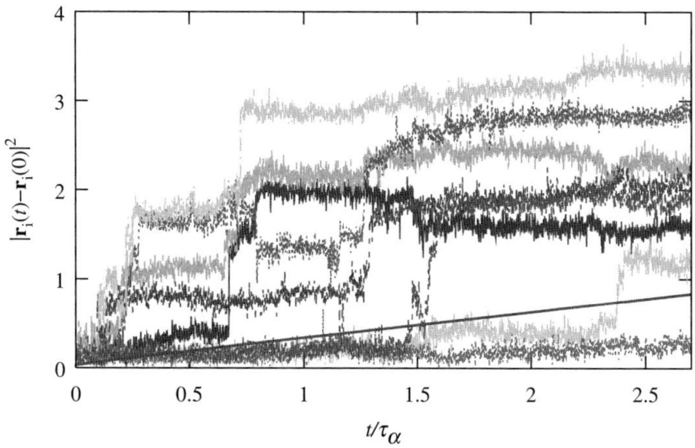

Fig. 3.3 Time-resolved squared displacements of individual particles in a simple model of a glass-forming liquid composed of Lennard-Jones particles (Berthier and Kob, 2007). The average is shown as a smooth full line and time is expressed in units of structural relaxation time τ_α. Trajectories are composed of long periods of time during which particles vibrate around well-defined positions, separated by rapid jumps that are statistically widely distributed in time, underlying the importance of dynamic fluctuations.

by fluctuations around the averaged dynamical behavior to understand structural relaxation in glassy materials.

Remaining at the single-particle level, these fluctuations can be characterized through the (time-dependent) distribution of particle displacements. This is the self-part of the van-Hove function, defined as

$$G_s(\mathbf{r},t) = \left\langle \frac{1}{N} \sum_{i=1}^{N} \delta(\mathbf{r} - [\mathbf{r}_i(t) - \mathbf{r}_i(0)]) \right\rangle . \tag{3.4}$$

For an isotropic Gaussian diffusive process, one has $G_s(\mathbf{r},t) \sim \exp\left(-\frac{|\mathbf{r}|^2}{4D_s t}\right)$. While simple liquids are well described by such a distribution, simulations of glassy systems instead reveal strong deviations from Gaussian behavior on the time scales relevant for structural relaxation (Kob *et al.*, 1997). In particular, they reveal "broad" tails in the distributions that are much wider than expected from the Gaussian approximation. These tails are in fact well described by an exponential, rather than Gaussian, decay in a wide time window comprising the structural relaxation,

$$G_s(\mathbf{r},t) \sim \exp\left(-\frac{|\mathbf{r}|}{\lambda(t)}\right), \tag{3.5}$$

which is a direct consequence of the intermittent motion shown in Fig. 3.3 (Chaudhuri *et al.*, 2007). These tails reflect the existence of a population of particles that moves distinctively further than the rest and appears therefore to be much more mobile. This observation implies that relaxation in a viscous liquid differs qualitatively from that of a normal liquid where diffusion is close to Gaussian, and that a non-trivial statistics of single-particle displacements exists in materials with glassy dynamics.

Another influential phenomenon that was related early on to the existence of dynamic heterogeneity is the decoupling of self-diffusion (D_s) and viscosity (η). In the high-temperature liquid, self-diffusion and viscosity are related by the Stokes–Einstein relation (Hansen and McDonald, 1986), $D_s\eta/T = const.$ For a large particle moving in a fluid the constant is equal to $1/(6\pi\mathrm{R})$ where R is the particle radius. Physically, the Stokes–Einstein relation means that two different measures of the relaxation time, R^2/D_s and $\eta R^3/T$, lead to the same time scale up to a constant factor. In supercooled liquids this phenomenological law breaks down, as shown in Fig. 3.4 for ortho-terphenyl (Mapes *et al.*, 2006). It is commonly found that D_s^{-1} does not increase as fast as η so that, at T_g, the product $D_s\eta$ has significantly increased as compared to its Stokes–Einstein value. The Stokes–Einstein "violation" factor is larger for more fragile liquids, and can be as high as 10^3. This phenomenon, although less spectacular than the overall change of viscosity, is a strong indication that different ways to measure relaxation times lead to different answers and, thus, is a strong hint of the existence of broad distributions of relaxation time scales (Stillinger and Hodgdon, 1994; Tarjus and Kivelson, 1995).

Indeed, a natural explanation of this effect is that different observables probe the underlying distribution of relaxation times in different ways (Ediger, 2000). For example, the self-diffusion coefficient of tracer particles is dominated by the more mobile particles, whereas the viscosity or other measures of structural relaxation probe the time scale needed for every particle to move. An unrealistic but instructive example

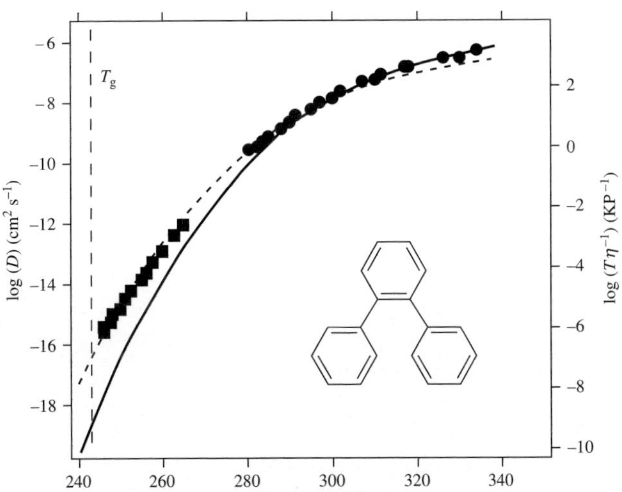

Fig. 3.4 Decoupling between viscosity (full line) and self-diffusion coefficient (symbols) in supercooled ortho-terphenyl (Mapes *et al.*, 2006). The dashed line shows a fit with a "fractional" Stokes–Einstein relation, $D_s \sim (T/\eta)^\zeta$ with $\zeta \sim 0.82$ instead of the expected value $\zeta = 1$.

is a model where there is a small, non-percolative subset of particles that are blocked forever, coexisting with a majority of mobile particles. In this case, the structure never fully relaxes but the self-diffusion coefficient is non-zero because of the mobile particles. Although unrealistic since all particles move in a viscous liquid, this example shows how different observables are likely to probe different moments of the distribution of time scales, as explicitly shown within several theoretical frameworks (Tarjus and Kivelson, 1995; Jung *et al.*, 2004; Hedges *et al.*, 2007; Heuer, 2008).

3.2.3 Early studies of dynamic heterogeneity

The phenomena described above, although certainly an indication of spatio-temporal fluctuations, do not allow one to study how these fluctuations are correlated in space. However, this is a fundamental issue both from the experimental and theoretical points of view, as discussed in the introduction. To discriminate between different explanations of glassy behavior, it would be useful to know: How large are the regions that are faster or slower than the average? How does their size depend on temperature? Are these regions compact or fractal?

These important questions were first addressed in pioneering works using four-dimensional NMR (Tracht *et al.*, 1998; Reinsberg *et al.*, 2001), and by directly probing fluctuations at the nanoscopic scale using microscopy techniques. In particular, Vidal Russel and Israeloff used atomic force microscopy techniques (Vidal Russell and Israeloff, 2000) to measure the polarization fluctuations in a volume of size of few tens of nanometers in a supercooled polymeric liquid (PVAc) close to T_g. In this spatially resolved measurement, the hope is to probe a small enough number of dynamically correlated regions, and to detect their dynamics. Indeed, the time signals shown in Ref. (Vidal Russell and Israeloff, 2000) show a very intermittent dynamics, switching between moments with intense activity, and moments with no dynamics at all, suggesting that extended regions of space indeed transiently behave as fast and slow regions. A much smoother signal would have been measured if dynamically correlated "domains" were not present.

Spatially resolved and NMR experiments are quite difficult. They give undisputed information about the typical lifetime of the dynamic heterogeneity, but their determination of a dynamic correlation length scale is rather indirect, and has been performed on a small number of liquids in narrow temperature windows. Nevertheless, an agreed consensus is that these experiments reveal the existence of a non-trivial dynamic correlation length emerging at the glass transition, where it reaches a value of the order of 5 to 10 molecule diameters (Ediger, 2000).

More recently, studies of systems for which single-particle dynamical data is available have led to direct observation of clusters of mobile particles, such as those shown in Fig. 3.1. This is possible in materials where the particles are big enough to be visualized through microscopy, such as colloids, or with a camera, such as granular materials. In early studies, mobile and immobile particles were usually identified by using a threshold on their displacement within a given time interval, dividing the particles into subpopulations. This then allows characterization of correlations within the populations, as shown in Fig. 3.5. For example, ratios of radial distribution

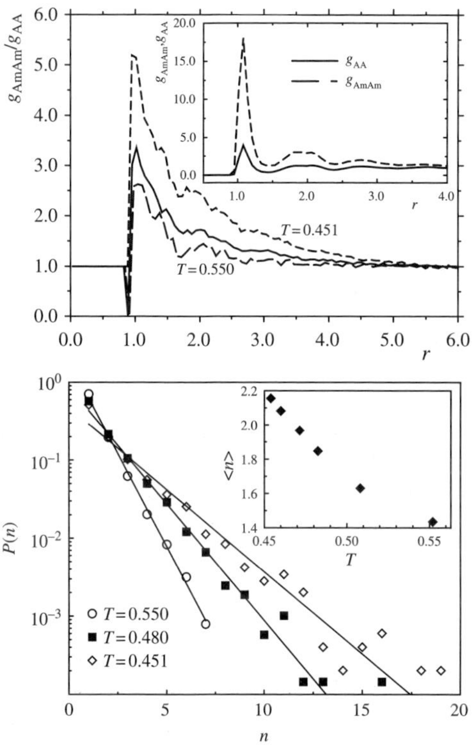

Fig. 3.5 Early measurements of dynamical heterogeneity from computer simulation. (Top) The main figure shows the probability that a particle is within the active population, given that it is a distance r from another active particle (Kob *et al.*, 1997). The increasing correlation between active particles is apparent on decreasing the temperature. (Bottom) The length distribution of string-like clusters of active particles (Donati *et al.*, 1998).

functions give the probability that particles in the vicinity of a mobile particle are themselves mobile. Other work concentrated for instance on the morphology and size of clusters of mobile particles, as shown in Fig. 3.5.

Results such as those of Fig. 3.5 gave clear evidence of large fluctuations and dynamical heterogeneity. However, unambiguous identification of a mobile population of particles proved difficult, with most distributions of mobility showing broad but unimodal distributions. Similarly, the identification of connected clusters of mobile particles introduces further ambiguity into the data processing, with their size and shape depending quite strongly on the definition of the mobile population. To achieve unambiguous and system-independent definitions of length scales and correlation volumes, it is useful to define correlation functions that do not involve separating particles into distinct populations, nor the identification of connected clusters. Four-point functions are a natural choice in this regard, and will be discussed in the next chapter.

3.2.4 Higher-order correlations: four-point functions

3.2.4.1 Definitions

In the previous section, we considered the probability that a particle is within some mobile population, given that it has a mobile particle a distance r away (recall Fig. 3.5). While the measurement of such a probability requires the identification of a mobile population, there is a straightforward alternative that contains similar information. We first define a continuously varying "mobility" $c_i(t,0)$ that indicates how far or how much particle i moves between times $t=0$ and t. Then, given two particles at separation r, one can measure the degree to which their mobilities are correlated. To this end, it is convenient to define a "mobility field" through (Bennemann *et al.*, 1999; Donati *et al.*, 1999; Glotzer *et al.*, 2000)

$$c(\mathbf{r};t,0) = \sum_i c_i(t,0)\delta(\mathbf{r}-\mathbf{r}_i). \tag{3.6}$$

Then, the spatial correlations of the mobility are naturally captured through the correlation function (Dasgupta *et al.*, 1991)

$$G_4(r;t) = \langle c(\mathbf{r};t,0)c(\mathbf{0};t,0)\rangle - \langle c(\mathbf{r};t,0)\rangle^2, \tag{3.7}$$

which depends only on the single time t and the single distance $r=|\mathbf{r}|$ as long as the average is taken at equilibrium in a translationally invariant system. The analogy with fluctuations in critical systems becomes clear in Eq. (3.7) if one considers the mobility field $c(\mathbf{r};t,0)$ as playing the role of the order parameter for the transition, characterized by non-trivial fluctuations and correlations near the glass transition.

Often, the mobility $c_i(t,0)$ is itself a two-point function. For example, to measure mobility on a length scale $2\pi/q$, one might consider $o_i(q,t) = e^{i\mathbf{q}\cdot\mathbf{r}_i(t)}$ and $c_i(t,0) = o_i(\mathbf{q},t)o_i(-\mathbf{q},0)$. In this case, $o_i(\mathbf{q},t)$ is related to a Fourier component of the density of the system, and the average of $c_i(t,0)$ is the self-part of the intermediate scattering function $F(\mathbf{q},t)$ defined in Eq. (3.3). Moving from a particle observable $o_i(\mathbf{q},t)$ to a field $o(\mathbf{r};\mathbf{q},t)$, one arrives at

$$G_4(r;t) = \langle o(\mathbf{r};\mathbf{q},t)o(\mathbf{r};-\mathbf{q},0)o(\mathbf{0};\mathbf{q},t)o(\mathbf{0};-\mathbf{q},0)\rangle - \langle o(\mathbf{r};\mathbf{q},t)o(\mathbf{r};-\mathbf{q},0)\rangle^2. \tag{3.8}$$

This correlation function is quartic in the operator o, so it is known as a "four-point function". It measures correlations on a length scale r, associated with motion between time zero and time t; it depends additionally on the length scale q^{-1} used in the definition of the particle mobility $c_i(t,0)$. Since structural relaxation typically involves particle motion over a distance comparable to the particle size R, one typically chooses $q \sim 1/R$ and studies the remaining t and r dependences.

This definition of a real-space correlation function of the mobility represents a vital advance in the characterization of dynamical heterogeneity. In particular, it allows the language of field theory and critical phenomena to be used in studying dynamical fluctuations in glassy systems. By analogy with critical phenomena, if there is a single dominant length scale ξ_4 then one expects that for large r, the correlation function decays as

$$G_4(r;t) \sim \frac{A(t)}{r^p} e^{-r/\xi_4(t)}, \tag{3.9}$$

with p an exponent whose value is discussed below. It is also natural to define the susceptibility associated with the correlation function

$$\chi_4(t) = \int \mathrm{d}r\, G_4(r;t). \tag{3.10}$$

If the pre-factor $A(t)$ were known, the susceptibility $\chi_4(t)$ could be used to extract the typical number of particles involved in correlated motion. That is, $\chi_4(t)$ may be interpreted as the size of the correlated clusters in Fig. 3.1.

Further, $\chi_4(t)$ can also be obtained from the fluctuations of the total mobility $C(t,0) = \int \mathrm{d}^d\mathbf{r}\, c(\mathbf{r};t,0)$, through

$$\chi_4(t) = N[\langle C(t,0)^2\rangle - \langle C(t,0)\rangle^2]. \tag{3.11}$$

In practice, this formula allows an efficient measure of the degree of dynamical heterogeneity, at least in computer simulations and in experiments where the dynamics can be spatially and temporally resolved. As long as the observable $c(\mathbf{r};t,0)$ is chosen appropriately, this observable can be measured in a wide variety of systems, and serves as a basis for fair comparisons of the extent of dynamical heterogeneity.

3.2.4.2 The spin-glass perspective: four-point functions in space and time

It is interesting to note that four-point functions have their origin in spin-glass physics, where they were used to investigate the onset of long-ranged amorphous order. The key insight is due to Edwards and Anderson (Edwards and Anderson, 1975). In a spin-glass, one considers a set of N localized degrees of freedom, the spins, which we denote by s_x, with x denoting the position in space. The spins interact by quenched random couplings. Then, two-point correlation functions between spins such as $\sum_x \langle s_x s_{x+r}\rangle$ typically vanish for $r \neq 0$, since sites separated by a distance r may be either correlated or anti-correlated, with equal probability. The Edwards–Anderson solution is to consider instead the static spin-glass correlation $\chi_{\mathrm{SG}} = N^{-1}\sum_x \langle s_x s_{x+r}\rangle^2$ that receives a positive contribution for both correlated and anti-correlated sites.

If one then considers a four-point dynamic function such as

$$G_4(r;t) = \frac{1}{N}\sum_x \langle s_x(0)s_{x+r}(0)s_x(t)s_{x+r}(t)\rangle, \tag{3.12}$$

then this approaches the static spin-glass correlation at long times. In spin-glasses, spontaneous symmetry breaking occurs into an amorphous solid with long-ranged order at a second-order spin-glass transition. Near such a phase transition, $G_4(r;t)$ obeys critical scaling behavior similar to that shown by static correlation functions near familiar phase transitions, and the corresponding susceptibility $\chi_4(t)$ diverges.

In structural glasses, the possibility of a static transition to an amorphous ordered state remains highly controversial. However, four-point functions such as $G_4(r;t)$ and

$\chi_4(t)$ can be usefully employed in the fluid state to characterize dynamical fluctuations, regardless of their possible connection to any critical point.

3.2.4.3 Four-point susceptibilities in molecular, colloidal and granular glasses

The function $\chi_4(t)$ has been measured in computer simulations of many different glass-forming liquids, by molecular dynamics, Brownian and Monte Carlo simulations, see e.g. (Bennemann *et al.*, 1999; Donati *et al.*, 1999; Glotzer *et al.*, 2000; Franz *et al.*, 1999; Lačević *et al.*, 2003; Berthier, 2004; Vogel and Glotzer, 2004; Berthier, 2007; Parisi, 1999; Parsaeian and Castillo, 2008). An example is shown in Fig. 3.6 for a Lennard-Jones liquid. The qualitative behavior is similar in all cases (Franz and Parisi, 2000; Toninelli *et al.*, 2005; Berthier *et al.*, 2007*a*): as a function of time $\chi_4(t)$ first increases, it has a peak on a time scale that tracks the structural relaxation time scale and then it decreases. As mentioned above, the decrease of $\chi_4(t)$ at long times constitutes a major difference with spin glasses. In a spin glass, χ_4 would be a monotonically increasing function of time whose long-time limit coincides with the static spin-glass susceptibility. Physically, the difference is that the spin-glass state that emerges at the transition is critical or marginally stable, i.e. characterized by singular static responses.

The peak value of $\chi_4(t)$ measures the volume over which the structural relaxation processes are correlated. Therefore, the most important result obtained from data such as presented in Fig. 3.6 is the temperature evolution of the peak height, which is found to increase when the temperature decreases and the dynamics slows down. From such data, one gets direct evidence that the approach to the glass transition is accompanied by the development of increasingly long-ranged spatial correlations of the dynamics.

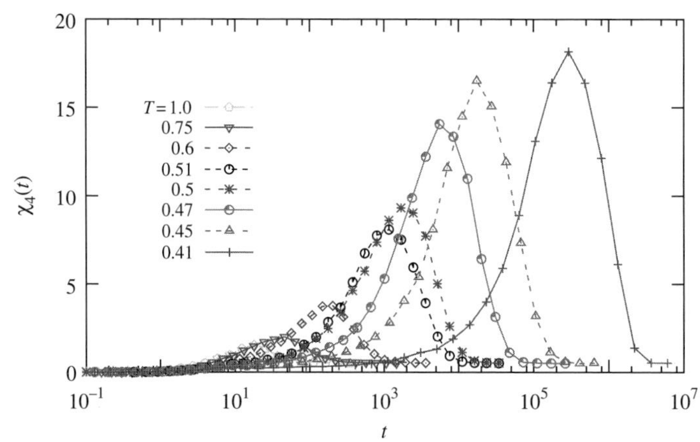

Fig. 3.6 Time dependence of $\chi_4(t)$ quantifying the spontaneous fluctuations of the self-intermediate scattering function in a molecular dynamics simulation of a Lennard-Jones supercooled liquid (Berthier, 2004). For each temperature, $\chi_4(t)$ has a maximum, which shifts to larger times and has a larger value when T is decreased, revealing the increasing length scale of dynamic heterogeneity in supercooled liquids approaching the glass transition.

Note that if the dynamically correlated regions were compact, the peak of χ_4 would be proportional to ξ_4^d (in d spatial dimensions), thus directly relating $\chi_4(t)$ measurements to that of the relevant length scale ξ_4 of dynamic heterogeneity.

In experiments, direct measurements of $\chi_4(t)$ have been made in colloidal (Weeks *et al.*, 2007) and granular materials (Dauchot *et al.*, 2005; Keys *et al.*, 2007) close to the colloidal and granular glass transitions, and also in foams (Mayer *et al.*, 2004) and gels (Duri and Cipelletti, 2006), because dynamics is more easily spatially and temporally resolved in those cases. The results obtained in all these cases are again broadly similar to those shown in Fig. 3.6, both for the time dependence of $\chi_4(t)$ and its evolution upon a change of the relevant variable controlling the dynamics.

A major issue is that obtaining information on the behavior of $\chi_4(t)$ and $G_4(r;t)$ from experiments on molecular systems is difficult. In molecular liquids, it remains a difficult task to resolve temporally the dynamics at the nanometer scale. Such measurements are, however, important because numerical simulations and experiments on colloidal and granular systems can typically only be performed for relaxation times spanning at most 5–6 decades. On the other hand, in molecular liquids, up to 14 decades can be measured, and extrapolation of simulation data all the way to the experimental glass transition is fraught with difficulty. Indirect estimates of $\chi_4(t)$ from experiments will be discussed below.

These results are also relevant because many theories of the glass transition assume or predict, in one way or another, that the dynamics slows down because there are increasingly large regions over which particles have to relax in a correlated or cooperative way, see Section 3.3.2 and Ref. (Tarjus, 2010). However, this length scale remained elusive for a long time. Measures of the spatial extent of dynamic heterogeneity, in particular $\chi_4(t)$ and $G_4(r;t)$, seem to provide the long-sought evidence of this phenomenon. This in turn suggests that the glass transition can indeed be considered as a form of critical phenomenon involving growing time scales and length scales. This is an important progress towards the understanding of the glass-transition phenomenon, even though a clear and conclusive understanding of the relationship between dynamic length scales and relaxation time scales is still the focus of intense research.

3.2.4.4 Dependence of $\chi_4(t)$ on the observable and probe length scale

As discussed above, one may define a four-point function $G_4(r;t)$ starting from any suitable mobility $c(\mathbf{r};t,0)$. Indeed, many candidates have been considered. It is not our intention to give a detailed list, but a few comments are in order.

A natural choice for $\chi_4(t)$ is to start from Eq. (3.11) and to take $C(t,0) = \rho_{\mathbf{q}}(t)\rho_{-\mathbf{q}}(0)$ as the autocorrelation of a single Fourier component of the density. In this case, the average of $C(t,0)$ is the intermediate scattering function $F(\mathbf{q},t)$ of Eq. (3.3). In computational studies, it is often more convenient to construct instead $C(t,0)$ from a simple sum over particles. That is, one defines the single-particle mobility

$$f_i(\mathbf{q},t,0) \equiv e^{i\mathbf{q}\cdot(\mathbf{r}_i(t)-\mathbf{r}_i(0))}, \tag{3.13}$$

whose average is the self-part of $F(\mathbf{q},t)$. The real-space four-point function is then given by Eq. (3.7), and the definition of $\chi_4(t)$ follows. These two definitions of $\chi_4(t)$

are not equivalent. Differences between them were discussed in Ref. (Lačević *et al.*, 2003), where it was concluded that they contain similar information.

Physically speaking, the key point is that as particle i moves away from its initial position $r_i(0)$, the function $f_i(\mathbf{q};t,0)$ decays from a value of unity, reaching zero when the particle has moved a distance of order $(\pi/|\mathbf{q}|)$. Once the particle has moved further than this, the oscillations in the cosine function imply that averages of f_i give numbers close to zero. Based on this physical interpretation, other choices for $c_i(t,0)$, including step functions, or smoothly decaying functions were used (Franz *et al.*, 1999; Bennemann *et al.*, 1999; Lačević *et al.*, 2003; Berthier, 2004). As expected on physical grounds, constructing four-point functions based on these choices for $c_i(t,0)$ again leads to qualitatively similar behaviors.

Yet another choice is to use a function $c_i(t,0)$ that depends not just on the positions at time zero and time t, but also on the whole history of the particle between these times. In particular, one may take a "persistence" function that takes a value of unity if the particle remains within a distance a of its initial position for all times between 0 and t; otherwise it takes the value zero. Again, one observes a broadly similar behavior (Chandler *et al.*, 2006).

All these choices for the definition of the local mobility $c_i(t,0)$ involve a "probe length scale", which is fixed by the choice of measurement, in contrast to the sought dynamic length scale $\xi_4(t)$, which should be a physical property of the system. While the specific form of $c(\mathbf{r};t,0)$ is typically unimportant for the qualitative behavior of four-point functions, the quantitative results do depend on the probe length scale.

Typically, if the probe length scale a is of the order of the particle diameter or smaller, $c(\mathbf{r};t,0)$ measures local motion, and this is often the scale on which heterogeneity is most apparent. As the probe length scale is increased, contributions to $\chi_4(t)$ come from pairs of particles that remain correlated over distances comparable to a and, typically, such correlations weaken as a increases, reducing $\chi_4(t)$ (Chandler *et al.*, 2006). Similarly, reducing a also reduces $\chi_4(t)$ as short-scale motion corresponding to thermal vibrations are also typically uncorrelated. Therefore, $\chi_4(t)$ is usually maximal for a probe length scale comparable to the particle size, and it is fixed to a constant when comparing data at different temperatures or densities. An alternative choice is to adjust the probe length scale a at different state points such that $\chi_4(t;a)$ reaches its absolute maximum, this can be very important for some systems like granular media close to the rigidity transition where the maximum is reached for values of the probe length far below the particle size (Lechenault *et al.*, 2008*a*,*b*; Heussinger *et al.*, 2010).

3.2.4.5 Real-space measurements and associated structure factors

We concluded above that a growing peak in $\chi_4(t)$ "directly" reveals the growth of a dynamic correlation length scale as the glass transition is approached. This can only be correct if the assumptions made above for the scaling form of $G_4(r;t)$ are correct. Dynamic length scales should in principle be obtained by direct measurements of a spatial correlation function (Doliwa and Heuer, 2000; Lačević *et al.*, 2002).

However, in contrast to $\chi_4(t)$, detailed measurements of $G_4(r;t)$ are technically more challenging as dynamic correlations must now be resolved in space over large distances with a very high precision, and so there is much less data to draw on. From the point of view of numerical simulations where many measurements of χ_4 were reported, the main limitation to properly measure ξ_4 is the system size. This might seem surprising as typical numbers extracted for the correlation length scale ξ_4 are modest, growing, say, from 1 to at most 5–10 in most reports. However, this modest number hides the fact that the correlation function only decays to zero for distances r that are several times larger than ξ_4. Given that $G_4(r;t)$ is accurately measured up to $r = L/2$ in a periodic system of linear size L, going to a few ξ_4 (say, five times), when $\xi_4 \sim 5$ requires systems containing at least $N \sim L^3 \sim (2 \times 5 \times 5)^3 \sim 10^5$ particles in three dimensions, assuming the density is near $\rho \approx 1$. Such large system sizes are not easily studied at low temperatures when relaxation times get very large, even with present-day computers. However, these studies are of vital importance in that they allow the dynamical length scale $\xi_4(t)$ to be measured directly. Moreover, insights from such studies can then be used when inferring the behavior of $\xi_4(t)$ from measurements of $\chi_4(t)$.

Some representative data are shown in Fig. 3.7. They are obtained for a lattice gas model with kinetic constraints, where measurements of $G_4(r;t)$ are somewhat easier than in molecular dynamics simulations. As discussed above, the idea is that for large r, $G_4(r;t) \simeq A(t)r^{-p}F(r/\xi_4(t))$ that then yields

$$\chi_4(t) \simeq A(t)\xi_4(t)^{d-p}. \tag{3.14}$$

Often, one estimates the pre-factor $A(t)$ to be equal to $G_4(0,t)$, which is simply the variance of the local quantity $c(r;0,t)$. However, the accuracy of this estimate is hard to assess without explicit evaluation of $G_4(r,t)$. In the example shown in Fig. 3.7,

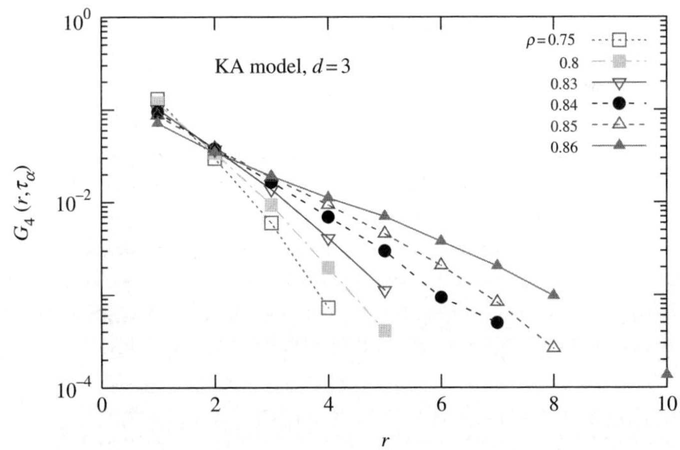

Fig. 3.7 Four-point correlation function $G_4(r; t = \tau_\alpha)$ measured in computer simulations of the Kob–Andersen kinetically constrained lattice gas in three dimensions (Berthier, 2003). The dynamics slows down when density ρ increases, and the slower spatial decay of G_4 directly reveals increasingly longer-ranged dynamic correlations accompanying the glass transition.

for instance the scaling between χ_4 and ξ_4 in Eq. (3.14) is well obeyed, and careful examination of $G_4(r;t)$ suggests that $p \approx 1$ and A is indeed a constant of order 1.

While this is a subtle situation that requires each case to be considered individually, the work in this domain is broadly consistent with $\chi_4(t)/G_4(0,t)$ representing the number of particles involved in heterogeneous relaxation. Note that these issues will also be relevant for discussion of other correlations and susceptibilities in later sections. Therefore, truly "direct" measurements indeed confirm that the increase of the peak of $\chi_4(t)$ corresponds, as expected, to a growing dynamic length scale $\xi_4(t)$ (Doliwa and Heuer, 2000; Lačević *et al.*, 2002; Bennemann *et al.*, 1999; Berthier, 2004; Berthier *et al.*, 2007*a*). As a result of subtleties related to the difference between four-point correlations in spin-glasses and structural glasses, an early study of four-point functions (Dasgupta *et al.*, 1991) drew an opposite conclusion, but the data of that study are in fact consistent with the now-established picture of a growing length scale.

We also note, in passing, that the power p may have more than one interpretation. Typically, one assumes that a typical cluster has size $\xi_4(t)$ and contains $\xi_4(t)^{d-p}$ particles, so that $d-p$ is interpreted as a fractal dimension. However, an alternative would be that clusters are all compact, but that the distribution of their sizes is rather broad. This uncertainty reflects the fact that four-point functions involve averages over many clusters, so that they do not resolve details of cluster structure.

Instead of direct inspection of $G_4(r;t)$, it is often convenient to consider its Fourier transform $S_4(k;t) = \int \mathrm{d}^d r e^{i\mathbf{k}\cdot\mathbf{r}} G_4(r;t)$ (Bennemann *et al.*, 1999; Donati *et al.*, 1999; Glotzer *et al.*, 2000; Lačević *et al.*, 2002; Berthier, 2004; Yamamoto and Onuki, 1998*a*,*b*; Flenner and Szamel, 2010). In particular, this allows data for different times and different temperatures to be combined into a scaling analysis that can yield the temperature dependence of ξ_4, leaving an uncertain pre-factor. This approach was taken for example in Refs. (Yamamoto and Onuki, 1998*b*; Berthier *et al.*, 2007*a*; Flenner and Szamel, 2010). Typically, simulation data result in length scales between 1 and 5 particle diameters.

So far, we have considered circularly averaged $G_4(r;t)$ and $S_4(k;t)$. However, if the probe function $c_i(t,0)$ is anisotropic, as in Eq. (3.13), four-point functions $G_4(r;t)$ then also depend on the angle between the separation vector $\mathbf{r}$ and the probe vector $\mathbf{q}$. Several papers (Weeks *et al.*, 2007; Flenner and Szamel, 2009; Doliwa and Heuer, 2000) have investigated this issue and found that motion in longitudinal directions is indeed more strongly correlated than motion in transverse directions, such that $G_4(r;t)$ is truly an anisotropic function. These findings add further to the difficulty of extracting the length scale ξ_4 from direct measurements of four-point structure factors.

3.2.5 Experimental estimates of multipoint susceptibilities

Although readily accessible in numerical simulations, the fluctuations of $C(t,0)$ that give access to $\chi_4(t)$ are in general very small and impossible to measure directly in experiments, except when the range of the dynamic correlation is macroscopic, as in granular materials (Marty and Dauchot, 2005) or in soft glassy materials where it can reach the micrometer and even millimetre range (Mayer *et al.*, 2004; Duri and Cipelletti, 2006). To access $\chi_4(t)$ in molecular liquids, one should perform time-resolved dynamic measurements probing very small volumes, with a linear size of the order of a few nanometers. Although possible, such experiments remain to be performed with the required accuracy.

It was recently realized that simpler alternative procedures exist. The central idea underpinning these results is that induced dynamic fluctuations are in general more easily accessible than spontaneous ones, and both types of fluctuations can be related to one another by fluctuation–dissipation theorems. The physical motivation is that while four-point correlations offer a direct probe of the dynamic heterogeneities, other multipoint correlation functions give very useful information about the microscopic mechanisms leading to these heterogeneities.

For example, one might expect that the slow part of a local enthalpy (or energy, density) fluctuation $\delta h_x(t=0)$ at position x and time $t=0$ triggers or eases the dynamics in its surroundings, leading to a systematic correlation between $\delta h_x(t=0)$ and $c(x+r;t,0)$. This physical intuition suggests the definition of a family of three-point correlation functions that relate thermodynamic or structural fluctuations to dynamical ones. Interestingly, and crucially, some of these three-point correlations are both experimentally accessible and give bounds or approximations to the four-point dynamic correlations (Berthier *et al.*, 2005, 2007*a*,*b*).

Based on this insight, one may obtain a lower bound on $\chi_4(t)$ using the Cauchy–Schwartz inequality $\langle \delta H(0)\delta C(t,0)\rangle^2 \leq \langle \delta H(0)^2\rangle \langle \delta C(t,0)^2\rangle$, where $H(t)$ denotes the enthalpy at time t, and $\delta X = X - \langle X\rangle$ denotes the fluctuating part of the observable X. By using a fluctuation–dissipation relation valid when the energy is conserved by the dynamics, the previous inequality can be rewritten as (Berthier *et al.*, 2005):

$$\chi_4(t) \geq \frac{k_{\mathrm{B}}T^2}{c_{\mathrm{P}}} \left[\chi_T(t)\right]^2, \tag{3.15}$$

where the multipoint response function $\chi_T(t)$ is defined by

$$\chi_T(t) = \left.\frac{\partial \langle C(t,0)\rangle}{\partial T}\right|_{N,P} = \frac{N}{k_{\mathrm{B}}T^2} \langle \delta H(0)\delta C(t,0)\rangle, \tag{3.16}$$

and c_{P} is the specific heat per particle (at constant pressure).

In this way, the experimentally accessible response $\chi_T(t)$ that quantifies the sensitivity of average correlation functions $\langle C(t,0)\rangle$ to an infinitesimal temperature change, can be used in Eq. (3.15) to yield a lower bound on $\chi_4(t)$. From Eq. (3.16), it is clear that χ_T is directly related to the covariance of enthalpy and dynamic fluctuations, and thus captures the part of dynamic heterogeneity that is triggered by enthalpy fluctuations.

Detailed numerical simulations and theoretical arguments (Berthier *et al.*, 2007*a*,*b*) strongly suggest that the right-hand side of Eq. (3.15) actually provides a good estimate of $\chi_4(t)$ in supercooled liquids, and not just a lower bound. Similar estimates exist considering density as a perturbing field instead of the temperature. These are useful when considering colloidal or granular materials where the glass transition is mostly controlled by the packing fraction. The quality of the corresponding lower bound was tested experimentally on granular packings close to the jamming transition (Lechenault *et al.*, 2008*b*), and numerically for colloidal hard spheres (Brambilla *et al.*, 2009).

Using this method, Dalle-Ferrier *et al.* (Dalle-Ferrier *et al.*, 2007) have been able to estimate the evolution of the peak value of χ_4 for many different glass formers in the entire supercooled regime. In Fig. 3.8 we show some of these results as a function of the relaxation time scale. The value on the y-axis, a bound on the peak of χ_4, is a proxy for the number of molecules, $N_{\rm corr,4}$ in a cluster of mobile or immobile particles. As discussed briefly above, $\chi_4(t)$ is expected to be equal to $N_{\rm corr,4}$, up to a proportionality constant $A(t)$ that is not known from experiments, probably explaining why the high-temperature values of $N_{\rm corr,4}$ are smaller than one. Figure 3.8 also indicates that $N_{\rm corr,4}$ grows faster when τ_α is not very large, close to the onset of slow dynamics, and a power-law relationship between $N_{\rm corr,4}$ and τ_α is good in this regime ($\tau_\alpha/\tau_0 < 10^4$). The growth of $N_{\rm corr,4}$ with τ_α becomes much slower closer to the glass-transition temperature T_g, where a change of 6 decades in time corresponds to a mere increase of a factor about 4 of $N_{\rm corr,4}$, suggesting logarithmic rather than power-law growth of dynamic correlations. A similar crossover towards a very slow growth of dynamic correlations is reported for colloidal hard spheres (Brambilla *et al.*, 2009) and model glasses (Berthier *et al.*, 2007*a*), and is observed in numerical simulations even if the dynamic length scale ξ_4 is directly estimated (Flenner and Szamel, 2010). The consequences of such an effect for theories of the glass transition are discussed

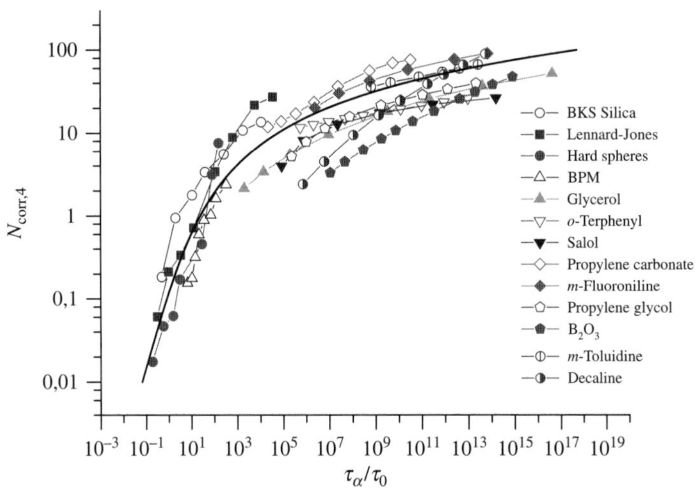

Fig. 3.8 Dynamic scaling relation between the number of dynamically correlated particles, $N_{\rm corr,4}$, and relaxation time scale, τ_α, for a number of glass formers (Dalle-Ferrier *et al.*, 2007), determined using the bound provided by Eq. (3.15). For all systems, dynamic fluctuations increase when the glass transition is approached. The full line through the data (Dalle-Ferrier *et al.*, 2007) suggests a crossover from algebraic, $N_{\rm corr,4} \sim \tau_\alpha^z$, to logarithmic, $N_{\rm corr,4} \sim \exp(\tau_\alpha^\psi)$, growth of dynamic correlations with increasing τ_α.

below. Bearing in mind all the caveats discussed above (unknown pre-factors, quality of the bound, etc.), the experimental data compiled in Fig. 3.8 do appear to confirm that dynamic fluctuations and correlation length scales do grow when the molecular liquids approach their glass transitions.

3.2.6 Four-point susceptibilities: some caveats

The above story about the four-point susceptibility looks quite enticing. We have essentially argued that dynamical heterogeneity should be quantified by the spatial correlations of the mobility. This correlation function is *a priori* hard to measure in molecular glasses, but a divine surprise occurs: using rather trivial mathematics, its spatial integral is found to be bounded from below by a quantity that is much easier to measure. Is this too good to be true? What is the physics underpinning this "easy" bound?

The answer is that four-point correlation functions pick up a contribution that depends both on the statistical ensemble used (i.e. NVE vs. NVT) and on the dynamics (i.e. Brownian vs. Newtonian). Using general scaling arguments based on a dominant length scale ξ_4, one can show that in Fourier space the four-point correlation function has the following structure (in the $k \to 0$ limit) (Berthier *et al.*, 2007*a*,*b*):

$$S_4(k;\tau_\alpha) = \xi_4^s \hat{H}_1(k\xi_4) + g(k)\left[\xi_4^s \hat{H}_2(k\xi_4)\right]^2, \tag{3.17}$$

where s is a certain exponent (related to p above through $2s = d - p$) and $\hat{H}_{1,2}$ are certain scaling functions that behave similarly at small and large arguments. The wavevector-dependent function $g(k)$ carries the dependence on the statistical ensemble and dynamics. In particular, $g(k=0) = 0$ when all conserved quantities are strictly fixed, as in the NVE ensemble.[2] Similarly, $g(k)$ is very different for Brownian dynamics (for which energy is not conserved) or for Newtonian dynamics (that conserves energy). So one has to face the uneasy truth that $\chi_4(t) \equiv S_4(k=0;t)$ depends on many microscopic and macroscopic details, although physically these should not affect the dynamic correlation length ξ_4. It is definitely not so easy to directly relate χ_4 to a dynamic correlation length. Intuitively, conserved variables play a role in "transmitting" the information about mobility from one region to another.

From the above expression, one sees that $S_4(k;t)$ mixes up two contributions that one would like to disentangle so as to extract the relevant scaling contribution from the first term only. It turns out that this first term is proportional to a three-point response function, that measures the change of the dynamics induced by a perturbation some distance away, and that we will discuss in more detail in Section 3.4. It is this three-point response, not G_4, that is the fundamental object carrying information about dynamical heterogeneities, and from which G_4 is constructed. The space integral of the three-point response function is $\xi_4^s \hat{H}_1(0)$; physically it represents the response of the dynamics to a uniform shift of an external parameter, such as the temperature or the density. Hence, $\chi_T(t)$ as defined in the previous section is a three-point function

[2]Actually, fast degrees of freedom can give subleading contributions in some cases.

at zero wavevector. This explains the physical nature of the lower bound on $\chi_4(t)$: due to the contribution of the energy as a conserved quantity, χ_4 has two contributions: one proportional to χ_T and one proportional to χ_T^2, the latter being precisely the lower bound of the previous section.

This discussion leads to the following caveats that we alluded to above: (a) the identification of a correlation volume from $\chi_4(t)$ alone is not warranted in general. The information contained in $S_4(k;\tau_\alpha)$ is needed to unambiguously relate the growth of $\chi_4(t)$ to a growing length scale;[3] (b) extracting information from S_4 can be difficult due to interference effects between the two terms and (c) three-point response functions are the fundamental building bricks for dynamic correlations, and are not soiled with problems related to conserved variables or statistical ensembles.

3.3 Theoretical discussion

3.3.1 Recent progress based on four-point functions

In the previous section, we have summarized some of the properties of four-point functions, their advantages for calculating the extent of dynamical heterogeneity, and some direct and indirect measurements of these quantities.

There are many subtleties associated with these measurements, but the same broad picture is observed in a variety of systems and is robust to the precise measurement used. Essentially, as relaxation times increase, the four-point susceptibility increases, suggesting the presence of a growing dynamic correlation length. Where the real-space function $G_4(r;t)$ can also be measured, this confirms more directly the increase of such a length scale, typically in the range between 1 and 10 molecular diameters. The fundamental, unavoidable conclusion seems to be that glassy behavior is not a purely local "caging" of particles by their neighbors, but indeed a genuine collective phenomenon.

Having established the existence of a growing correlation length, several questions arise. From a theoretical perspective, we are familiar with the idea, borrowed from equilibrium critical phenomena, that when correlation length scales get large, microscopic features of the system become unimportant, and "universal" behavior emerges. Whether realistic glassy systems have length scales that are large enough for such a universal description remains unclear. Many analyses in this spirit have nevertheless been attempted, as we shall discuss shortly. It is likely that in order to reach a good quantitative agreement a careful treatment of pre-asymptotic effects will have to be performed.

A second fundamental point concerns the microscopic mechanisms that give rise to the correlations revealed by four-point functions. Many model systems can demonstrate the presence of increasing time scales, coupled with increasing susceptibilities

[3]Related to this point, it is worth mentioning the case of purely Arrhenius systems, which are considered to be non-cooperative systems. Still, the lower bound based on χ_T proves that $\chi_4(\tau_\alpha)$ diverges in these systems as least as T^{-2} as the temperature goes to zero (Dalle-Ferrier *et al.*, 2007). The physical interpretation of such an apparent growth of the range of dynamical correlations is still unclear.

$\chi_4(t)$ and length scales ξ_4. Predictions from different theoretical frameworks of the form of four-point functions are discussed in the next section, and we evaluate some of the theories in the light of existing results.

3.3.2 Models of the glass transition and their predictions of dynamic heterogeneity

We now turn to perhaps the most fundamental question in this area: what are the dominant mechanisms by which structural relaxation takes place in glassy materials? We give a quick survey of the dominant pictures for molecular glasses, their predictions for four-point functions, and the extent to which these are borne out.

3.3.2.1 Mode-coupling theory

The mode-coupling theory (MCT) of the glass transition (Götze and Sjögren, 1995) was historically derived from liquid-state theory. Starting from exact microscopic equations of motion for the density field in a liquid, several uncontrolled approximations are then performed to yield a closed set of dynamical equations for intermediate scattering functions. These equations give rise to a dynamic singularity at some finite temperature, T_c, where relaxation times diverge in an algebraic manner. Additionally, very precise quantitative predictions can be made about the specific form of intermediate scattering functions, suggesting a very rich behavior of time-correlation functions, which do resemble the behavior observed experimentally and reported in Fig. 3.2. As is well known these predictions only apply over a modest time window of about 2–3 decades in the moderately supercooled regime, but dramatically break down nearer to the glass transition (Götze, 2008).

Another dramatic failure of the traditional formulation of the theory, more relevant to the present contribution, is its inability to accurately predict the shape of the van Hove distribution function described above, the resulting wavevector dependence of the self-intermediate scattering function (especially at low wavevector), and the corresponding decoupling between self-diffusion constant and the viscosity.

Following this historical route, therefore, it is not obvious whether the MCT dynamic singularity is accompanied by non-trivial dynamic fluctuations. This also means that the theory is not easily interpreted in physical terms. Recently, mode-coupling theory was reformulated in such a way that both these issues were greatly clarified. Using a field-theoretic formulation, it is possible to perform consistent mode-coupling approximations to get analytical predictions for both averaged two-time dynamic correlation functions and for the dynamic fluctuations around the averaged behavior, i.e. for $\chi_4(t)$ (Franz and Parisi, 2000; Berthier *et al.*, 2007*b*; Biroli and Bouchaud, 2004). In a subsequent recent move, "inhomogeneous mode-coupling" predictions for the shape and scaling form of four-point spatial correlation functions $G_4(r;t)$ and its Fourier transform $S_4(k;t)$ were finally obtained (Biroli *et al.*, 2006). Thus, overall, mode-coupling theory is now able to make an impressive set of very detailed predictions for a very large family of spatio-temporal correlation functions for any given liquid, starting from the form of the microscopic interaction between the particles.

A few numerical simulations have been presented to test these new predictions. First, the temporal evolution of the four-point susceptibility $\chi_4(t)$ was compared to mode-coupling predictions. Just as time-correlation functions decay within MCT in a two-step process similar to the data presented in Fig. 3.2, $\chi_4(t)$ is predicted to grow with time with two distinct power laws, $\chi_4 \sim t^a$ and $\chi_4 \sim t^b$ in the time regimes, respectively, corresponding to the approach to, and departure from, the plateau. These two power-law regimes have been successfully identified in numerical work, with numerical values for the exponents a and b that are in "reasonable" agreement with numbers predicted by MCT (Berthier *et al.*, 2007*b*; Berthier, 2007; Berthier and Kob, 2007).

The peak of the four-point susceptibility is predicted to diverge algebraically at the critical temperature. This prediction was observed numerically to hold over a similar (restricted) temperature window as for the averaged relaxation time τ_α itself. This finding implies that the peak of χ_4, when plotted as a function of τ_α, follows a power-law scaling, $\chi_4 \sim \tau_\alpha^z$, where z is predicted to be a non-universal critical exponent. Returning to the data compilation in Fig. 3.8, we remark that the data obtained in the moderately supercooled regime do indeed approximately follow an initial growth that is consistent with the MCT prediction, while clearly breaking down at lower temperatures.

Finally, predictions for the detailed shape of four-point correlation functions, in particular for $S_4(k;t)$, were recently confronted by numerical results, with inconclusive results. While a first paper (Stein and Andersen, 2008) reports excellent agreement with MCT predictions both for the wavevector dependence of $S_4(k;t)$ and its evolution with temperature, a more recent report (Karmakar *et al.*, 2009) claims that disagreements with theoretical predictions arise when larger system sizes are included in the numerical analysis. This ongoing debate illustrates the fact mentioned above that even for the modest correlation length scales characterizing relaxation in supercooled liquids, very large system sizes are needed to unambiguously and accurately measure four-point functions. Clearly, more work is needed to clarify the status of the large body of MCT predictions regarding four-point functions.

3.3.2.2 Facilitation picture and kinetically constrained models

In the facilitation picture of supercooled liquids, structural relaxation is thought to originate from propagation of localized excitations, or "defects", throughout the system (Glarum, 1960). The idea is that the local structure at position x changes rapidly when a defect visits the neighborhood of x. Then, one makes the further hypothesis that defects are sparse, and most of the system is in fact immobile, such that defects can be considered as uncorrelated, independent objects. In practice, very little is known about the nature, origin, or even the existence of such defects in real liquids.

Nevertheless, it is clear within this picture that dynamics is highly heterogeneous in space, and temporally intermittent. Physically, one expects a dynamic correlation length scale ξ_4 to emerge, which should basically correspond to the linear size of the region explored independently by a given defect. This also means that the time dependence of $\xi_4(t)$, and thus of $\chi_4(t)$ can directly be connected, in this view, to how fast the defects move. In the simplest approximation where defects are diffusive objects, one would expect $\xi_4(t) \sim \sqrt{t}$ for $t < \tau_\alpha$.

In recent years, these ideas have been pursued quite extensively, based on the proposal (Garrahan and Chandler, 2003) that facilitation is essential for explaining the heterogeneous dynamics of supercooled liquids. The theory has been developed primarily through studies of systems called kinetically constrained models (Ritort and Sollich, 2003). Numerous distinct lattice models belong to this family, which can be distinguished by the set of microscopic rules governing the dynamics of the localized defects, the existence or absence of conservation laws, the topology of the lattice, etc. The simplest models, such as the (one-spin facilitated) Fredrickson–Andersen (FA) model, in fact reproduce nearly exactly the scaling relations mentioned above for four-point functions (Toninelli *et al.*, 2005; Chandler *et al.*, 2006; Whitelam *et al.*, 2004, 2005).

Given the large number of distinct models, and the fact that models are postulated instead of being derived as approximate, coarse-grained representations of liquids, it is not clear how one should compare their behavior to numerical results obtained for realistic liquid models. This issue is discussed in a recent review (Chandler and Garrahan, 2010). Thus, it is perhaps better to interpret this diversity as being suggestive of the different types of behavior one can possibly encounter in liquids. On the other hand, from the theoretical point of view, having well-defined, relatively simple, statistical models defined on the lattice is very appealing as very many detailed and quantitative results can be obtained by exploiting tools from statistical mechanics. Thus, kinetically constrained models can also be viewed as "toy supercooled liquids".

In this regard, the study of the dynamically heterogeneous behavior of kinetically constrained models has been a very active field of research in recent years. Many models have been investigated and a large number of time-correlation functions (two-point, four-point, persistence functions) have been analyzed, suggesting possible behaviors for dynamic susceptibilities (Berthier *et al.*, 2007*a*; Chandler *et al.*, 2006) or decoupling phenomena (Jung *et al.*, 2004; Pan *et al.*, 2005). We refer to the chapter by Sollich, Toninelli, and Garrahan for further details and references on this topic.

At the qualitative level, it is obvious that all models are characterized by rapidly growing time scales and length scales, and are thus interesting models to study dynamic heterogeneity. However, models with diffusive point defects (like the simplest of Fredrickson–Andersen models), do not compare well with the real liquids that have been studied so far. In three dimensions, they predict simple exponential relaxation and no decoupling phenomena (Jack *et al.*, 2006). The dependence of χ_4 on time and temperature are also characterized by scaling laws that have not been observed in numerical and experimental results (Toninelli *et al.*, 2005; Berthier *et al.*, 2005). However, the Arrhenius scaling of their relaxation time indicates a relation between these models and "strong" liquids (Garrahan and Chandler, 2003; Whitelam *et al.*, 2005) and there are comparatively few results on dynamical heterogeneity for such materials. In particular, while such models predict rather large dynamical length scales in strong materials (Garrahan and Chandler, 2003; Berthier and Garrahan, 2005), these have not yet been observed.

On the other hand, more complicated models where defects move subdiffusively or cooperatively seem to be more appropriate representations of "fragile" liquids, which have a non-Arrhenius scaling of relaxation time with temperature. Such kinetically constrained models exhibit stretched exponential relaxation as in Fig. 3.2, decoupling phenomena (Jung *et al.*, 2004) similar to the results in Fig. 3.4, realistic form of four-point structure factors as in Fig. 3.7, or dynamic length scales that grow very slowly (Garrahan and Chandler, 2003), in qualitative agreement with the experimental results shown in Fig. 3.8.

3.3.2.3 Adam–Gibbs and the mosaic picture

The idea that relaxation events in glasses are collective and involve the simultaneous motion of several particles dates back at least to Adam and Gibbs, who provided an argument to relate the size of these "cooperatively rearranging regions" (CRR) to the configurational entropy of the supercooled liquid. A lot of the work on dynamical heterogeneities is in fact motivated by the Adam–Gibbs picture and attempts to determine the size of these CRR (Binder and Kob, 2005).

The Adam–Gibbs picture was later put on more solid ground in the context of the random first-order transition (RFOT) theory of glasses (Kirkpatrick *et al.*, 1989; Lubchenko and Wolynes, 2006). RFOT suggests that supercooled liquids can be thought of as a mosaic of "glass nodules" or "glassites" with a spatial extension $\ell^*(T)$ limited by the configurational entropy. Regions of size smaller than $\ell^*(T)$ are ideal glasses: they cannot relax, even on very long time scales, because the number of states towards which the system can escape is too small to compete with the energy that blocks the system in a given favorable configuration. Regions of size greater than ℓ^* are liquid in the sense that they explore with time an exponentially large number of unrelated configurations, and all correlation functions go to zero. The relaxation time of the whole liquid is therefore the relaxation time of glassites of size ℓ^*. This relaxation occurs through collective activated events that sweep a region of size ℓ^*, which are the CRR regions of the Adam–Gibbs theory. The crucial assumption of RFOT is that thermodynamics alone fixes the value of ℓ^*, whereas the relaxation time τ_α involves the height of the activation barriers on scale ℓ^*, which is assumed to grow as a power law, $\ell^{*\psi}$, where ψ is a certain exponent (Bouchaud and Biroli, 2004).

Note that when an activated event takes place within a glassite of size ℓ^*, the boundary conditions of the nearby region change. There is a substantial probability that this triggers, or facilitates, an activated event there as well, possibly inducing an "avalanche" process that extends over the dynamic correlation length scale $\xi > \ell^*$. The dynamics on length scales less than ℓ^* is, within RFOT, inherently cooperative, but the relation between the dynamic correlation length ξ, defined for example through three- or four-point point correlation functions and the mosaic length ℓ^* is at this stage an important open problem (see (Dalle-Ferrier *et al.*, 2007; Capaccioli *et al.*, 2008)). If ξ is of the order of a few glassite lengths ℓ^*, then one expects that $\chi_4(\tau_\alpha)$ should grow as ℓ^*

to some power. Assuming activated scaling, $\ln \tau_\alpha \sim \ell^{*\psi}$, finally leads to $\chi_4 \sim (\ln \tau_\alpha)^z$, instead of a power-law relation predicted by MCT or by non-cooperative KCMs. The crossover towards this logarithmic behavior is not incompatible with the data (Xia and Wolynes, 2000), see Section 3.2.5 above. In fact, the details of the crossover between the MCT region and the RFOT region are still very mysterious (see (Biroli and Bouchaud, 2009) for a recent discussion), but some claims have been made about the evolution of the shape of the dynamically correlated regions, that should morph from stringy, fractal objects in the MCT region to compact blobs at lower temperatures (Stevenson *et al.*, 2006). It would be interesting to devise some experimental protocol to test these predictions.

3.4 Beyond four-point functions: other tools to detect dynamical correlations

So far, we have discussed how four-point functions can be used to estimate dynamical length scales, and we have stated that these are typically found to be in the range 1–10 molecular diameters. However, we have also noted that (a) the four-point susceptibility estimate of the dynamical correlation volume may lead to erroneous results (see Section 3.2.6) and (b) a variety of different theories of the glass transition are broadly consistent with the above estimate.

To make further progress, it seems that more adapted and discriminating observables will be required. In fact, there are a wealth of methods that have been used to characterize dynamical heterogeneity, of which we discuss just a few, and we refer to other chapters in this book for details. Here, we mainly emphasize the questions that can be addressed by different methods, and give an overview of the relationships between some of the methods that have been developed in different contexts.

3.4.1 Non-linear susceptibilities

In standard critical phenomena, diverging two-point correlations lead to singular linear responses. It is therefore quite natural to conjecture that increasing dynamical correlations should also lead to anomalous responses of some kind. Spin-glass theory provides, again, an interesting insight. As discussed in Section 3.2.4.2 the spin-glass transition (at zero external field) is signaled by the divergence of the four-point static correlation function $\chi_{\rm SG}$. It can be easily established that close to the transtion, $\chi_{\rm SG}$ is related to the third-order non-linear magnetic response at zero frequency $\chi_3(\omega = 0)$ (Fischer and Hertz, 1991):

$$\chi_3(\omega = 0) = -\frac{\chi_{\rm SG} - 2/3}{(k_{\rm B}T)^2}. \tag{3.18}$$

Thus, although linear responses are blind to the development of spin-glass long-range order, the non-linear magnetic response is not. Actually, it diverges at the transition and, hence, is a direct experimental probe, contrary to $\chi_{\rm SG}$, which can instead only be measured in numerical simulations.

The analogy with spin-glasses discussed in Section 3.2.4.2 therefore suggests that glasses should also display increasingly non-linear responses approaching the glass transition, as first argued in (Bouchaud and Biroli, 2005). Theoretically, this can be substantiated by some general scaling arguments and by a mode-coupling calculation. These are described in (Tarzia *et al.*, 2010); we will just briefly summarize them in the following. The starting point is to rewrite the generic third-order non-linear response $\chi_3(t)$ in terms of the second-order change R_2 of the linear response R:

$$\chi_3(t) = \int_{-\infty}^{t} dt_1 dt_2 dt_3 \frac{\delta P(t)}{\delta E(t_1)\delta E(t_3)\delta E(t_2)} E(t_1)E(t_2)E(t_3) \tag{3.19}$$

$$= \int_{-\infty}^{t} dt_1 dt_2 dt_3 E(t_1) \frac{\delta R(t,t_1)}{\delta E(t_2)\delta E(t_2)} E(t_2)E(t_3) = \int_{-\infty}^{t} dt_1 E(t_1) R_2(t,t_1).$$

Note that we will focus on the dielectric non-linear response (so P is the electric polarization) but the generalization to other perturbing fields is straightforward.[4] It is easy to understand, at least at low frequency, why R_2 and therefore the non-linear susceptibility have a singular behavior. In fact, within an adiabatic approximation, one finds that the linear response in the steady state created by a slowly alternating field is:

$$R_{\rm eq}(t-t', E\cos(\omega t)), \tag{3.20}$$

where $R_{\rm eq}(t-t', E)$ is the equilibrium response function with a *static* field E. Since we are interested in the small-E behavior, we can expand the above expression up to second order in E, this yields:

$$R_2(t,t_1) = R_{\rm eq}(\tau, E\cos(\omega t)) - R_0(\tau) \approx \frac{E^2\cos^2(\omega t)}{2} \frac{\partial^2 R_{\rm eq}(t-t_1,E)}{\partial E^2}\bigg|_{E=0}, \tag{3.21}$$

where $R_0(\tau)$ is the unperturbed equilibrium response function, and the derivative is computed with respect to a constant external field. The second term is expected to give a singular contribution because close to the glass transition a small applied field E is roughly equivalent to a shift of the order of E^2 of the glass-transition temperature (see below), and τ_α significantly varies when the temperature is changed by a small amount close to T_g. More precisely, by taking into account the $E \to -E$ symmetry one can rewrite $\partial^2 R_{\rm eq}/\partial E^2$ as $2\partial R_{\rm eq}/\partial\Theta$, where $\theta = E^2$. Using the time–temperature superposition, one finds:

$$\frac{\partial R_{\rm eq}(\tau,E)}{\partial\Theta}\bigg|_{E=0} \simeq \kappa \frac{\partial R_{\rm eq}(\tau,0)}{\partial T}, \tag{3.22}$$

[4] Depending on the perturbing field, the symmetry $E \to -E$ will hold or not. In the latter case, the first non-linear response is the quadratic one, in the former the quadratic vanishes by symmetry and one has to focus on the third-order one.

where $\kappa = \frac{\partial\tau_\alpha/\partial\Theta}{\partial\tau_\alpha/\partial T}$. Using that $R_{\rm eq}(\tau,0)$ is the Fourier transform of the linear susceptibility $\chi_1(\omega)$ and plugging the previous expressions in Eq. (3.19) (and after some algebra detailed in (Tarzia *et al.*, 2010)) one finds the following result:

$$\chi_3(\omega) \approx \kappa \frac{\partial \chi_1(2\omega)}{\partial T}, \tag{3.23}$$

which is expected to hold at low enough frequency, at least when the deviations from time–temperature superposition are weak. κ is expected to be a slowly varying function of temperature, a constant in first approximation (Tarzia *et al.*, 2010). In this expression, $\partial\chi_1(\omega)/\partial T$ is akin to the three-point susceptibility χ_T defined in Section 3.2.5. Thus, Eq. (3.23) is an important result since it establishes a relationship with the linear dynamical responses that have been used to evaluate dynamical correlations, and it also proves that supercooled liquids should respond in an increasingly non-linear way approaching the glass transition since, as we have discussed before, χ_T and therefore $\frac{\partial\chi_1(2\omega)}{\partial T}$ increase approaching the glass transition.

The above general heuristic arguments can be supplemented by more microscopic ones based on MCT, which provides quantitative predictions on the critical behavior of χ_3. Although the corresponding results are restricted to the low-temperature regime where MCT is believed to apply, they are nevertheless guidelines for the general behavior of χ_3. We sketch the evolution of the absolute value of χ_3 with frequency in Fig. 3.9.

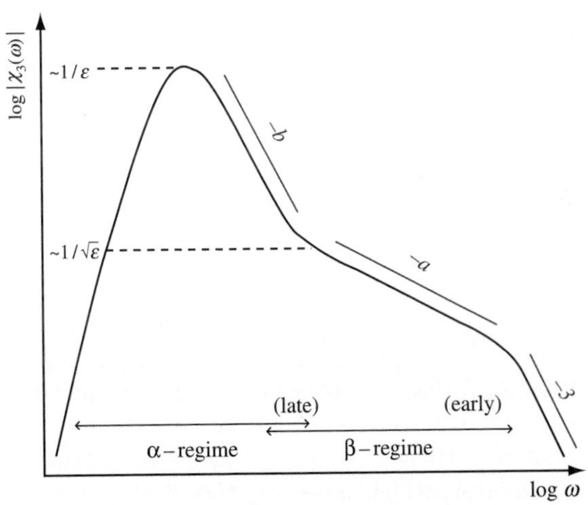

Fig. 3.9 Sketch of $\log|\chi_3(\omega)|$ as a function of $\log\omega$, showing different frequency regimes and crossovers (Tarzia *et al.*, 2010): $\omega\tau_\alpha \ll 1$, $\omega\tau_\alpha \sim 1$, $\tau_\beta/\tau_\alpha \ll \omega\tau_\beta \ll 1$ ($\epsilon = T - T_c$), $\omega\tau_\beta \gg 1$, $\omega\tau_0 \sim 1$. Note that the low-frequency limit is non-zero but much smaller than the peak value for T close to T_c.

- In the α-regime, i.e. $\omega \sim 1/\tau_\alpha \sim \epsilon^{1/2a+1/2b}/\tau_0$, the absolute value of $\chi_3(\omega)$ grows with decreasing ω and reaches its maximum, of height of order $1/\epsilon$, after which it decreases as $\omega^{-b}\tau_\beta$ at large ω. In this regime, one has the scaling form: $\chi_3(\omega) = \frac{1}{\epsilon}\mathcal{G}(\omega\tau_\alpha)$.
- At the crossover between the early α-regime and late β-regime the absolute value of $\chi_3(\omega)$ is of order $1/\sqrt{\epsilon}$.
- In the β-regime, i.e. $\omega \sim 1/\tau_\beta \sim \epsilon^{1/2a}/\tau_0$, the absolute value of $\chi_3(\omega)$ decreases as $\omega^{-b}\tau_\beta$ at small ω and as $\omega^{-a}\tau_\beta$ at large ω. In this regime one has a scaling form $\chi_3(\omega) = \frac{1}{\sqrt{\epsilon}}\mathcal{F}(\omega\tau_\beta)$.

Exponents a and b are the well-known critical exponents of MCT introduced above; τ_0 is a microscopic relaxation time and $\epsilon = (T - T_c)/T_c$ is the distance from the mode-coupling critical temperature T_c. We remark that the existence of the peak and the decrease at low frequency is a non-trivial prediction since it is in contrast with the (trivial) non-linear response of uncorrelated Brownian dipoles (Déjardin and Kalmykov, 1999) and with spin glasses. The decrease with an exponent three at high frequency sketched in Fig. 3.9 is instead trivial and present also for independent dipoles (Déjardin and Kalmykov, 1999).

Recent experiments of third-order non-linear dielectric responses of supercooled glycerol have indeed shown, for the first time, that these theoretical expectations are qualitatively correct (Crauste-Thibierge *et al.*, 2010). We refer to the chapter by Alba-Simionesco *et al.* for a presentation of these experimental results and their discussion.

We conclude this section by emphasizing that perturbing fields, other than electric fields, are expected to lead to similar results. Studying non-linear responses seems to us a very promising route to follow in order to probe the glassy state in a new way. A particularly interesting case worth studying corresponds to non-linear mechanical responses of colloidal glassy liquids. In this case, the values of the perturbing field that affect the sample are of the order of Pa, thus much smaller than the ones affecting the measuring apparatus, which are of the order of GPa. Thus, within a very good approximation, the only non-linear output signal is from the sample itself. This is not the case for dielectric measurements, which are therefore very difficult since one has to be able to filter out the trivial non-linear part due to the amplifiers, etc. This was the main difficulty in the experiment reported in (Crauste-Thibierge *et al.*, 2010).

3.4.2 Inhomogeneous dynamical susceptibilities

We have seen in Section 3.2.5 that the variation of a dynamical correlator with respect to an external parameter (e.g., χ_T) is a way to obtain estimates of the number of dynamically correlated particles. As a natural generalization, one can study the variation of a local correlator, which measures the dynamics around the position $\mathbf{x}$, induced by a perturbation at certain other point $\mathbf{z}$. By summing over all $\mathbf{z}$ one obtains again a global dynamical response such as χ_T, since this then corresponds to computing the variation with respect to a uniform shift of the external parameter.

This new, spatially dependent, dynamical response function is akin to G_4 and allows one to probe the spatial structure of dynamic heterogeneity and to measure directly a dynamical correlation length ξ. The physical reason is that *spontaneous* dynamical fluctuations measured by the 4-point function and *induced* dynamical fluctuations measured by this new type of response function are intimately related. Accelerating or slowing down the dynamics at one given point (by adding an external potential) must perturb the dynamics over a length scale ξ if the dynamics are indeed correlated over this distance.

Let us define more precisely this new dynamical response. Consider the change in the local dynamical structure factor $F(\mathbf{x}, \mathbf{y}, t)$ due to an extra, spatially varying, external potential $U(\mathbf{z})$. Note that this observable can always be decomposed in Fourier modes: $\frac{\delta F(\mathbf{x},\mathbf{y},t)}{\delta U(\mathbf{z})}|_{U=0} = \int d\mathbf{k}d\mathbf{q}e^{-i\mathbf{q}\cdot(\mathbf{x}-)+i\mathbf{k}\cdot(-\mathbf{z})}\chi_{\mathbf{k}}(\mathbf{q}, t)$, where $\chi_{\mathbf{k}}(\mathbf{q}, t) \propto \frac{\delta F(\mathbf{q},\mathbf{q}+\mathbf{k},t)}{\delta U(\mathbf{k})}|_{U=0}$ is the response of the dynamical structure factor to a static external perturbation in Fourier space. For a perturbation localized at the origin, $U(\mathbf{z}) = U_0\delta(\mathbf{z})$, one finds $\delta F(\mathbf{q}, \mathbf{y}, t) = U_0 \int d\mathbf{k}e^{i\mathbf{k}\cdot}\chi_{\mathbf{k}}(\mathbf{q}, t)$. This susceptibility is also related to a 3-point density correlation function in the absence of the perturbation. It is very important to note the very different role played by $\mathbf{q}$ (the standard wavevector) and $\mathbf{k}$ (the wavevector of the perturbation): only the latter is sensitive to dynamic correlations.[5]

The susceptibility $\chi_{\mathbf{k}}(\mathbf{q}, t)$ is interesting for experimental reasons, at least in colloids, since it could be measured by using optical tweezer techniques. From a fundamental point of view, it provides a very useful way to characterize dynamic heterogeneity since it is not affected by complications due to conservation laws and the type of dynamics, contrary to χ_4 and G_4. Therefore, we expect that extracting spatial information and, especially, a precise estimate of ξ should be cleaner by using $\chi_{\mathbf{k}}(\mathbf{q}, t)$ (see the discussion in Section 3.2.6).

Finally, another advantage of $\chi_{\mathbf{k}}(\mathbf{q}, t)$ is that precise quantitative predictions have been obtained within MCT by analytical arguments (Biroli *et al.*, 2006), which were later confirmed by numerical analysis (Szamel and Flenner, 2009) and complementary approaches (Szamel, 2008). The critical behavior of $\chi_{\mathbf{k}}(\mathbf{q}, t)$ approaching the MCT transition temperature is the following (we use the same MCT notation introduced previously):

- In the β-regime, i.e for times of the order of $\tau_\beta = \epsilon^{-1/2a}$, one finds

$$\chi_{\mathbf{k}}(\mathbf{q}, t) \propto \frac{1}{\sqrt{\epsilon} + \Gamma k^2} g_\beta\left(\frac{k^2}{\sqrt{\epsilon}}, t\epsilon^{1/2a}\right), \tag{3.24}$$

where the proportionality constant depends on q. The scaling function $g_\beta(k^2/\sqrt{\epsilon}, t\epsilon^{1/2a})$ is regular for $k = 0$, thus implying that the $\mathbf{k} = 0$ value diverge as $1/\sqrt{\epsilon}$. For large values of $u = t\epsilon^{1/2a}$ one finds that $g_\beta(k^2/\sqrt{\epsilon}, t\epsilon^{1/2a})$ equals $\Xi(\Gamma k^2/\sqrt{\epsilon})u^b$, with Ξ a certain regular function.

[5] We note that compared to Refs. (Berthier *et al.*, 2007*a*; Biroli *et al.*, 2006), the notations $\mathbf{q}$ and $\mathbf{k}$ have been inverted.

- In the α-regime, i.e. for times of the order of $\tau_\alpha = \epsilon^{-1/2a-1/2b}$, one finds

$$\chi_{\mathbf{k}}(\mathbf{q},t) = \frac{\Xi(\Gamma k^2/\sqrt{\epsilon})}{\sqrt{\epsilon}(\sqrt{\epsilon}+\Gamma k^2)}\, g_{\alpha,q}\left(\frac{t}{\tau_\alpha}\right), \tag{3.25}$$

  where Ξ is the same function defined previously. It has the properties: $\Xi(0) \neq 0$ and $\Xi(v \gg 1) \sim 1/v$ such that $\chi_{\mathbf{k}}$ behaves as k^{-4} for large $k\epsilon^{-1/4}$, independently of ϵ. Also, $g_{\alpha,q}(u \ll 1) \propto u^b$, as to match the β regime, and $g_\alpha(u \gg 1, k) \to 0$.

Note that the spatial scaling variable is $k^2/\epsilon^{-1/2}$ in both the α and the β regimes. The physical consequence is that there exists a unique diverging dynamic correlation length $\xi \sim \sqrt{\Gamma}|\epsilon|^{-1/4}$ that rules the response of the system to a space-dependent perturbation within MCT. The analysis of the early β regime where $t \ll |\epsilon|^{-1/2a}$ shows that this length in fact first increases as $t^{a/2}$ and then saturates at ξ. Interestingly, this suggests that while keeping a fixed extension ξ, the (fractal) geometrical structures carrying the dynamic correlations significantly "fatten" between τ_β and τ_α, where more compact structures are expected, as perhaps suggested by the results of (Appignanesi *et al.*, 2006).

Up to now, there are no simulations or experiments measuring a spatial dynamic response such as $\chi_{\mathbf{k}}(\mathbf{q},t)$. Hopefully, these will be performed in the future. As discussed previously, we do believe that this new observable is a simpler and more direct measure of dynamical correlations than G_4. Furthermore, quantitative results beyond scaling can be obtained within MCT. Thus, one could consider comparing the MCT predictions for dynamical heterogeneities to numerical and experimental result in a stringent way, as it has been done for the intermediate scattering function, see e.g. (Kob and Andersen, 1995).

3.4.3 Structure and dynamics: Is dynamic heterogeneity connected to the liquid structure?

One of the most frequently asked questions in studies of dynamical heterogeneity is whether the observed fluctuations might be structural in origin. Such questions have attracted sustained interest. For example, in early numerical work on dynamic heterogeneity, immobile regions were discussed in connection with compositional fluctuations in fluid mixtures (Hurley and Harrowell, 1995). Thirteen years later, some form of local crystalline order is invoked to account for slow domains in numerical work (Kawasaki *et al.*, 2007).

It should be noted that this chapter, and perhaps even this whole book about "dynamic heterogeneity" would not exist in this form if the question of the connection between structure and dynamics had been satisfactorily answered. In that case, indeed, research would be dedicated to understanding the development of structural correlations at low temperatures in supercooled liquids, and to developing tools to measure, quantify and analyze such static features.

A key advance in connecting structural properties to dynamical heterogeneity has been the development of the so-called "iso-configurational ensemble" (Widmer-Cooper *et al.*, 2004). In this approach, one calculates a traditional ensemble average in two stages. First, one averages the particles' velocities, keeping their initial positions fixed. (If the dynamics are stochastic, this step also contains an average over random noises.) Averaging a local dynamical observable such as $c(r;t,0)$ in this way, one arrives at an "iso-configurational average" $\langle c(r;t,0)\rangle_{\text{iso}}$, which still depends on the position r through the fixed initial particle positions. This average is therefore able to reveal the influence of the structure of the initial configuration on the dynamical behavior at that point. To return to a traditional ensemble average, one carries out an average over the initial particle positions in a second step.

The right panel of Fig. 3.1 represents the spatial dependence of the iso-configurationally averaged single-particle mobility in a two-dimensional mixture of soft disks (Widmer-Cooper *et al.*, 2004). The fact that this image is not uniform demonstrates that part of the dynamic heterogeneity has a structural origin. This raises two different questions. First, can one predict from structural measurements the pattern produced by the iso-configurational average in Fig. 3.1? Secondly, how much of the "real" dynamic heterogeneity is actually preserved by the iso-configurational average and has thus a genuine structural origin?

Harrowell and coworkers have provided detailed answers to the first question (Widmer-Cooper *et al.*, 2004; Widmer-Cooper and Harrowell, 2006, 2005, 2007). Statistical analysis of iso-configurational ensembles has been very useful in assessing the statistical significance of correlations between mobility and the local energy, composition or free volume. They have recently made the point that strong correlations exist between vibrational properties of the liquid and isoconfigurational mobilities (Widmer-Cooper *et al.*, 2008). They have also made vivid the distinction between the existence of a statistical correlation between structural and dynamical fluctuations, and the much more demanding notion of a causal link between the two, that is, of a correlation that is strong enough that prediction of the mobility can be made based on a given structural information (Widmer-Cooper and Harrowell, 2006, 2005). These two notions are very often confused in the dynamic heterogeneity literature.

The example of enthalpy fluctuations is useful in this respect. The fact that the four-point susceptibility $\chi_4(t)$ can be quantitatively estimated with good accuracy from a three-point susceptibility such as $\chi_T(t) \propto \langle \delta H(0)\delta C(t,0)\rangle$ provides evidence of a strong correlation between enthalpy and dynamic fluctuations. However, enthalpy fluctuations are not good predictors of dynamic heterogeneity, presumably because they contain short-ranged and short-lived fluctuations that do not correlate well with slow dynamics. Indeed, suitably filtered enthalpy fluctuations correlate very strongly with dynamic heterogeneity (Matharoo *et al.*, 2006).

We finally return to the second question: is dynamic heterogeneity truly captured by iso-configurational averages, and thus does it fully originate from the structure? The response is more subtle than expected as it depends on which observable, and more precisely on which length scale, it is analyzed. We mentioned above that dynamic heterogeneity primarily revealed itself through the intermittent single-particle dynamics (Fig. 3.3) leading to broad distributions of single-particle displacements with broad tails. These features almost completely disappear after the iso-configurational average is performed (Berthier and Jack, 2007). In other words, the distinction between mobile and immobile particles is mostly dynamical in nature, suggesting that the quest for a connection between the static and dynamic properties of glass formers at the particle level is in vain.

Nevertheless, mobility fluctuations do display interesting spatial correlations, as illustrated in Fig. 3.1. This suggests that the distinction between fast and slow domains remains consistent in the iso-configurational ensemble. This observation can be quantified by measuring a "restricted" four-point function

$$\chi_4^{\rm iso}(t) = N\left\langle \langle C(t,0)^2\rangle_{\rm iso} - \langle C(t,0)\rangle_{\rm iso}^2 \right\rangle_{\text{initial cond.}} . \tag{3.26}$$

While $\chi_4(t)$ measures the total strength of dynamic heterogeneity, $\chi_4^{\rm iso}(t)$ makes use of the iso-configurational ensemble and first records the strength of dynamic heterogeneity at fixed initial conditions, the average over initial conditions being performed afterwards. In the case where iso-configurational mobility (and thus the image in Fig. 3.1) is uniform, one has $\chi_4^{\rm iso}(t) = \chi_4(t)$, since the average over initial conditions is trivial in this case. More generally, a large contribution of $\chi_4^{\rm iso}(t)$ to $\chi_4(t)$ indicates that the dynamic fluctuations captured by $\chi_4(t)$ are inherently dynamical in origin and do not originate in the liquid structure. Numerical measurements in molecular dynamics simulations indicate that the opposite is true and $\chi_4^{\rm iso}(t)$ contributes less and less to $\chi_4(t)$ as temperature decreases (Berthier and Jack, 2007). This suggests that the search of a causal link between structure and mobility does make sense, at least on large length scales. Interestingly, the vibrational properties investigated in Ref. (Widmer-Cooper *et al.*, 2008) as relevant structural indicators of dynamic heterogeneity are a suitable candidate, since the vibrational spectrum is a collective property.

3.4.4 Point to set correlations: Emergence of amorphous long-range order?

As discussed in the previous section, it is quite natural to ask what structural features (if any) might be responsible for the growth of dynamical correlations. One possibility is that actually there exists a static growing length that drives the increase of dynamical correlations. As we discussed in the introduction, simple static correlations are rather featureless when approaching the glass transition. However, a new length called the "point to set" length was recently introduced (Bouchaud and Biroli, 2004; Mézard and Montanari, 2006). It is naturally devised to probe the growth of static amorphous long-range order (Bouchaud and Biroli, 2004; Mézard and Montanari, 2006) and has been shown to grow close to the glass transition (Biroli *et al.*, 2008). This is reviewed in the chapter by Semerjian and Franz.

The basic idea is to measure how much boundary conditions affect the behavior of the system, far away from the boundaries themselves. This is the usual way to test for the emergence of long-range order in statistical mechanics. However, for standard phase transitions, the appropriate boundary conditions are known from the outset.

For example, in the case of ferromagnetic transitions, one can fix the boundary spins mostly in the up direction and check whether this leads to a positive magnetization for spins in the bulk. The difficulty in the case of glasses, for which one would like to test the presence of long-range amorphous order, is that the boundary conditions one has to use look just as random as the amorphous configuration one wants to select. The way out is to let the system itself choose the boundary conditions: the procedure is to take an equilibrated configuration α, freeze all particles outside a cavity of radius $\mathcal{R}$ and then recompute the thermodynamics for the particles inside the cavity, that now are subjected to a typical equilibrium boundary condition. One can then study a suitably defined average overlap $q(\mathcal{R})$ between the new thermalized configurations at the center of the cavity and the reference state α, as a function of $\mathcal{R}$. The quantity $q(\mathcal{R})$ is called a "point-to-set" correlation (Mézard and Montanari, 2006). The characteristic length scale over which $q(\mathcal{R})$ drops to zero is called the point to set length. The increase of this length is a clear signal that the system is developing long-range static order, and in the case of glasses, amorphous long-range order. This point-to-set length is precisely the size of the glassites within the RFOT theory of glasses, see Section 3.3.2.3.

This topic is discussed in detail by Franz and Semerjian, to which we refer for a presentation of the general theoretical and numerical results. Here, we simply mention that this point to set length has been shown to grow in numerical simulations of supercooled liquids (Biroli *et al.*, 2008). Furthermore, it was proved that it must diverge whenever the relaxation time does so (Montanari and Semerjian, 2006).

Therefore, an important open question is whether the correlation length picked up by the dynamical correlators discussed above are actually just a consequence of hidden static correlations or if they are instead quite unrelated to them. In the first case, the study of dynamical correlation will still be very valuable because it provides an easier way to probe static correlations. Whatever is the correct answer, it would lead to a substantial progress in our understanding of the glass transition and would help us in pruning down the correct theory. We do hope that numerical and experimental studies will be devoted to this important problem in the future.

3.4.5 Large deviations and space-time thermodynamics

It is clear from its definition in Eq. (3.7) that the correlation $G_4(r, t)$ is the covariance of the mobility $c(r; 0, t)$ at two nearby spatial points. Similarly, the three-point functions of the previous section are covariances of $c(r; 0, t)$ with the local energy, enthalpy or free volume. Of course, not all information about the mobility is contained in such covariances: one might consider higher moments of these functions or indeed the joint distribution of mobilities at all points. However, the inherent difficulty of characterizing the distribution of an entire mobility field requires physical intuition in choosing which observable to measure.

A natural first choice for such a scheme is to consider the fluctuations of the spatially averaged correlation function $C(t, 0)$, beyond the Gaussian level. Such

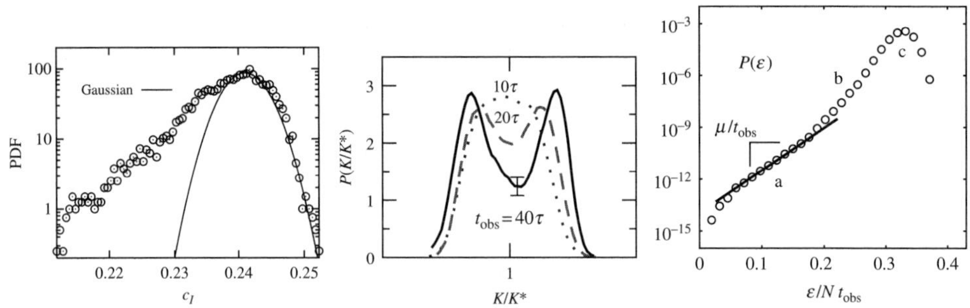

Fig. 3.10 Distributions of dynamical observables. (Top) Distribution of the correlation function in a coarsening foam (Duri *et al.*, 2005). The skewed distribution arises from rare trajectories with more motion than average. (Middle) The distribution of the "dynamical action" ε in a kinetically constrained model (Merolle *et al.*, 2005). The action ε measures the amount of motion in a trajectory in the same spirit as the activity K defined in Eq. (3.27). The distribution is left-skewed and non-convex, indicating a population of trajectories with low mobility. (Bottom) Distribution of the activity K in a model of Lennard-Jones particles, with a biasing field s in place, as discussed in the text. Two peaks are evident when the t_{obs} is large, revealing the presence of two distinct dynamical phases (Hedges *et al.*, 2009).

measurements are possible in experiments such as those of Duri *et al.* (Duri *et al.*, 2005) (see Fig. 3.10) and in computer simulations of a variety of models (Chamon *et al.*, 2004). They have typically been considered in out-of-equilibrium situations but this is not essential (see also below). Typically, such distributions are skewed and non-Gaussian, and it is natural to connect the asymmetry of the distribution with dynamical heterogeneity. For example, even on time scales t much greater than the structural relaxation time, there is a substantial probability that regions of the system have persisted in an immobile state for all times between 0 and t. This enhances the probability of observing a larger than average value for $C(t,0)$. The opposite behavior may occur on short times: while typical regions have not relaxed, cooperative motion in some regions enhances the probability of observing a smaller than average of $C(t,0)$.

In making this connection, it seems reasonable that regions where $C(t,0)$ is large possess rather stable structure at the molecular level, while regions where $C(t,0)$ is small correspond to relatively unstable local structure. This fact has recently been exploited in computational studies that probe trajectories where relaxation is much slower than average. To identify such trajectories, it is useful to define a measure of dynamical activity, for systems of N particles evolving over an observation time t_{obs}. For example, one may take (Hedges *et al.*, 2009)

$$K(N, t_{\text{obs}}) = \sum_{i=1}^{N} \sum_{j=1}^{t_{\text{obs}}/\Delta t} |\boldsymbol{r}_i(t_j) - \boldsymbol{r}_i(t_j - \Delta t)|^2, \tag{3.27}$$

where $\boldsymbol{r}_i(t)$ is the position of particle i at time t and the $t_j = j\Delta t$ are equally spaced times. For large N and $t_{\rm obs}$, the distribution of K becomes sharply peaked about its average, $\langle K \rangle$.

In general, for large N and $t_{\rm obs}$, one expects a the equilibrium distribution of K to have the form

$$P(K) \simeq \exp[-Nt_{\rm obs}f(K/Nt_{\rm obs})] \tag{3.28}$$

where the function $f(k)$ resembles a free-energy density: it gives the probability of observing a substantial deviation between the measured K and its average $\langle K \rangle$. (The variance of K was also considered in Ref. (Merolle *et al.*, 2005). In general, this quantity contains different information from four-point functions, such as $\chi_4(k,t)$ although $\chi(t_{\rm obs})$ and $\chi_4(k,t)$ may sometimes be related through scaling arguments.[6]) In some kinetically constrained models (Merolle *et al.*, 2005), the distribution $P(K)$ has a characteristic shape, skewed towards small activity, with an apparently exponential tail, as shown in the central panel of Fig. 3.10. Further, on estimating $f(k)$ from this plot, there is a range of K over which $f(k)$ is non-convex (that is, $f''(K) < 0$). The behavior of $f(k)$ away from its minimum describes the properties of rare trajectories in the system and their relevance for the liquid behavior is not clear *a priori*. However, the key motivation of this study was to formulate a thermodynamic approach to the statistical properties of trajectories of glassy systems (Garrahan and Chandler, 2002). Within such a framework, non-convexity of $f(K)$ has a direct interpretation as a "dynamical phase transition" in the system.

This interpretation is most easily seen by taking the Legendre transform of $f(K)$ to obtain a new "dynamical free energy"

$$\psi(s) = -\min_k[sk + f(k)], \tag{3.29}$$

which describes the response of the system to a field s that biases the system towards trajectories with small (or large) activity K. In particular, the effect of the bias is to change the average of K from $\langle K \rangle$ to $K(s) = -Nt_{\rm obs}\frac{\rm d}{{\rm d}s}\psi(s)$. Then, a non-convex form for $f(K)$ results in a jump singularity of $K(s)$ for a specific biasing field $s = s^*$. While the field s has no simple physical interpretation, one may view it as a mathematically convenient trick for sampling the distribution $P(K)$.

Turning to the results of this formalism, the key point is that glassy systems may exhibit singular responses to the field s, leading to "ideal glass" states that are characterized by values of K that is much smaller than its equilibrium average $\langle K \rangle$. The existence of these phase transitions has been proven in simple models (Garrahan

[6]In the notation of Eq. (3.7), one may write $K = \sum_i \sum_j c_i(t_j, t_j + \Delta t)$ so that the variance of K contains terms like $g_{ijmn} = \langle c_i(t_j, t_j + \Delta t) c_m(t_n, t_n + \Delta t)\rangle$. Assuming that g_{ijmn} depends on scaling variables such as $|r_i - r_j|/\xi_4$ and $(t_j - t_n)/\tau_\alpha$, one may connect the variance of K to the dynamical length scale ξ_4 and time scale τ_α. Such connections are analogous to the relation (3.14) between $\chi_4(t)$ and $\xi_4(t)$, but there is considerable freedom in the scaling ansatz for g_{ijmn}, which seems to prevent a more direct connection between the variance of K and $\chi_4(t)$.

et al., 2007; Jack and Garrahan, 2010) and numerical results for Lennard-Jones model liquids are also consistent with the existence of such a transition (Hedges *et al.*, 2009). In particular, if the field s is chosen to lie at the putative phase-transition point, then one may construct the distribution of K in the presence of the field s, $P_s(K) \propto P(K)e^{-sK}$. In the presence of a phase transition, $P_s(K)$ has two peaks, which correspond to distinct active (liquid) and inactive (glass) states. An example is shown in the right panel of Fig. 3.10.

To summarize then, dynamical heterogeneity is concerned with distributions of dynamical quantities, most often through their means and covariances. However, the tails of these distributions can reveal information about possible new phases in the system, whose structure is very stable and whose relaxation times are very long. This leads to the hypothesis that the nature of the dynamically heterogeneous fluid state should be interpreted in terms of coexistence between and active liquid and inactive "ideal glass" states.

3.5 Open problems and conclusions

The aim of this chapter was to review the recent progress in the quantitative analysis of dynamical heterogeneities. We showed that the introduction of four-point correlation functions played an important role, both conceptually and operationally, by providing a precise quantitative measure to characterize dynamical heterogeneities. These four-point correlations have now been measured or estimated in numerical simulations of schematic and realistic models of glass formers, and experimentally on molecular glasses, colloids and granular assemblies close to jamming. They have also been investigated theoretically within simplified models or within the mode-coupling approximation, and have indeed been shown to be critical as the glass transition is approached. These four-point correlations are the natural counterpart, for glass or spin-glass transitions, of the standard two-point correlations that diverge close to a usual second-order phase transition. These can be considered to be breakthroughs that have significantly improved our understanding of the microscopic mechanisms leading to glass formation, and that have already spilled over to many different scientific communities.

However, it soon became clear that four-point correlation functions are not a panacea. It was for example not anticipated that these functions would be delicately sensitive to details such as the choice of statistical ensemble or the microscopic dynamics. Secondly, although these four-point objects give valuable information, they are not powerful enough to answer more precise questions about the geometry of the structures that carry dynamical heterogeneities, or about the nature of the relaxation events (continuous vs. activated). Along the same line of thought, the relation between the dynamical correlation length extracted from these four-point points and the intuitive (but not so clearly defined) notion of cooperative relaxation is at this stage quite elusive. How many different "dynamical" length scales does one expect in general?

We have seen that the study of three-point response functions and non-linear susceptibilities allows one to bypass some of the difficulties inherent to four-point functions. We note that the experimental and numerical situation on that front is much less developed, and should be encouraged. The very recent measurement of the non-linear dielectric properties of glycerol (Crauste-Thibierge *et al.*, 2010) is a remarkable exception.

Higher-order correlation functions might also contain interesting quantitative information about dynamical heterogeneities, but this subject is at this stage totally unexplored. It was recently suggested (Lechenault *et al.*, 2010) that six-point functions might provide a way to measure intermittent dynamics and identify activated events. Skewness (or kurtosis) might indeed detect that the dynamics is intermittent, as one expects if "activated" events dominate, with a few rare events decorrelating the system completely, while most events decorrelate only weakly. More work in that direction would certainly be worthwhile.

Finally, we have not touched upon the problem of out-of-equilibrium dynamical heterogeneities, in particular in the aging regime. This is clearly a very interesting topic, for which experimental efforts are underallocated, although results in this regime might be able to discriminate between theories. We refer to (Parisi, 1999; Castillo *et al.*, 2003; Parsaeian and Castillo, 2008; Vollmayer-Lee *et al.*, 2002; Vollmayer-Lee and Baker, 2006; El Masri *et al.*, 2010; Courtland and Weeks, 2003) and the chapter on aging in this book for interesting lines of research on this issue in spin glasses.

References

Adam, G. and Gibbs, J. H. (1965). *J. Chem. Phys.*, **43**, 139.

Appignanesi, G. A., Rodríguez Fris, J. A., Montani, R. A., and Kob, W. (2006). *Phys. Rev. Lett.*, **96**, 057801.

Bennemann, C., Donati, C., Baschnagel, J., and Glotzer, S. C. (1999). *Nature*, **399**, 246.

Berthier, L. (2003). *Phys. Rev. Lett.*, **91**, 055701.

Berthier, L. (2004). *Phys. Rev. E*, **69**, 020201(R).

Berthier, L. (2007). *Phys. Rev. E*, **76**, 011507.

Berthier, L., Biroli, G., Bouchaud, J.-P., Cipelletti, L., El Masri, D., L'Hôte, D., Ladieu, F., and Pierno, M. (2005). *Science*, **310**, 1797.

Berthier, L., Biroli, G., Bouchaud, J.-P., Kob, W., Miyazaki, K., and Reichman, D. R. (2007*a*). *J. Chem. Phys.*, **126**, 184503.

Berthier, L., Biroli, G., Bouchaud, J.-P., Kob, W., Miyazaki, K., and Reichman, D. R. (2007*b*). *J. Chem. Phys.*, **126**, 184504.

Berthier, L. and Garrahan, J. P. (2005). *J. Phys. Chem. B*, **109**, 3578.

Berthier, L. and Jack, R. L. (2007). *Phys. Rev. E*, **76**, 041509.

Berthier, L. and Kob, W. (2007). *J. Phys.: Condens. Matter*, **19**, 205130.

Binder, K. and Kob, W. (2005). *Glassy Materials, and Disordered Solids*. Singapore, World Scientific.

Biroli, G. and Bouchaud, J.-P. (2004). *Europhys. Lett.*, **67**, 21.

Biroli, G. and Bouchaud, J.-P. (2009). *arXiv:0912.2542*.

Biroli, G., Bouchaud, J.-P., Cavagna, A., Grigera, T. S., and Verrocchio, P. (2008). *Nature Phys.*, **4**, 771.

Biroli, G., Bouchaud, J.-P., and Miyazaki, K. (2006). *and D. R. Reichman Phys. Rev. Lett.*, **97**, 195701.
Bouchaud, J.-P. and Biroli, G. (2004). *J. Chem. Phys.*, **121**, 7347.
Bouchaud, J.-P. and Biroli, G. (2005). *Phys. Rev. B*, **72**, 064204.
Brambilla, G., El Masri, D., Pierno, M., Petekidis, G., Schofield, A. B., Berthier, L., and Cipelletti, L. (2009). *Phys. Rev. Lett.*, **102**, 085703.
Capaccioli, S., Ruocco, G., and Zamponi, F. (2008). *J. Phys. Chem. B*, **112**, 10652.
Castillo, H., Chamon, C., Cugliandolo, L. F., Iguain, J. L., and Kennett, M. P. (2003). *Phys. Rev. B*, **68**, 134442.
Chamon, C., Charbonneau, P., Cugliandolo, L. F., and Reichman, D. R. (2004). *J. Chem. Phys.*, **121**, 10120.
Chandler, D. and Garrahan, J. P. (2010). *Annu. Rev. Phys. Chem.*, **61**, 191.
Chandler, D., Garrahan, J. P., Jack, R. L., Maibaum, L., and Pan, A. C. (2006). *Phys. Rev. E*, **74**, 051501.
Chaudhuri, P., Berthier, L., and Kob, W. (2007). *Phys. Rev. Lett.*, **99**, 060604.
Courtland, R. E. and Weeks, E. R. (2003). *J. Phys.: Condens. Matter*, **15**, S359.
Crauste-Thibierge, C., Brun, C., Ladieu, F., L'Hôte, D., Biroli, G., and Bouchaud, J.-P. (2010). *Phys. Rev. Lett.*, **104**, 165703.
Dalle-Ferrier, C., Thibierge, C., Alba-Simionesco, C., Berthier, L., Biroli, G., Bouchaud, J.-P., Ladieu, F., L'Hôte, D., and Tarjus, G. (2007). *Phys. Rev. E*, **76**, 041510.
Dasgupta, C., Indrani, A. V., Ramaswamy, S., and Phani, N. K. (1991). *Europhys. Lett.*, **15**, 307.
Dauchot, O., Marty, G., and Biroli, G. (2005). *Phys. Rev. Lett.*, **95**, 265701.
Debenedetti, P. G (1996). *Metastable Liquids.* Princeton, Princeton University Press.
Déjardin, J. L. and Kalmykov, Yu. P. (1999). *Phys. Rev. E*, **61**, 1211.
Doliwa, B. and Heuer, A. (2000). *Phys. Rev. E*, **61**, 6898.
Donati, C., Douglas, J. F., Kob, W., Plimpton, S. J., Poole, P. H., and Glotzer, S. C. (1998). *Phys. Rev. Lett.*, **80**, 2338.
Donati, C., Glotzer, S.C., and Poole, P.H. (1999). *Phys. Rev. Lett.*, **82**, 5064.
Duri, A., Bissig, H., Trappe, V., and Cipelletti, L. (2005). *Phys. Rev. E*, **72**, 051401.
Duri, A. and Cipelletti, L. (2006). *Europhys. Lett.*, **76**, 972.
Ediger, M. D. (2000). *Annu. Rev. Phys. Chem.*, **51**, 99.
Edwards, S. F. and Anderson, P. W. (1975). *J. Phys. F: Metal Phys.*, **5**, 965.
El Masri, D., Berthier, L., and Cipelletti, L. (2010). *Phys. Rev. E*, **82**, 031503 (2010).
Fernandez, L. A., Martin-Mayor, V., and Verrocchio, P. (2006). *Phys. Rev. E*, **73**, 020501.
Fischer, K. H. and Hertz, J. A. (1991). *Spin Glasses.* Cambridge, Cambridge University Press.
Flenner, E. and Szamel, G. (2009). *Phys. Rev. E*, **79**, 051502.
Flenner, E. and Szamel, G. (2010). *Phys. Rev. Lett.* **105**, 217801 (2010).
Franz, S., Donati, C., Parisi, G., and Glotzer, S. C. (1999). *Philos. Mag. B*, **79**, 1827.
Franz, S. and Parisi, G. (2000). *J. Phys.: Condens. Matter*, **12**, 6335.
Garrahan, J. P. and Chandler, D. (2002). *Phys. Rev. Lett.*, **89**, 035704.
Garrahan, J. P. and Chandler, D. (2003). *Proc. Nat. Acad. Sci. USA*, **100**, 9710.

Garrahan, J. P., Jack, R. L., Lecomte, V., Pitard, E., van Duijvendijk, K., and van Wijland, F. (2007). *Phys. Rev. Lett.*, **98**, 195702.

Glarum, S. H. (1960). *J. Chem. Phys.*, **33**, 1371.

Glotzer, S. C., Novikov, V., and Schroder, T. B. (2000). *J. Chem. Phys.*, **112**, 509.

Götze, W. (2008). *Complex Dynamics of Glass-forming Liquids: A Mode-coupling Theory.* Oxford, Oxford University Press.

Götze, W. and Sjögren, L. (1995). *Rep. Prog. Phys.*, **55**, 241.

Hansen, J. P. and McDonald, I. R. (1986). *Theory of Simple Liquids.* London, Academic.

Hedges, L. O., Jack, R. L., Garrahan, J. P., and Chandler, D. (2009). *Science*, **323**, 16665.

Hedges, L. O., Maibaum, L., Chandler, D., and Garrahan, J. P. (2007). *J. Chem. Phys.*, **127**, 211101.

Heuer, A. (2008). *J. Phys.: Condens. Matter*, **20**, 373101.

Heussinger, C., Berthier, L., and Barrat, J.-L. (2010). *EPL*, **90**, 20005.

Hurley, M. M. and Harrowell, P. (1995). *Phys. Rev. E*, **52**, 1694.

Jack, R. L. and Garrahan, J. P. (2010). *Phys. Rev. E*, **81**, 011111.

Jack, R. L., Mayer, P., and Sollich, P. (2006). *J. Stat. Mech.*, P03006.

Jung, Y., Garrahan, J.P., and Chandler, D. (2004). *Phys. Rev. E*, **69**, 061205.

Karmakar, S., Dasgupta, C., and Sastry, S. (2009). *Proc. Natl. Acad. Sci. USA*, **106**, 3675.

Kawasaki, T., Araki, T., and Tanaka, H. (2007). *Phys. Rev. Lett.*, **99**, 215701.

Kegel, W. K. and van Blaaderen, A. (2000). *Science*, **287**, 290.

Keys, A. S., Abate, A. R., Glotzer, S. C., and Durian, D. J. (2007). *Nature Phys.*, **3**, 260.

Kirkpatrick, T. R., Thirumalai, D., and Wolynes, P. G. (1989). *Phys. Rev. A*, **40**, 1045.

Kob, W. and Andersen, H. C. (1995). *Phys. Rev. E*, **51**, 4626.

Kob, W., Donati, C., Plimpton, S. J., Poole, P. H., and Glotzer, S. C. (1997). *Phys. Rev. Lett.*, **79**, 2827.

Lačević, N., Starr, F. W., Schroder, T. B., and Glotzer, S. C. (2003). *J. Chem. Phys.*, **119**, 7372.

Lačević, N., Starr, F. W., Schroder, T. B., Novikov, V. N., and Glotzer, S. C. (2002). *Phys. Rev. E*, **66**, 030101.

Lechenault, F., Candelier, R., Dauchot, O., Bouchaud, J.-P., and Biroli, G. (2010). *Soft Matter*, **6**, 3059.

Lechenault, F., Dauchot, O., Biroli, G., and Bouchaud, J.-P. (2008*a*). *EPL*, **83**, 46003.

Lechenault, F., Dauchot, O., Biroli, G., and Bouchaud, J.-P. (2008*b*). *EPL*, **83**, 46002.

Lubchenko, V. and Wolynes, P.G. (2006). *Ann. Rev. Phys. Chem.*, **58**, 235.

Lunkenheimer, P. and Loidl, A. (2002). *Chem. Phys.*, **284**, 205.

Mapes, M. K., Swallen, S. F., and Ediger, M. D. (2006). *J. Chem. Phys.*, **124**, 054710.

Marty, G. and Dauchot, O. (2005). *Phys. Rev. Lett.*, **94**, 015701.

Matharoo, G. S., Razul, M. S. G., and Poole, P. H. (2006). *Phys. Rev. E*, **74**, 050502.

Mayer, P., Bissig, H., Berthier, L., Cipelletti, L., Garrahan, J. P., Sollich, P., and Trappe, V. (2004). *Phys. Rev. Lett.*, **93**, 115701.

Menon, N. and Nagel, S. R. (1995). *Phys. Rev. Lett.*, **74**, 1230.
Merolle, M., Garrahan, J. P., and Chandler, D. (2005). *Proc. Natl. Acad. Sci. USA*, **102**, 10837.
Mézard, M. and Montanari, A. (2006). *J. Stat. Phys.*, **124**, 1317.
Mézard, M., Parisi, G., and Virasoro, M. (1988). *Spin Glass Theory, and Beyond.* Singapore, World Scientific.
Montanari, A. and Semerjian, G. (2006). *J. Stat. Phys.*, **125**, 23.
Pan, A. C., Garrahan, J. P., and Chandler, D. (2005). *Phys. Rev. E*, **72**, 041106.
Parisi, G. (1999). *J. Chem. Phys. B*, **103**, 4128.
Parsaeian, A. and Castillo, H. E. (2008). *Phys. Rev. E*, **78**, 060105(R).
Reinsberg, S. A., Qiu, X. H., Wilhelm, M., Spiess, H. W., and Ediger, M. D. (2001). *J. Chem. Phys.*, **114**, 7299.
Richert, R. (2002). *J. Phys.: Condens. Matter*, **14**, R703.
Richert, R., Israeloff, N., Alba-Simionesco, C., Ladieu, F., and L'Hôte, D. (2010). *Chapter 5*.
Ritort, F. and Sollich, P. (2003). *Adv. Phys.*, **52**, 219.
Sillescu, H. (1999). *J. Non-Cryst. Solids*, **243**, 81.
Stein, R. S. L. and Andersen, H. C. (2008). *Phys. Rev. Lett.*, **101**, 267802.
Stevenson, J. D., Schmalian, J., and Wolynes, P. G. (2006). *Nature Phys.*, **2**, 268.
Stillinger, F. H. and Hodgdon, J. A. (1994). *Phys. Rev. E*, **50**, 2064.
Szamel, G. (2008). *Phys. Rev. Lett.*, **101**, 205701.
Szamel, G. and Flenner, E. (2009). *Phys. Rev. E*, **79**, 021503.
Tarjus, G. (2011). *Chapter 2*, this book.
Tarjus, G. and Kivelson, D. (1995). *J. Chem. Phys.*, **103**, 3071.
Tarzia, M., Biroli, G., Bouchaud, J.-P., and Lefèvre, A. (2010). *J. Chem. Phys.*, **132**, 054501.
Toninelli, C., Wyart, M., Berthier, L., Biroli, G., and Bouchaud, J.-P. (2005). *Phys. Rev. E*, **71**, 041505.
Tracht, U., Wilhelm, M., Heuer, A., Feng, H., Schmidt-Rohr, K., and Spiess, H. W. (1998). *Phys. Rev. Lett.*, **81**, 2727.
Vidal Russell, E. and Israeloff, N. E. (2000). *Nature*, **408**, 695.
Vogel, M. and Glotzer, S. C. (2004). *Phys. Rev. E*, **70**, 061504.
Vollmayer-Lee, K. and Baker, E. A. (2006). *Europhys. Lett.*, **76**, 1130.
Vollmayer-Lee, K., Kob, W., Binder, K., and Zippelius, A. (2002). *J. Chem. Phys.*, **116**, 5158.
Weeks, E. R., Crocker, J. C., Levitt, A. C., Schofield, A., and Weitz, D. A. (2000). *Science*, **287**, 627.
Weeks, E. R., Crocker, J. C., and Weitz, D. A. (2007). *J. Phys.: Condens. Matter*, **19**, 205131.
Whitelam, S., Berthier, L., and Garrahan, J.P. (2004). *Phys. Rev. Lett.*, **92**, 185705.
Whitelam, S., Berthier, L., and Garrahan, J.P. (2005). *Phys. Rev. E*, **71**, 026128.
Widmer-Cooper, A. and Harrowell, P. (2005). *J. Phys.: Condens. Matter*, **17**, S4025.
Widmer-Cooper, A. and Harrowell, P. (2006). *Phys. Rev. Lett.*, **96**, 185701.
Widmer-Cooper, A. and Harrowell, P. (2007). *J. Chem. Phys.*, **126**, 154503.

Widmer-Cooper, A., Harrowell, P., and Fynewever, H. (2004). *Phys. Rev. Lett.*, **93**, 135701.

Widmer-Cooper, A., Perry, H., Harrowell, P., and Reichman, D. R. (2008). *Nature Phys.*, **4**, 711.

Wuttke, J., Petry, W., and Pouget, S. (1996). *J. Chem. Phys.*, **105**, 5177.

Xia, X. Y. and Wolynes, P. G. (2000). *Proc. Natl. Acad. Sci. USA*, **97**, 2990.

Yamamoto, R. and Onuki, A. (1998*a*). *Phys. Rev. Lett.*, **81**, 4915.

Yamamoto, R. and Onuki, A. (1998*b*). *Phys. Rev. E*, **58**, 3515.

4
Glassy dynamics and dynamical heterogeneity in colloids

Luca Cipelletti and Eric R. Weeks

Abstract

Concentrated colloidal suspensions are a well-tested model system that has a glass transition. Colloids are suspensions of small solid particles in a liquid, and exhibit glassy behavior when the particle concentration is high; the particles are roughly analogous to individual molecules in a traditional glass. Because the particle size can be large (100–1000 nm), these samples can be studied with a variety of optical techniques including microscopy and dynamic light scattering. Here, we review the phenomena associated with the colloidal glass transition, and in particular discuss observations of spatial and temporally heterogeneous dynamics within colloidal samples near the glass transition. Although this chapter focuses primarily on results from hard-sphere-like colloidal particles, we also discuss other colloidal systems with attractive or soft repulsive interactions.

4.1 Colloidal hard spheres as a model system for the glass transition

4.1.1 The hard-sphere colloidal glass transition

When some materials are rapidly cooled, they form an amorphous solid known as a glass. This transition to a disordered solid is the glass transition (Götze and Sjögren, 1992; Stillinger, 1995; Ediger *et al.*, 1996; Angell *et al.*, 2000). As the temperature of a molecular glass-forming material is decreased the viscosity rises smoothly but rapidly, with little apparent change in the microscopic structure (Ernst *et al.*, 1991; van Blaaderen and Wiltzius, 1995). Glass formation may result from dense regions of well-packed molecules or a decreasing probability of finding mobile regions. As no structural mechanisms for this transition have been found, many explanations rely on dynamic mechanisms. Some theoretical explanations focus on the idea of dynamical heterogeneities (Götze and Sjögren, 1992; Sillescu, 1999; Ediger, 2000;

Adam and Gibbs, 1965). The underlying concept is that, for any molecule to move, all molecules within a surrounding region must "cooperate" in their movement. As the glass transition is approached the sizes of these regions grow, causing the rise in macroscopic viscosity (Adam and Gibbs, 1965). The microscopic length scale characterizing the size of these regions could potentially diverge, helping to explain the macroscopic viscosity divergence. However, it is also possible that these regions could grow but not be directly connected to the viscosity divergence. Additionally, it is not completely clear if the viscosity itself diverges or simply becomes too large to measure (Hecksher *et al.*, 2008). While the existence of dynamical heterogeneities in glassy systems has been confirmed in a wide variety of systems, the details of this conceptual picture remain in debate (Sillescu, 1999; Ediger, 2000; Glotzer, 2000; Ngai, 1999; Richert, 2002; Cipelletti and Ramos, 2005).

Colloidal suspensions are composed of microscopic-sized solid particles in a liquid, and are a useful model system for studying the glass transition. In terms of interparticle interaction, the simplest colloids are those in which the particles interact as hard spheres, i.e. the interparticle potential arises solely due to excluded volume effects (Pusey and van Megen, 1986). Hard spheres are a useful theoretical model for glass-forming systems due to their simplicity (Bernal, 1964). Clearly, attractive interactions between atoms and molecules are responsible for dense phases of matter. But given dense states of matter, repulsive interactions play the dominant role in determining the structure. Hard spheres are useful simulation models for crystals, liquids, and glasses, although it is still debated whether a purely repulsive interparticle potential is sufficient to reproduce the glass transition in general (Berthier and Tarjus, 2009).

The control parameter for hard-sphere systems is the concentration, expressed as the fraction, φ, of the sample volume occupied by the particles. Most colloidal hard-sphere systems act like a glass for φ larger than an "operational" glass-transition volume fraction $\varphi_g \approx 0.58$. The transition is the point where particles no longer diffuse through the sample on experimentally accessible time scales; for $\varphi < \varphi_g$ spheres do diffuse at long times, although the asymptotic diffusion coefficient D_∞ decreases sharply as the concentration increases (Bartsch *et al.*, 1993; van Megen *et al.*, 1998; Kasper *et al.*, 1998). The transition at φ_g occurs even though the spheres are not completely packed together; in fact, the density must be increased to $\varphi_{RCP} \approx 0.64$ for "random-close-packed" spheres (O'Hern *et al.*, 2003; Bernal, 1964; Torquato *et al.*, 2000; Donev *et al.*, 2004; O'Hern *et al.*, 2004) before the spheres are motionless. In simulations, a collection of same-sized (i.e. monodisperse) hard spheres almost always crystallizes. A binary mixture of spheres or indeed a distribution of sizes is therefore needed to frustrate crystallization and enable access to the glass transition, both numerically and experimentally (Henderson and van Megen, 1998; Zaccarelli *et al.*, 2009*a*).

The glass transition for suspensions of nearly hard-sphere colloids (Pusey and van Megen, 1986; van Megen and Pusey, 1991; van Blaaderen and Wiltzius, 1995; van Megen and Underwood, 1993; Bartsch *et al.*, 1993; Bartsch, 1995; Mason and Weitz, 1995) is comparable in many respects to the hard-sphere glass transition studied in simulations and theory (Speedy, 1998). Macroscopically, a colloidal liquid flows like a viscous fluid, whereas the colloidal glass does not flow easily, like a paste (Segrè *et al.*,

1995*b*; Cheng *et al.*, 2002). For colloidal samples with low polydispersity, samples can crystallize for $\varphi < \varphi_g$ (with crystallization nucleating in the interior of the sample), while for $\varphi > \varphi_g$, crystals only nucleate at flat sample boundaries such as the walls of the container (Pusey and van Megen, 1986). Microscopically, the glass-transition point is identified as the point where $D_\infty \to 0$ (Bartsch *et al.*, 1993; van Megen *et al.*, 1998; Kasper *et al.*, 1998).

Note that there are some questions about the colloidal glass transition. First, prior measurements disagree about the nature of the viscosity divergence in colloidal glasses (Segrè *et al.*, 1995*b*; Cheng *et al.*, 2002). This may be due to difficulties in reconciling measurements of the volume fraction φ (de Schepper *et al.*, 1996; Segrè *et al.*, 1996). Secondly, it has been seen that a glassy colloidal suspension can crystallize under microgravity conditions (Zhu *et al.*, 1997), and potentially also when in density-matching solvents (Kegel, 2000; Simeonova and Kegel, 2004), suggesting that the apparent glass transition at $\varphi_g = 0.58$ is an artifact of gravity. The interpretation of these observations is unclear. One possibility is that this is merely heterogeneous nucleation at the walls, as seen before (Pusey and van Megen, 1986). Another possibility is that these samples were more monodisperse than most, and a slightly larger polydispersity ($> 5\%$) may be necessary to induce a glass transition at $\varphi_g = 0.58$ (Henderson and van Megen, 1998; Meller and Stavans, 1992; Auer and Frenkel, 2001; Schope *et al.*, 2007). In practice, most experimental samples always have a polydispersity of at least 5%. Recent simulations suggest that the relationship between polydispersity, crystallization, and glassy dynamics is more complex than perhaps was previously appreciated (Zaccarelli *et al.*, 2009*a*). Overall, we note that the interpretations of the various observations described in this paragraph are often still controversial.

Colloidal glasses are quite similar to molecular glasses:

- Both are microscopically disordered (van Blaaderen and Wiltzius, 1995).
- Both are macroscopically extremely viscous near their transition point (Segrè *et al.*, 1995*b*; Cheng *et al.*, 2002).
- Colloidal glasses have a non-zero elastic modulus at zero frequency, which is absent in the liquid phase (Mason and Weitz, 1995).
- Colloidal glasses are out of equilibrium and show aging behavior—their properties depend on the time elapsed since preparation (Courtland and Weeks, 2003; Cianci *et al.*, 2006; Cianci and Weeks, 2007; Simeonova and Kegel, 2004), similar to polymer glasses (Hodge, 1995; McKenna, 2003) and other molecular glasses (Angell *et al.*, 2000; Castillo and Parsaeian, 2007).
- Colloidal glasses exhibit dynamical heterogeneity (Kegel and van Blaaderen, 2000; Weeks *et al.*, 2000), similar to that seen in simulations (Donati *et al.*, 1998; Glotzer, 2000; Doliwa and Heuer, 1998) and experiments on molecular glasses (Ediger, 2000; Sillescu, 1999; Vidal Russell and Israeloff, 2000).
- The colloidal glass transition is sensitive to finite-size effects (Nugent *et al.*, 2007), similar to molecular glasses (Alcoutlabi and McKenna, 2005) and polymers (Roth and Dutcher, 2005).

One difference between colloids and molecules is that colloidal particles move via Brownian motion, whereas the latter move ballistically at very short time scales; several simulations indicate that this difference is unimportant for the long-time dynamics that is of the most interest (Gleim *et al.*, 1998; Szamel and Flenner, 2004; Höfling *et al.*, 2008). Likewise, hydrodynamic interactions between particles influence their motion on short time scales, but do not modify the pairwise interaction potential (which remains hard-sphere-like), suggesting that they should not be relevant for the long-time dynamics (Brambilla *et al.*, 2009).

One advantage colloids have over traditional molecular glass formers is that their time scales are significantly slower, with relaxation taking $O(1\text{–}1000\,\text{s})$, allowing easy study of the relaxation processes. A second advantage of colloids is that their large size $[O(1\,\mu\text{m})]$ allows for measurement using optical microscopy or dynamic light scattering, as will be discussed in Section 4.2.

Experiments have one chief advantage over simulations, in that they more easily avoid finite-size effects. Near the glass transition, dynamical length scales can be large ($\sim$ 4–5 particle diameters (Weeks *et al.*, 2007; Doliwa and Heuer, 2000)) and finite-size effects on structure and dynamics may extend to even larger length scales ($\sim$ 20 particle diameters (Nugent *et al.*, 2007)). Microscope sample chambers typically contain $\sim 10^9$ particles, and light-scattering cuvettes contain even more.

The most popular "hard-sphere" colloid is colloidal PMMA (poly-methyl-methacrylate), sterically stabilized to minimize interparticle attraction (Pusey and van Megen, 1986; Weeks *et al.*, 2000; Dinsmore *et al.*, 2001; Antl *et al.*, 1986). The particles can be placed in density-matching solvents to inhibit sedimentation (see Section 4.1.2), and/or solvents that match their index of refraction to enable microscopy (Dinsmore *et al.*, 2001) and light scattering (see Sections 4.2.2 and 4.2.3). In some of these solvents, the particles pick up a slight charge and thus have a slightly soft repulsive interaction in addition to the hard-sphere core. Despite this charge the spheres behave similarly to hard spheres with some phase transitions shifted to slightly lower φ (Gasser *et al.*, 2001). Salt can be added to the samples to screen the charges and shift the interaction back to more hard-sphere like (Yethiraj and van Blaaderen, 2003; Royall *et al.*, 2003).

A second popular colloid is colloidal silica, which is relatively easy to fabricate (Stober, 1968). These particles are suspended in water (or a mixture of water and glycerol), avoiding the organic solvents that are required for PMMA colloids. Because they are in water, repulsion due to charge is the primary mechanism preventing flocculation; adding salt can screen the charges and cause flocculation (which is often irreversible). Silica colloids are hard to density match (their density varies but is larger than $2\,\text{g/cm}^3$), and also it is hard to match their refractive index. This latter constraint makes microscopy difficult except at lower volume fractions (Mohraz *et al.*, 2008).

Several glass-transition theories have been applied to the colloidal glass transition. The colloidal glass transition appears to be well described by mode-coupling theory up to $\varphi \approx 0.58$ (van Megen and Pusey, 1991; van Megen and Underwood, 1993, 1994; Götze and Sjögren, 1991; Schweizer and Saltzman, 2003, 2004; Saltzman and Schweizer, 2006*a*,*b*), although other glass-transition theories successfully capture many features of the colloidal glass transition as well (Ngai and Rendell, 1998; Liu and

Oppenheim, 1996; Tokuyama and Oppenheim, 1995; Tokuyama, 2007). (See the discussion in Section 4.3.1 on the strengths and weaknesses of mode-coupling theory as applied to colloids.)

4.1.2 Experimental challenges in studying colloidal hard spheres

As argued in the previous section, colloidal hard spheres are a good model system for investigating the glass transition. However, several experimental challenges have to be faced. Probably, the most serious problem is that of a precise determination of the volume fraction. In optical microscopy, φ can be obtained by counting the number of particles in a given volume and using the particle size as obtained, e.g., from electron microscopy. It should be noted that an error of just 1% in the radius a of the particles results in a 3% error in φ. Additionally, one has to take into account the thickness of the stabilizing layer (e.g. the grafted polymer for PMMA particles or the counterion cloud for silica particles in a polar solvent), which is often difficult to measure precisely. Since typical values of the thickness are on the order of 10 nm (Pusey, 1991), this contribution can be relevant for the small particles used in light scattering ($a \sim 100$–300 nm), while it is less important for the micrometer-sized particles used in microscopy. Other methods to determine φ include precise density or refractive-index measurements (Phan *et al.*, 1996). For PMMA, these methods require special care, since the particles can be swollen by organic solvents such as tetralin or brominated solvents, which change their density and refractive index compared to those of the bulk material.

Because of these difficulties, the absolute volume fraction is often determined indirectly by comparing the phase behavior or the dynamical behavior of the sample to theoretical and numerical predictions. Samples sufficiently monodisperse ($\sigma = \sqrt{\overline{a^2} - \overline{a}^2}/\overline{a}^2 < 5-8\%$) crystallize for $0.494 < \varphi < 0.545$ (Pusey, 1991). The absolute volume fraction can then be calibrated by matching the experimentally determined freezing volume fraction with $\varphi_\mathrm{f} = 0.494$ as determined by simulations (Hoover and Ree, 1968). However, some uncertainty is still left, because the exact value of the freezing fraction depends on σ (Bolhuis and Kofke, 1996; Fasolo and Sollich, 2004; van Megen and Underwood, 1994; Zaccarelli *et al.*, 2009*a*; Pusey *et al.*, 2009). Moreover, this method cannot be applied to more polydisperse suspensions that do not crystallize over several months or years. Alternatively, φ may be calibrated against predictions for the volume fraction dependence of the low shear viscosity (Pusey, 1991; Poon *et al.*, 1996) or the short-time self-diffusion coefficient (Beenakker and Mazur, 1983; Tokuyama and Oppenheim, 1994) in the dilute regime. For samples where both the calibration against φ_f and that using the short-time self-diffusion coefficient are possible, the two methods appear to be consistent (van Megen and Underwood, 1989; Segrè *et al.*, 1995*a*). In summary, while relative values of φ can be measured very precisely (down to 10^{-4} using an analytical balance), absolute values are typically affected by an uncertainty of a few %. This should always be kept in mind when comparing sets of data obtained in different experiments.

For colloids, the equivalent of $T = 0$ in a molecular system is random close packing, the volume fraction $\varphi_\mathrm{RCP} \sim 0.64$ where osmotic pressure diverges and all motion ceases because no free volume is left. Clearly, knowledge of the precise location of φ_{RCP} is very important to discriminate between theories that predict a glass transition below random close packing (for example the mode-coupling theory (Gotze, 1999) or thermodynamic glass-transition theories (Cardenas *et al.*, 1999; Parisi and Zamponi, 2005)) and scenarios like jamming, where no arrest is predicted below φ_{RCP}. However, the location of random close packing is still highly debated (Berthier and Witten, 2009; Xu *et al.*, 2009; Kamien and Liu, 2007; Chaudhuri *et al.*, 2010) and its very existence is challenged, based on the argument that one can always trade order for packing efficiency (Donev *et al.*, 2007), up to $\varphi = \pi/\sqrt{18} \approx 0.7405$, the packing fraction of a (monodisperse) hard-sphere crystal. Experimentally, it is difficult to measure φ_{RCP} because of the uncertainty on absolute volume fractions discussed above and because applying the high pressure (e.g. by centrifugation) needed to approach it may result in the compression of the stabilizing layer. Additionally, both experiments (Clusel *et al.*, 2009) and simulations (Schaertl and Sillescu, 1994; Berthier and Witten, 2009) show that φ_{RCP} depends on σ.

Some of the experimental challenges posed by hard spheres stem from the very same features that make them a valuable model system: their relatively large time- and length scales. The microscopic time in a colloidal system is the Brownian time τ_{B}, i.e. the time required by a particle to diffuse over its own size in a diluted system, defined as

$$\tau_{\mathrm{B}} \equiv a^2/6D = \pi\eta a^3/k_{\mathrm{B}}T\,. \tag{4.1}$$

Here, η is the solvent viscosity, T the absolute temperature, k_{B} Boltzman's constant and D is the diffusion coefficient for a sphere of radius a, given by the Stokes–Einstein–Sutherland formula (Einstein, 1905; Sutherland, 1905):

$$D = k_{\mathrm{B}}T/6\pi\eta a\,. \tag{4.2}$$

The time scale τ_{B} is typically on the order of $10^{-3} - 1$ s. The largest relaxation time that can be measured in concentrated systems is on the order of 10^5 s; thus, the accessible dynamical range covers at most 8 decades, as opposed to 15 decades in molecular glass formers. Additionally, owing to their relatively large size and due to the mismatch between their density and that of the solvent in which they are suspended, colloidal particles experience gravitational forces that can modify their phase behavior (Zhu *et al.*, 1997; Pusey *et al.*, 2009) and dynamical properties (El Masri *et al.*, 2009; Simeonova and Kegel, 2004). The relevant parameter to gauge the importance of sedimentation is the inverse Péclet number, $\mathrm{Pe}^{-1} = \frac{3k_{\mathrm{B}}T}{4\Delta\rho g a^4}$, defined as the ratio of the gravitational length to the particle radius (g is the acceleration of gravity and $\Delta\rho$ the density mismatch). For diluted suspensions, gravity becomes relevant as Pe^{-1} approaches (from above) unity; for concentrated suspensions, sedimentation effects may set in at even higher values of Pe^{-1}, since gravitational stress is transmitted over increasingly larger distances as the sample becomes more solid-like. For example, sedimentation effects have been reported to alter the dynamics of PMMA particles in organic solvents ($\Delta\rho \sim 0.3\,\mathrm{g/cm^3}$) for $\mathrm{Pe}^{-1} = 44.1$ (Simeonova and Kegel, 2004) or even for $\mathrm{Pe}^{-1} = 1350$ (El Masri *et al.*, 2009). Density-matching solvents can mitigate these effects, although matching closely both the index of refraction (as required for optical observations) and the density of the particles without altering their hard-sphere behavior has been proved difficult for PMMA particles (Royall *et al.*, 2003) and impossible for other systems such as silica spheres.

4.2 Experimental methods for measuring both the average dynamics and dynamical heterogeneity

4.2.1 Main features of optical microscopy and dynamic light scattering

Optical microscopy and dynamic light scattering are the main techniques to probe both the average dynamics and its spatio-temporal fluctuations in dense colloidal suspensions. Each of them comes with specific advantages and limitations. Optical microscopy is unsurpassed in providing detailed information on the structure and the dynamics at the single-particle level. The same quantities introduced in theory and simulations to characterize the dynamics can be precisely measured (e.g. the mean-square displacement or the intermediate scattering function for the average dynamics, and the dynamical susceptibility, χ_4, and the spatial correlation of the dynamics, g_4, for its fluctuations, see Chapter 2 and Section 4.2.4). Additionally, direct visualization of the sample allows any experimental problem to be readily detected, such as particle aggregation, sedimentation, or wall effects. Finally, techniques such as optical or magnetic tweezing (Grier, 2003; Amblard *et al.*, 1996) allow one to manipulate single particles and thus to measure the microscopic response of the system to a local perturbation (Habdas *et al.*, 2004).

Dynamic light scattering probes a very large number of particles simultaneously, yielding very good averages. Moreover, particles used in light scattering are usually smaller than those for optical microscopy ($a = 100 - 500$ nm as opposed to $a = 0.5 - 1.5\,\mu$m), which has a twofold advantage. First, the microscopic time τ_B (see Eq. (4.1)) is significantly reduced, since $\tau_B \sim a^2/D \sim a^3$, thus increasing substantially the experimentally accessible dynamical range. Secondly, gravitational effects are much less of concern, since $\mathrm{Pe}^{-1} \sim a^{-4}$. Finally, as we will discuss in Section 4.2.4, recent developments allow dynamical heterogeneity to be probed by dynamic light scattering, although not at the level of microscopic detail afforded by optical microscopy. These methods extend the possibilities of light scattering by adding features characteristic of imaging techniques. Quite in a symmetric way, very recent microscopy methods such as dynamic differential microscopy (Cerbino and Trappe, 2008) have extended imaging techniques by adding the capability of measuring the intermediate scattering function $f(q, \tau)$ defined below in Section 4.2.3.

4.2.2 Optical and confocal microscopy

Given the large size of many colloidal systems (particle radius $a \sim 100 - 1000$ nm), optical microscopy is a useful tool for observing these systems. First, these sizes are comparable to the wavelength of light, thus rendering them visible. Secondly, the time scales of their motion are often slow enough for video cameras to follow their motion.

This can be seen by considering how quickly particles diffuse. Colloidal particles undergo Brownian motion due to thermal fluctuations, as discussed in Section 4.1.2. The diffusion coefficent D (Eq. (4.2)) is related to the mean square displacement of particles as:

$$\langle \Delta x^2 \rangle = 2D\Delta t, \tag{4.3}$$

where Δt is the time scale over which the displacements are taken. A particle of diameter $2a = 1$ μm in water ($\eta = 1$ mPa s) at room temperature diffuses approximately 1 μm in 1 s. This motion is easy to study with a conventional video camera and a microscope; most video cameras take data at 30 images per second, thus they allow one to follow the Brownian motion of colloidal particles of these sizes. Of course, particles that are 10 times smaller move 1000 times faster, by Eq. (4.1). In practice, often one can choose to study larger colloidal particles (Weeks *et al.*, 2000) or use careful data-analysis techniques to learn information about smaller-sized particles that aren't directly imaged (Simeonova and Kegel, 2004). Additionally, when studying dense colloidal samples, the time scales increase simply because of the glassy dynamics.

Two main methods of microscopy have been used to study dense colloidal samples: conventional optical microscopy and confocal microscopy. First, there is the possibility of using a conventional light-microscope technique such as "bright-field microscopy." These techniques typically depend on slight differences between the index of refraction of the colloidal particles and the solvent (Inoué and Spring, 1997). A limitation is that these differences also scatter light; each particle acts like a tiny lens. This ultimately limits how deeply into a sample one can observe. (This is the same phenomenon that makes milk appear white, even though composed of transparent components; the different components all have different indices of refraction. Snow is white for a similar reason, due to the contrast in index of refraction between the ice crystals and air.) A further limitation of conventional optical microscopy is that the images are limited to a plane, although this can be fine for quasi-two-dimensional samples (Marcus *et al.*, 1999).

A related conventional technique is fluorescence microscopy. Here, the particles can be precisely index matched with the solvent. However, the particles also contain a fluorescent dye. In fluorescence microscopy, the particles are illuminated with short-wavelength light. The dye molecules absorb this light, and radiate slightly longer-wavelength (lower-energy) light, which is imaged by the camera. Special filters and mirrors are used to direct the light appropriately from the light source to the sample, and from the sample to the camera. While this method avoids the problem of light scattering off of different parts of the sample, in dense samples it can still be a problem that too much of the sample fluoresces at the same time, thus giving a large background illumination. Trying to observe bright particles on a bright background thus limits fluorescence microscopy of dense samples. One way to overcome this is to only dye a few tracer particles.

Fluorescence microscopy has one significant limitation: photobleaching. After dye molecules absorb the excitation light, but before they emit light, they can chemically react with oxygen present in the sample to form a non-fluorescent molecule. This only happens when they are excited, so photobleaching happens in direct proportion to the illumination light. Photobleaching manifests itself as the image becoming gradually darker. This can be a useful technique for studying local diffusion in samples, a technique known as "fluorescent recovery after photobleaching" (Axelrod *et al.*, 1976). Intense light is used to photobleach a region of the sample, and then low-intensity light is used to monitor the recovery of fluorescence as non-bleached particles diffuse back into the region. With this method, the diffusivity of the particles can be measured, which has been used to study the behavior of colloidal glasses (Simeonova and Kegel, 2004).

An extension of fluorescence microscopy is confocal microscopy, sometimes termed laser scanning optical microscopy. Here, a laser is used to excite fluorescence in dye added to a sample. Typically, the laser beam is reflected off two scanning mirrors that raster the beam in the x and y directions on the sample. Any resulting fluorescent light is sent back through the microscope, and becomes descanned by the same mirrors. A mirror directs the fluorescent light onto a detector, usually a photomultiplier tube.

One additional modification is necessary to make a confocal microscope: before reaching the detector, the fluorescent light is focused onto a screen with a pinhole. All of the light from the focal point of the microscope passes through the pinhole, while any out of focus fluorescent light is blocked by this screen. This spatial filtering technique blocks out the background fluorescence light, allowing the particles to be viewed as bright objects on a dim background.

This ability to reject out-of-focus fluorescent light directly results in the main strength of confocal microscopy, the ability to take three-dimensional pictures of samples. By rejecting out-of-focus light, a crisp two-dimensional image can be obtained, as shown in Fig. 4.1. The sample (or objective lens) can be moved so as to focus at a different height z within the sample, and a new 2D image obtained. By collecting a stack of 2D images at different heights z, a 3D image is built up. The time to scan one 2D image can range from 10 ms to several seconds, depending on the details of the confocal microscope and the desired image size and quality. The time to scan a 3D image depends on the 2D scan speed and the desired number of pixels in the z-direction; reasonable 3D images can be acquired in 2–20 s depending on the microscope (Nugent *et al.*, 2007; Weeks *et al.*, 2000). Finally, we mention coherent anti-Stokes Raman scattering (CARS) microscopy (Kaufman and Weitz, 2006), a technique that allows one to image in 3D colloidal samples with spatial and temporal resolutions comparable to those of confocal microscopy.

Once a series of images has been acquired (2D or 3D), the next step is typically tracking the individual particles within the images. Within each image, a computer can determine the positions of all of the particles. If the particles do not move large distances between subsequent images, then they can be easily tracked (Crocker and Grier, 1996). Specifically, they need to move less between images than their typical interparticle spacing. With confocal microscopy, particles can be tracked in three dimensions (Besseling *et al.*, 2009; Dinsmore *et al.*, 2001). This then is the same type of data that is analyzed from computer simulations. Simulations, of course,

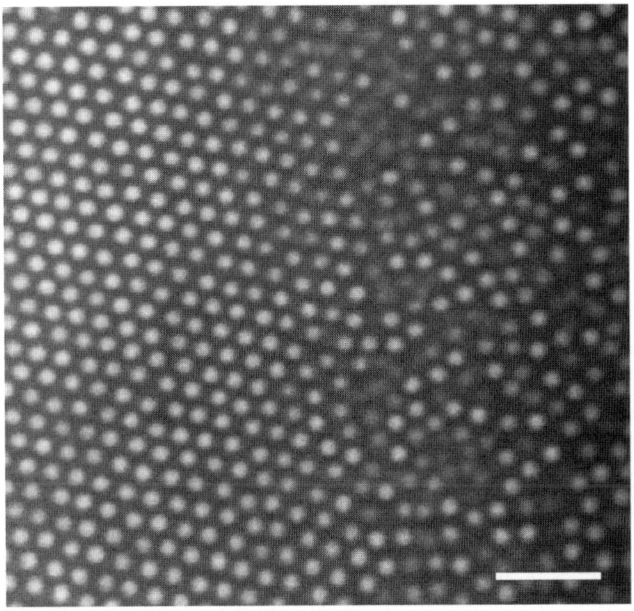

Fig. 4.1 Image of 2-μm diameter colloidal particles taken with a confocal microscope. The scale bar is 10 μm. The sample is in coexistence between a colloidal crystal and a liquid, see Ref. (Hernández-Guzmán and Weeks, 2009) for details. Taken by Jessica Hernández-Guzmán and Eric R. Weeks.

have many advantages, including a tunability of particle interaction, the ability to study either Brownian or ballistic dynamics, and more precise and instantaneous control over parameters such as temperature and pressure. The experiments have an advantage that typically the boundaries are far away: while perhaps a few thousand particles might be viewed, they are embedded in a much larger sample with millions of particles.

4.2.3 Dynamic light scattering

Dynamic light scattering (DLS) (Berne and Pecora, 1976), also termed photon correlation spectroscopy, probes the temporal fluctuations of the refractive index of a sample. In colloidal systems, scattering arises from a mismatch between the index of refraction of particles and that of the solvent, so that DLS probes particle-density fluctuations. Experimentally, one measures $g_2(\tau) - 1$, the time autocorrelation function of the temporal fluctuations of the intensity scattered at a wavevector $q = 4\pi/\lambda \sin(\theta/2)$. Here, θ is the scattering angle and λ is the wavelength in the solvent of the incoming light, usually a laser beam. Under single-scattering conditions, the intensity autocorrelation function is directly related to the intermediate scattering function $f(q, \tau)$ (ISF, sometimes also referred to as the dynamic structure factor):

$$f(q,\tau) \equiv \left\langle N^{-1} \sum_{j,k} \exp\left\{-i\mathbf{q}\cdot[\mathbf{r}_j(t+\tau)-\mathbf{r}_k(t)]\right\} \right\rangle$$

$$= \sqrt{\beta^{-1}[g_2(\tau)-1]} = \sqrt{\beta^{-1}\left[\frac{< I(t+\tau)I(t) >_t}{< I(t) >_t^2} - 1\right]}, \tag{4.4}$$

where $\beta \leq 1$ depends on the collection optics, $I(t)$ is the time-varying scattered intensity, $\mathbf{r}_j$ the position of the jth particle, and N the number of particles in the scattering volume. Note that $f(q,\tau)$ decays significantly when $\Delta\mathbf{r}(\tau) = \mathbf{r}(t+\tau) - \mathbf{r}(t)$ is of the order of q^{-1}: depending on the choice of θ, DLS probes motion on length scales ranging from tens of nm to tens of μm. Another important point to be noticed is that DLS usually probes collective motion, since the sum in Eq. (4.4) extends over all pairs of particles. However, there are ways to measure the self-part of the ISF by making the contribution of the $j \neq k$ terms vanish from Eq. (4.4). This may be accomplished by choosing the scattering vector q in such a way that $S(q) = 1$, where $S(q)$ is the static structure factor (Pusey *et al.*, 1982; Pusey, 1978). Alternatively, one can use optically polydisperse suspensions, as in Ref. (van Megen and Underwood, 1989), where silica and PMMA particles of nearly the same radius but different refractive index were mixed. When matching the average refractive index of the colloids, the measured ISF contains only contributions from the self-terms, $i = k$. Note that optically composite particles that are polydisperse in size are usually also optically polydisperse. This is the case, e.g., of PMMA colloids stabilized by a polymer layer with a refractive index different from that of the core, when the core is polydisperse in size.

The nature of the averages indicated by the brackets in Eq. (4.4) is an important issue. In the definition of the ISF, the average is over an ensemble of statistically equivalent particle configurations, while operationally g_2 is averaged over time. Therefore, ergodicity is required for Eq. (4.4) to hold. Additionally, in order to reduce noise to an acceptable level, g_2 has to be averaged over at least $10^3 - 10^4\ \tau_\alpha$, with τ_α the relaxation time of the ISF. These requirements are often impossible to meet for supercooled or glassy colloidal systems, where τ_α can be as large as hundreds of thousands of seconds. To overcome these difficulties, various schemes have been proposed, among which the most popular is probably the "multispeckle" method (Wong and Wiltzius, 1993; Bartsch *et al.*, 1997). In a multispeckle experiment, the phototube or avalanche photodiode used in regular DLS is replaced by a multielement detector, typically a CCD or CMOS camera sensor. The collection optics is chosen in such a way that each pixel of the detector corresponds to a different speckle (Goodman, 2007), i.e. to a slightly different scattering vector. Because distinct speckles carry statistically independent information, the time average can be replaced in part by an average over pixels:

$$g_2(\tau) - 1 = \left\langle \frac{< I_p(t+\tau)I_p(t) >_p}{< I_p(t+\tau) >_p < I_p(t) >_p} - 1 \right\rangle_t . \tag{4.5}$$

Here I_p indicates the intensity measured by the pth pixel, $< \cdots >_t$ and $< \cdots >_p$ denote averages over time and pixels, respectively. The set of pixels is chosen in such a way that they correspond to nearly the same magnitude of the scattering vector q. The number of pixels is typically of order $10^4 - 10^5$, so that time averaging only needs to extend over a few τ_α. This approach allows very slow and non-stationary dynamics to be probed effectively.

Although many DLS experiments have been carried out on a variety of glassy colloidal systems, the single-scattering conditions required by this technique are probably more the exception than the rule. For mildly turbid suspensions, smart detection schemes, most of which were pioneered by K. Schätzel (Schätzel, 1991), allow the rejection of multiply scattered photons, thereby efficiently suppressing artifacts due to multiple scattering. Popular implementations of this concept include the so-called two-color and 3D apparatuses (see Ref. (Pusey, 1999) for a review). For very turbid samples, where photons are scattered a large number of times before leaving the sample and the contribution of single scattering is negligible, an alternative formalism has been developed, termed diffusing wave spectroscopy (DWS) (Weitz and Pine, 1993). In a DWS experiment the intensity correlation function $g_2 - 1$ is related to the mean-squared displacement, $< \Delta r^2(\tau) >$, rather than to the ISF, as in DLS. Another important difference is the probed length scale, which in DWS typically covers the range 0.1–100 nm, much smaller than in DLS. Finally, DWS experiments typically probe the self-motion of the particles, rather than their collective relaxation.

An alternative way to tackle multiple scattering is provided by X-photon correlation spectroscopy (XPCS). Modern synchrotron sources deliver X-ray radiation that is coherent enough to perform the same kind of experiments as with a laser beam in DLS. Because colloidal systems scatter X-rays much less efficiently than visible light, in most cases XPCS measurements can be safely performed in the single-scattering regime without adjusting the refractive index of the solvent. While long-term beam stability is often still an issue, several XPCS studies on the slow dynamics of colloidal systems have been published in recent years (Bandyopadhyay *et al.*, 2004; Chung *et al.*, 2006; Robert *et al.*, 2006; Trappe *et al.*, 2007; Wandersman *et al.*, 2008; Herzig *et al.*, 2009; Duri *et al.*, 2009*a*).

4.2.4 Time- and space-resolved dynamic light scattering

In a traditional DLS experiment, the detector is placed in the far-field, so that it collects light scattered by a macroscopic region, typically of volume 1 mm^3 or more. Additionally, the intensity correlation function $g_2 - 1$ has to be extensively averaged over time. Because of these averages over both time and space, no information can be *a priori* extracted on dynamical heterogeneity, e.g. on the spatial and temporal fluctuations of the dynamics. In recent years, however, novel light-scattering methods have been proposed to overcome these limitations, providing either spatially averaged but temporally resolved data (time-resolved correlation, TRC (Cipelletti *et al.*, 2003)), or both spatially and temporally resolved measurements (photon correlation imaging, PCI (Duri *et al.*, 2009*b*)).

In a TRC experiment, one uses a CCD or CMOS detector to calculate a two-time correlation function $c_I(t, \tau)$ defined by

$$c_I(t,\tau) = \frac{< I_p(t+\tau) I_p(t) >_p}{< I_p(t+\tau) >_p < I_p(t) >_p} - 1\,. \tag{4.6}$$

Note that the usual intensity correlation function $g_2(\tau) - 1$ defined in Eq. (4.5) is the temporal average of $c_I(t,\tau)$. Because the detector is typically placed in the far-field, c_I is a temporally resolved but spatially averaged correlation function. Figure 4.2 shows an example of TRC data and their relationship to $g_2 - 1$ for a diluted Brownian suspension (Duri *et al.*, 2005) and for a colloidal gel (Duri and Cipelletti, 2006). When plotted as a function of time t for a fixed time delay τ, the data for the Brownian suspension are essentially constant. Indeed, for this system the dynamics are homogeneous and time-translational invariant, so that the evolution of the system and hence the degree of correlation over a fixed time lag τ does not depend on t. The small fluctuations around the mean value are due to the statistical noise of the measurement, associated with the finite number of pixels over which c_I is averaged (Duri *et al.*, 2005). For the colloidal gel, by contrast, c_I exhibits significant temporal fluctuations, indicative of heterogeneous dynamics. Sudden drops of c_I measured for short lags, as in the top trace of Fig. 4.2c, are indicative of a sudden rearrangement event that

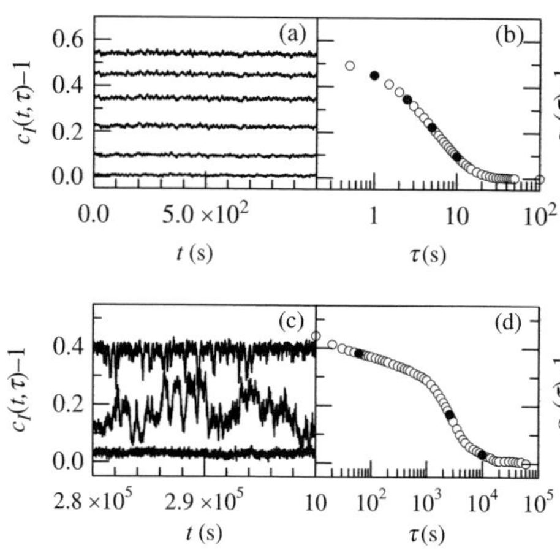

Fig. 4.2 (a) Degree of correlation $c_I(t,\tau)$ for a diluted suspension of Brownian particles: the dynamics are stationary and homogeneous, as seen by the very small fluctuations of c_I, due uniquely to measurement noise. From top to bottom, $\tau = 0, 1, 2.5, 5, 10$ and 700 s. (b) Intensity correlation function $g_2 - 1$ obtained by averaging over time the data in (a). The solid circles correspond to the time delays for which c_I is shown in (a). (c), (d): degree of correlation and intensity correlation function for a colloidal gel (Duri and Cipelletti, 2006) (see Section 4.5.2). In (c), $t = 0$ when the gel is formed. Note the large fluctuations of c_I, due to the heterogeneous nature of the dynamics. Individual events are discernible in the top trace ($\tau = 60\,\text{s}$), while the large fluctuations of the middle trace are due to the superposition of a fluctuating number of events ($\tau = 2600\,\text{s}$). For the bottom trace, $\tau = 10\,000\,\text{s}$.

has led to a loss of correlation between the intensity patterns recorded at times t and $t+\tau$. As the rearrangement event ceases, the degree of correlation recovers its typical level. At longer time lags (middle trace in Fig. 4.2c), c_I has a highly fluctuating behavior, because several events may occur during the probed lag. By contrast, almost no fluctuations are observed at very long lags (bottom trace in Fig. 4.2c), since a large number of events has occurred for all pairs of images, leading to a full decorrelation.

Various way of characterizing the fluctuations of c_I have been proposed, including analyzing the probability distribution function and the moments of its fluctuations, or their temporal autocorrelation (Duri *et al.*, 2005). Here, we focus on the variance

$$\chi(\tau) = \mathrm{var}(c_I) \equiv \left\langle [c_I(t,\tau) - < c_I(t,\tau) >_t)]^2 \right\rangle_t , \tag{4.7}$$

which is the analog in light scattering of the dynamical susceptibility χ_4 discussed in detail in Chapter 2. Intuitively, one understands that large fluctuations of c_I must be associated to "rare", large events: if the rearrangements were very localized, many such events would be necessary to significantly decorrelate the light scattered by a macroscopic sample volume. Assuming independent events, the resulting spatial average would yield a smooth c_I trace. More precisely, χ can be shown to be proportional to the volume integral of the spatial correlation of the dynamics, as discussed for χ_4 in Chapter 2. An example of the scaling of χ with the size of the events for a coarsening foam, where events can be unambiguously identified, is discussed in (Mayer *et al.*, 2004) (see also Chapter 5). A few differences exist between χ in light-scattering experiments and χ_4 in simulations or real-space measurements. Contrary to χ_4, χ is not normalized with respect to the number N of particles (compare Eq. (4.7) to Eq. (11) in Chapter 3), since usually N is not known precisely in light scattering. Accordingly, typical values reported for χ are much smaller than those for χ_4. Moreover, correction methods (Duri *et al.*, 2005) to remove the contribution of the statistical noise to χ are often used: using these corrections, one has $\chi = 0$ for homogeneous dynamics.

While time-resolved correlation (TRC) experiments probe temporal fluctuations of the dynamics, they still lack spatial resolution. In photon correlation imaging (PCI), by contrast, spatial resolution is achieved by modifying the collection optics. As shown in (Duri *et al.*, 2009*b*) and sketched in Fig. 4.3 for $\theta = 90°$, a lens is used to image the scattering volume onto a CCD or CMOS detector, while a diaphragm limits the range of q vectors accepted by the detector. Under these conditions, each pixel of the sensor is illuminated by light issued from a small region of the sample and scattered in a small solid angle associated with the same, well-defined scattering vector. The images are analyzed in the same way as for TRC, except that they are divided into regions of interest (ROIs): a local degree of correlation $c_I(\boldsymbol{r}, t, \tau)$ is calculated for each ROI, by averaging the intensity correlation function over a small set of pixels centered around $\boldsymbol{r}$. The spatial correlation of the dynamics can then be measured by comparing the temporal evolution of $c_I(\boldsymbol{r}, t, \tau)$ for ROIs separated by a distance $\Delta\boldsymbol{r}$.

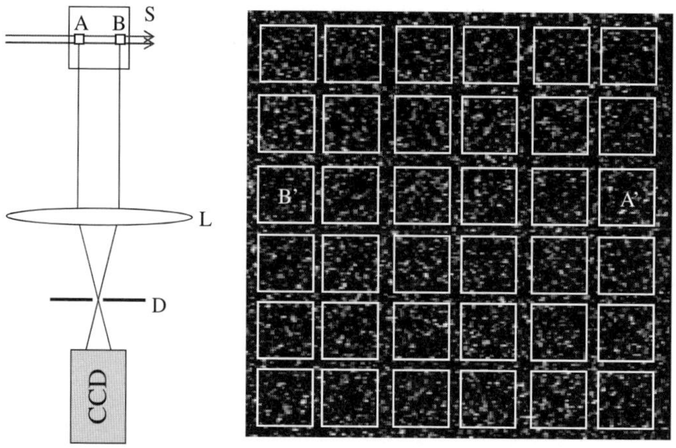

Fig. 4.3 Left: scheme of the photon correlation imaging apparatus for a scattering angle $\theta = 90°$. The lens L makes an image of the sample S onto the CCD detector. The diaphragm D, placed in the focal plane of L, selects only light rays scattered at $\theta \approx 90°$. Right: typical CCD image recorded by a PCI apparatus. The overlaid boxes indicate the grid of ROIs for which local degrees of correlation, $c_I(\boldsymbol{r}, t, \tau)$, are calculated. The size L of the ROIs has been exaggerated for the sake of clarity; typically, L is of the order of 10–20 pixels, corresponding to $\sim 20 - 100$ μm in the sample. ROIs A' and B' correspond to two distinct regions A and B in the sample.

More specifically, we define (Duri *et al.*, 2009*b*)

$$g_4(\Delta\boldsymbol{r}, \tau) = B(\tau) \left\langle \frac{\langle \delta c_I(\boldsymbol{r}, t, \tau)\delta c_I(\boldsymbol{r} + \Delta\boldsymbol{r}, t, \tau)\rangle_t}{\sqrt{\mathrm{var}[\delta c_I(\boldsymbol{r}, t, \tau)]\mathrm{var}[\delta c_I(\boldsymbol{r} + \Delta\boldsymbol{r}, t, \tau)]}} \right\rangle_{\boldsymbol{r}}, \tag{4.8}$$

where $\delta c_I = c_I - < c_I >_t$ are the temporal fluctuations of the local dynamics and $B(\tau)$ is a normalizing coefficient chosen so that $g_4(\Delta\boldsymbol{r}, \tau) \to 1$ as $\Delta\boldsymbol{r} \to 0$. This is the analog, albeit at a coarse-grained level, of the spatial correlation of the dynamics calculated in numerical and experimental work where particle trajectories are accessible (see Chapters 3 and 5). In most cases, the dynamics are isotropic and g_4 is averaged over all orientations of $\Delta\boldsymbol{r}$. It is important to distinguish the length scale over which the dynamics are probed from the spatial resolution with which the local dynamics can be measured. The former is dictated by the inverse scattering vector. Depending on the scattering angle, typical values range from a fraction of a μm up to tens of μm. The latter is determined by the size of the ROIs and the magnification with which the sample is imaged. Typical values are in the range 20–100 μm. As a final remark, we note that the differential dynamic microscopy method (Cerbino and Trappe, 2008) mentioned at the end of Section 4.2.1 could be easily adapted to calculate g_4. This would improve the spatial resolution as compared to that of PCI, thanks to the larger magnification typically used in a microscope.

4.3 Average dynamics and dynamical heterogeneity in the supercooled regime

4.3.1 Structural relaxation time

The average dynamics of colloidal hard spheres in the supercooled regime ($\varphi < \varphi_g$) has been thoroughly studied in a series of works on PMMA-based systems (Pusey and van Megen, 1987; van Megen and Underwood, 1994; van Megen *et al.*, 1998; Brambilla *et al.*, 2009; El Masri *et al.*, 2009). Figure 4.4 shows typical ISFs measured for a variety of volume fractions at a scattering vector $q = 2.5/a$ ($a = 100$ nm) (Brambilla *et al.*, 2009, 2010), below the first peak of the static structure factor [we quote here the more precise determination of a reported in (Brambilla *et al.*, 2010), slightly smaller than that in (Brambilla *et al.*, 2009)]. These experiments are performed close to the best index-matching conditions for a PMMA sample with size polydispersity $\sigma = 12.2\%$ (Brambilla *et al.*, 2010). Under these conditions the sample is optically polydisperse, as discussed in Section 4.2.3; thus, the self-part of the ISF, f_s, is probed (El Masri *et al.*, 2009). At low volume fractions, the decay of f_s is well fitted by a single exponential, as expected for diluted Brownian particles. As φ increases, the ISFs develop a two-step relaxation. The initial decay depends weakly on φ and corresponds to the motion of a particle in the cage formed by its neighbors. The final decay corresponds to the relaxation of the cage; its characteristic time, τ_α, increases by almost 7 decades in the range of φ investigated, where all samples equilibrate.

Figure 4.5 shows $\tau_\alpha(\varphi)$, as obtained by fitting the final relaxation of the ISF to a stretched exponential:

$$f_s = B \exp[-(\tau/\tau_\alpha)^\beta], \tag{4.9}$$

with $\beta \approx 0.56$ in the glassy regime. In the range $0.517 < \varphi < 0.585$, corresponding to about three decades in relaxation time, the volume fraction dependence of τ_α agrees very well with the critical law predicted by MCT (Gotze, 1999):

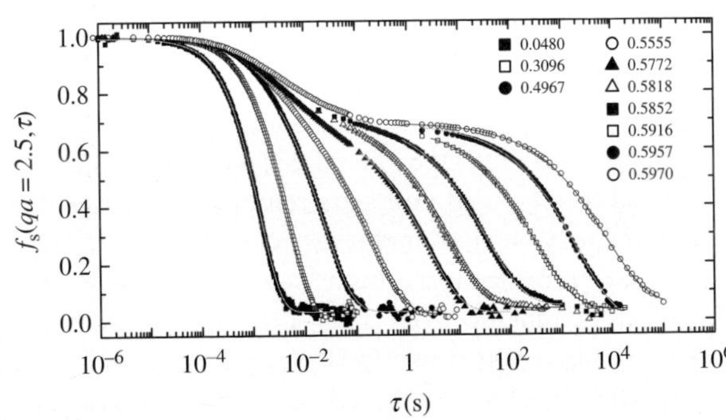

Fig. 4.4 Intermediate scattering functions for colloidal hard spheres (Brambilla *et al.*, 2009, 2010). Data are labeled by the volume fraction φ. The lines are stretched exponential fits to the final decay of $f_s(q, \tau)$. Adapted from (Brambilla *et al.*, 2009) with permission.

$$\tau_\alpha = \tau_0 \left(\frac{\varphi_c}{\varphi_c - \varphi} \right)^\gamma , \tag{4.10}$$

with $\gamma = 2.6$ and $\varphi_c = 0.59$. This is shown in Fig. 4.6a, where τ_α is plotted against $(1 - \varphi/\varphi_c)^{-1}$. Deviations are observed at low φ (as expected, since here MCT does not apply) and, more importantly, at the highest volume fractions that could be probed, where τ_α grows much slower than expected from MCT. This suggests that the divergence predicted by MCT is in fact avoided, as confirmed unambiguously by the fact that equilibrium ISFs could be measured up to $\varphi = 0.598$, above the critical packing fraction $\varphi_c = 0.59$ obtained from the MCT fit (Brambilla *et al.*, 2009; El Masri *et al.*, 2009). Figure 4.5 also shows data taken from previous works on PMMA systems (van Megen and Underwood, 1994; van Megen *et al.*, 1998), which exhibit a similar (apparent) algebraic divergence of τ_α. Note, however, that in these works data at very high φ could not be obtained, possibly explaining why the crossover between

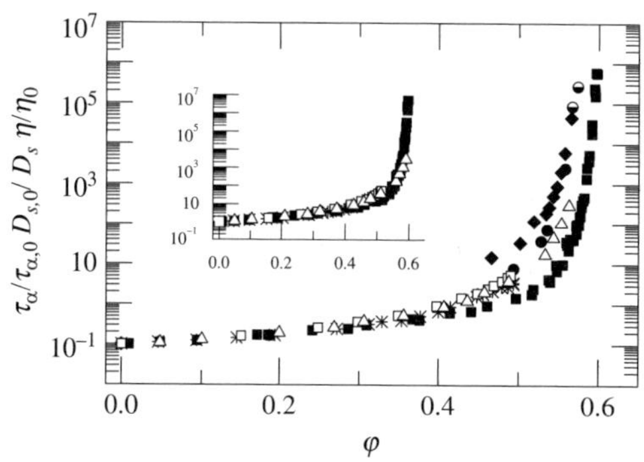

Fig. 4.5 Main panel: comparison of the volume fraction dependence of the structural relaxation time, self-diffusion coefficient and low shear viscosity as determined in various works on PMMA colloidal hard spheres with size polydispersity ranging from $\approx 4\%$ to 12.2%. All quantities are normalized with respect to their value in the $\varphi \to 0$ limit. Solid squares (Brambilla *et al.*, 2009), solid circles (van Megen and Underwood, 1994) and solid diamonds (van Megen *et al.*, 1998) are τ_α data obtained by DLS. Semifilled symbols indicate data where only a partial decay of $f(q, \tau)$ could be measured. Stars are self-diffusion data obtained by DLS (Segrè *et al.*, 1995*c*). Open squares and open triangles are viscosity data from Refs. (Segrè *et al.*, 1995*c*) and (Cheng *et al.*, 2002), respectively. Inset: same data as a function of scaled volume fraction, where scaling factors ranging from 1 to 1.05 were chosen so as to superimpose all curves for $\varphi \leq 0.2$. [The data sets from Refs. (van Megen and Underwood, 1994; van Megen *et al.*, 1998), for which no low-φ data are available, are not included in this plot]. A reasonably good collapse is obtained also for $\varphi > 0.2$, suggesting that the main source of discrepancy between the various data sets lies in the uncertainty on the absolute volume-fraction determination.

the MCT regime and the high φ "activated" regime reported in (Brambilla *et al.*, 2009; El Masri *et al.*, 2009) was not observed previously.

A similar crossover is observed in molecular glass formers (Donth, 2001): the reduced dynamical range accessible in colloids is responsible for the difficulty in pinpointing it, since typically φ_c is close to φ_g, the volume fraction above which equilibrium dynamics become too slow to be experimentally accessible. The analogy with molecular glass formers suggests that $\tau_\alpha(\varphi)$ may be fitted by a Vogel–Fulcher–Tammann (VFT)-like form:

$$\tau_\alpha = \tau_\infty \exp\left[\frac{A}{(\varphi_0 - \varphi)^\delta}\right], \tag{4.11}$$

where the exponent δ has been introduced for the sake of generality ($\delta = 1$ for the VFT law). Although a very good fit of the data of Ref. (Brambilla *et al.*, 2009) is obtained for $\delta = 1$, a somehow better fit is obtained for $\delta = 2$ and $\varphi_0 = 0.637$. The quality of the fit thus obtained can be appreciated in Fig. 4.6b, where the data of Ref. (Brambilla *et al.*, 2009) are plotted using reduced variables so that Eq. (4.11) reduces to a straight line. A crucial question is whether φ_0 should be identified with φ_{RCP}. Experimentally, this question is still open, due to the difficulties in measuring precisely the volume fraction at random close packing. Numerical simulations for a binary mixture of hard spheres (Berthier and Witten, 2009), however, show that $\varphi_0 = 0.641 < \varphi_{\mathrm{RCP}} \leq 0.664$ supporting the thermodynamic glass-transition scenario and implying that in colloidal hard spheres the glass and the jamming transitions are distinct phenomena (Krzakala and Kurchan, 2007; Berthier and Witten, 2009). This viewpoint is still highly debated (see, e.g., (Xu *et al.*, 2009; Kamien and Liu, 2007)).

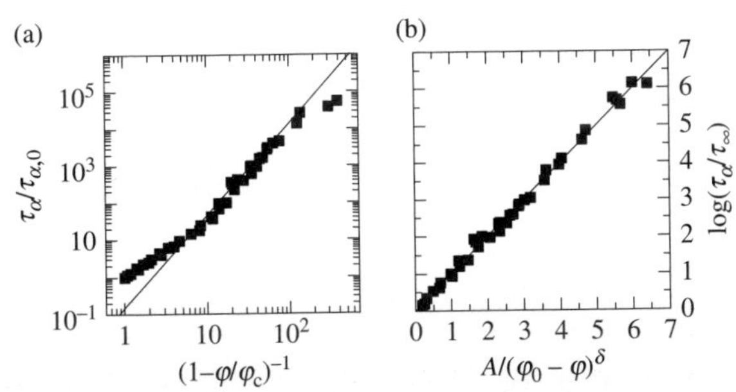

Fig. 4.6 (a): Double logarithmic plot of the structural relaxation time τ_α as a function of $(1 - \varphi/\varphi_c)^{-1}$, where the critical volume fraction φ_c is obtained from a fit to Eq. (4.10) (adapted from (Brambilla *et al.*, 2009) with permission). In this representation, the MCT law is a straight line with slope $\gamma = 2.5$ (solid line). Data at $\varphi > \varphi_c$ are not represented in this plot. (b): data from Ref. (Brambilla *et al.*, 2009) for $\varphi > 0.41$, plotted using reduced variable such that the generalized VFT law, Eq. (4.11) with $\varphi_0 = 0.637$ and $\delta = 2$, corresponds to the straight line shown in the plot (reproduced from (Brambilla *et al.*, 2009) with permission).

As a final remark on the structural relaxation in colloidal hard spheres, we note that one may wonder whether size polydispersity may have a significant impact on the dynamical behavior (van Megen and Williams, 2010; Brambilla *et al.*, 2010) and in particularly on the location of the (apparent) MCT divergence. Although no systematic experiments have been performed to address this issue (Williams and van Megen, 2001), computer simulations (El Masri *et al.*, 2009; Pusey *et al.*, 2009; Zaccarelli *et al.*, 2009*a*) have shown that φ_c is essentially unaffected by σ in the range $3\% < \sigma < 12\%$ typically explored in experiments.

4.3.2 Viscosity

Viscosity measurements are an alternative way to probe the average dynamics of hard-sphere suspensions. Several works have been devoted to the φ dependence of the viscosity, η, in the limit of a vanishingly small applied shear, see, e.g., Ref. (Cheng *et al.*, 2002) and references therein. These measurements are quite delicate, since the applied stress has to be very small (as low as a fraction of a mPa) in order to avoid non-linear effects, and because the resulting shear rate becomes extremely small beyond $\varphi \approx 0.5$, limiting measurements in the deeply supercooled regime. Additionally, the comparison between results obtained for different samples is affected by the uncertainties in the absolute volume fraction discussed above. Cheng *et al.* (Cheng *et al.*, 2002) show that data from various groups collapse reasonably well on a master curve when the absolute volume fractions are scaled by a factor up to about 1.03 to account for polydispersity. Figure 4.5 shows viscosity data from Refs. (Cheng *et al.*, 2002; Segrè *et al.*, 1995*c*) together with the DLS data discussed in Section 4.3.1. While there is some discrepancy between DLS and viscosity data, these differences are likely to be due, to a great extent, to uncertainties in φ and to the different methods used in the determination of the absolute volume fraction. Indeed, data sets for which data points at low φ are available can be scaled reasonably well onto a master curve by correcting φ using scaling factors between 1 and 1.05, so as to superimpose the growth of the viscosity or the relaxation time at low volume fraction ($\varphi \leq 0.2$) (see the inset of Fig. 4.5). The relationship between η and τ_α is discussed in more detail in Ref. (Segrè *et al.*, 1995*c*), where it is found that the low shear rate viscosity and the structural relaxation time measured by the collective ISF at the peak of the structure factor agree remarkably well up to $\varphi \sim 0.5$, while the relaxation time for the self-part of the ISF is somewhat lower.

The nature of the divergence of η is still a matter of debate. In their early work (Phan *et al.*, 1996), Russel, Chaikin and coworkers reported that $\eta(\varphi)$ is fitted well by the Krieger–Dougherty equation, a critical law of the same form as Eq. (4.10) with $\gamma = 2$, yielding $\varphi_c = 0.577$, consistent with the critical packing fraction obtained from MCT fits of DLS data. However, in their subsequent analysis of data on a larger range of viscosity (Cheng *et al.*, 2002), they report that a VFT-like fit (e.g. of the form of Eq. (4.11) with $\delta = 1$) yields better results. They find $\varphi_0 = 0.625$, close to random close packing, and significantly higher than φ_c. It should, however, be recalled that viscosity measurements are feasible only up to volume fractions lower than φ_c

(e.g. $\varphi \leq 0.562$ in Ref. (Cheng *et al.*, 2002)), making it difficult to draw unambiguous conclusions on the nature of the divergence of η, and in particular on the existence of a divergence at $\varphi_c \approx 0.58$.

4.3.3 Dynamical heterogeneity

The first microscopy experiment that examined the motion of supercooled colloidal particles was by Kasper, Bartsch, and Sillescu in 1998 (Kasper *et al.*, 1998). They devised a clever colloidal system primarily composed of refraction-index-matched particles made from crosslinked poly-*t*-butylacrylate. They then added a small concentration of tracer particles that had non-index-matched cores of polystyrene coated with shells of poly-*t*-butylacrylate. Using dark-field microscopy, they could observe the motion of the tracer particles. They observed that particles exhibited caged motion, as described above in Section 4.3.1. That is, a particle would diffuse within some local region, trapped in a cage formed by its neighbors, and then occasionally exhibit a quicker motion to a new region. This was useful evidence that the dynamics are temporally heterogeneous, and the first direct experimental visualization of caged motion. Averaging over all of the tracer particles, they noted that the distribution of displacements was non-Gaussian, likely linked to the cage trapping and cage rearrangements. This experiment exploited an inherent size polydispersity of about 8% to prevent crystallization. The chief limitations of the experiment were that the observations were limited to two-dimensional slices of the three-dimensional sample, and also that only isolated tracer particles were observed, rather than every particle.

The next published experiment, by Marcus, Schofield, and Rice, had different tradeoffs (Marcus *et al.*, 1999). They used a very thin sample chamber to study a quasi-two-dimensional colloidal suspension; the spacing between the walls of their sample chamber was approximately 1.2 particle diameters. Because of the relative ease of studying a thin sample, they did not need tracer particles, but rather could follow the motion of every particle within the field of view. Like Ref. (Kasper *et al.*, 1998), they observed cage trapping and cage rearrangements, and found a non-Gaussian distribution of displacements. Being able to see all of the particles, they also noted that the cage-rearrangement motions were spatially heterogeneous, with groups of particles exhibiting string-like motions. The string-like motions were quite similar to those seen in simulations (Kob *et al.*, 1997; Donati *et al.*, 1998; Hurley and Harrowell, 1996). Their results clearly showed a connection between the non-Gaussian behavior and the spatially heterogeneous dynamics, in that the non-Gaussian displacements were due to the particles involved in the cage rearrangements.

The experiment of Kegel and van Blaaderen in 2000 used confocal microscopy to improve upon the prior experimental limitations (Kegel and van Blaaderen, 2000). They observed the motion of core–shell colloidal particles, in a fully three-dimensional sample (although their observations were limited to two-dimensional images to maximize the imaging rate). Their colloidal samples were well characterized as having hard-sphere interactions. In this experiment, they again observed string-like regions of high mobility, related to the non-Gaussian distribution of displacements. This was the first experiment to directly visualize spatially heterogeneous dynamics in a

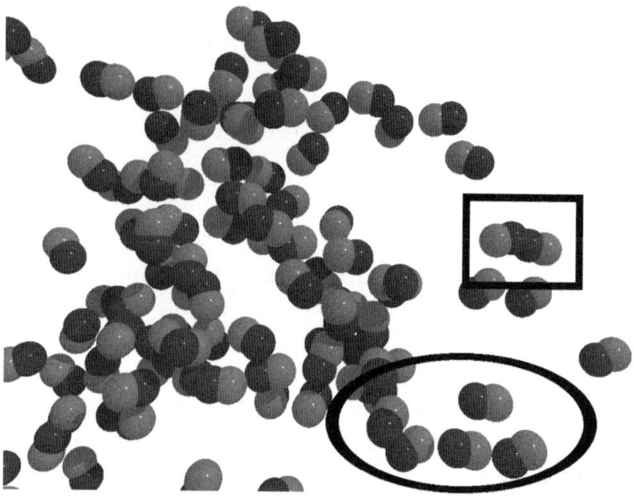

Fig. 4.7 Picture depicting the current positions of the most mobile colloidal particles (light colored) and the direction they are moving in (dark colored). The particles are drawn at 75% of their correct size, and for clarity only the most mobile particles are shown. Many particles move in similar directions to their neighbors, as shown by those within the oval. However, some particles also move in opposite directions to their neighbors, for example closing gaps between them, as highlighted by the rectangle. This is a sample with $\varphi = 0.56$, data taken from Ref. (Weeks *et al.*, 2000). The particles have radius $a = 1.18$ μm and the time scale used to determine the mobility is $\Delta t = 1000$ s.

three-dimensional sample, confirming what simulations had already suggested, that string-like motion is not an artifact of two-dimensional systems (Marcus *et al.*, 1999; Kob *et al.*, 1997; Donati *et al.*, 1998; Hurley and Harrowell, 1996; Perera and Harrowell, 1999).

Shortly after Ref. (Kegel and van Blaaderen, 2000), Weeks *et al.* published a similar experiment using confocal microscopy to study three-dimensional colloidal samples (Weeks *et al.*, 2000). Utilizing a faster confocal microscope than Ref. (Kegel and van Blaaderen, 2000), they were able to observe displacements in three dimensions. They too saw string-like motion, although also noted that some particles were moving in non-string-like ways termed "mixing" (Weeks and Weitz, 2002); see Fig. 4.7. A limitation of this experiment is that the particles were later discovered to be slightly charged, rather than being ideal hard spheres (Gasser *et al.*, 2001). The three-dimensional observations enabled the fractal nature of the regions of mobile particles to be measured as $d_{\mathrm{f}} = 1.9 \pm 0.4$, similar to simulations (Donati *et al.*, 1999). The sizes of the mobile regions increased dramatically as the colloidal glass transition was approached.

Subsequently, the data of Ref. (Weeks *et al.*, 2000) has been reanalyzed to highlight other features. The idea of caging was quantified in Ref. (Weeks and Weitz, 2002), finding a result similar to Kasper *et al.* (Kasper *et al.*, 1998), that caging manifested itself as an anti-correlation of particle displacements in time. That is, if a particle pushed against the "walls" of its cage (formed by its neighbors), then the particle was likely to subsequently be pushed backwards. The original work of Ref. (Weeks *et al.*, 2000) highlighted mobile regions using a slightly arbitrary definition of which particles were mobile. Later analysis used spatial correlation functions to avoid defining a particular subset of particles as mobile (Weeks *et al.*, 2007). These correlation functions found length scales for particle mobility that grew by a factor of 2 as the glass transition was approached, finding the largest length scale for a supercooled liquid of approximately $8a$ in terms of the particle radius a.

More recent work has taken the study of dynamical heterogeneities in colloidal suspensions in new directions. One notable set of experiments uses superparamagnetic particles and an external magnetic field to control the glassiness of a sample *in situ* (König *et al.*, 2005; Ebert *et al.*, 2009; Mazoyer *et al.*, 2009). Other experiments study the colloidal glass transition in confinement (Nugent *et al.*, 2007), or glassy samples as they are sheared (Besseling *et al.*, 2007; Schall *et al.*, 2007; Chen *et al.*, 2010).

Dynamical heterogeneity in the equilibrium regime has also been probed by DLS. Attempts to directly measure χ_4 or g_4 using the time- and space-resolved methods discussed in Section 4.2.4 have, so far, failed, because these techniques lack the resolution needed to detect dynamical heterogeneity on the length scale of a few particles. By contrast, χ_4 can be measured indirectly using the theory developed in Chapter 3 (see Section 3.2.5 therein). For colloidal hard spheres the following relation holds (Berthier *et al.*, 2005, 2007*a*,*b*):

$$\chi_4(q,\tau) = \chi_4(q,\tau)|_\varphi + \rho k_B T \kappa_T [\varphi \chi_\varphi(q,\tau)]^2, \tag{4.12}$$

where ρ is the number density, κ_T the isothermal compressibility (taken from the Carnahan–Starling equation of state), $\chi_4(q,\tau)|_\varphi$ denotes the value taken by $\chi_4(q,\tau)$ in a system where density is strictly fixed, and $\chi_\varphi(q,\tau) \equiv \partial f_s(q,\tau)/\partial\varphi$. Only the second term in the r.h.s. of Eq. (4.12) can be accessed experimentally. Numerical simulations, where both terms in the r.h.s. of Eq. (4.12) can be calculated (Brambilla *et al.*, 2009), show that the first term can be neglected in the deep supercooled regime: $\chi_4(q,\tau) \approx \rho k_B T \kappa_T [\varphi \chi_\varphi(q,\tau)]^2$ when the latter term is larger than unity. Experimentally, χ_φ can be obtained either by numerical differentiation using two ISFs measured at close enough volume fractions (see Fig. 4.8a), or by using the chain rule in the r.h.s. of Eq. (4.9):

$$\frac{\partial f_s}{\partial \varphi} = \frac{\partial f_s}{\partial B}\frac{\partial B}{\partial \varphi} + \frac{\partial f_s}{\partial \tau_\alpha}\frac{\partial \tau_\alpha}{\partial \varphi} + \frac{\partial f_s}{\partial \beta}\frac{\partial \beta}{\partial \varphi}, \tag{4.13}$$

where the partial derivatives with respect to volume fraction of the coefficients B, τ_α and β defined in Eq. (4.9) are obtained by fitting their φ dependence by smooth polynomials (Dalle Ferrier *et al.*, 2007; Brambilla *et al.*, 2009).

Figure 4.8a shows $\chi_4(q = 2.5a, \tau)$ together with the ISFs used to obtain it by finite difference (Berthier *et al.*, 2005), for the same system as in Ref. (Brambilla *et al.*, 2009). The dynamical susceptibility has the characteristic peaked shape observed in a variety of glassy systems (see also Chapters 3 and 5), with the maximum of the

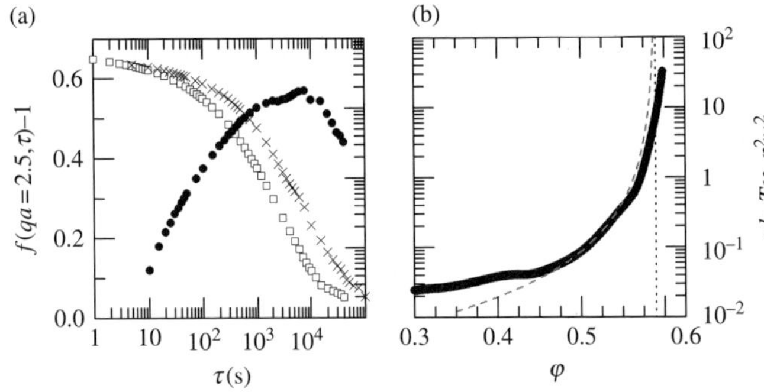

Fig. 4.8 (a), left axis: ISFs at two nearby volume fractions ($\varphi = 0.5953$ and 0.5970 for the open squares and the crosses, respectively). The solid circles are χ_φ as obtained by finite difference from the two ISFs (right axis, same scale as in (b)). (b): height of the peak of χ_4, χ^*, as a function of φ (adapted from (Brambilla *et al.*, 2009) with permission). Here, χ_4 has been estimated using the chain rule, as explained in the text. The dashed line is a MCT fit to the dynamical susceptibility. The vertical dotted line indicates the location of the divergence predicted by MCT but avoided by the data.

fluctuations, $\chi^* \equiv \chi_4(\tau^*)$, occurring on a time scale τ^* comparable to the structural relaxation time τ_α. As discussed in detail in Chapters 3 and 5, χ^* is of order N_{corr}, the number of particles that undergo correlated rearrangements. Figure 4.8b shows χ^*, obtained for the same system but using the chain rule (4.13), as a function of φ. The amplitude of dynamical fluctuations increases with volume fraction, supporting theoretical scenarios where a growing dynamical length accompanies the divergence of the relaxation time on approaching a glass transition. The dashed line shows the predictions of advanced mode-coupling theories, $\chi^* \sim (\varphi_c - \varphi)^{-2}$ (Berthier *et al.*, 2007*a*,*b*; Biroli and Bouchaud, 2004). As observed for the average dynamics, the data agree with MCT only over a limited range of volume fractions; in particular, MCT overpredicts the growth of χ^* at high φ, where the data remain finite across the critical volume fraction φ_c. Therefore, the analysis of dynamical fluctuations confirms the crossover from an MCT regime to an activated regime at high volume fractions inferred from the average dynamics.

4.4 Average dynamics and dynamical heterogeneity in non-equilibrium regimes

4.4.1 The glassy regime

Thus far, we have focused on dynamical heterogeneities in supercooled colloidal liquids; these systems will crystallize after some time (if the particles are sufficiently monodisperse). For polydisperse systems, the dynamics can be stationary (unless the sample has been recently sheared or stirred), since the sample is in (metastable) equilibrium. In contrast, glasses (volume fraction $\varphi > \varphi_g$) are out of equilibrium, and their properties depend on time, a phenomenon termed "aging." In particular, consider a low volume fraction colloidal suspension that is centrifuged to rapidly increase its volume fraction to the point where it becomes a colloidal glass at some time $t_w = 0$. The motion of particles within this sample depends on the time t_w since this formation, called the waiting time or simply the age of the sample. Initially, particles can move relatively rapidly, but as the system ages, particle motion slows (van Megen *et al.*, 1998). Particles take longer to move the same distance that was covered quickly at an earlier age. Equivalently, the cages formed by a particle's neighbors are more long-lasting, and given that these samples have a high volume fraction, particles spend almost all of their time tightly confined within their cages.

In the earliest microscopy experiments (Kasper *et al.*, 1998), there was evidence of slow motion within glassy samples. While the mean square displacement of the particles grew extremely slowly, it did grow. This provided some evidence that particles were not completely frozen, but rather moved to new positions as the sample aged.

More direct observations were obtained via confocal microscopy in Ref. (Weeks *et al.*, 2000). In that experiment, Weeks *et al.* observed that while mobile regions of particles grew larger as $\varphi \to \varphi_g$ (in the supercooled state), in glassy samples the mobile regions were quite small. This implied that spatial dynamical heterogeneities were less important for colloidal glasses. This result was revised in 2003 by Courtland and Weeks, who observed larger clusters of mobile particles in an aging colloidal glass, again using confocal microscopy (Courtland and Weeks, 2003). The key difference was in the data analysis. In a glassy sample, most of the particle motion is Brownian motion within the cages. Occasionally, particles have cage rearrangements, but this is hard to distinguish from the Brownian motion as the distances particles move during these rearrangements gets very small (Weeks and Weitz, 2002; Courtland and Weeks, 2003). Courtland and Weeks were able to observe the cage rearrangements by low-pass filtering the raw trajectories, thus smoothing out the Brownian motion and making the cage rearrangements clearer. They found that there was indeed spatially heterogeneous dynamics comparable to the behavior of supercooled colloids, but surprisingly they did not see any dependence on the aging time t_w. Similar results were also seen in a later confocal microscopy study of a binary colloidal glass (Lynch *et al.*, 2008).

One important consideration for studies of the aging of colloidal glasses is the protocol for achieving the initial glassy state. In traditional molecular glasses, a sample is quenched from a liquid state by rapidly reducing the temperature. In a colloidal glass, the analogous quench would be to rapidly increase the volume fraction, for example by centrifugation. However, in the experiments described above (Courtland and Weeks, 2003; Lynch *et al.*, 2008), the samples were instead prepared at a constant volume fraction, and then shear-melted by stirring them with an embedded stir bar. This is sometimes termed "shear-rejuvenation" as the stirring will make a well-aged colloidal glass look like a young colloidal glass, that is, it resets $t_w = 0$. However, there is evidence that these two protocols give different glassy states in molecular glass formers (McKenna, 2003).

Motivated by this, two groups have found ways to quench a colloidal glass *in situ* from a low volume fraction state to a high volume fraction state. One method uses an external magnetic field to control the effective interparticle attraction in a two-dimensional colloidal sample composed of superparamagentic particles (Assoud *et al.*, 2009). This technique has not been used to study aging explicitly, but so far has focused on slow crystallization after a rapid quench (Assoud *et al.*, 2009). Another method uses the sample temperature to control swelling in hydrogel particles, again in a quasi-two-dimensional experiment (Yunker *et al.*, 2009). Experiments using soft swellable particles are described in Section 4.5.1.

Dynamical heterogeneity in glassy colloidal samples has also been probed by DWS. We recall that this light-scattering technique is sensitive to motion on very small length scales, down to a fraction of a nm (see Section 4.2.3), a highly desirable feature for glassy systems where particles hardly move. Reference (Ballesta *et al.*, 2008*b*) discusses both the average dynamics and its temporal fluctuations in very dense suspensions of relatively large particles ($a \approx 10\ \mu$m). After initializing the sample by shaking it vigorously, the dynamics slow down until a pseudo-stationary regime is attained, where all measurements are performed. This regime does not correspond to a true equilibrium state, but rather to a regime where the very local dynamics probed in this experiment do not evolve significantly on the time scale of the experiments (up to a few days).

Figure 4.9 shows the volume-fraction dependence of various parameters characterizing the dynamics in the pseudo-stationary regime. As shown in Fig. 4.9a, the average relaxation time of the intensity correlation function, as obtained from a fit $g_2(t,\tau) - 1 = B \exp[-(\tau/\tau_0(t))^{\beta(t)}]$, increases smoothly with φ, seemingly diverging at $\varphi = \varphi_{\max} = 0.752$, presumably the random close-packing volume fraction for this quite polydisperse system. Figure 4.9b shows the φ dependence of the average stretching exponent, $\overline{\beta}$, which increases from about 0.9 to 1.3 as random close packing is approached. Figure 4.9c shows the height, χ^*, of the peak of the dynamic susceptibility χ defined in Eq. (4.7). Quite surprisingly, χ^* is non-monotonic: it first increases with φ, reaches a maximum, but eventually drops dramatically close to random close packing.

The interpretation of these data requires special care, since the usual DWS formalism has been developed for spatially and temporally homogeneous dynamics. By contrast, Ref. (Ballesta *et al.*, 2008*b*) discusses a general model for DWS for heterogeneous dynamics, based on an extension of the formalism originally proposed by Durian *et al.* for DWS measurements of the spatially localized, temporally intermittent dynamics of a foam (Durian *et al.*, 1991). According to the model of (Ballesta *et al.*, 2008*b*), the non-monotonic behavior of $\chi^*(\varphi)$ results from the competition between two contrasting effects. On the one hand, the size ξ of the regions that undergo a rearrangement increases continuously with volume fraction, leading to enhanced temporal fluctuations as observed in supercooled samples (see Section 4.3.3). On the other hand, as φ increases the particle displacement associated with each of these events is increasingly restrained, due to crowding. Thus, an increasingly large number of rearrangement events is required to decorrelate the scattered light at high φ, leading to reduced fluctuations. While simple simulations of a DWS experiment in a medium undergoing rearrangements described by this scenario reproduce the experimental data (Ballesta *et al.*, 2008*b*), a direct measurement of $\xi(\varphi)$ would be necessary to confirm it. This would allow one to better understand analogies and differences with

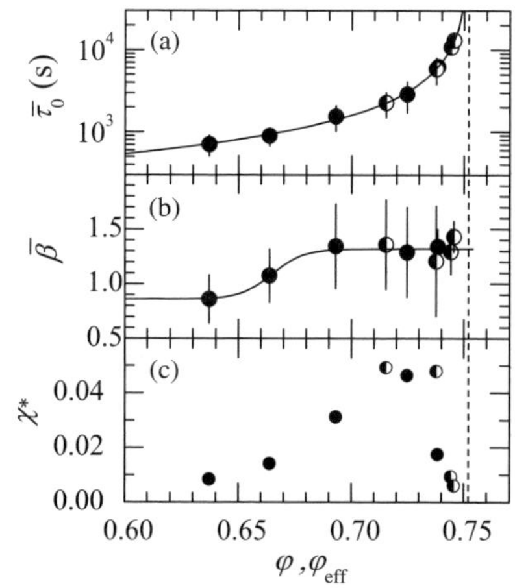

Fig. 4.9 Volume-fraction dependence of the dynamics and its temporal fluctuations for a concentrated suspension of colloids as probed by DWS. Solid symbols refer to fresh samples, semifilled symbols to aged samples that have been rejuvenated mechanically. See (Ballesta *et al.*, 2008*b*) for more details. (a) Average relaxation time, the line is a critical-law fit to the data, $\overline{\tau_0} \sim (\varphi_{\max} - \varphi)^{-y}$ with $\varphi_{\max} = 0.752$ and $y = 1.5$. Here and in the other panels the vertical dashed line indicates the location of $\varphi_{\max}$. (b) Stretching exponent obtained by fitting the final relaxation of the intensity correlation function to a stretched exponential. (c) Height χ^* of the peak of the dynamical susceptibility, $\chi(\tau)$. Note the abrupt drop of χ^* on approaching $\varphi_{\max}$. In (a) and (b), the bars indicate the standard deviation of the temporal distribution of the relaxation time and the stretching exponent, respectively. Adapted from Ref. (Ballesta *et al.*, 2008*b*) with permission.

the experiments on driven grains (Lechenault *et al.*, 2008) mentioned in Chapter 5, where a non-monotonic behavior of *both* χ^* and ξ has been reported.

4.4.2 Dynamical heterogeneity under shear

Aging systems are out of equilibrium; another type of non-equilibrium system is a driven system. Of particular interest are samples that are sheared. The importance of shear can be quantified by a Péclet number (see also Section 4.1.2). This is the ratio of the time scale for diffusion to the time scale for shear-induced motion. The shear induced time scale is given in terms of the strain rate as $1/\dot{\gamma}$. Recall that the Brownian time scale is $\tau_B = a^2/(6D)$ (Eq. (4.1)). Typically, when considering dense colloidal suspensions, the relevant diffusion constant is D_∞, the long-time diffusion constant, which varies with the volume fraction. The other option would be to use D_0, as given by the Stokes–Einstein–Sutherland formula, Eq. (4.2), which is the diffusion constant in the $\varphi \to 0$ limit. If D_0 is used, the Péclet number is termed the bare Péclet number, Pe, and if D_∞ is used, it is termed the modified Péclet number, Pe*. Given that we wish to understand how particles move and rearrange, this is the long-time scale motion (D_∞), and combining the expressions above we find $\text{Pe}^* = a^2\dot{\gamma}/(6D_\infty)$.

For $\mathrm{Pe}^* < 1$, diffusion is the primary influence on particle motion, and the shear is only a small perturbation. For $\mathrm{Pe}^* > 1$, the shear-induced motion is expected to be more significant. The main consideration here is that for the shear rate to be significant, an experiment must have $\mathrm{Pe}^* > 1$, which is the case for the experiments described below in this subsection.

Simulations of supercooled fluids found that as Pe^* increases, the system becomes more liquid-like, and dynamically heterogeneous regions become smaller (Yamamoto and Onuki, 1997), similar to the idea of shear unjamming a sample (Liu and Nagel, 1998). As yet, this relationship between Pe^* and dynamical heterogeneity has been untested by colloidal experiments.

A complementary question is to ask what the nature of a shear-induced motion is, for a dense amorphous sample. For example, sheared crystalline materials respond by the internal motion of dislocation lines. Influential simulations by Falk and Langer found, for amorphous materials, that rearrangements occurred in localized regions, termed "shear-transformation zones" (Falk and Langer, 1998). In these regions, locally the stress builds up until it is released by a rapid local rearrangement event. These behaviors have been directly observed in an experiment by Schall *et al.*, where they used confocal microscopy to examine a colloidal glass as it was sheared between two parallel plates at low strains $\sim 4\%$ (Schall *et al.*, 2007). In their experiment, the shear-induced dynamical heterogeneities had a small spatial extent, although in some cases they appeared to relax the strain over a large region even though the farther-away particles did not move very far. Their activation energy was estimated to be $E \sim 16 k_{\mathrm{B}}T$, where $k_{\mathrm{B}}T$ is the thermal energy. Given the low strain, it is possible that these observations are of shear-induced aging effects, rather than shear flow.

A separate experiment examined the shear-induced motion of colloidal particles in supercooled colloidal liquids (Chen *et al.*, 2010). In these less-dense samples, localized rearranging regions were observed at high strain values. These samples were sheared between two parallel plates that moved back and forth with a triangle wave displacement curve. Strikingly, the shapes of these rearranging regions were isotropic on average, and showed no distinction between the velocity direction, velocity gradient direction, and the vorticity direction mutually perpendicular to the first two. Similar observations were made of a sheared colloidal glass (Besseling *et al.*, 2007): in those experiments, little difference was found in the effective diffusivity in the three directions, once the affine motion of the average shear profile was subtracted from the particle displacements. This study was the first direction observation of a sheared colloidal glass that directly imaged shear-induced dynamical heterogeneities (Besseling *et al.*, 2007). Note that the isotropic nature of plastic rearrangements observed in experiments is at odds with recent simulations of a 2D supercooled fluid of soft particles, where anisotropic structural rearrangements were observed (Furukawa *et al.*, 2009). These observations depended on careful analysis methods, which have not yet been applied to experimental data from colloids.

While not directly the same idea as shear-induced dynamical heterogeneity, it is important to note that many soft glassy systems—such as colloids—exhibit shear banding (Dhont, 1999; Dhont *et al.*, 2003; Fielding, 2007; Dhont and Briels, 2008; Fielding *et al.*, 2009). That is, when a large sample is sheared, sometimes the strain is localized near one of the boundaries. Within the "shear band" the sample is straining a significant amount and plastically deforming, while outside the shear band, the sample is nearly unstrained. As the stress must be continuous throughout the sample, this suggests that the material is acting as if it has two states, a low-viscosity state in the shear band where the sample flows, and an elastic state outside the shear band without flow. Shear bands have been noted in colloidal suspensions (Vermant, 2001; Chen *et al.*, 2010; Besseling *et al.*, 2007; Ballesta *et al.*, 2008; Dhont and Briels, 2008), colloidal gels (Moller *et al.*, 2008), worm-like micelles (Berret *et al.*, 1997; Olmsted, 1999; Cates and Fielding, 2006), foams (Debrégeas *et al.*, 2001; Lauridsen *et al.*, 2004; Janiaud *et al.*, 2006), and granular materials (Losert *et al.*, 2000; Utter and Behringer, 2008). Note that the experiments discussed in the previous paragraph (Besseling *et al.*, 2007; Chen *et al.*, 2010) were observations of a homogeneously shearing subregion within the shear band.

4.5 Beyond hard spheres

4.5.1 Soft particles

Colloidal particles with soft repulsive potential interactions can be obtained in several ways. One possibility is to exploit the softness of an electrostatic or magnetic repulsive potential. Another possibility is to modify the synthesis of the particles, e.g. in star polymers or microgel particles, where the degree of softness is governed by the number of arms and the degree of cross-linking, respectively. Finally, systems based on the self-assembly of amphiphilic molecules can form soft spheres, e.g. with amphiphilic diblock copolymers.

For soft systems, a nominal volume fraction, φ_{nom}, is often defined, based on the size of isolated particles and their number concentration; since particles can be squeezed, φ_{nom} can exceed unity. At low concentration, soft spheres behave similarly to hard spheres, provided that an effective size that takes into account the range of the soft repulsion is used. This is shown, e.g., by viscosity measurements (Roovers, 1994; Buitenhuis and Förster, 1997; Senff *et al.*, 1999), where the increase of the relative viscosity with concentration is essentially indistinguishable from that of hard spheres. At higher concentration, however, the behavior deviates significantly (Roovers, 1994; Buitenhuis and Förster, 1997; Sessoms *et al.*, 2009; Mattsson *et al.*, 2009*a*): the growth of the viscosity or the relaxation time with φ_{nom} is much gentler and samples with $\varphi_{\mathrm{nom}} > \varphi_{\mathrm{RCP}}$ may still be fluid, since particle deformations allow for structural relaxation.

In the recent years, aqueous solutions of poly-N-isopropylacrylamide (PNiPAM) microgels have become one of the most popular soft systems. Not only can their softness be controlled during the synthesis (by varying the amount of cross-linking), but it can also be tuned by varying the temperature, T. Under appropriate conditions, a decrease of a few degrees of T results in a $\sim 20\%$ growth of the particle radius (Senff *et al.*, 1999) and in an increased softness. The influence of the softness of concentrated PNiPAM particles on their dynamics has been studied in detail by DLS in Ref. (Mattsson *et al.*, 2009*a*). Quite remarkably, the authors find that softness correlates with fragility, defined here as the slope of $\log \tau_\alpha$ vs. concentration, ζ, at the glass transition, in analogy with the definition for molecular glass formers (Donth, 2001), where $1/T$ replaces ζ. Very soft particles have an Arrhenius-like behavior, $\tau_\alpha \sim \exp(C\zeta)$, harder particles have a larger fragility (steeper increase of $\tau_\alpha(\zeta)$), and hard spheres are the most fragile system, with the steepest increase of τ_α on approaching the glass transition. Thus, soft colloids appear as a promising system for understanding the origin of fragility in glasses, a long-standing problem.

A similar system has been investigated by photon correlation imaging (PCI) in Ref. (Sessoms *et al.*, 2009). Both χ and g_4 have been studied as a function of φ_{nom} in measurements of the collective dynamics at low q. The amplitude of temporal fluctuations of the dynamics is found to increase monotonically with φ_{nom}, while the range of spatial correlations of the dynamics has a non-monotonic behavior, a maximum being observed at $\varphi_{\text{nom}} \approx \varphi_{\text{RCP}}$. Remarkably, spatial correlations of the dynamics here extend over several mm, in analogy with what was reported for other jammed materials (see Section 4.5.2).

The T dependence of the size of PNiPAM particles provides a convenient way to quench them rapidly in a glassy state, by preparing a relatively concentrated—yet fluid—sample, which is then cooled by a few degrees, thereby swelling the particles to a quenched glassy state in a fraction of a second. This protocol was used in a quasi-two-dimensional experiment (Yunker *et al.*, 2009), where observations of aging over nearly six decades in aging time t_{w} were possible, due to the rapid quench rate. The authors observed spatial dynamical heterogeneity; particles occasionally underwent irreversible rearrangements. While the average size of the rearranging regions remained approximately constant during aging, similar to the prior study of 3D hard-sphere samples (Ref. (Courtland and Weeks, 2003)), they identified an increase in the size of a particular class of rearranging regions as the sample aged. Specifically, the domain size of rearranging particles surrounding irreversible rearrangements increased during aging. The largest clusters of rearranging regions involved approximately 100 particles, a size much smaller than the range of dynamical correlations measured in (Sessoms *et al.*, 2009). However, the technique used to identify rearranging particles limited the correlation size, making direct quantitative comparisons difficult. Additionally, they saw a relation between the local structure and the particle motion, which agrees with some prior work on hard spheres (Cianci *et al.*, 2006).

4.5.2 Attractive particles

When colloidal particles experience attractive forces, arrested phases may be obtained also at a volume fraction lower than that required for glassy dynamics to be observed in hard spheres. The distinction is often made between colloidal gels (up to $\varphi \sim 0.3$) and attractive glasses (over $\varphi \sim 0.5$) (Trappe and Sandkuhler, 2004). Very recent work by Zaccarelli and Poon further refines the classification of concentrated attractive systems (Zaccarelli *et al.*, 2009*b*), based on the pre-dominance of either caging or bonding. Concentrated, attractive glasses (Eckert and Bartsch, 2002; Pham *et al.*, 2002) have a structure similar to that of repulsive HS systems, while the structure of diluted gels depends on the strength of the interactions: highly attractive systems tend to form string-like, fractal structures, while gel strands are thicker when the interparticle potential well at contact is close to $k_{\mathrm{B}}T$ (Campbell *et al.*, 2005; Dibble *et al.*, 2006). The mechanism leading to gel formation is still intensely debated; recent work, still controversial, points to the role of an underlying fluid–fluid phase transition that is arrested once the dense phase becomes too concentrated (Lu *et al.*, 2008) (see Ref. (Zaccarelli, 2007) for a recent review on colloidal gels). Experimentally, attractive systems are typically realized either by screening the Coulomb repulsion in charge-stabilized systems, thereby exposing the particles to short-range, attractive van der Waals forces (Russel *et al.*, 1992), or by means of the depletion force (Asakura and Oosawa, 1958; Vrij, 1976) induced by adding to the suspension smaller particles, often polymer coils (Poon, 2002).

Dynamical heterogeneity in weak gels has been explored mainly by confocal microscopy (Gao and Kilfoil, 2007; Dibble *et al.*, 2008) and simulations (Puertas *et al.*, 2004; Charbonneau and Reichman, 2007). The general picture emerging from these works is that DH is closely related to structural heterogeneity. The probability distribution function of the particles displacement (van Hove function) has typically a non-Gaussian shape, with "fat" tails corresponding to fast particles whose displacement is anomalously large (Gao and Kilfoil, 2007; Dibble *et al.*, 2008; Chaudhuri *et al.*, 2008). These particles are located at the boundaries of the thick strands constituting the gel, while the particles buried within the strands are the least mobile. A detailed analysis of particle mobility as a function of the number of their neighbors (Dibble *et al.*, 2008) confirms this picture. It should be noted that such a structural origin of DH is in contrast with hard-sphere systems, where no clear connection between DH and structural quantities could be established so far. The influence of structure on dynamical heterogeneity in attractive systems has also been highlighted in a series of simulation papers by Coniglio and coworkers (see e.g. Ref. (Fierro *et al.*, 2008)).

The length scale dependence of dynamical heterogeneity in weak gels has been explored both numerically (Charbonneau and Reichman, 2007) and in XPCS experiments (Trappe *et al.*, 2007). In these works, the peak χ^* of the dynamical susceptibility χ_4 has been measured as a function of scattering vector q. χ^* has a non-monotonic behavior, the largest dynamical fluctuations being observed on a length scale of the order of the range of the attractive interparticle potential. This has to be contrasted with the case of repulsive systems, where the maximum of χ^* typically occurs around the interparticle distance (Charbonneau and Reichman, 2007; Dauchot *et al.*, 2005). On a more technical level, it is worth noting that Ref. (Trappe *et al.*, 2007) has demonstrated that modern synchrotron sources and X-ray detectors are now sufficiently advanced to allow for measurements of dynamical heterogeneities. The activity in this field is thus growing rapidly (Trappe *et al.*, 2007; Wandersman *et al.*, 2008; Herzig *et al.*, 2009; Duri *et al.*, 2009*a*; Wochner *et al.*, 2009) and there is hope that eventually X-ray scattering experiments may probe dynamical heterogeneity in molecular glass formers and not only for colloidal systems.

Optical microscopy studies of the dynamics of strong gels are difficult, due to the restrained and very slow motion of particles in these systems (Dibble *et al.*, 2008). By contrast, scattering techniques have been successfully applied to characterize the slow dynamics of tenuous, fractal-like gels made of particles tightly bound by van der Waals forces (Cipelletti *et al.*, 2000; Duri and Cipelletti, 2006; Duri *et al.*, 2009*b*). Time-resolved correlation (TRC, see Section 4.2.4) experiments show that the dynamics are due to intermittent rearrangement events where particles move over relatively small distances, of the order of a fraction of a μm (Duri and Cipelletti, 2006). Quite surprisingly, each of these events affects a macroscopic portion of the sample: the spatial correlation of the dynamics measured by photon correlation imaging (PCI, see Section 4.2.4) hardly decays over several millimeters, indicating that spatial correlations of the dynamics are limited essentially only by the system size (Duri *et al.*, 2009*b*). Once averaged over both time and space, the intensity correlation function $g_2(q,\tau)-1$ measured by "regular" multispeckle DLS exhibits a peculiar q dependence of both the relaxation time and the stretching exponent p obtained by fitting g_2-1 to a stretched exponential, $g_2(q,\tau)-1 \sim \exp[-(\tau/\tau_r)^p]$ (Cipelletti *et al.*, 2000). As shown in Fig. 4.10a, p is larger than one; accordingly, the relaxation has been termed a "compressed" exponential, as opposed to the stretched exponential relaxations often observed in glassy systems ($p<1$). Moreover, $\tau_r \sim q^{-1}$, as opposed to $\tau_r \sim q^{-2}$ as for diffusive motion.

In Refs. (Cipelletti *et al.*, 2000; Bouchaud and Pitard, 2002) these dynamics were interpreted as ultraslow ballistic motion due to the slow evolution of a strain field set by internal dipolar stresses. A more refined model has been proposed in Ref. (Duri and Cipelletti, 2006), taking into account the results from time- and space-resolved light-scattering experiments that highlight the discontinuous nature of the relaxation process. The model is based on the following assumptions: (i) the dynamics are due to individual rearrangement events that are random in time (Poissonian

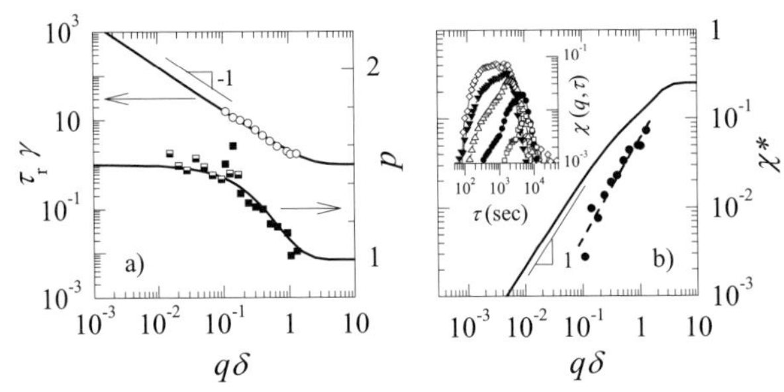

Fig. 4.10 Average dynamics and dynamical heterogeneity in a strong gel (adapted from (Duri and Cipelletti, 2006) with permission). (a): q dependence of the relaxation time τ_r of the intensity correlation function (left axis, open circles) and of the stretching exponent (right axis, solid and semifilled squares). The data are normalized with respect to the parameters of the model (see also the text): $\delta = 250$ nm is the average particle displacement due to one single rearrangement event and $\gamma = (960\ \text{s})^{-1}$ is the event rate. Note that the motion is, on average, ballistic-like ($\tau_r \sim q^{-1}$) and that the decay of g_2-1 is steeper than exponential ("compressed" exponential, $p \geq 1$). The lines are the predictions of the model. (b) Inset: dynamical susceptibility $\chi(q,\tau)$ for various q. Main plot: q dependence of the height, χ^*, of the peak of $\chi(\tau)$. The dashed line is a power-law fit with an exponent 1.13 ± 0.11, the solid line is the model.

statistics); (ii) each event affects the whole scattering volume (as indicated by PCI); (iii) the displacement field induced by one single event is that due to the long-range elastic deformation of the gel under the action of dipolar stresses (in order to account for $p > 1$); (iv) on the length scales probed by the scattering experiments, the displacement due to successive events occurs along the same direction (i.e. the motion is, on average, ballistic-like, as implied by the scaling $\tau_r \sim q^{-1}$). The model contains just two adjustable parameters: the rate of the events in the scattering volume, γ, and the average particle displacement resulting from one single event, δ. The model captures well the q dependence of both p and τ_r as observed in the average dynamics, as shown in Fig. 4.10a. It also captures correctly the growing trend for the q dependency of the amplitude of dynamical heterogeneity, although it overestimates their magnitude by about a factor of two, as seen in Fig. 4.10b. Physically, the growth of dynamical heterogeneity with increasing q can be understood as the result of the competition between the length scale, q^{-1}, over which the dynamics is probed, and the typical particle displacement, δ, due to a rearrangement event. At very large q, $q\delta \geq 1$, one single event is sufficient to fully decorrelate $g_2 - 1$. In this regime, the instantaneous relaxation time depends on the time between successive events, which is a fluctuating quantity due to the Poissonian nature of the events. This yields very large fluctuations of the dynamics. As q decreases, an increasing number of events is required to decorrelate $g_2 - 1$, thus leading to smoother dynamics.

Quite intriguingly, the main features of the average dynamics reported for the strongly attractive gels discussed above have been also found in a large variety of glassy soft materials (for a review, see e.g. Ref.(Cipelletti and Ramos, 2005)). Models inspired by Ref. (Duri and Cipelletti, 2006) have been used also to describe the dynamics of nanoparticles embedded in molecular systems approaching the glass transition (Caronna *et al.*, 2008; Guo *et al.*, 2009). Moreover, photon correlation imaging experiments (Sessoms *et al.*, 2009; Maccarrone *et al.*, 2010) on several systems (from concentrated soft spheres to repulsive glasses of charged clays and humidity-sensitive biofilms) suggest that system-spanning correlations of the dynamics may be a ubiquitous feature of jammed materials. This is in stark contrast with supercooled colloidal hard spheres, where spatial correlations of the dynamics extend over a few particle sizes at most, as discussed in Section 4.3.3. It is likely that the predominantly elastic behavior of jammed materials is responsible for the ultralong spatial correlations of the dynamics observed in those systems, since the strain field set by a local rearrangement can propagate over large distances before being appreciably damped. Indeed, internal stress relaxation is often invoked as the origin of theses dynamics (Cipelletti *et al.*, 2000; Bouchaud and Pitard, 2002), although a complete understanding of the physical mechanisms underlying this peculiar yet general relaxation behavior is still lacking.

4.6 Perspectives and open problems

While the average dynamics of glassy colloidal systems has been intensively studied since the 1980s, experiments on dynamical heterogeneities started only about thirteen years ago, spurred by advances in microscopy and light-scattering methods and stimulated by numerical works. A (partial) list of what we have learned in the past years on both the average dynamics and dynamical heterogeneity of glassy colloids includes:

- Microscopy experiments have allowed us to observe directly what caged motion and cage rearrangements look like. Data have been used to quantify the meaning of "caging" (Weeks and Weitz, 2002).
- Scattering experiments have shed new light on the dynamics of concentrated hard spheres, for which equilibrium dynamics above the ergodic–non-ergodic transition predicted by mode-coupling theory have been reported (Brambilla *et al.*, 2009).
- A variety of microscopy and light-scattering experiments agree on both the existence and the magnitude of dynamical heterogeneity in several colloidal systems, both 2D (Marcus *et al.*, 1999; König *et al.*, 2005; Ebert *et al.*, 2009; Mazoyer *et al.*, 2009) and 3D (Kasper *et al.*, 1998; Kegel and van Blaaderen, 2000; Weeks *et al.*, 2000), with both hard (Kegel and van Blaaderen, 2000; Weeks *et al.*, 2000) and soft (König *et al.*, 2005; Yunker *et al.*, 2009; Sessoms *et al.*, 2009) repulsive interactions. Along with simulations (Doliwa and Heuer, 1998; Donati *et al.*, 1998; Glotzer, 2000; Yamamoto and Onuki, 1998), this provides nice evidence that dynamical heterogeneities are ubiquitous in glassy systems and not artifacts of one particular colloidal system, one particular experimental technique, or one particular simulation method.
- Experiments on colloidal hard spheres have provided some of the first experimental quantitative evidence that the length scale of dynamical heterogeneities increases when approaching a glass transition (Weeks *et al.*, 2000; Berthier *et al.*, 2005; Weeks *et al.*, 2007).
- Light-scattering experiments on deeply jammed attractive or soft repulsive systems (Ballesta *et al.*, 2008; Sessoms *et al.*, 2009; Duri *et al.*, 2009*b*; Maccarrone *et al.*, 2010) have unveiled a richer-than-expected scenario, with ultralong-ranged spatial correlations of the dynamics not observed so far in simulations.

In spite of these advances, several questions remain open, making dynamical heterogeneity an exciting field of research:

- Is there a structural origin of dynamical heterogeneity? While for diluted, attractive systems this has been shown to be the case, for concentrated, repulsive particles a clear answer is still lacking. Progress in this area will likely require the measurement of non-conventional structural quantities, e.g. the "point-to-set" correlation function introduced in numerical works (Biroli *et al.*, 2008) and discussed in Chapter 3.
- What is the behavior of dynamical heterogeneity in colloidal glasses and its relationship with aging? While recent work on soft particles suggests that the slowing down of the dynamics during aging may be associated with a growth of spatial correlations of the dynamics (Yunker *et al.*, 2009), this was not seen in experiments with harder spheres (Courtland and Weeks, 2003); this question requires further exploration.

- Is the non-monotonic φ dependence of dynamical fluctuations observed in some systems (Ballesta *et al.*, 2008*b*; Sessoms *et al.*, 2009) a general feature? Although a somewhat similar behavior has been reported for granular systems (Lechenault *et al.*, 2008), the explanation proposed for colloids and grains are different: can these contrasting views be reconciled?
- Recent work (Duri *et al.*, 2009*b*; Sessoms *et al.*, 2009; Maccarrone *et al.*, 2010) on jammed soft materials suggests that the relatively short-ranged correlation of the dynamics of supercooled hard spheres may be the exception rather than the rule, since system-size dynamical correlations are observed in those materials. What is the physical origin of these correlations? Why do the numerical simulations not capture these correlations?
- Most work on colloidal systems has been devoted to hard-sphere-like systems. A general understanding of the role of the interaction potential on the slow dynamics and dynamical heterogeneity is still lacking. Recent experiments show that softer colloids have strikingly different behaviors than hard colloids (Mattsson *et al.*, 2009*b*), although these results are not fully understood.

Finally, we remark that numerical simulations can nowadays probe a range of relaxation times comparable to that explored by experiments (see e.g. (Brambilla *et al.*, 2009)). On the one hand, this calls for a more rigorous approach to the design of new experiments, since the question of what can be learned from experiments that simulations can not address needs to be asked. On the other hand, this opens the exciting possibility to compare in great detail numerical and experimental results, thereby allowing one to identify the physical mechanisms that are relevant in determining the slow relaxation of glassy colloidal systems.

Acknowledgments

The work of ERW was supported by NSF Grant No. CHE-0910707. The work of LC was supported by grants from ACI, ANR, CNRS, CNES, Région Languedoc-Roussillon, Institut Universitaire de France. LC acknowledges many discussions and fruitful collaborations with L. Berthier, G. Biroli, V. Trappe, and D.A. Weitz. ERW acknowledges helpful discussions and fruitful collaborations with J. C. Crocker and D. A. Weitz.

References

Adam, Gerold and Gibbs, Julian H. (1965). *J. Chem. Phys.*, **43**(1), 139–146.

Alcoutlabi, Mataz and McKenna, Gregory B. (2005, April). *J. Phys.: Condens. Matter*, **17**(15), R461–R524.

Amblard, François, Yurke, Bernard, Pargellis, Andrew, and Leibler, Stanislas (1996). *Rev. Sci. Instrum.*, **67**(3), 818–827.

Angell, C. A., Ngai, K. L., McKenna, G. B., McMillan, P. F., and Martin, S. W. (2000). *J. Appl. Phys.*, **88**(6), 3113–3157.

Antl, L., Goodwin, J. W., Hill, R. D., Ottewill, R. H., Owens, S. M., Papworth, S., and Waters, J. A. (1986, January). *Colloids Surf.*, **17**(1), 67–78.

Asakura, S. and Oosawa, F. (1958). *J. Polym. Sci.*, **33**(126), 183–192.

Assoud, Lahcen, Ebert, Florian, Keim, Peter, Messina, René, Maret, Georg, and Löwen, Hartmut (2009). *Phys. Rev. Lett.*, **102**(23), 238301.

Auer, Stefan and Frenkel, Daan (2001, October). *Nature*, **413**(6857), 711–713.

Axelrod, D., Koppel, D., Schlessinger, J., Elson, E., and Webb, W. (1976, September). *Biophys. J.*, **16**(9), 1055–1069.

Ballesta, P., Besseling, R., Isa, L., Petekidis, G., and Poon, W. C. K. (2008*a*). *Phys. Rev. Lett.*, **101**(25), 258301.

Ballesta, Pierre, Duri, Agnes, and Cipelletti, Luca (2008*b*, June). *Nature Phys.*, **4**(7), 550–554.

Bandyopadhyay, R., Liang, D., Yardimci, H., Sessoms, D. A., Borthwick, M. A., Mochrie, S. G. J., Harden, J. L., and Leheny, R. L. (2004, Nov). *Phys. Rev. Lett.*, **93**(22), 228302.

Bartsch, E. (1995, December). *J. Non-Cryst. Solids*, **192-193**, 384–392.

Bartsch, E., Frenz, V., Baschnagel, J., Schärtl, W., and Sillescu, H. (1997). *J. Chem. Phys.*, **106**(9), 3743–3756.

Bartsch, E., Frenz, V., Moller, S., and Sillescu, H. (1993, December). *Physica A*, **201** (1-3), 363–371.

Beenakker, C. and Mazur, P. (1983, July). *Physica A: Stat. Theor. Phys.*, **120**(3), 388–410.

Bernal, J. D. (1964, July). *Proc. Roy. Soc. London. Series A*, **280**(1382), 299–322.

Berne, B. J. and Pecora, R. (1976). *Dynamic Light Scattering.* Wiley, New York.

Berret, Jean F., Porte, Grégoire, and Decruppe, Jean P. (1997, Feb). *Phys. Rev. E*, **55**(2), 1668–1676.

Berthier, L., Biroli, G., Bouchaud, J. P., Cipelletti, L., Masri, D. El, L'Hote, D., Ladieu, F., and Pierno, M. (2005, December). *Science*, **310**(5755), 1797–1800.

Berthier, L., Biroli, G., Bouchaud, J. P., Kob, W., Miyazaki, K., and Reichman, D. R. (2007*a*). *J. Chem. Phys.*, **126**(18), 184504.

Berthier, L., Biroli, G., Bouchaud, J. P., Kob, W., Miyazaki, K., and Reichman, D. R. (2007*b*). *J. Chem. Phys.*, **126**(18), 184503.

Berthier, L. and Tarjus, G. (2009, Oct). *Phys. Rev. Lett.*, **103**(17), 170601+.

Berthier, Ludovic and Witten, Thomas A. (2009, Aug). *Phys. Rev. E*, **80**(2), 021502.

Besseling, R., Isa, L., Weeks, E. R., and Poon, W. C. K. (2009, February). *Adv. Colloid Interf. Sci.*, **146**(1-2), 1–17.

Besseling, R., Weeks, Eric R., Schofield, A. B., and Poon, W. C. K. (2007). *Phys. Rev. Lett.*, **99**(2), 028301.

Biroli, G. and Bouchaud, J. P. (2004). *Europhys. Lett.*, **67**(1), 21–27.

Biroli, G., Bouchaud, J. P., Cavagna, A., Grigera, T. S., and Verrocchio, P. (2008, August). *Nature Phys.*, **4**(10), 771–775.

Bolhuis, Peter G. and Kofke, David A. (1996, Jul). *Phys. Rev. E*, **54**(1), 634–643.

Bouchaud, J. P. and Pitard, E. (2002, November). *Eur. Phys. J. E - Soft Matter*, **9**(3), 287–291.

Brambilla, G., El Masri, D E. M., Pierno, M., Berthier, L., Cipelletti, L., Petekidis, G., and Schofield, A. B. (2009, Feb). *Phys. Rev. Lett.*, **102**(8), 085703.

Brambilla, G., Masri, D. El, Pierno, M., Berthier, L., Cipelletti, L., Petekidis, G., and Schofield, A. (2010, Apr). *Phys. Rev. Lett.*, **104**(16), 169602.

Buitenhuis, Johan and Förster, Stephan (1997). *J. Chem. Phys.*, **107**(1), 262–272.

Campbell, Andrew I., Anderson, Valerie J., van Duijneveldt, Jeroen S., and Bartlett, Paul (2005, May). *Phys. Rev. Lett.*, **94**(20), 208301.

Cardenas, Miguel, Franz, Silvio, and Parisi, Giorgio (1999). *J. Chem. Phys.*, **110**(3), 1726–1734.

Caronna, Chiara, Chushkin, Yuriy, Madsen, Anders, and Cupane, Antonio (2008, Feb). *Phys. Rev. Lett.*, **100**(5), 055702.

Castillo, H. E. and Parsaeian, A. (2007). *Nature Phys.*, **3**(1), 26–28.

Cates, M. E. and Fielding, S. M. (2006). *Adv. Phys.*, **55**(7/8).

Cerbino, Roberto and Trappe, Veronique (2008, May). *Phys. Rev. Lett.*, **100**(18), 188102.

Charbonneau, P. and Reichman, D. R. (2007, Sep). *Phys. Rev. Lett.*, **99**(13), 135701.

Chaudhuri, Pinaki, Berthier, Ludovic, and Sastry, Srikanth (2010, Apr). *Phys. Rev. Lett.*, **104**(16), 165701.

Chaudhuri, Pinaki, Gao, Yongxiang, Berthier, Ludovic, Kilfoil, Maria, and Kob, Walter (2008). *J. Phys.: Condens. Matter*, **20**(24), 244126.

Chen, Dandan, Semwogerere, Denis, Sato, Jun, Breedveld, Victor, and Weeks, Eric R. (2010, Jan). *Phys. Rev. E*, **81**(1), 011403.

Cheng, Zhengdong, Zhu, Jixiang, Chaikin, Paul M., Phan, See-Eng, and Russel, William B. (2002, April). *Phys. Rev. E*, **65**(4), 041405.

Chung, B., Ramakrishnan, S., Bandyopadhyay, R., Liang, D., Zukoski, C. F., Harden, J. L., and Leheny, R. L. (2006, Jun). *Phys. Rev. Lett.*, **96**(22), 228301.

Cianci, Gianguido C., Courtland, Rachel E., and Weeks, Eric R. (2006, September). *Solid State Comm.*, **139**(11–12), 599–604.

Cianci, Gianguido C. and Weeks, Eric R. (2007). In *Reports of the Institute of Fluid Science*, Volume 19, pp. 51–56. Tohoku University.

Cipelletti, Luca, Bissig, H., Trappe, V., Ballesta, P., and Mazoyer, S. (2003, December). *J. Phys.: Condens. Matter*, **15**, S257–S262.

Cipelletti, Luca, Manley, S., Ball, R. C., and Weitz, D. A. (2000, Mar). *Phys. Rev. Lett.*, **84**(10), 2275–2278.

Cipelletti, Luca and Ramos, Laurence (2005, February). *J. Phys.: Condens. Matter*, **17**(6), R253–R285.

Clusel, Maxime, Corwin, Eric I., Siemens, Alexander O. N., and Brujic, Jasna (2009, July). *Nature*, **460**(7255), 611–615.

Courtland, Rachel E. and Weeks, Eric R. (2003, January). *J. Phys.: Condens. Matter*, **15**(1), S359–S365.

Crocker, John C. and Grier, David G. (1996, April). *J. Colloid Interf. Sci.*, **179**(1), 298–310.

Dalle Ferrier, C., Thibierge, C., Simionesco, C. Alba, Berthier, L., Biroli, G., Bouchaud, J. P., Ladieu, F., L'Hôte, D., and Tarjus, G. (2007, Oct). *Phys. Rev. E*, **76**(4), 041510.

Dauchot, O., Marty, G., and Biroli, G. (2005, Dec). *Phys. Rev. Lett.*, **95**(26), 265701.
de Schepper, I. M., Cohen, E. G. D., and Verberg, R. (1996, July). *Phys. Rev. Lett.*, **77**(3), 584.
Debrégeas, G., Tabuteau, H., and di Meglio, J. M. (2001, October). *Phys. Rev. Lett.*, **87**(17), 178305.
Dhont, Jan and Briels, Wim (2008, April). *Rheologica Acta*, **47**(3), 257–281.
Dhont, Jan K. G. (1999, Oct). *Phys. Rev. E*, **60**(4), 4534–4544.
Dhont, Jan K. G., Lettinga, M. Pavlik, Dogic, Zvonimir, Lenstra, Tjerk A. J., Wang, Hao, Rathgeber, Silke, Carletto, Philippe, Willner, Lutz, Frielinghaus, Henrich, and Lindner, Peter (2003). *Faraday Disc.*, **123**, 157–172.
Dibble, Clare J., Kogan, Michael, and Solomon, Michael J. (2006, Oct). *Phys. Rev. E*, **74**(4), 041403.
Dibble, Clare J., Kogan, Michael, and Solomon, Michael J. (2008, May). *Phys. Rev. E*, **77**(5), 050401.
Dinsmore, A. D., Weeks, E. R., Prasad, V., Levitt, A. C., and Weitz, D. A. (2001). *App. Opt.*, **40**(24), 4152–4159.
Doliwa, B. and Heuer, A. (1998, June). *Phys. Rev. Lett.*, **80**(22), 4915–4918.
Doliwa, B. and Heuer, A. (2000, Jun). *Phys. Rev. E*, **61**(6), 6898–6908.
Donati, Claudio, Douglas, Jack F., Kob, Walter, Plimpton, Steven J., Poole, Peter H., and Glotzer, Sharon C. (1998, Mar). *Phys. Rev. Lett.*, **80**(11), 2338–2341.
Donati, Claudio, Glotzer, Sharon C., and Poole, Peter H. (1999, Jun). *Phys. Rev. Lett.*, **82**(25), 5064–5067.
Donev, Aleksandar, Stillinger, Frank H., and Torquato, Salvatore (2007). *J. Chem. Phys.*, **127**(12), 124509.
Donev, Aleksandar, Torquato, Salvatore, Stillinger, Frank H., and Connelly, Robert (2004, Oct). *Phys. Rev. E*, **70**(4), 043301.
Donth, E. (2001). *The Glass Transition*. Springer, Berlin.
Duri, A., Autenrieth, T., Stadler, L. M., Leupold, O., Chushkin, Y., Grübel, G., and Gutt, C. (2009*a*, Apr). *Phys. Rev. Lett.*, **102**(14), 145701.
Duri, Agnès, Bissig, Hugo, Trappe, Véronique, and Cipelletti, Luca (2005, Nov). *Phys. Rev. E*, **72**(5), 051401.
Duri, A. and Cipelletti, L. (2006). *Europhys. Lett.*, **76**(5), 972–978.
Duri, A., Sessoms, D. A., Trappe, V., and Cipelletti, L. (2009*b*, Feb). *Phys. Rev. Lett.*, **102**(8), 085702.
Durian, D. J., Weitz, D. A., and Pine, D. J. (1991, May). *Science*, **252**(5006), 686–688.
Ebert, F., Dillmann, P., Maret, G., and Keim, P. (2009). *Rev. Sci. Instrum.*, **80**(8), 083902.
Eckert, T. and Bartsch, E. (2002, Aug). *Phys. Rev. Lett.*, **89**(12), 125701.
Ediger, M. D. (2000). *Annu. Rev. Phys. Chem.*, **51**(1), 99–128.
Ediger, M. D., Angell, C. A., and Nagel, S. R. (1996, January). *J. Phys. Chem.*, **100**(31), 13200–13212.
Einstein, A. (1905). *Annalen der Physik (Leipzig)*, **17**, 549–560.
El Masri, D., Brambilla, G., Pierno, M., Petekidis, G., Schofield, A. B., Berthier, L., and Cipelletti, L. (2009, July). *J. Stat. Mech.: Theory Exp.*, **2009**(07), P07015.

Ernst, Richard M., Nagel, Sidney R., and Grest, Gary S. (1991, April). *Phys. Rev. B*, **43**(10), 8070–8080.

Falk, M. L. and Langer, J. S. (1998, Jun). *Phys. Rev. E*, **57**(6), 7192–7205.

Fasolo, Moreno and Sollich, Peter (2004, Oct). *Phys. Rev. E*, **70**(4), 041410.

Fielding, Suzanne M. (2007). *Soft Matter*, **3**(10), 1262–1279.

Fielding, S. M., Cates, M. E., and Sollich, P. (2009). *Soft Matter*, **5**(12), 2378–2382.

Fierro, A., Gado, E. Del, Candia, A. De, and Coniglio, A. (2008, April). *J. Stat. Mech.: Theory Exp.*, **2008**(04), L04002.

Furukawa, Akira, Kim, Kang, Saito, Shinji, and Tanaka, Hajime (2009, Jan). *Phys. Rev. Lett.*, **102**(1), 016001.

Gao, Y. and Kilfoil, M. L. (2007, Aug). *Phys. Rev. Lett.*, **99**(7), 078301.

Gasser, U., Weeks, Eric R., Schofield, Andrew, Pusey, P. N., and Weitz, D. A. (2001, April). *Science*, **292**(5515), 258–262.

Gleim, Tobias, Kob, Walter, and Binder, Kurt (1998, November). *Phys. Rev. Lett.*, **81**(20), 4404–4407.

Glotzer, Sharon C. (2000, September). *J. Non-Cryst. Solids*, **274**(1-3), 342–355.

Goodman, J. W. (2007). *Speckle Phenomena in Optics: Theory and Applications*. Roberts and Company, Englewood Greenwood Village.

Gotze, Wolfgang (1999, March). *J. Phys.: Condens. Matter*, **11**(10A), A1–A45.

Götze, W. and Sjögren, L. (1991, May). *Phys. Rev. A*, **43**(10), 5442–5448.

Götze, W. and Sjogren, L. (1992). *Rep. Prog. Phys.*, **55**(3), 241–376.

Grier, David G. (2003, August). *Nature*, **424**(6950), 810–816.

Guo, Hongyu, Bourret, Gilles, Corbierre, Muriel K., Rucareanu, Simona, Lennox, R. Bruce, Laaziri, Khalid, Piche, Luc, Sutton, Mark, Harden, James L., and Leheny, Robert L. (2009, Feb). *Phys. Rev. Lett.*, **102**(7), 075702.

Habdas, P., Schaar, D., Levitt, A. C., and Weeks, E. R. (2004). *Europhys. Lett.*, **67**(3), 477–483.

Hecksher, Tina, Nielsen, Albena I., Olsen, Niels B., and Dyre, Jeppe C. (2008). *Nature Phys.*, **4**(9), 737–741.

Henderson, S. I. and van Megen, W. (1998, January). *Phys. Rev. Lett.*, **80**(4), 877–880.

Hernández-Guzmán, Jessica and Weeks, Eric R. (2009, September). *Proc. Nat. Acad. Sci.*, **106**(36), 15198–15202.

Herzig, E. M., Robert, A., van 't Zand, Cipelletti, L., Pusey, P. N., and Clegg, P. S. (2009, Jan). *Phys. Rev. E*, **79**(1), 011405.

Hodge, Ian M. (1995, March). *Science*, **267**(5206), 1945–1947.

Höfling, Felix, Munk, Tobias, Frey, Erwin, and Franosch, Thomas (2008). *J. Chem. Phys.*, **128**(16), 164517.

Hoover, William G. and Ree, Francis H. (1968). *J. Chem. Phys.*, **49**(8), 3609–3617.

Hurley, M. M. and Harrowell, Peter (1996). *J. Chem. Phys.*, **105**(23), 10521–10526.

Inoué, Shinya and Spring, Kenneth R. (1997, August). *Video Microscopy: The Fundamentals (The Language of Science)* (2nd edn). Springer, Berlin.

Janiaud, E., Weaire, D., and Hutzler, S. (2006, Jul). *Phys. Rev. Lett.*, **97**(3), 038302.

Kamien, Randall D. and Liu, Andrea J. (2007, Oct). *Phys. Rev. Lett.*, **99**(15), 155501.

Kasper, A., Bartsch, E., and Sillescu, H. (1998, September). *Langmuir*, **14**(18), 5004–5010.

Kaufman, Laura J. and Weitz, David A. (2006). *J. Chem. Phys.*, **125**(7), 074716.
Kegel, W. K. (2000). *Langmuir*, **16**(3), 939–941.
Kegel, Willem K. and van Blaaderen, Alfons (2000, January). *Science*, **287**(5451), 290–293.
Kob, Walter, Donati, Claudio, Plimpton, Steven J., Poole, Peter H., and Glotzer, Sharon C. (1997, Oct). *Phys. Rev. Lett.*, **79**(15), 2827–2830.
König, H., Hund, R., Zahn, K., and Maret, G. (2005, November). *Euro. Phys. J. E*, **18**(3), 287–293.
Krzakala, Florent and Kurchan, Jorge (2007, Aug). *Phys. Rev. E*, **76**(2), 021122.
Lauridsen, John, Chanan, Gregory, and Dennin, Michael (2004, Jul). *Phys. Rev. Lett.*, **93**(1), 018303.
Lechenault, F., Dauchot, O., Biroli, G., and Bouchaud, J. P. (2008). *Europhys. Lett.*, **83**(4), 46003.
Liu, Andrea J. and Nagel, Sidney R. (1998, November). *Nature*, **396**(6706), 21–22.
Liu, Cliff Z. and Oppenheim, Irwin (1996, January). *Phys. Rev. E*, **53**(1), 799–802.
Losert, W., Bocquet, L., Lubensky, T. C., and Gollub, J. P. (2000). *Phys. Rev. Lett.*, **85**(7), 1428–1431.
Lu, Peter J., Zaccarelli, Emanuela, Ciulla, Fabio, Schofield, Andrew B., Sciortino, Francesco, and Weitz, David A. (2008, May). *Nature*, **453**(7194), 499–503.
Lynch, Jennifer M., Cianci, Gianguido C., and Weeks, Eric R. (2008). *Phys. Rev. E*, **78**(3), 031410.
Maccarrone, S., Brambilla, G., Pravaz, O., Duri, A., Ciccotti, M., Fromental, J. M., Pashkovski, E., Lips, A., Sessoms, D., Trappe, V., and Cipelletti, L. (1995). *Soft Matter*, **6**, 5514.
Marcus, Andrew H., Schofield, Jeremy, and Rice, Stuart A. (1999, Nov). *Phys. Rev. E*, **60**(5), 5725–5736.
Mason, T. G. and Weitz, D. A. (1995, Oct). *Phys. Rev. Lett.*, **75**(14), 2770–2773.
Mattsson, Johan, Wyss, Hans M., Fernandez-Nieves, Alberto, Miyazaki, Kunimasa, Hu, Zhibing, Reichman, David R., and Weitz, David A. (2009*a*, November). *Nature*, **462**(7269), 83–86.
Mattsson, Johan, Wyss, Hans M., Fernandez-Nieves, Alberto, Miyazaki, Kunimasa, Hu, Zhibing, Reichman, David R., and Weitz, David A. (2009*b*, November). *Nature*, **462**(7269), 83–86.
Mayer, P., Bissig, H., Berthier, L., Cipelletti, L., Garrahan, J. P., Sollich, P., and Trappe, V. (2004, Sep). *Phys. Rev. Lett.*, **93**(11), 115701.
Mazoyer, S., Ebert, F., Maret, G., and Keim, P. (2009). *Europhys. Lett.*, **88**(6), 66004.
McKenna, Gregory B. (2003). *J. Phys.: Condens. Matter*, **15**(11), S737–S763.
Meller, Amit and Stavans, Joel (1992, June). *Phys. Rev. Lett.*, **68**(24), 3646–3649.
Mohraz, Ali, Weeks, Eric R., and Lewis, Jennifer A. (2008). *Phys. Rev. E*, **77**(6), 060403(R).
Moller, P. C. F., Rodts, S., Michels, M. A. J., and Bonn, Daniel (2008). *Phys. Rev. E*, **77**(4), 041507.
Ngai, K. L. (1999, December). *J. Phys. Chem. B*, **103**(48), 10684–10694.
Ngai, K. L. and Rendell, R. W. (1998, February). *Phil. Mag. B*, **77**(2), 621–631.

Nugent, Carolyn R., Edmond, Kazem V., Patel, Hetal N., and Weeks, Eric R. (2007). *Phys. Rev. Lett.*, **99**(2), 025702.

O'Hern, Corey S., Silbert, Leonardo E., Liu, Andrea J., and Nagel, Sidney R. (2003, Jul). *Phys. Rev. E*, **68**(1), 011306.

O'Hern, Corey S., Silbert, Leonardo E., Liu, Andrea J., and Nagel, Sidney R. (2004, Oct). *Phys. Rev. E*, **70**(4), 043302.

Olmsted, P. (1999, April). *Curr. Opin. Colloid Interf. Sci.*, **4**(2), 95–100.

Parisi, Giorgio and Zamponi, Francesco (2005). *J. Chem. Phys.*, **123**(14), 144501.

Perera, Donna N. and Harrowell, Peter (1999). *J. Chem. Phys.*, **111**(12), 5441–5454.

Pham, K. N., Puertas, A. M., Bergenholtz, J., Egelhaaf, S. U., Moussaid, A., Pusey, P. N., Schofield, A. B., Cates, M. E., Fuchs, M., and Poon, W. C. K. (2002, April). *Science*, **296**(5565), 104–106.

Phan, See-Eng, Russel, William B., Cheng, Zhengdong, Zhu, Jixiang, Chaikin, Paul M., Dunsmuir, John H., and Ottewill, Ronald H. (1996, Dec). *Phys. Rev. E*, **54**(6), 6633–6645.

Poon, W. C. K. (2002). *J. Phys.: Condens. Matter*, **14**(33), R859–R880.

Poon, Wilson C. K., Meeker, S. P., Pusey, P. N., and Segrè, P. N. (1996, November). *J. Non-Newton. Fluid Mech.*, **67**, 179–189.

Puertas, Antonio M., Fuchs, Matthias, and Cates, Michael E. (2004). *J. Chem. Phys.*, **121**(6), 2813–2822.

Pusey, P. (1999, June). *Curr. Opin. Colloid Interf. Sci.*, **4**(3), 177–185.

Pusey, P. N. (1978). *J. Phys. A: Math. General*, **11**(1), 119–135.

Pusey, Peter N. (1991), pp. 763–943. Les Houches session LI, North Holland, Amsterdam.

Pusey, P. N., Fijnaut, H. M., and Vrij, A. (1982). *J. Chem. Phys.*, **77**(9), 4270–4281.

Pusey, P. N. and van Megen, W. (1986, March). *Nature*, **320**(6060), 340–342.

Pusey, P. N. and van Megen, W. (1987, Nov). *Phys. Rev. Lett.*, **59**(18), 2083–2086.

Pusey, P. N., Zaccarelli, E., Valeriani, C., Sanz, E., Poon, Wilson C. K., and Cates, Michael E. (2009, December). *Phil. Trans. Roy. Soc. A*, **367**(1909), 4993–5011.

Richert, Ranko (2002). *J. Phys.: Condens. Matter*, **14**(23), R703–R738.

Robert, A., Wandersman, E., Dubois, E., Dupuis, V., and Perzynski, R. (2006). *Europhys. Lett.*, **75**(5), 764–770.

Roovers, Jacques (1994, September). *Macromolecules*, **27**(19), 5359–5364.

Roth, Connie B. and Dutcher, John R. (2005, October). *J. Electroanal. Chem.*, **584**(1), 13–22.

Royall, C. P., Leunissen, M. E., and Blaaderen, A. Van (2003). *J. Phys.: Condens. Matter*, **15**(48), S3581–S3596.

Russel, W. B., Saville, D. A., and Schowalter, W. R. (1992, March). *Colloidal Dispersions (Cambridge Monographs on Mechanics)*. Cambridge University Press, Cambridge.

Saltzman, Erica J. and Schweizer, Kenneth S. (2006*a*). *J. Chem. Phys.*, **125**(4), 044509.

Saltzman, Erica J. and Schweizer, Kenneth S. (2006*b*). *Phys. Rev. E*, **74**(6), 061501.

Schaertl, W. and Sillescu, H. (1994, December). *J. Stat. Phys.*, **77**(5), 1007–1025.

Schall, Peter, Weitz, David A., and Spaepen, Frans (2007, December). *Science*, **318**(5858), 1895–1899.

Schätzel, Klaus (1991). *J. Mod. Opt.*, **38**(9), 1849–1865.

Schope, H. J., Bryant, G., and van Megen, W. (2007). *J. Chem. Phys.*, **127**, 084505.

Schweizer, Kenneth S. and Saltzman, Erica J. (2003). *J. Chem. Phys.*, **119**(2), 1181–1196.

Schweizer, K. S. and Saltzman, E. J. (2004, December). *The J. Phys. Chem. B*, **108**(51), 19729–19741.

Segrè, P. N., Behrend, O. P., and Pusey, P. N. (1995*a*, Nov). *Phys. Rev. E*, **52**(5), 5070–5083.

Segrè, P. N., Meeker, S. P., Pusey, P. N., and Poon, W. C. K. (1995*b*, Jul). *Phys. Rev. Lett.*, **75**(5), 958–961.

Segrè, P. N., Meeker, S. P., Pusey, P. N., and Poon, W. C. K. (1995*c*, Jul). *Phys. Rev. Lett.*, **75**(5), 958–961.

Segrè, P. N., Meeker, S. P., Pusey, P. N., and Poon, W. C. K. (1996, July). *Phys. Rev. Lett.*, **77**(3), 585.

Senff, H., Richtering, W., Norhausen, Ch, Weiss, A., and Ballauff, M. (1999, January). *Langmuir*, **15**(1), 102–106.

Sessoms, David A., Bischofberger, Irmgard, Cipelletti, Luca, and Trappe, Véronique (2009, December). *Phil. Trans. Roy. Soc. A*, **367**(1909), 5013–5032.

Sillescu, Hans (1999, February). *J. Non-Cryst. Solids*, **243**(2-3), 81–108.

Simeonova, Nikoleta B. and Kegel, Willem K. (2004, Jul). *Phys. Rev. Lett.*, **93**(3), 035701.

Speedy, Robin J. (1998). *Molec. Phys.*, **95**(2), 169–178.

Stillinger, Frank H. (1995, March). *Science*, **267**(5206), 1935–1939.

Stober, W. (1968, January). *J. Colloid Interf. Sci.*, **26**(1), 62–69.

Sutherland, W. (1905). *Phil. Mag.*, **9**, 781–785.

Szamel, G. and Flenner, E. (2004, September). *Europhys. Lett.*, **67**(5), 779–785.

Tokuyama, Michio (2007, May). *Physica A*, **378**(2), 157–166.

Tokuyama, Michio and Oppenheim, Irwin (1994, Jul). *Phys. Rev. E*, **50**(1), R16–R19.

Tokuyama, Michio and Oppenheim, Irwin (1995, June). *Physica A*, **216**(1–2), 85–119.

Torquato, S., Truskett, T. M., and Debenedetti, P. G. (2000, Mar). *Phys. Rev. Lett.*, **84**(10), 2064–2067.

Trappe, V., Pitard, E., Ramos, L., Robert, A., Bissig, H., and Cipelletti, L. (2007, Nov). *Phys. Rev. E*, **76**(5), 051404.

Trappe, V. and Sandkuhler, P. (2004, April). *Curr. Opin. Colloid Interf. Sci.*, **8**(6), 494–500.

Utter, Brian and Behringer, R. P. (2008). *Phys. Rev. Lett.*, **100**(20), 208302.

van Blaaderen, A. and Wiltzius, P. (1995, November). *Science*, **270**, 1177–1179.

van Megen, W., Mortensen, T. C., Williams, S. R., and Müller, J. (1998, November). *Phys. Rev. E*, **58**(5), 6073–6085.

van Megen, W. and Pusey, P. N. (1991, May). *Phys. Rev. A*, **43**(10), 5429–5441.

van Megen, W and Underwood, S. M. (1989). *J. Chem. Phys.*, **91**, 552–559.

van Megen, W. and Underwood, S. M. (1993, Jan). *Phys. Rev. E*, **47**(1), 248–261.

van Megen, W. and Underwood, S. M. (1994, May). *Phys. Rev. E*, **49**(5), 4206–4220.

van Megen, W. and Williams, Stephen R. (2010, Apr). *Phys. Rev. Lett.*, **104**(16), 169601.

Vermant, J. (2001, November). *Curr. Opin. Colloid Interf. Sci.*, **6**(5-6), 489–495.

Vidal Russell, E. and Israeloff, N. E. (2000, December). *Nature*, **408**(6813), 695–698.

Vrij, A. (1976). *Pure Appl. Chem.*, **48**(4), 471–483.

Wandersman, E., Duri, A., Robert, A., Dubois, E., Dupuis, V., and Perzynski, R. (2008). *J. Phys.: Condens. Matter*, **20**(15), 155104.

Weeks, Eric R., Crocker, J. C., Levitt, Andrew C., Schofield, Andrew, and Weitz, D. A. (2000, January). *Science*, **287**(5453), 627–631.

Weeks, Eric R., Crocker, John C., and Weitz, D. A. (2007). *J. Phys.: Condens. Matter*, **19**(20), 205131.

Weeks, Eric R. and Weitz, D. A. (2002, Aug). *Phys. Rev. Lett.*, **89**(9), 095704.

Weitz, D. A. and Pine, D. J. (1993). In *Dynamic Light Scattering* (ed. W. Brown), pp. 652–720. Clarendon Press, Oxford.

Williams, S. R. and van Megen, W. (2001). *Phys. Rev. E*, **64**(4), 041502.

Wochner, Peter, Gutt, Christian, Autenrieth, Tina, Demmer, Thomas, Bugaev, Volodymyr, Ortiz, Alejandro D., Duri, Agnès, Zontone, Federico, Grübel, Gerhard, and Dosch, Helmut (2009, July). *Proc. Nat. Acad. Sci.*, **106**(28), 11511–11514.

Wong, Apollo P. Y. and Wiltzius, P. (1993). *Rev. Sci. Inst.*, **64**(9), 2547–2549.

Xu, Ning, Haxton, Thomas K., Liu, Andrea J., and Nagel, Sidney R. (2009, Dec). *Phys. Rev. Lett.*, **103**(24), 245701.

Yamamoto, R. and Onuki, A. (1997). *Europhys. Lett. (EPL)*, **40**(1), 61–66.

Yamamoto, Ryoichi and Onuki, Akira (1998). *Phys. Rev. E*, **58**(3), 3515–3529.

Yethiraj, Anand and van Blaaderen, Alfons (2003, January). *Nature*, **421**(6922), 513–517.

Yunker, Peter, Zhang, Zexin, Aptowicz, Kevin, and Yodh, A. (2009, Sep). *Phys. Rev. Lett.*, **103**(11), 115701.

Zaccarelli, Emanuela (2007, August). *J. Phys.: Condens. Matter*, **19**(32), 323101+.

Zaccarelli, E., Valeriani, C., Sanz, E., Poon, W. C. K., Cates, M. E., and Pusey, P. N. (2009*a*, Sep). *Phys. Rev. Lett.*, **103**(13), 135704.

Zaccarelli, E., Valeriani, C., Sanz, E., Poon, W. C. K., Cates, M. E., and Pusey, P. N. (2009*b*, Sep). *Phys. Rev. Lett.*, **103**(13), 135704.

Zhu, Jixiang, Li, Min, Rogers, R., Meyer, W., Ottewill, R. H., Sts-73, Russel, W. B., and Chaikin, P. M. (1997, June). *Nature*, **387**(6636), 883–885.

5

Experimental approaches to heterogeneous dynamics

Ranko RICHERT, Nathan ISRAELOFF,
Christiane ALBA-SIMIONESCO,
François LADIEU and Denis L'HÔTE

Abstract

Dynamic heterogeneity refers to the independence of fast and slow modes in a system displaying dispersive relaxation and is not an alternative model to the dynamics in complex materials, but rather a required approach for rationalizing an increasing number of experimental results on supercooled liquids. The last twenty years has seen the advent of numerous experimental techniques aimed at studying the heterogeneous nature of viscous liquids. The methods either demonstrate heterogeneity *per se*, or provide information on the time- and length scale involved in this concept. This chapter reviews the experimental techniques that focus on molecular and polymeric glass formers, the challenges involved, and the results obtained in this relatively new field. Most approaches have particular strengths and weaknesses, so that only the synergistic efforts of combining numerous approaches will advance our understanding of the heterogeneous nature of dynamics significantly.

5.1 Introduction

(R. RICHERT)
Practically all relaxation phenomena in disordered materials are non-exponential regarding the time dependence with which they approach equilibrium after some perturbation (Richert and Blumen, 1994; Jäckle, 1986). Deviations from an exponential time-correlation function are particularly obvious for supercooled liquids, i.e. glass-forming liquids at temperatures somewhere between their melting temperature T_{m} and glass-transition temperature T_{g} (Ediger *et al.*, 1996; Angell *et al.*, 2000). In this range, the average relaxation times are in the 100 ns to 100 s range, and the viscosities

are accordingly high. The recurring observation of dispersive relaxation patterns has led to a variety of attempts to describe the time-correlation function in question, $\phi(t)$. Gaussian distributions of logarithmic relaxation times (Richert, 1985), algebraic waiting-time distributions (Phillips, 1996), laws based upon fractal dimensionalities (Jurjiu *et al.*, 2002), and fractional differential equation approaches (Sokolov *et al.*, 2002) are among the models employed for rationalizing non-exponential decays. In the time domain, the most widely used relaxation function is the Kohlrausch–Williams–Watts (KWW) or stretched exponential decay (Kohlrausch, 1854; Williams and Watts, 1970),

$$\phi(t) = \phi_0 \exp\left[-\left(\frac{t}{\tau_{\mathrm{KWW}}}\right)^{\beta_{\mathrm{KWW}}}\right] . \tag{5.1}$$

Here, a single parameter β with $0 < \beta \leq 1$ captures the dispersive character of $\phi(t)$. It turned out that this empirical approach describes a large number of observations, regardless of the chemical nature of the sample and of the experimental technique used for measuring $\phi(t)$ (Fourkas *et al.*, 1997). Values of β approaching unity imply that a single relaxation time constant τ is almost sufficient for describing the correlation function, whereas low values of β indicate that there can be orders of magnitude separating the slowest and fastest τ involved.

The degree of stretching often correlates with other properties of the supercooled liquid. The deviation from an exponential relaxation pattern has been found to correlate with the fragility m, a quantity used to gauge the deviation from an Arrhenius-type activation trace (Angell, 1991),

$$m = \left.\frac{\partial \log_{10} < \tau >}{\partial (T_{\mathrm{g}}/T)}\right|_{T=T_{\mathrm{g}}} . \tag{5.2}$$

However, the results will depend on the metric used for the relaxation-time dispersion (Nielsen *et al.*, 2009). According to Böhmer *et al.* (Böhmer *et al.*, 1993), materials with high fragility m are commonly those with low values of β, i.e. a strongly non-Arrhenius temperature dependence would indicate a pronounced relaxation-time dispersion. The fragile (opposed to "strong") cases are also known to exhibit the more pronounced secondary relaxations of the Johari–Goldstein type (Johari and Goldstein, 1970) and translation–rotation decoupling in the viscous regime (Ngai, 1999).

With the experimental validation of the stretched exponential patterns with $0.3 < \beta < 0.9$ as an almost universal phenomenon in liquids, it is tempting to seek a rationale for the non-exponential behavior. According to the work of Palmer, Stein, Abrahams, and Anderson (Palmer *et al.*, 1984), the simplest route to non-exponential dynamics is to "postulate a statistical distribution of relaxation times τ across different atoms, clusters, or degrees of freedom". Then, the correlation function would take the form

$$\phi(t) = \phi_0 \int_0^{\infty} g(\tau) \exp -(t/\tau) \mathrm{d}\tau = \phi_0 \sum_{i=0}^{N} g_i \exp -(t/\tau_i) , \tag{5.3}$$

with an appropriate probability density of relaxation times, $g(\tau)$, or weights, g_i. Palmer *et al.* dismiss this picture saying "However, this approach is microscopically arbitrary, and does not explain the universality of Kohlrausch's law. It is also normally associated with a picture of *parallel* relaxation, in which each degree of freedom x_i relaxes independently with characteristic time τ_i", and a serial relaxation scheme is proposed instead. The parallel scheme is now referred to as heterogeneous relaxation, whereas the serial picture is a realization of a homogeneous scenario. As has been emphasized previously (Richert and Blumen, 1994; Richert, 1993) and indicated in Fig. 5.1, both views are equally capable of justifying a non-exponential two-time correlation function in the regime of linear responses. Therefore, more sophisticated experimental tools are required for a decisive answer. The aim of the following sections is to provide an overview of the techniques used to discriminate heterogeneous from homogeneous dynamics and provide details on the length- and time scales involved in this issue.

It should be noted that the concept of heterogeneity is particularly non-trivial for single-component liquids, where perhaps the chemical uniformity has led to anticipating homogeneous dynamics. In the case of binary systems or more complicated

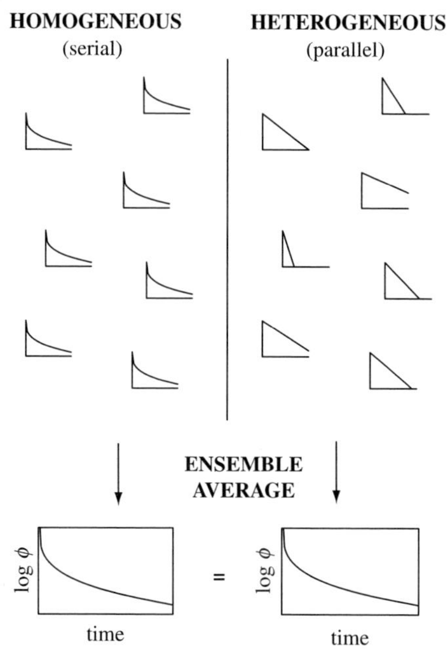

Fig. 5.1 Comparison of two extremes regarding the nature of dispersive relaxations. Left: Homogeneous or serial case, where each relaxing unit equally displays the non-exponential relaxation pattern. Right: Heterogeneous or parallel scenario, in which each domain relaxes independently and exponentially with its distinct relaxation time constant. By virtue of the ensemble average inevitable in most experiments, the resulting observed two-point correlation functions in the linear response regime will be identical for the two different pictures of microscopic relaxation behavior.

compositions, dynamic heterogeneity is readily expected on the basis of concentration fluctuations (Duvvuri and Richert, 2004). The origin of heterogeneity for simple neat molecular liquids is considered a much less obvious feature, as emphasized in these reviews (Ediger, 2000; Böhmer, 1998*a*,*b*; Richert, 2002; Sillescu, 1999; Glotzer, 2000).

For situations in which the relaxation is entirely characterized by a single time constant, i.e. exponential or Debye-type dynamics, the question of heterogeneous versus homogeneous dynamics ceases to make sense. Liquids tend to approach Debye behavior at sufficiently high temperatures, where the relaxation time falls below a threshold of around 1 ps (Wang and Richert, 2007*a*). Single-exponential relaxation is also observed for some liquids in their viscous regime, either the very strong cases or, more notably, the dielectric relaxation of many monohydroxy alcohols and several other associating liquids (Hansen *et al.*, 1997; Murthy and Tyagi, 2002; Wang and Richert, 2004*a*; Huth *et al.*, 2007).

5.2 Techniques based on spectral selectivity

Many techniques aimed at approaching the nature of relaxations, including the original clear cut evidence for heterogeneous dynamics based on nuclear magnetic resonance (NMR), involve spectral selectivity. If a subensemble regarding the time constants (e.g. the slower or faster than average modes) can be modified without affecting the remainder of the spectral range equally, then some degree of independence among the modes and thus heterogeneity is demonstrated (Böhmer *et al.*, 1998*a*). The nature of the spectrally selective modification will vary with the experimental technique used. However, common to the experiments outlined in this section is that they do not allow us to derive a length scale of heterogeneity.

A note of caution applies to relaxation measurements involving pulse excitation of a finite duration instead of a step in the generalized force. While shorter excitation pulse durations will lead to faster response curves, this is not a signature of spectral selectivity. Instead, Boltzmann's superposition principle (i.e. linear response) is all that is required to account for such observations (Böhmer *et al.*, 1995).

5.2.1 Nuclear magnetic resonance

(R. RICHERT)

The first unambiguous evidence for heterogeneous dynamics is based upon reduced 4-dimensional NMR detection of segmental reorientation in poly(vinylacetate) at a temperature $T_g + 20$ K (Schmidt-Rohr and Spiess, 1991). The orientation of probe nuclei, e.g. ^{2}H or ^{13}C, is detected via their Larmor frequency and its dependence on the orientation of the molecule with respect to an external homogeneous magnetic field. In a two-time stimulated-echo experiment, a pair of radio-frequency (RF) pulses with time separation t_p gives rise to an echo signal for a time denoted t. See Fig. 5.2 for a schematic outline of this technique. At the end of this mixing period a third RF pulse leads to a reversal of the phase evolution and after the time t_p an echo is recorded. The amplitude F_2 of this echo signal is proportional to the number of spins that have not changed orientation by a certain amount during the mixing time

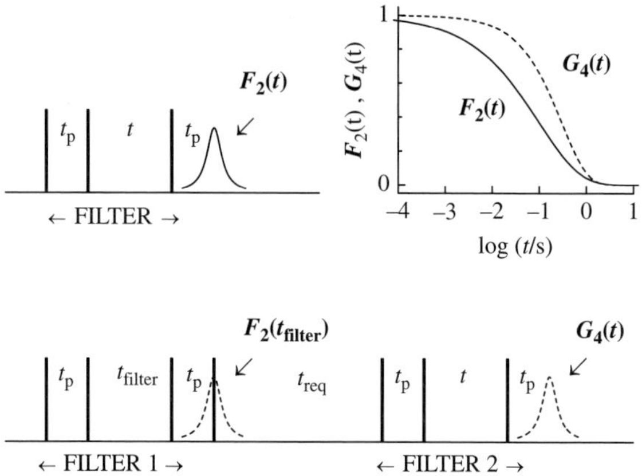

Fig. 5.2 Illustration of the NMR spin-echo technique that facilitates spectral filtering. The shorter pulse sequence leading to $F_2(t)$ acts as a filter, where the echo amplitude represents the fraction of spins that have not reoriented within the time interval t. In the longer sequence, the same type of filter is used to prepare a subensemble of spins that have not moved within the time t_{filter}, and the remaining sequence (filter 2) measures the reorientation of those spins that have passed filter 1. As indicated in the graph, $G_4(t)$ is found to be slower and more exponential than $F_2(t)$.

t. Thereby, $F_2(t)$ can be interpreted as a two-point orientation correlation function of segmental dynamics. The spins contributing to the echo are those that have not lost a certain amount of correlation during t, and are thus selectively the slower than average ones.

The higher-dimensional NMR techniques exploit the fact that the spins selected by the above method are now available for further investigations. The sequence leading to F_2 is now used as a filter with efficiency $\text{FE} = 1 - F_2(t_{\text{filter}})$, i.e. variation of t_{filter} will determine the fraction of slower modes that will enter the remainder of the sequence. After a re-equilibration interval of duration t_{req}, a further pulse sequence measures the orientation correlation function $F_4(t)$ of the fraction FE that have been slowest prior to t_{req}. For the special case $t_{\text{req}} = 0$, $F_4(t)$ is denoted $G_4(t)$. In the case of heterogeneous dynamics, this subensemble is expected to display slower and more exponential relaxation, which is what is being indicated in Fig. 5.2 and observed in real systems. Details of the technique are provided elsewhere (Böhmer *et al.*, 2001; Schmidt-Rohr and Spiess, 1994; Heuer *et al.*, 1995; Kuebler *et al.*, 1997). Variation of t_{req} (for a fixed $t = t_{\text{filter}}$) facilitates measuring the exchange time τ_{ex}. This exchange time or "heterogeneity lifetime" is the time it takes for a slow mode to attain an average relaxation time constant again, eventually leading to $F_4(t)$ approaching $F_2(t)$ for a sufficiently long t_{req}.

Based on quantitative analyses of data from the above techniques, the main results regarding molecular liquids and the segmental motion in polymers are as follows: orientational motion on time scales in the 1 to 100 ms range is heterogeneous in the sense that the overall dispersion originates from a superposition of distinct exponential modes. The exchange time for these modes is practically equal to the average relaxation time of the selected slower subensemble, as expressed by the ratio (Heuer, 1997),

$$Q = \frac{\tau_{\text{ex}}}{\tau_\alpha}, \tag{5.4}$$

with $Q \approx 1$. This result had been obtained for a variety of polymeric and molecular systems (Kuebler *et al.*, 1997; Böhmer *et al.*, 1996, 1998*b*; Heuer *et al.*, 1999; Hinze, 1998; Qi *et al.*, 2000, 2003).

5.2.2 Deep photobleaching

(R. RICHERT)
Optical anisotropy of chromophores in solution is a typical approach to the dynamics of a liquid (Horng *et al.*, 1997). What is actually measured is the single-particle rotation correlation function of the solute, usually present at very low concentration. Typically, a polarized laser pulse generates an optical anisotropy regarding the excited molecules, and the emission intensity is monitored at parallel and perpendicular angles relative to the excitation. From such data, the second-order projection $P_2(\cos\theta)$ of the three-dimensional isotropic rotational motion can be derived (Nickel, 1989). Naturally, such measurements are limited by the lifetime of the excited state. Ediger and coworkers have devised a photobleaching technique that facilitates the measurement of probe rotation at much longer times, aimed at matching the time scale of a supercooled liquid or polymer in the viscous regime near T_g (Cicerone and Ediger, 1993). Here, probe molecules have been carefully selected so that their rotation times and dispersion are practically equal to the structural (α) relaxation of the solvent (Cicerone *et al.*, 1995*a*).

The technique of photobleaching uses a polarized laser to remove a subset of chromophores by photo-oxidation. Initially, chromophores are isotropically oriented, but the polarized beam will preferentially bleach those chromophores whose transition dipoles for absorption are parallel to the polarization direction of the laser beam, leaving an anisotropic ensemble of chromophores. The return to an isotropic state can be monitored by optical spectroscopy without time limit. Bleaching a small fraction (few per cent) of the probes leads to the equivalent of a two-point correlation function for the probe rotation. Spectral selectivity is introduced by performing a "deep bleach", where about half of the probes are undergoing a photochemical change (Cicerone and Ediger, 1995). In this case, it turns out that the probes surviving such a deep bleach are somewhat slower than the original ensemble, and several explanations for this spectral selectivity have been offered (Cicerone and Ediger, 1995). Regardless of the origin of this bleaching selectivity, the observation that spectral selectivity is possible is an indication of heterogeneity.

Following a deep bleach, a sequence of low-depth photobleaching experiments can serve to measure how the slower subensemble reverts back to the original distribution of rotation times that was present prior to the deep bleach. This sequence of measurements is illustrated in Fig. 5.3. The main observation in these experiments is that the

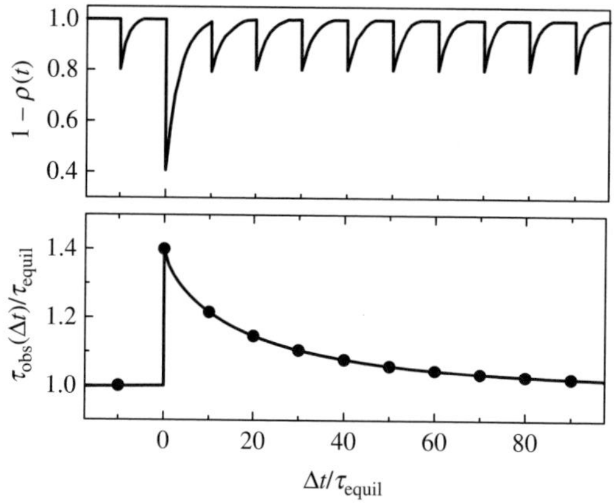

Fig. 5.3 Schematic illustration of the deep photobleaching technique. Each dip in the top panel indicates a reduction in isotropic alignment by photobleaching. Each recovery of isotropy defines the observed time scale τ_{obs} of reorientation. The deep bleach at time zero eliminates the faster probes from the equilibrium ensemble. According to the lower graph, it takes long compared with the equilibrium reorientation time τ_{equil} to revert to the equilibrium state in which $\tau_{obs} = \tau_{equil}$.

time required to return to the overall equilibrium behavior is longer that the average time scale of probe rotation itself. For tetracene in *o*-terphenyl, such measurements around $T_g = 243$ K led to the following values for the exchange to relaxation time ratio Q: $\tau_{ex} = 6.5 \times \tau_\alpha$ for $\tau_\alpha = 13$ s at $T_g + 4$ K and $\tau_{ex} = 540 \times \tau_\alpha$ for $\tau_\alpha = 125$ s at $T_g + 1$ K, where τ_α ($= \tau_{equil}$) reflects the equilibrium average rotation time of the probe (Wang and Ediger, 1999). Similar behavior has been reported for polystyrene (Wang and Ediger, 2000). According to photobleaching results, the values of Q are not only much larger compared with the NMR results, but also display a pronounced temperature dependence in this very viscous regime that is not accessible by but consistent with the 4D-NMR technique.

5.2.3 Hole-burning techniques

(R. RICHERT)
In (non-resonant) hole-burning experiments, modes in a certain spectral range are modified regarding their relaxation time constant rather than removing them from the ensemble (Schiener *et al.*, 1996). These modifications are the result of energy being absorbed from an external time-dependent field of sufficient amplitude. Electric (Schiener *et al.*, 1997; Richert and Böhmer, 1999; Duvvuri and Richert, 2003; Richert, 2001*a*; Kircher *et al.*, 2001), magnetic (Chamberlin, 1999), as well as mechanical shear (Shi and McKenna, 2005) fields have been used. The frequencies applied in the hole-burning range from mHz to kHz, implying that structural modes (opposed to vibrational modes) are absorbing the energy. The thermal coupling of these slow degrees of freedom to the heat bath is many orders of magnitude below that of thermal

conductivity (Chamberlin *et al.*, 1997). Therefore, the effect of this energy absorption can be expected to differ qualitatively from a simple increase in temperature. We shall describe the hole-burning technique in terms of the dielectric variant, because it is the most widely used approach. The magnetic and mechanical counterpart experiments are performed in an analogous fashion.

The basic idea of the technique is to transfer energy to a subensemble of relaxation times and thereby modify their response behavior (Schiener *et al.*, 1996). To this end, a large time-dependent electric field is applied to the sample and energy is transferred irreversibly from the field to those modes whose loss profile overlaps significantly with the power spectrum of the field. The highest spectral selection is expected for a narrow power spectrum, and thus a sinusoidal field is employed in most cases. Such a field is applied for a few cycles, followed by a variable waiting time after which a field or charge step is applied. The step is designed to be well within the linear response regime and is used to probe the polarization response in the time domain. The polarization observed after the step consists of a response to the step and an often larger component originating from the response to the preceding high-amplitude sine wave. In order to eliminate the latter contribution, a field cycle is performed in which the signs of the sine wave and/or the step are altered and the pure step response is extracted from combining the individual results. In the case of a field step, the dielectric displacement is recorded, which is proportional to the dielectric function $\epsilon(t)$ (Schiener *et al.*, 1997). If a charge step is used instead, the electric field is recorded, which is proportional to the electric modulus $M(t)$ (Wagner and Richert, 1997). The relation between the two quantities is given by their steady-state equation $M = 1/\epsilon$ (Howell *et al.*, 1974), and the decay of $M(t)$ always faster than its $\epsilon(t)$ counterpart (Jäckle and Richert, 2008).

The typical variables in this technique are the amplitude E_0, frequency ν, and number of cycles n of the field applied, the waiting time t_{wait} between pump and probe process. The aim is to compare two polarization traces $\varphi(t)$ and $\varphi^\star(t)$ obtained without and with the preceding large-amplitude burn field ($\varphi = \epsilon$ or $\varphi = M$). As depicted in Fig. 5.4, homogeneous dynamics would result if all molecules absorbed energy equally and the effect of the high field were the same as an increase in temperature. In such a case, the result would be a uniform shift of the correlation function along the $\log t$ scale (assuming time–temperature superposition for these small changes in the net T). If, however, modes are independent, as in the heterogeneous situation, only a certain range of relaxation time constants will shift towards lower values, and the modification of the polarization response is limited on the time scale. The differences between the high-field modified $\varphi^\star(t)$ traces and the equilibrium $\varphi(t)$ response are strongly exaggerated in Fig. 5.4 and usually seen only if the difference $\Delta_V\varphi(t) = \varphi^\star(t) - \varphi(t)$ is examined. As is easily seen from the curves in Fig. 5.4, this "vertical" difference will display a peaked curve in both the homogeneous and heterogeneous scenario. More decisive is the "horizontal" difference $\Delta_H\varphi(t)$, usually calculated as $\Delta_V\varphi(t)/(-d\varphi(t)/d\ln t)$. As shown in the inset, the $\Delta_H\varphi(t)$ curves now differ qualitatively, with a peak appearing only in the case of heterogeneous dynamics. The quantity $\Delta_H\varphi(t)$ can be viewed as a gauge for the extent of shifts in $\ln\tau$, and translates into an effective temperature scale for the modes affected by energy absorption.

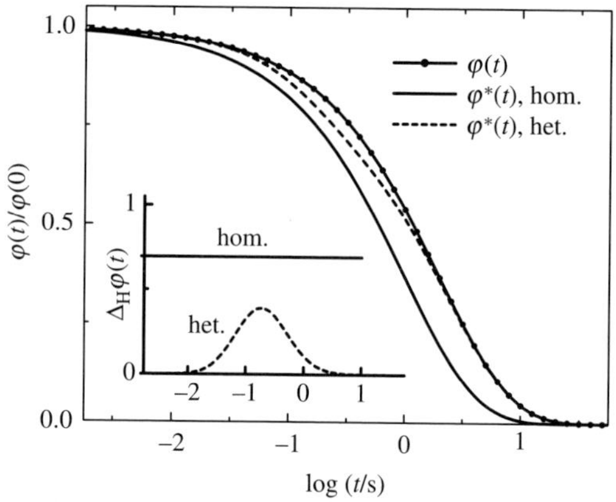

Fig. 5.4 Polarization decays in equilibrium, $\varphi(t)$, and subsequent to energy absorption, $\varphi^\star(t)$, for the heterogeneous and the homogeneous case. The function φ represents either ϵ or M, depending on the experimental situation. Effects are exaggerated over those observed in real systems. The inset displays the horizontal differences between $\varphi(t)$ and $\varphi^\star(t)$, which will show a peak only in the case of heterogeneous dynamics.

An alternative approach to the spectrally selective energy absorption and resulting acceleration of the dynamics of a subensemble is by high-field impedance spectroscopy (Richert and Weinstein, 2006; Weinstein and Richert, 2007*a*) or its time-resolved variant (Wang and Richert, 2007*b*). Compared with the pump-wait-probe protocol of hole burning, quasi-steady-state impedance uses the same sinusoidal high-amplitude field to transfer energy to certain modes and to measure the modification at the frequency ν in terms of the frequency-domain dielectric function, e.g. as the field-induced change of the loss, $\epsilon''(E_0 \gg 0) - \epsilon''(E_0 \approx 0)$. Effectively, the pump and probe components are combined, thus eliminating the requirement of the time-consuming phase cycle and leading to effects that are significantly larger and better resolved. However, probing is limited to the frequency used for the pump process and the time resolution is lost.

A more sophisticated high-field method is outlined in Fig. 5.5, which combines the impedance technique with time-resolution capabilities (Huang and Richert, 2008*a*). Here, the field amplitude is switched from low to high or vice versa and the modification can be resolved in time by subjecting each period of the voltage and current trace to a Fourier analysis. From such data, the loss factor can be obtained from the phase-angle difference $\Delta\varphi$ of current and voltage, $\tan\delta = \pi/2 - \Delta\varphi$. A typical change in the loss factor $\tan\delta$ derived from such an experiment is indicated in the lower part of Fig. 5.5, which shows the relative field-induced change of $\tan\delta$. The steady-state amplitude of $\Delta\ln(\tan\delta)$ and its time evolution depend on frequency in a characteristic manner that is incompatible with homogeneous dynamics.

Hole burning and the related steady-state and time-resolved impedance techniques all have in common that the findings can be rationalized only on the basis of independent degrees of freedom regarding both structural and thermal relaxation

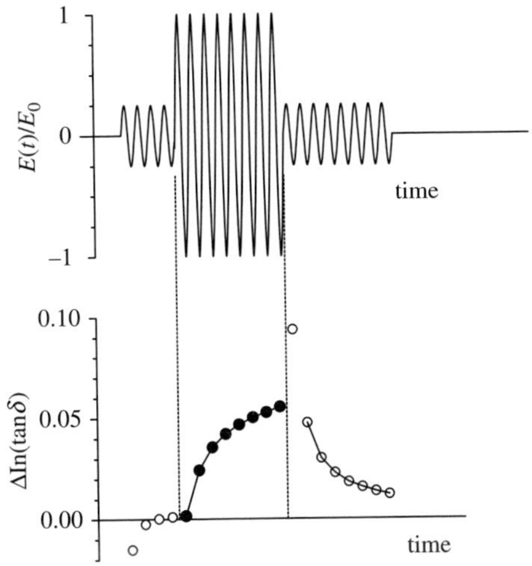

Fig. 5.5 Typical field pattern for recording the dielectric properties during the addition and removal of significant power absorbed from the field. An oscilloscope records both voltage across the sample, $V(t)$, and current through the sample, $I(t)$. The time-resolved phase difference $\Delta\varphi$ between $V(t)$ and $I(t)$ is determined by subjecting each period to a Fourier analysis. The lower graph indicates a typical response in terms of the relative change of $\tan\delta$. The signature of heterogeneity is that the rate of initial rise of $\tan\delta$ changes linearly with the frequency used.

times (Richert and Weinstein, 2006; Weinstein and Richert, 2007*b*). A large number of observations pertaining to the effects of energy absorption and the concomitant change in relaxation times have been rationalized on a quantitative level (Schiener *et al.*, 1997; Chamberlin *et al.*, 1997; Huang and Richert, 2008*a*; Jeffrey *et al.*, 2003; Blochowicz and Rössler, 2005; Weinstein and Richert, 2005). The complete calculation needed to match a steady-state high-field spectrum or the time dependence at a given frequency can be involved (Huang and Richert, 2008*b*, 2009), so only the basic ingredients are mentioned here. Regarding its dynamics, the sample is envisioned to consist of dynamically distinct domains, where each domain is characterized by a single relaxation time constant τ. We consider the balance of power or energy flow $p = dQ/dt$. The gain term is given by a generalization of Joule's law, $p_{\text{gain}} = \epsilon_0 E_0^2 \epsilon''(\omega) V\omega/2$, averaged over one cycle, where E_0 is the peak field, $\epsilon''(\omega)$ is the loss of the domain at the angular frequency ω, and $V = Ad$ is the sample volume. The loss term is given by the thermal relaxation of the configurational temperature T_{cfg} towards the temperature T of the heat bath, $p_{\text{loss}} = -\rho C_{\text{p}}(T_{\text{cfg}} - T)/\tau$, where ρC_{p} is the heat capacity and τ is the relaxation time of the domain. The net effect of energy absorption increasing the configurational temperature T_{cfg} is that the mode in question adjusts its relaxation time to the value it would have at $T_{\text{cfg}} = T$ in equilibrium. The overall apparent activation energy E_{A} can be employed to determine how sensitive $\ln\tau$ changes in response to an offset of T_{cfg} relative to T. An important aspect of this model is that thermal and structural (dielectric) relaxation times not only share the same distribution but are identical within each domain for simple liquids.

These experiments demonstrate the heterogeneous nature of the relaxation associated with the technique used (dielectric, magnetic, or mechanical), along with the correlated heterogeneity of thermal relaxation. Thus far, hole burning has not contributed towards measuring the time scale τ_{ex} of rate exchange.

5.2.4 Guest/host rotation correlation

(R. RICHERT)

Consider a glass-forming liquid doped with a small concentration of chromophores that can be excited with a polarized laser pulse. As in the photobleach case of Section 5.2.2, the solutes are considered to match the solvent molecules in size so that the rotation of the probes reflects the structural dynamics of the liquid. In order to simplify the discussion of optical anisotropy, it is further assumed that the optical transition dipoles for absorption and emission are parallel, i.e. the fundamental anisotropy angle of the molecule is zero. In this situation the detection angle can be used as a spectral filter regarding the rotation time constant for the following reason. If molecular motion is frozen, the emission at an angle parallel to the excitation polarization will be higher in intensity relative to perpendicular detection, because the anisotropy persists. It follows that with probe rotation, the parallel (VV) intensity will be dominated by the probes that rotate slowly, while the intensity at a perpendicular (VH) angle will preferentially originate from those probes that have reoriented prior to emission (Horng *et al.*, 1997; Richert, 2000).

Having established that VV and VH detection geometries can serve as a spectral filter regarding the rotation of the probe, we now add a solvation dynamics experiment to be performed for both the VV and VH cases. Solvation dynamics refers to a technique that uses chromophores to measure the time-resolved solvent reorganization via the Stokes shift correlation function. To this end, the solutes are excited electronically by a laser pulse, and their emission profiles are detected as a function of time. In the typical case of dipolar solvation, the excitation-induced change in the solutes dipole moment initiates a reorientation of solvent dipoles towards the newly established equilibrium (Richert, 2000; Bagchi and Chandra, 1991; Fleming and Cho, 1996*a*). In the course of this solvent response, the energy level of the excited state is lowered, while that of the ground state is raised. The net effect is a reduction of the emission energy on the time scale of the solvent motion. It is important to realize that the dynamics of the solvent immediately surrounding the probe are revealed by evaluating the emission energy as a function of time, regardless of the accompanying changes is the emission intensities. Therefore, the application of VV and VH geometries does not interfere with the solvation experiment. As the energy scale is not relevant here, the red-shift of the peak emission energies $\nu_{\mathrm{p}}(t)$ is normalized to its limiting values at times zero and infinity to yield the Stokes shift correlation function $C(t)$.

The main goal of combining the spectral selection regarding probe rotation with solvation is to determine whether a suitable probe tracks the environmental heterogeneity of the liquid. The probe is selected so that its rotation correlation time matches the structural time scale of the solvent. In this case, the rotation time dispersion will be similar to the relaxation time dispersion of the primary solvent mode (see Section 5.6.2). The question is whether a faster or slower than average rotating probe can be shown to correlate with a faster or slower solvent environment. If so, the probes are subject to the spatial fluctuations of local viscosities and thus mirror the heterogeneity of the solvent. Such an experiment has been performed using quinoxaline as rotational probe and as chromophore for the triplet-state solvation study in 2-ethyl-1-hexanol (Wang and Richert, 2004*b*). Clearly in that system near its glass transition, the time scales of probe rotation and solvent response or effective viscosity are locally correlated. Figure 5.6 represents what such a measurement would display for the case of a molecule with zero fundamental anisotropy, i.e. for parallel transition dipoles for absorption and emission.

The efficiency of such an anisotropy-based filter is limited, with the average relaxation times for the two subensembles being within a factor of two. Therefore, one might question whether the relatively small difference between $C_{\|}(t)$ and $C_{\perp}(t)$ is due to the local correlation of solute/solvent dynamics. In order to clarify this, the excitation-wavelength dependence of the fundamental anisotropy of quinoxaline was exploited. Thus, the experiment could be repeated with a different laser wavelength that shifted the fundamental anisotropy κ of the probe from $\kappa < \mu$ to $\kappa > \mu$, where $\mu = 57.74°$ is the magic angle. This swaps the spectral filter character from selecting the slow to the fast subensembles and the order of $C_{\|}(t)$ and $C_{\perp}(t)$ is observed to reverse accordingly (Wang and Richert, 2004*b*). This observation provides strong

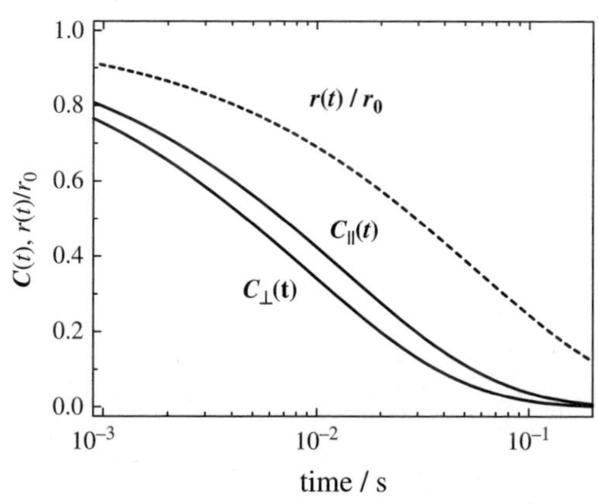

Fig. 5.6 Normalized rotation correlation function of a probe molecule in a viscous solvent (dashed). For the same probe, the solid lines represent Stokes shift correlation functions $C(t)$ for parallel ($C_{\|}(t)$, VV) and perpendicular ($C_{\perp}(t)$, VH) detection relative to the excitation polarization. In VV (VH) geometry, probes that rotate more slowly (rapidly) are emphasized. Therefore, the difference in VV and VH solvent response indicates a correlation between probe rotation and solvent response time. All curves are similarly non-exponential.

confirmation of the correlated dynamics rationale for the probe and its immediate solvent environment.

5.2.5 Photochromic probes

(R. RICHERT)
Photochromic molecules are species that can interconvert between two isomeric forms that differ considerably in their optical absorption behavior. Spiropyrans, fulgides, and azobenzenes and their derivatives have been studied regarding the isomerization reaction in different media (Smets *et al.*, 1983). In many cases, the stable isomer "S" is colorless in the visible wavelength range and the reaction towards the metastable and colored form "M" requires ultraviolet photoexcitation. The $S \leftarrow M$ backreaction occurs thermally in most cases, but irradiation in the visible can accelerate the discoloration considerably (Richert, 1988*a*). In liquid solutions the kinetics are purely exponential, as expected from the first-order nature of these reactions. In viscous and solid matrices, however, the overall kinetics are dispersive, while the molecular $S \leftrightarrow M$ type simplicity of the reaction scheme is retained (Richert and Heuer, 1997).

Realizing that these isomerization reactions are sterically demanding, their dispersive nature in polymer matrices is readily explained by a heterogeneous distribution of environments, leading to a distribution of first-order reaction rates. To a good level of approximation, the logarithmic rates $\epsilon = -\ln k$ can be represented by a gaussian probability density with width σ and center at $\epsilon_0 = -\ln k_0$. As the glass transition of the matrix is approached from below, the width σ decreases and the kinetics approach the exponential behavior observed in fluid solution. This "line-narrowing" effect regarding the reaction-rate dispersion has been evaluated on the basis of rate exchange that occurs at a rate of $k_{\text{ex}} = 1/\tau_{\text{ex}}$. With n representing the time-dependent concentration density at a logarithmic rate ϵ and M representing the concentration of all metastable molecules with $M(t = 0) = 1$, the effect of exchange could be cast in the form (Richert, 1988*b*)

$$\frac{\partial n(\epsilon, t)}{\partial t} = -kn(\epsilon, t) - k_{\text{ex}} n(\epsilon, t) + k_{\text{ex}} n(\epsilon, 0) \int_{-\infty}^{\infty} n(\epsilon, t) d\epsilon$$

$$M(t) = \int_{-\infty}^{\infty} n(\epsilon, t) d\epsilon$$

$$n(\epsilon, 0) = \frac{1}{\sigma\sqrt{2\pi}} \exp\left[-\frac{(\epsilon - \epsilon_0)^2}{2\sigma^2}\right]. \tag{5.5}$$

Analogous to the environmental fluctuation approach by Anderson and Ullman (Anderson and Ullman, 1967), the decay of $M(t)$ becomes more exponential as k_{ex} increases above the average reaction rate, $< k >$.

A different route to assessing rate exchange on the basis of photochromic reactions has been reported by Grebenkin and coworkers (Grebenkin and Bol'shakov, 2007). Because the overall rate of an azobenzene derivative $S \leftarrow M$ backreaction can be controlled by irradiation intensity, the average reaction rate can be tuned across the exchange rate. This would manifest itself as a transition from dispersive to more exponential overall kinetics. Two molecular glass formers, n-butanol and o-terphenyl have been subject to this technique (Grebenkin and Syutkin, 2007; Grebenkin, 2008).

A further variant of this idea is to interrupt the reaction by switching off the irradiation altogether for certain time intervals. This method requires that the thermal backreaction is negligible compared to the photoinduced pathway. In such a case, a spectral filter can be applied by an incomplete $S \rightarrow M$ conversion, which selects the faster subensemble of the M states. After this preparation, the backreaction can be measured either immediately or after a waiting time, during which the reaction itself is halted but exchange will continue to occur. Based on this technique, Grebenkin has concluded on the exchange time scale being only somewhat slower than the structural relaxation time of the o-terphenyl liquid in a temperature range around T_g (Grebenkin, 2008).

In an analogous fashion, a "low-pass" filter can be realized with a photoinduced isomerization reaction, as outlined in Fig. 5.7. Here, an equilibrium population of M states would have been generated for time $t < 0$ by sufficiently long UV irradiation. The backreaction is then initiated by switching on the irradiation that promotes the $S \leftarrow M$ process. After a certain time or amplitude, $M(t = 270s) = 0.5$ in the case of Fig. 5.7, the irradiation is switched of for various times, here 200 and 1000 s. Without exchange occurring during the dark interval the decay would continue as in the absence of a dark phase. By contrast, a sufficiently long off-time would have reverted the system back to the equilibrium ensemble by virtue of rate exchange. Accordingly, the decay following $\Delta t_{\mathrm{on}} = 200$ s is slower and more exponential than the one after

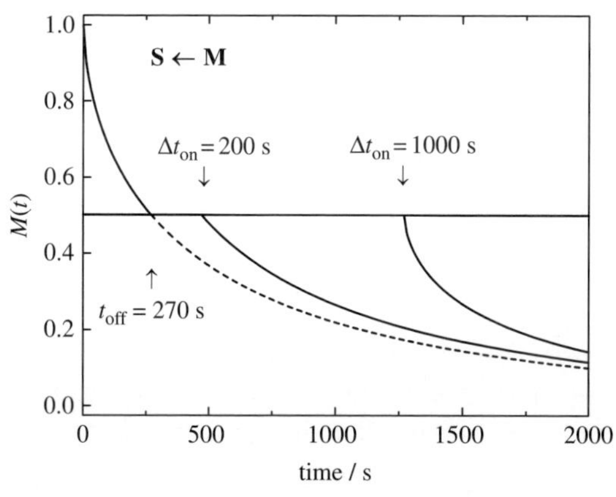

Fig. 5.7 Illustration of spectral filtering with photoinduced isomerization reactions of photochromic probes in viscous matrices. The decay starting at $M(0) = 1$ and continued as dashed line represents the decay of M in the course of the $S \leftarrow M$ reaction under irradiation. The solid curves are the decays expected for interrupting irradiation at t_{off} for the duration Δt_{on}. For $\Delta t_{\mathrm{on}} < 1/k_{\mathrm{ex}}$ the remaining decay is simply delayed in time, for $\Delta t_{\mathrm{on}} > 1/k_{\mathrm{ex}}$ the ensemble has re-equilibrated and the decay will be a replica of the uninterrupted case.

$\Delta t_{\rm on} = 1000$ s. Effectively, such an experiment would generate a four-point correlator that is very similar to the NMR case described above.

5.3 Spatially selective techniques

Spectral selectivity demonstrates the independence of fast and slow modes associated with a parallel relaxation scheme. While it is natural to assume that spatial separation of distinct modes is responsible for their independence, the experiments discussed above do not provide significant insight into such a spatial aspect of heterogeneity. More specifically, the length scales involved are of interest. The following approaches address this issue, pre-dominantly by eliminating the macroscopic ensemble average inherent in the majority of relaxation experiments.

5.3.1 Local polarization fluctuations

(N. ISRAELOFF)
Response functions (susceptibility) or fluctuations measured locally in complex materials can reveal important features of the microscopic dynamics (Weeks *et al.*, 2002; Vidal Russell and Israeloff, 2000). Both offer the tantalizing possibility of seeing and directly characterizing nanoscale dynamical heterogeneities in glass-forming materials. Via the fluctuation–dissipation relation (FDR), the two measured quantities, noise and response, should be related, in equilibrium. Local moduli such as rheological properties in soft glassy materials, are regularly inferred from the fluctuations in light scattering (Mason and Weitz, 1995). On the other hand, local deviations from the FDR have been predicted for aging glassy systems (Castillo *et al.*, 2002). Recent theory of glasses and dynamical heterogeneities has also emphasized space-time trajectories of the spontaneous dynamics (Garrahan and Chandler, 2002). Here, we describe quantitative measurements and imaging of local dielectric susceptibility and polarization noise near $T_{\rm g}$, in polymer films.

Broadened, non-exponential response functions, can be produced by a collection of quasi-independent modes with nearly exponential dynamics, but differing relaxation times. With spatially selective probes, sensitive to mesoscopic-scale regions, deviations from bulk response may be observed. These deviations may be used to infer properties of the dynamical heterogeneities, such as their size and lifetimes, and possible interactions. Do we need to probe length scales, L, that are comparable to the heterogeneity length, ξ, in order to see mesoscopic effects? This depends somewhat on the model of the heterogeneous dynamics, but for the simplest models we can make a reasonable estimate.

If measuring a frequency-dependent response function, say dielectric susceptibility, $\epsilon(f) = \epsilon'(f) - i\epsilon''(f)$, then both spatial and dynamical selectivity (see previous sections) can be applied. For exponential local modes, which have Debye-like peaks contributing to $\epsilon''(f)$, we can determine a density, $n(f)$, of dynamical heterogeneities whose Debye peaks lie within a factor of e of frequency f. From Kramers–Kronig, it can be shown that $n(f) \sim \epsilon''(f)/\xi^3\epsilon'(0)$. Then, the probed number of distinct domains is $N(f) = n(f)L^3 = \epsilon''(f)L^3/\xi^3\epsilon'(0)$. $N(f)$ reflects the number of active regions with similar peak frequencies. It is the statistical fluctuation in this number that would produce deviations from bulk response. When $N(f)$ is small, of order 10 or less, significant statistical fluctuations in $\epsilon''(f)$ should, in principle, be observable. Simulations indicate about 10% fluctuations in $\epsilon''(f)$ should be found for $N(f) = 10$. These effects should therefore be easiest to observe in the high-frequency tail of the response where $\epsilon''(f)$ is small. Suppose $\epsilon''(f)/\epsilon'(0) \sim 0.01$. This would typically occur about three decades above the α-peak frequency. In this case, we might expect to see mesoscopic effects when $L = 10\ \xi$. Estimates from NMR give $\xi \approx 2 - 3$ nm (Tracht *et al.*, 1998). Thus, we need probes with $20 - 30$ nm resolution, in order to have hope of seeing these effects. This turns out to be close to the resolution of nanodielectric spectroscopy and polarization noise methods discussed below. A challenge may arise from the *time scale* of these fluctuations, i.e. the lifetime of the heterogeneities, τ_{ex}. If τ_{ex} is too short compared with the averaging time needed for the measurement, the effects will be lost. Simple heterogeneous models, and aging and NMR experiments, suggest that τ_{ex} should be of order τ_α.

Nanodielectric spectroscopy. A variation of atomic force microscopy, electric force microscopy (EFM) involves oscillating a small silicon cantilever with a sharp metal-coated tip at its resonance frequency in ultrahigh vacuum. The EFM tip is held a distance, z, above a dielectric sample, and biased with voltage, V, relative to a conducting substrate on which the sample is coated. The bias provides an electrostatic force between the tip and sample surface, which shifts the cantilever resonance frequency by $\delta f(V)$, which is then measured by means of a precision phase-locked loop (PLL). Polymer films typically 1 μm thick, which have a strong dielectric response, such as poly(vinylacetate) (PVAc) and poly(methylmethacrylate) (PMMA) have been studied thus far. The tip electrostatic force can be calculated from the charging energy on the tip capacitance, $U = CV^2/2$ and $F = -dU/dz$. The bias-induced shift in cantilever resonance frequency, $\delta f(V)$, can be obtained from the force gradient dF/dz that supplies a fractional change in the cantilever spring constant, k. Surface potentials and surface charges due for example to polarization of the sample, both act to produce a dc offset, V_{p}, to the applied bias. The electrostatic component of δf is therefore:

$$\delta f = \frac{1}{2}\frac{f_0}{k}\frac{dF}{dz} = -\frac{1}{4}\frac{f_0}{k}\frac{d^2C}{dz^2}(V - V_{\mathrm{p}})^2\,. \tag{5.6}$$

Because the signal is proportional to the force derivative on the sharp tip, it is most sensitive to sample polarization near the surface. For a conical tip with typical 30 nm tip radius, a depth of about 20 nm below the surface is probed. This is comparable to the thickness of polymer films in which dramatic changes in T_{g} have been observed and attributed to confinement effects.

With a dc bias, the changes in V_{p} due to dielectric relaxation or spontaneous polarization fluctuations have been studied in PVAc and PMMA films (Vidal Russell and Israeloff, 2000; Vidal Russell *et al.*, 1998; Walther *et al.*, 1998*a*,*b*), however, contributions from other tip–surface interactions were not excluded by this approach. By applying a low-frequency oscillating bias, $V = V_0 \sin(\omega t)$, and using a lock-in amplifier to detect the oscillation in δf, the electrostatic contributions can be isolated. The term in the expanded Eq. (5.6) proportional to $2V_0V_\mathrm{p}\sin(\omega t)$ due to sample polarization, will be detected by the lock-in as a signal, δf_ω, and can be used to study time-dependent polarization due to noise or relaxation (see next section). The $V_0^2\sin^2(\omega t)$ term can be detected as a 2ω lock-in signal, $\delta f_{2\omega}$. This signal turns out to be related to the local dielectric susceptibility (Crider *et al.*, 2007). By studying the amplitude and phase of the $\delta f_{2\omega}$ signal as a function of frequency (0.1 – 100 Hz) and temperature, one sees curves very similar in shape to bulk dielectric susceptibility for a polymer PVAc (Fig. 5.8). By applying a dc plus ac bias, dielectric relaxation is seen in the δf_ω signal. These can be Fourier transformed to effectively extend the frequency range easily down to 10^{-3} Hz. These curves can be accounted for in a model with local $\epsilon'(f)$ and $\epsilon''(f)$ as input parameters. However, it was found that the locally measured peaks in $\epsilon''(f)$ were shifted to lower temperatures as compared with bulk by about 5 K, indicating surface mobility was considerably higher than bulk (Crider *et al.*, 2008). For 30-nm tips, the probed volume is estimated to be about $\Omega = 5 \times 10^4$ nm^3, which means $L/\xi \sim 12$, however, thus far, no clear mesoscopic fluctuations in the nanodielectric response have been found with this approach. This may indicate that the simple picture of dynamical heterogeneities as similar-sized fast and slow compact regions is not applicable in this situation, or that the statistical fluctuations of these faster than average regions occur on time scales short compared with τ_α. In other words, the system appears homogeneous already at the given time and spatial averaging of this experiment.

Polarization noise. Polarization fluctuations in a volume, Ω, with dielectric susceptibility, χ, have the form $< \delta P^2 > \sim k_\mathrm{B}T\chi/\Omega$ as a consequence of the fluctuation–dissipation relation. Thus, in probing a small volume, fluctuations should become all the more apparent. In fact, such noise can be seen readily in the polarization as measured by the δf_ω signal described above (Crider and Israeloff, 2006). In

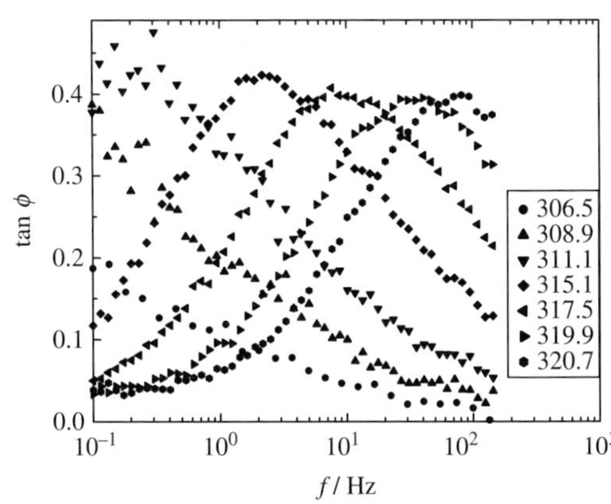

Fig. 5.8 Nanodielectric spectroscopy of PVAc. Tangent of the phase angle vs. frequency in measurements of δf_{2w} for several temperatures near T_g, as indicated. The curves are comparable in shape to $\tan\delta = \epsilon''(f)/\epsilon'(f)$ measured in bulk.

equilibrium, this should contain the same information about the dynamics that is found in the susceptibility. The slow time-dependent fluctuations can be quantified by their autocorrelation function $A(t)$ and this in turn relates to the glassy time-dependent part of the susceptibility, $\Delta\chi(t)$, by $\Delta\chi(t) = [A(0) - A(t)]/(k_B T)$. A similar expression can be found (Oukris and Israeloff, 2010) for the measured quantities, $\langle \delta V_p^2 \rangle$ and $\Delta\chi_{exp}(t) = \Delta V_p(t)/V_{dc}$: $\Delta\chi_{exp}(t) = C_{eff}[A(0) - A(t)]/(k_B T)$, where $C_{eff} \sim< C >^2 /C$, and C is the sample contribution to the tip capacitance. C_{eff} calculated from an image charge model and finite-element calculation of tip capacitance for the experiment give estimates for $C_{eff} = 6.3\ (\pm 1.8) \times 10^{-18}$ F. In Fig. 5.9 the measured correlation function, $A(t)$ of the noise and $\Delta\chi_{exp}(t)$ are shown for one temperature in PVAc. Also, in the inset of Fig. 5.9 the two quantities are plotted vs. each other for two temperatures near T_g for PVAc. First, it is clear that a straight line is obtained with nearly the same slope for the two temperatures, indicating that the noise and relaxation do indeed contain the same time-dependence information. Secondly, from the slopes a $C_{eff} = 8.45(\pm 1) \times 10^{-18}$ F is extracted. The value is nearly the same for different temperatures and in reasonable agreement with the calculated value. This approach has been used recently to study quenched samples during aging, where deviations from straight-line behavior were observed (Oukris and Israeloff, 2010) indicating global and possible local violations of the FDR.

The noise can also be used to produce spatio-temporal images of the dynamics (Crider and Israeloff, 2006). In this case the surface potential is repeatedly measured along a one-dimensional spatial line. These space-time images of the surface polarization are striking, in that they clearly demonstrate the spatio-temporal aspects of glassy dynamics, e.g. the correlations seen along the time axis are distinctly longer at the lower temperatures. Some patterns appeared similar to those produced by the

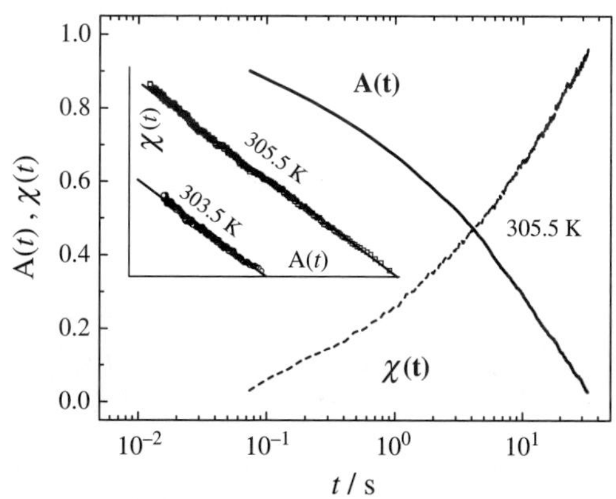

Fig. 5.9 Correlation function of polarization noise ($A(t)$, solid curve) and dielectric response function ($\chi(t)$, dashed curve) versus time for $T = 305.5$ K for PVAc. The inset displays the correlation of the two quantities by plotting the response χ versus correlation function A for two temperatures in equilibrium. For $T = 303.5$ K: $d\chi/dA = 2020\ (\pm 250)$, $C_{eff} = 8.47\ (\pm 1) \times 10^{-18}$ F, and for $T = 305.5$ K: $d\chi/dA = 1990\ (\pm 250)$, $C_{eff} = 8.42\ (\pm 1) \times 10^{-18}$ F.

Garrahan–Chandler picture (Garrahan and Chandler, 2002) of activated dynamics, but on much larger 50-nm length scales, thus we should be cautious about interpreting them this way. Space-time images of fluctuations offer a tremendous opportunity to study high-order space-time correlation functions, that are believed to be important measures of growing correlation lengths, however, a reduced probe size would be most useful.

5.3.2 Single-molecule rotation and translation studies

(R. RICHERT)

The possibility of observing single molecules as pioneered by Moerner (Moerner, 1994) opens new routes to studying heterogeneous dynamics. The attractive feature is that the otherwise necessary ensemble averaging can be circumvented and rotational and translational motion can be monitored in time. Rotational behavior is usually recorded in terms of the reduced linear dichroism, $A = (I_{\|} - I_{\perp})/(I_{\|} + I_{\perp})$, where the observed quantity is often a complicated projection of the actual three-dimensional reorientation (Fourkas, 2001; Hinze *et al.*, 2004). Long-time trajectories of $A(t)$ can be recorded, and the autocorrelation function of these $A(t)$ traces provides anisotropy decays, as illustrated in Fig. 5.10. Often, only a fraction of the recorded $A(t)$ curve is required for defining a correlation time τ, and thus the time scale of rotation can be determined many times in sequence for a single molecule. From such an analysis, leading to τ_1 through τ_4 in Fig. 5.10, the overall relaxation dispersion can be obtained from

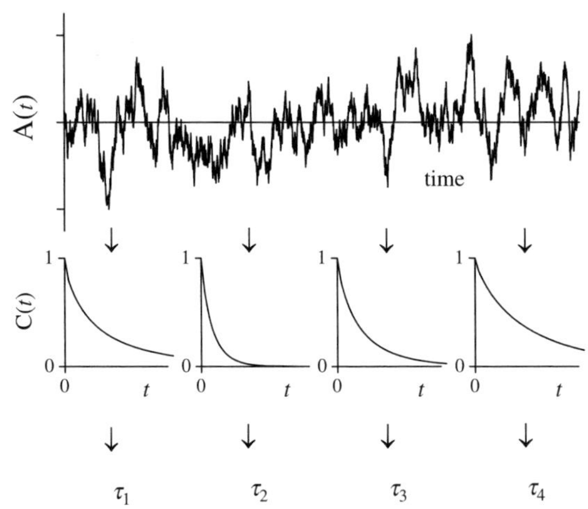

Fig. 5.10 Illustration of how heterogeneous dynamics are observed on the basis of linear dichroism trajectories $A(t)$. For time intervals sufficient to obtain meaningful decays for the autocorrelation functions $C(t)$, characteristic values such as average time constant τ and stretching exponent β can be evaluated. A time sequence of $C(t)$ curves will provide information on the long-time average dispersion (equal to the ensemble average in the ergodic state) and rate exchange for a single-molecular probe.

long-time statistics and the rate exchange can be evaluated directly from the temporal fluctuation of the τs.

Single-molecule experiments require fluorescent dye molecules with exceptionally high quantum yield and photochemical stability (Kulzer and Orrit, 2004), which limits the choice of appropriate probe molecules considerably. Single-molecule studies of heterogeneous dynamics include both polymeric and molecular glass-forming systems. Examples are measurements of rhodamine B in poly(vinylacetate) ($\tau_g/\tau_h = 80$, $Q < 80$) (Adhikari *et al.*, 2007), rhodamine 6G in *o*-terphenyl ($\tau_g/\tau_h = 20$, $Q = 300$) (Deschenes and Vanden Bout, 2002), rhodamine 6G in poly(methylacrylate) ($Q = 10$) (Schob *et al.*, 2004), and N,N'-bis(2,5-di-tertbutylphenyl)- 3,4,9,10-perylenedicarboximide in glycerol ($\tau_g/\tau_h = 80$, $Q > 10^6$) (Zondervan *et al.*, 2007). As far as is known, the guest (probe rotation time) to host (α relaxation time) relaxation time ratios τ_g/τ_h and the derived reduced time scales of rate exchange $Q = \tau_{ex}/\tau_\alpha$ are included in this compilation. The values for Q are taken from the original papers and may be at variance with the definition in Eq. (5.4). The above single-molecule studies all report clear signatures of heterogeneous dynamics.

The above examples of single-molecule rotation in viscous materials have in common that the values of Q reported are significantly higher than what was concluded from higher-dimensional NMR with $Q \approx 1$. Another common feature, and unlike the photobleaching case of Section 5.2.2, the single-molecule probes used have hydrodynamic volumes that are much larger than those of the molecules or polymer segments of the host material. Accordingly, the probe rotation is significantly slower than the structural relaxation of the liquid itself, i.e. $\tau_g/\tau_h \gg 1$. As detailed in Section 5.6.2, rotational probes with $\tau_g/\tau_h > 8$ are safely in the hydrodynamic range as rate exchange has averaged out the heterogeneity on the time scale of the slower probe rotation. In single-molecule experiments, such a situation does not imply an exponential decay of the reduced linear dichroism (Fourkas, 2001), and only stretching exponents below about 0.85 should be considered a sign of heterogeneous probe dynamics. An explanation for the apparently long exchange times $Q \gg 1$ could be the following. The probe is subject to an already time-averaged environment and, as a result, significant excursions from the current time constant become relatively rare events. That this time averaging affects conclusions regarding Q and that $\tau_g \approx \tau_h$ is a desirable working condition for future single-molecule experiments has been pointed out by Bingemann and coworkers (Adhikari *et al.*, 2007).

Single-molecule translational dynamics have been observed by von Borczyskowski and collaborators in terms of broad distributions of diffusion constants in thin films, where the dispersion is attributed to the distance from the interface or anisotropic motion (Schuster *et al.*, 2003, 2004). An analogous experiment for a bulk system has been performed on the rhodamine 6G in poly(methylacrylate) system mentioned as an example of rotational dynamics above (Schob *et al.*, 2004). In this latter case, no sign of heterogeneous diffusivity is observed, and the probe-to-segment size mismatch is again a likely cause for efficient averaging. Similarly, enhanced diffusivity near T_g is considered a consequence of heterogeneous dynamics and disappears for larger diffusion probes, see Section 5.6.1.

5.3.3 4D cross-polarization NMR

(R. RICHERT)

The multidimensional NMR technique outlined in Section 5.2.1 is capable of revealing the independence of fast and slow modes in the system, but spatial scales ξ of heterogeneity have remained inconsequential. The technique of 4D cross-polarization (CP) NMR facilitates a measurement of the length scale involved in heterogeneous dynamics by adding spin diffusion to the filtering capabilities (Tracht *et al.*, 1998, 1999). CP transfers the spin polarization from ^{13}C to ^{1}H nuclei and vice versa, with the main feature that the spin diffusion for the ^{1}H system is much more efficient compared with the ^{13}C counterpart. Therefore, this CP process can effectively be used to initiate and terminate spin diffusion within the ^{13}C subsystem. If spin diffusion is promoted for variable times t_{diff}, spatial ranges will be explored, whose extent is under experimental control. With this technique at hand, slow molecules are selected analogous to the 4D-NMR spin-echo technique discussed in Section 5.2.1, and after diffusion for a time t_{diff} the experiment determines whether the initially slow subensemble has undergone a change to shorter correlation times. If so, the conclusion is that diffusion across boundaries of dynamically distinct domains is responsible. Otherwise, the condition $\sqrt{(6Dt_{\mathrm{diff}})} < \xi$ holds and net diffusion has not crossed domain boundaries. How an adjustable t_{diff} can be exploited to assess ξ is depicted schematically in Fig. 5.11 for a small and a large t_{diff} case.

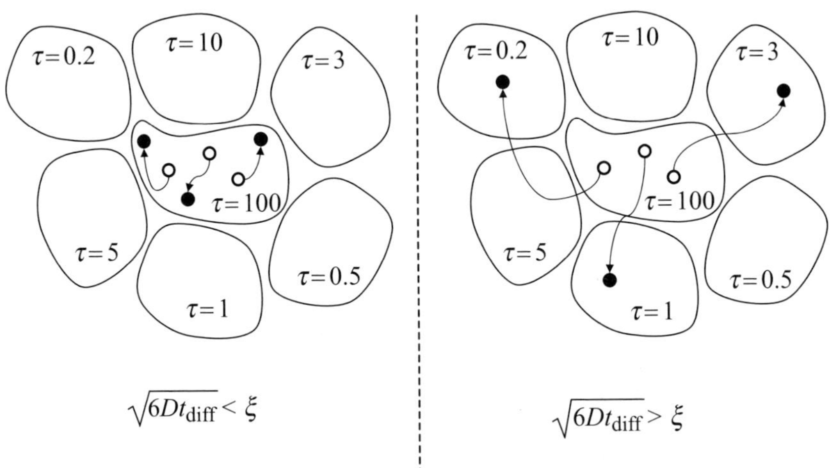

Fig. 5.11 Cartoon representation of how controlled diffusion can measure the length scales involved in heterogeneous dynamics. Diffusion is activated for a variable time t_{diff}, and the experiment determines whether correlation times after the diffusion differ significantly from the initial spectral selection, here $t_{\mathrm{sel}} = 100$ s ($\tau_{\mathrm{avg}} = 17$ s). The mean-squared displacement has to exceed the domain size in order for diffusion to alter the relaxation times through transfer into other domains with different time constants τ.

Several materials have been studied using this 4D CP NMR technique, and length scales between 2 and 4 nm have been found for poly(vinylacetate), D sorbitol, and *o*-terphenyl, whereas glycerol shows a smaller value of 1 nm (Reinsberg *et al.*, 2001, 2002; Qiu and Ediger, 2003). For non-spherical domains, the analysis will yield the shortest axis as length scale.

5.4 Using higher-order correlation functions

Most measurements of relaxation behavior determine a two-time autocorrelation function of a certain variable, e.g. dielectric polarization, orientation, magnetization, shear stresses, etc. Results of such measurements will not provide information of the heterogeneous versus homogeneous nature of the dynamics. As outlined in previous sections, four-time correlation functions can be designed to add spectral filtering and are thus capable of revealing details regarding heterogeneous dynamics. An alternative approach is based on higher-order correlation functions, e.g. four-point, where still only two points in time are involved. Examples are those where two points in time are combined with two points in frequency or in real space. The cases described below exploit such higher-order two-time correlations where no spectral filtering is involved.

Direct observations of space-time correlation functions are possible by evaluating particle trajectories obtained either from MD simulations (Lačević *et al.*, 2003*a*) or recorded in real time for colloidal systems (Marcus *et al.*, 1999). Colloids are covered in Chapter 3 of this book.

5.4.1 Solvation dynamics

(R. RICHERT)

When a chromophore in solution is excited electronically, the solute/solvent system is removed from its equilibrium, which is subsequently restored by the response of the solvent (Ware *et al.*, 1971; Maroncelli *et al.*, 1989). The typical source of significant interactions are of an electrostatic nature, as in the case of dipolar solvation, where the excitation induced change in the solutes dipole moment interacts with the dipoles in a polar solvent (Maroncelli, 1993; Fleming and Cho, 1996*b*). As a result of the solvent response, the emission energy is shifted towards lower energies. Accordingly, the experiments aimed at detecting solvation dynamics will measure the time-resolved optical emission bands. In the case of a linear response, the process of solvation dynamics can be represented by diffusion within a parabolic potential, known as the Ornstein–Uhlenbeck process (Risken, 1989). As a consequence of the linear response, the inhomogeneously broadened emission profiles are gaussian profiles at time zero and in the steady-state limit. Under favorable experimental conditions this profile of the electronic transition can be monitored in time. In such a case, not only the average but higher moments of the relaxing quantity (variance and/or asymmetry) are revealed (Richert and Wagener, 1991). While the true inhomogeneous line broadening is often obscured by vibrational modes in many fluorescence solvation studies, long-lived triplet probes and the concomitant low solvent temperature have facilitated the direct detection of gaussian bands of electronic origin (Richert, 2000).

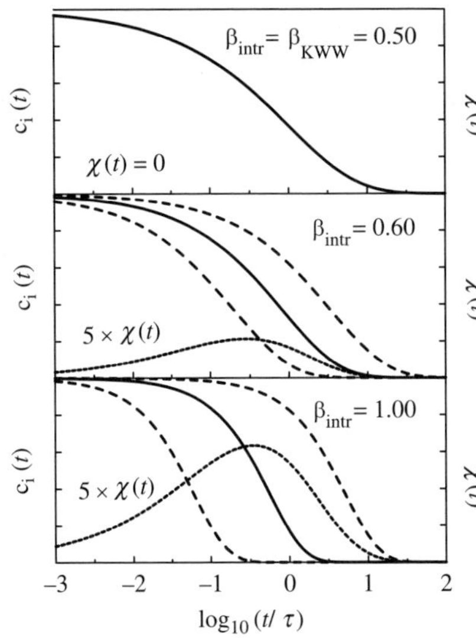

Fig. 5.12 Individual relaxation patterns $c_i(t)$ for a common ensemble-averaged decay $\phi(t) =< c_i(t) >$ of the KWW type with $\tau_{\mathrm{KWW}} = 1$ and $\beta_{\mathrm{KWW}} = 0.5$ and for various values of $\beta_{\mathrm{intr}} = 0.50$, 0.60, and 1.00. The decays $c_i(t)$ in each panel are traces of $\exp\left[-(t/\tau_i)^{\beta_{\mathrm{intr}}}\right]$ with $\ln(\tau_i) =< \ln(\tau_i) > \pm$ the standard deviation of $\ln(\tau_i)$. The dotted $\chi(t) =< c_i(t)^2 > - < c_i(t) >^2$ curves represent the variances of the site-specific decays. (Böhmer, 1998*b*; Wendt and Richert, 2000)

Figure 5.12 is meant to justify the interest in studying the time-dependent inhomogeneous linewidth σ or the variance σ^2 instead of the average emission energy alone. Consider the overall correlation function $C(t)$ as a superposition of local response functions $c_i(t)$, where i enumerates all relaxing units. As an example, a stretched exponential, $C(t) = \exp\left[-(t/\tau)^{\beta}\right]$, with exponent β is assumed for the ensemble averaged response. For the local responses, we also use stretched exponentials, but with an "intrinsic" exponent β_{intr} ($\geq \beta$) and site-specific relaxation time τ_i, leading to $c_i(t) = \exp\left[-(t/\tau_i)^{\beta_{\mathrm{intr}}}\right]$ (Richert, 1997; Richert and Richert, 1998). No rate exchange is considered at this point. In the case of homogeneous dynamics, we have $\beta_{\mathrm{intr}} = \beta$ and $\tau_i \equiv \tau$, and all $c_i(t)$ would be identical, as indicated in the top panel of Fig. 5.12. In the other extreme of a purely heterogeneous relaxation scenario, we have $\beta_{\mathrm{intr}} = 1$ and a wide distribution of τ_is. For this case, the bottom panel represents the distribution by three exponential $c_i(t)$ curves and the dotted line is the variance $\chi(t)$ of all $c_i(t)$ values evaluated as a function of time. The center panel displays an intermediate situation with $\beta_{\mathrm{intr}} = 0.6$ and a narrower distribution of τ_is.

Qualitatively, this variance gauges how much separated the faster and slower modes are on this normalized 0–1 ordinate scale. Naturally, homogeneous dynamics will lead to $\chi(t) \equiv 0$ and the amplitude of $\chi(t)$ will increase with overall relaxation time dispersion and with more exponential intrinsic decay patterns. Therefore, $\chi(t)$ is a reliable indicator of the extent of heterogeneity and was subsequently employed more generally as a non-gaussian parameter or dynamic susceptibility $\chi_4(t)$. The underlying correlator is effectively a four-point correlation function with two points in time and two points along the scale of the observable (real space, emission frequency, etc.).

By analyzing the gaussian emission lines in the course of a solvation dynamics experiments using the probe quinoxaline (QX) in the glass former 2-methyltetrahydrofuran (MTHF), both the average energy as well as the variance became available as a function of time and for various temperatures. The average energy decays according to a KWW law with exponent $\beta = 0.5$ for all seven temperatures used in that experiment, 91 K to 97 K (Wendt and Richert, 2000). Both temperature and time can be eliminated as explicit variables by plotting the dynamic susceptibility $\chi(t)$ versus $C(t)$ (Ito *et al.*, 2006). Here, these normalized quantities are determined from the peak frequency $\omega_p(t)$ and the inhomogeneous width $\sigma_{inh}(t)$ via

$$C(t) = \frac{\omega_p(t) - \omega_p(\infty)}{\omega_p(0) - \omega_p(\infty)} \quad \text{and} \quad \chi(t) = \frac{\sigma_{inh}^2(t) - \sigma_{inh}^2(\infty)}{\Delta^2} \tag{5.7}$$

where $\Delta = \omega_p(0) - \omega_p(\infty)$ is the amplitude of the response in terms of the red-shift. The line in Fig. 5.13 represents the relation $\chi(t) = C(2t) - C^2(t)$, which is the case expected for intrinsically exponential dynamics, i.e. $\beta_{intr} = 1$ (Richert, 2001*b*). Clearly, exponential local dynamics provide the best description, and $\beta_{intr} = 0.9$ would already show systematic deviations from the experimental data. Homogeneity corresponds to $\chi(t) \equiv 0$ and is thus qualitatively incompatible with the observations. Heterogeneity of solvent response times is also considered as the origin of the red-edge effect of fluorescence observed in ionic liquids (Hu and Margulis, 2006*a*,*b*).

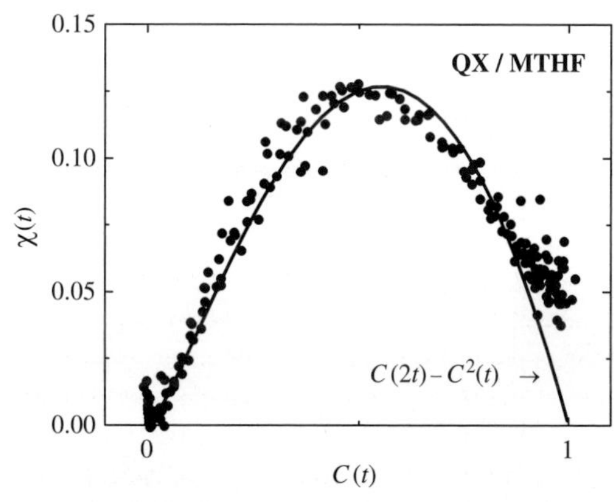

Fig. 5.13 Master plot of $\chi(t)$ (symbols) versus t/τ_{KWW} based on the $\sigma_{inh}(t)$ data of the system QX/MTHF at temperatures 91 K $\leq T \leq$ 97 K. The line is the predicted $\chi(t)$ curve using $\phi(t) = \exp\left[-(t/\tau_{KWW})^{0.5}\right]$ for $\beta_{intr} = 1.0$. Assuming $\beta_{intr} = 0.9$ would already lead to an inferior coincidence with experimental data (Wendt and Richert, 2000). The homogeneous case $\beta_{intr} = \beta_{KWW} = 0.5$ would lead to $\chi(t) \equiv 0$.

The above modeling of the dynamic susceptibility has disregarded rate exchange. Adding rate exchange to the spectral diffusion process can be accomplished by introducing a coupling or exchange term in the Fokker–Planck equations and solving these numerically. An alternative is to solve the equation of motion resulting from stochastic Langevin forces (Richert, 2001*c*).

For the Ornstein–Uhlenbeck-type process, the iteration in time for a single trajectory is given by

$$\omega(t + \Delta t) = \omega(t)\,[1 - k\Delta t] + g_{\mathrm{r}}\sigma_{\infty}\sqrt{2k\Delta t}\,, \tag{5.8}$$

with $\omega(0)$ taken at random from the initial gaussian (with width σ_∞ and centered at Δ) and g_{r} taken at random from a gaussian with $< g_{\mathrm{r}} >= 0$ and $< g_{\mathrm{r}}^2 >= 1$. In such a simulation, rate exchange can be introduced by letting $k = 1/\tau$ jump between two (or more) different values, k_i and k_j, where the persistence time t_{ex} in a state is determined by the exchange rate $k_{\mathrm{ex}} = 1/\tau_{\mathrm{ex}}$, equivalent to picking t_{ex} at random from the probability density $k_{\mathrm{ex}} \exp(-k_{\mathrm{ex}} t_{\mathrm{ex}})$. A typical trajectory result from a two-state model of this process is depicted in Fig. 5.14.

Such calculations show that rate exchange leads to a $\chi(t)$ that is greater than the static heterogeneity counterpart. Based upon the data of Fig. 5.13, the calculations showed that τ_{ex} has to exceed 9 times the average solvent response time scale in order to remain compatible with the experimental findings (Richert, 2001*c*). For a comparison with other results regarding the ratio $\tau_{\mathrm{ex}}/\tau_\alpha$, one has to realize that these solvent responses are about a factor of 6 faster than most standard metrics of the primary relaxation time τ_α. This notion leaves $\tau_{\mathrm{ex}} > 1.5\tau_\alpha$ for an estimate of the exchange process, simply implying that this particular experiment is not very sensitive to exchange.

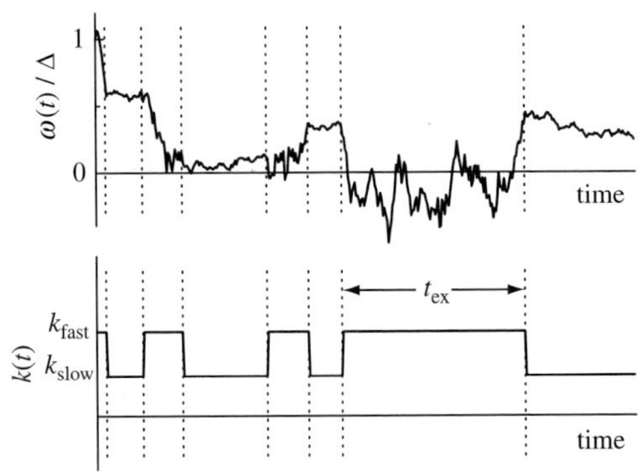

Fig. 5.14 A typical trajectory of the transition frequency $\omega(t)$ for an Ornstein–Uhlenbeck process subject to a fluctuating rate constant. The simulation has been carried out for two relaxation times $\tau_i = 0.3$ and $\tau_j = 10$ with $\tau_x = 2$ and equal probability for a site to reside in either state. For clarity, the initial frequency has been set to $\omega(0) = \Delta$ and a small $\sigma_\infty = 0.2\Delta$ has been used. The lower panel indicates the rate-exchange process for this trajectory. Dashed vertical lines mark the actual times of rate exchange.

5.5 Correlation volume estimates from dynamic susceptibilities

(CHR. ALBA-SIMIONESCO, F. LADIEU, D. L'HÔTE)

While four-point correlation function and dynamic susceptibilities provide insight into the heterogeneous nature of liquid dynamics, their direct experimental determination is limited to few techniques. This is why to investigate possible connections between high-order spatio-temporal correlators and other kinds of experimental observables is of great interest. The four-point correlator,

$$G_4(\mathbf{r}' - r, t) = <\delta\rho(\mathbf{r}, 0)\delta\rho(r, t)\delta\rho(r', 0)\delta\rho(r', t)> - <\delta\rho(r, 0)\delta\rho(r, t)>^2 \qquad (5.9)$$

measures the correlation between time evolutions at points $\mathbf{r}$ and $\mathbf{r}'$ of the density fluctuations $\delta\rho$. Note that it is possible to consider various kinds of correlations, e.g. orientational ones (Dalle-Ferrier *et al.*, 2007) that would be more convenient, e.g. for dielectric spectroscopy. In what follows, we shall see how measurable quantities, i.e. the linear dynamic susceptibility χ_{T} and the non-linear susceptibility χ_3 are related to this correlator or its normalized space integral $\chi_4(t) \propto \int d^3\mathbf{r}\, G_4(r, t)$, and more generally to the average number of dynamically correlated molecules N_{corr}. Bouchaud and Biroli (Bouchaud and Biroli, 2005) first established a connection between N_{corr} and the non-linear susceptibility χ_3. Later, Berthier *et al.* (Berthier *et al.*, 2005) have shown that information related to those contained in four-point susceptibilities can be derived from normalized correlation functions $\phi(t, T)$, if measured for various values of a control parameter, e.g. temperature T. An average number $N_{\mathrm{corr,T}}$ of dynamically correlated molecules can be defined, and the result reads

$$N_{\mathrm{corr,T}} = \sqrt{\frac{k_{\mathrm{B}}T^2}{\Delta c_{\mathrm{p}}}} \max_t \mid \chi_{\mathrm{T}} \mid, \qquad (5.10)$$

where Δc_{p} is the configurational heat capacity (the difference between the c_{p} of the liquid and the c_{p} of the glass) of the system under study, and χ_{T} is defined by

$$\chi_{\mathrm{T}} = \frac{\partial\phi(t, T)}{\partial T}. \qquad (5.11)$$

χ_{T} is a three-point dynamic susceptibility (Dalle-Ferrier *et al.*, 2007): When calculated through a fluctuation-dissipation relation in the NPT ensemble (which corresponds to most experimental situations), it reads

$$\chi_{\mathrm{T}}(t) = \frac{\rho}{k_{\mathrm{B}}T^2} \int d^3\mathbf{r} < \delta h(\mathbf{0}; 0)\delta c(\mathbf{r}; 0, t) >, \qquad (5.12)$$

where ρ is the density, $h(r, t)$ the local enthalpy density per molecule, δh its variation with respect to its average value, $c(\mathbf{r}; 0, t)$ a two-times correlator that characterizes the dynamics from time 0 to t at point $\mathbf{r}$. χ_{T} estimates the "size" of the correlation between an enthalpy fluctuation at point $\mathbf{0}$ and time 0 and the change in the dynamics at point $\mathbf{r}$ from time 0 to time t. Its time dependence is that it starts at zero, grows to a maximum reached for $t \sim \tau_\alpha$, then decreases to zero. Such time evolutions of the dynamical correlations have been investigated for various theoretical models (Berthier *et al.*, 2005; Donati *et al.*, 2002; Lačević *et al.*, 2003*b*; Berthier *et al.*, 2007*a*,*b*). The connection of χ_{T} with the four-point dynamic susceptibility $\chi_4(t)$ has been investigated thoroughly (Dalle-Ferrier *et al.*, 2007; Berthier *et al.*, 2005, 2007*a*); the main result being:

$$\chi_4^{\mathrm{NPT}} = \frac{k_{\mathrm{B}}T^2}{\Delta c_{\mathrm{p}}}(\chi_{\mathrm{T}}^{\mathrm{NPT}})^2 + \chi_4^{\mathrm{NPH}}, \tag{5.13}$$

where the superscripts indicate the statistical ensemble in which the susceptibility is considered. χ_4^{NPH} is a positive quantity, thus the square of χ_{T} gives a lower bound for χ_4. This lower bound should give a reasonable estimate of χ_4 since simulations support the fact that χ_4^{NPH} is small compared to $(k_{\mathrm{B}}T^2/\Delta c_{\mathrm{p}})\left(\chi_{\mathrm{T}}^{\mathrm{NPT}}\right)^2$ (Dalle-Ferrier *et al.*, 2007; Berthier *et al.*, 2005, 2007*a*,*b*). As a consequence, the average number of correlated molecules according to a four-point correlator (see eqn (5.9)), $N_{\mathrm{corr},4} = \max_t[\chi_4(t)]$ has $(N_{\mathrm{corr,T}})^2$ for lower bound and for estimate.

An underlying aspect of this discussion on the relationship between χ_4 and χ_{T} is that the supercooled liquid can be described in the framework of a "Newtonian" microscopical dynamics. The nature of the dynamics used to describe the physics of the system plays an important role in numerical simulations: Their results may depend on the "Newtonian", "Brownian" or "Monte Carlo" nature of the prescriptions governing the energy transfers between the elementary components of the system and their surroundings (Berthier *et al.*, 2007*a*). Newtonian simulations led to the realization that $\chi_4(t)$ mixes dynamical heterogeneities with another physical effect, i.e. correlations due to conservation laws. Energy fluctuations related to a dynamical heterogeneity motion must be related—because of energy conservation—to other energy fluctuations somewhere else in the system. This is why it is the square of χ_{T} that appears as a lower bound of χ_4. $\chi_4(t)$ and $N_{\mathrm{corr},4}$ might therefore not be the most effective tools to work with. On the contrary, χ_{T} and $N_{\mathrm{corr,T}}$ appear not to be affected by such a difficulty (Dalle-Ferrier *et al.*, 2007).

As the primary mechanism creating dynamic correlations can be attributed to the fact that any local perturbation coupled to the slow dynamics (energy, enthalpy, etc.) affects the dynamics over a certain volume growing as the glass transition is approached, it results that the mechanism inducing dynamic correlations is well captured by χ_{T} and $N_{\mathrm{corr,T}}$. It also appeared that an important result of various theoretical models or numerical simulations was that a unique length scale underlies the variations of both χ_{T} and χ_4 (Berthier *et al.*, 2005, 2007*a*,*b*). The size of the dynamical heterogeneities should thus be reasonably estimated by using the three-point dynamic susceptibility χ_{T}, easily deduced experimentally, rather than χ_4, not experimentally available so far.

Possible limitations of χ_{T} when considered as a tool to estimate the size of the dynamical heterogeneities are, however, worth noting. A first one is that the three-point correlation function appearing in the integral of eqn (5.12) may vary in such a way that the integral would give the variations of the magnitude of the correlations rather than that of the number of correlated molecules. In particular, the T dependence of the latter is given by $\chi_{\mathrm{T}}(T)$ only if the magnitude of the correlations between the enthalpy fluctuations and the molecules, positions or orientations, etc. are temperature independent. This point is evoked in Section II.C of Ref. (Berthier *et al.*, 2007*a*) (study of the quantity χ_0). It was confirmed in numerical simulations and in MCT (Berthier *et al.*, 2007*a,b*) that the magnitude of the correlations were indeed temperature independent, but this point remains questionable for real systems close to T_g. Another possible limitation of probing the number of correlated molecules with χ_T is the fact that for purely Arrhenius systems, i.e. $\tau_\alpha = \tau_0 \times \exp(\Delta/kT)$, where Δ is some constant activation energy and no collective processes should be expected, this procedure leads to a growth of the correlation size as T decreases. This is related to energy conservation and should not affect $N_{\mathrm{corr,T}}$ estimates for fragile liquids (Berthier *et al.*, 2007*a*). Because of those two possible limitations, it is important to compare the χ_T variations with those of another completely different estimator based on the non-linear susceptibility, χ_3 (Bouchaud and Biroli, 2005; Tarzia *et al.*, 2010; Crauste-Thibierge *et al.*, 2010).

The fact that the average number of dynamically correlated molecules N_{corr} can be obtained by using the dynamical susceptibility χ_T is of strong interest because it allows investigation of the dependences of N_{corr} as a function of the parameters governing the glass transition, such as the temperature or the relaxation time, etc. (Dalle-Ferrier *et al.*, 2007; Berthier *et al.*, 2005; Fragiadakis *et al.*, 2009; Capacioli *et al.*, 2008; Dalle-Ferrier *et al.*, 2008). Above all, it provides the first experimental evidence of a growing length scale over the whole T and τ_α ranges of the viscous slowing down.

In Section 5.5.1, we present how to obtain these informations by using the linear dynamic susceptibility χ_T. Section 5.5.2 is devoted to a completely different method based on the non-linear susceptibility χ_3.

5.5.1 Studies of the correlation volume estimated through the linear dynamic susceptibility χ_T

By using the above-mentioned Eqs. (5.10), (5.11), and (5.13), the behavior of $N_{\mathrm{corr,T}}$ and $N_{\mathrm{corr,4}}$ can be obtained from the temperature derivative of the (normalized) correlation function in time. In Fig. 5.15, the three-point dynamic susceptibility $\chi_T(t)$ is drawn as extracted from the intermediate scattering function at short time scale measured by neutron spin-echo experiments in the case of glycerol for temperatures between 302 K and 350 K (Dalle-Ferrier, 2009), well above T_g ($T_g = 185$–190 K). When the temperature decreases, the time at which $\chi_T(t)$ is maximum increases, and the corresponding maximum value, $\max_t \mid \chi_T(t) \mid$, increases. By using Eqs. (5.10) and (5.11), this leads to increasing values of $N_{\mathrm{corr,T}}$ when T decreases. In Ref. (Berthier *et al.*, 2007*a*), it was shown that $N_{\mathrm{corr,T}}$ and $N_{\mathrm{corr,4}}$ can also be estimated from any (normalized) linear response function in frequency space. This is of great practical importance since close to the glass-transition temperature T_g, linear response functions are often more easily accessible in frequency space.

In Fig. 5.16, the temperature derivative of the normalized frequency-dependent dielectric constant $\chi(\omega, T) = \mathrm{Re}[\epsilon(\omega) - \epsilon(\infty)]/[\epsilon(0) - \epsilon(\infty)]$ has been used to calculate $T\chi_T(\omega)$ for glycerol close to T_g. Using the maximum over ω of $\chi_T(\omega)$ yields an estimate

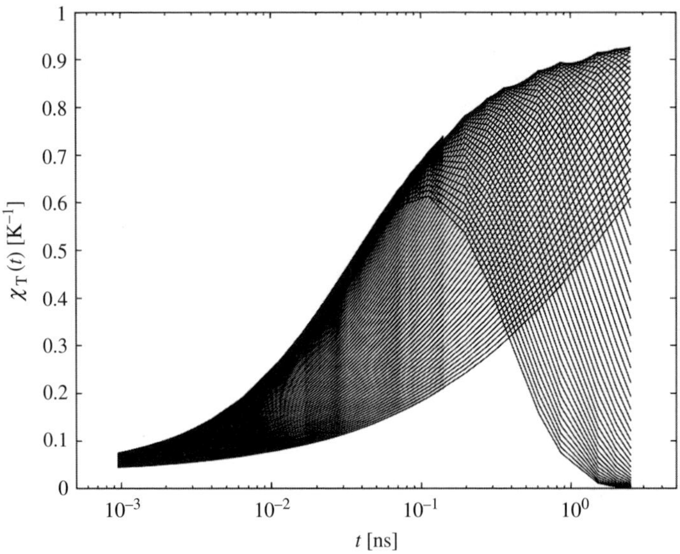

Fig. 5.15 The three-point dynamic susceptibility of glycerol versus time (measured by neutron spin-echo experiments), for a set of 48 temperatures separated by 1 K, between 302 K and 350 K. At each temperature, $\chi_T(t)$ shows a maximum around the average relaxation time τ_α. From (Dalle-Ferrier, 2009).

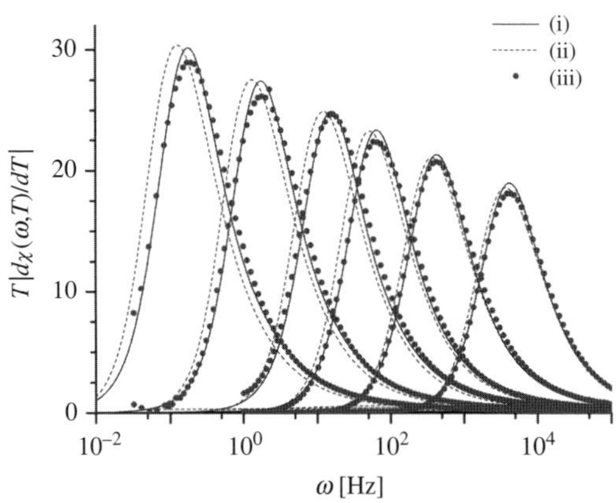

Fig. 5.16 The three-point dynamic susceptibility obtained from the linear response measured in the frequency domain. Here, the dielectric susceptibility of glycerol $\epsilon(\omega, T)$ was measured to give $T\partial\chi(\omega, T)/\partial T$, where $\chi(\omega, T) = \text{Re}[\epsilon(\omega, T) - \epsilon(\infty, T)]/[\epsilon(0, T) - \epsilon(\infty, T)]$, for six temperatures ranging from 196.6 K (extreme left curve) to 225.6 K (extreme right curve). Three different estimates (i)–(iii) of $T\partial\chi(\omega, T)/\partial T$ were used, yielding consistent results, with small deviations that are well understood. From (Dalle-Ferrier *et al.*, 2007).

of $N_{\text{corr,T}}$ that has the same main trend as that obtained with the maximum over t of $\chi_T(t)$: Although the magnitudes may be different, both estimates of $N_{\text{corr,T}}$ yield a similar increase when T decreases towards T_g.

In Fig. 5.17, $N_{\text{corr,T}}$ is plotted as a function of τ_α/τ_0, where τ_α is the α-relaxation time and τ_0 is a microscopic time, for a large number of glass-forming systems. For consistency, all the $N_{\text{corr,T}}$ values were obtained by using the maximum over time of the normalized correlation function (close to T_g the correlation functions in the time domain were obtained as Fourier transforms of the linear susceptibility data in the frequency domain). A common behavior is seen in Fig. 5.17, despite the fact that, depending on the available set of experimental data for each system, the $N_{\text{corr,T}}$ values were calculated from different kinds of experiments (dielectric spectroscopy, photon correlation spectroscopy, optical Kerr effect, neutron scattering). The most important feature is that for all the systems, $N_{\text{corr,T}}$ increases with τ_α, i.e. when the temperature of the supercooled liquid is decreased towards T_g. The number of dynamically correlated molecules thus increases as one approaches the glass transition for all the systems of Fig. 5.17.

In Fig. 5.17, the increase of $N_{\text{corr,T}}$ is quite fast when τ_α/τ_0 is not very large, typically when $\tau_\alpha/\tau_0 < 10^2$: This corresponds to the high-temperature part of the supercooled liquid dynamics, where the mode-coupling theory (MCT) can be considered. A power-law fit $N_{\text{corr,T}} \sim (\tau_\alpha)^{1/\gamma}$ yields $\gamma = 2$–3, compatible with the MCT predictions (Berthier *et al.*, 2007*b*; Biroli *et al.*, 2006). For larger τ_α, i.e. when τ_α/τ_0 is between 10^6 and 10^{12}, the increase of $N_{\text{corr,T}}$ is much weaker. This is expected from

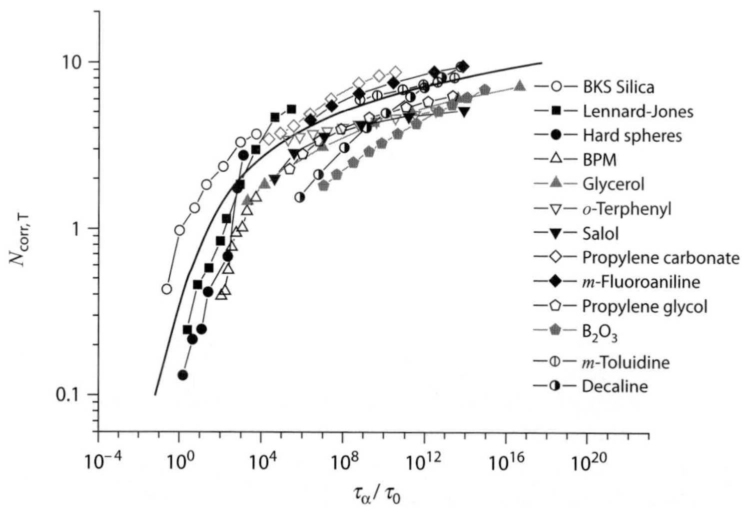

Fig. 5.17 Evolution of $N_{\text{corr,T}}$ as the glass transition is approached for 10 different liquids indicated in the figure (the microscopic time τ_0 was set to 1 ps), as well as for two simulation results (BKS model for silica (Berthier *et al.*, 2007*a*), and binary Lennard-Jones mixture). For comparison, the results for a colloidal hard-sphere system with $\tau_0 = 1$ ms are also included (Berthier *et al.*, 2005)). From (Dalle-Ferrier *et al.*, 2007).

the activation-based approaches, where activated events avoid the MCT divergence at T_c and are responsible for a slow logarithmic growth of length scales (Xia and Wolynes, 2000; Bouchaud and Biroli, 2004). The solid line in Fig. 5.17 is a phenomenological fit by a formula that contains these two regimes:

$$\tau_\alpha = A\left(\frac{N_{\mathrm{corr,T}}}{N_0}\right)^{\gamma}\exp\left[\left(\frac{N_{\mathrm{corr,T}}}{N_0}\right)^{\psi}\right], \tag{5.14}$$

where A, N_0, γ, and ψ are adjustable parameters.

The uncertainties about the pre-factors in Eqs. (5.10) and (5.11) and the normalization problems do not allow the determination of the absolute values of $N_{\mathrm{corr,T}}$, i.e. the real number of correlated molecules (Dalle-Ferrier *et al.*, 2008). A signature of this caveat is that the values of $N_{\mathrm{corr,T}}$ fall below 1 at high temperature (see Fig. 5.17). However, one can check that the main result, i.e. the temperature variation of $N_{\mathrm{corr,T}}$, does not significantly change when one replaces $\Delta c_{\mathrm{p}}(T_{\mathrm{g}})$ by $\Delta c_{\mathrm{p}}(T)$ or by $c_{\mathrm{p}}(T)$ in the pre-factor that relates $N_{\mathrm{corr,T}}$ to χ_{T} in eqn (5.10) (Dalle-Ferrier *et al.*, 2008). Indeed, the specific heat to be used in the calculation of $N_{\mathrm{corr,T}}$ may depend on whether the degrees of freedom involved are those lost at T_{g} or all those available in the system. We have seen that χ_{T} provides a lower bound for χ_4 (eqn (5.13)). Another aspect of the connection between the two quantities is the contribution of density variations to χ_4. The density becomes an important control parameter at high temperatures and should be considered through the following expression (Dalle-Ferrier *et al.*, 2007):

$$\chi_4^{\mathrm{NPT}} = \frac{k_{\mathrm{B}}T^2}{c_{\mathrm{v}}}\left(\chi_{\mathrm{T}}^{\mathrm{NVT}}\right)^2 + \rho^3 k_{\mathrm{B}} T \kappa_{\mathrm{T}}\left(\chi_{\rho}^{\mathrm{NPT}}\right)^2 + \chi_4^{\mathrm{NVE}}, \tag{5.15}$$

where $\chi_\rho^{\mathrm{NPT}}(t) = \left.\frac{\partial\phi}{\partial\rho}\right|_T$, χ_ρ characterizes the effect of the density ρ, c_{v} is the heat capacity at constant volume, κ_{T} is the isothermal compressibility and the superscripts indicate the considered statistical ensembles. Still corrections applied to take into account ρ variations would not change the T-dependence trend.

To summarize, all the above-mentioned limitations suggest that one should rather focus on the domain of the dramatic increase of the relaxation time. Accordingly the absolute numbers could be rescaled to one somehow above T_{c}, namely at the temperature T_{onset} above which no collective dynamics is expected.

A more rapid and rough way to get the temperature (or τ_α) variation of $N_{\mathrm{corr,T}}$ is to make the derivation of a Kohlrausch relaxation function in Eq. (5.11) assuming that the shape of the function remains unchanged, applying the so-called time–temperature superposition (TTS), and assuming that the stretching exponent β of the Kohlraush relaxation function is independent of the temperature. This leads to the following formula where the T dependence of $N_{\mathrm{corr,T}}$ is now directly expressed by the T dependence of the relaxation time, adjustable by any fitting procedure (using a VTF formula or any other functional form),

$$N_{\mathrm{corr,T}} = \frac{R}{\sqrt{\Delta C_{\mathrm{p}}}}\frac{\beta}{e}\left(\frac{d\ln(\tau_\alpha)}{d\ln(T)}\right). \tag{5.16}$$

Such a formula can also be applied when TTS is not verified by adding another term. It already gives the general trends of the increase of $N_{\mathrm{corr,T}}$ (see Fig. 5.17).

Finally, one should note that since Δc_{p} can be taken as a constant and replaced by $\Delta c_{\mathrm{p}}(T_{\mathrm{g}})$ or $c_{\mathrm{p}}(T_{\mathrm{g}})$, one expects, from Eq. (5.16), a direct relationship between the steepness of the $N_{\mathrm{corr,T}}$ temperature dependence and the fragility index. In fact, one finds no clear correlation between the temperature variations of $N_{\mathrm{corr,T}}$ and the so-called Angell plot yielding the fragility index (Angell, 1991), each system having a specific $\Delta c_{\mathrm{p}}(T_{\mathrm{g}})$ or $c_{\mathrm{p}}(T_{\mathrm{g}})$ or $\beta(T)$ exponent.

Recently, Capaccioli *et al.* (Capacioli *et al.*, 2008) proposed to analyze the $N_{\mathrm{corr}}(T)$ dependence in terms of "cooperative rearranging regions" (CRR) originally introduced by Adam and Gibbs (Adam and Gibbs, 1965). The authors start with the important assumption that $N_{\mathrm{corr,4}}$ is the average number of particles of the CRRs. This line of thought amounts to assuming that the linear size ξ of these CRRs is related to $N_{\mathrm{corr,4}}$ by $\xi^d = N_{\mathrm{corr,4}}/\rho$, where ρ is the molecules density and $d = 3$, leading to

$$\sigma_{\mathrm{CRR}}(T) = \frac{S_{\mathrm{c}}(T)}{k_{\mathrm{B}}} N_{\mathrm{corr,4}}(T), \tag{5.17}$$

where σ_{CRR} is the total configurational entropy of a CRR, and $S_{\mathrm{c}}(T)$ is the configurational entropy per molecule estimated from the excess entropy of the supercooled liquid with respect to the crystal.

The values of $S_{\mathrm{c}}(T)$ and $N_{\mathrm{corr,4}}(T)$ were computed in Ref. (Capacioli *et al.*, 2008) from various experimental sets of data (dielectric susceptibility in most cases, but also photon correlation spectroscopy as well as mechanical relaxation) for different liquids and polymers. For a given $\log(\tau_\alpha/\tau_0)$, the spreading of the $N_{\mathrm{corr,4}}$ values is larger than one order of magnitude, while the corresponding spreading of the σ_{CRR} values is only by about a factor 2. This suggests that the α-relaxation time is given by a material-dependent microscopic time τ_0 times a universal function of $\sigma_{\mathrm{CRR}}(T)$, namely $\log(\tau_\alpha(T)/\tau_0) = F[\sigma_{\mathrm{CRR}}(T)]$. As a result, the authors concluded that $N_{\mathrm{corr,4}}(T_{\mathrm{g}})$ and $S_{\mathrm{c}}(T_{\mathrm{g}})$ should be anti-correlated. Moreover, they observed that σ_{CRR} depends on the temperature T, contradicting the Adam–Gibbs theory, but satisfying a prediction from the random first-oder theory (RFOT) (Lubchenko and Wolynes, 2003, 2007):

$$\sigma_{\mathrm{CRR}}(T) \propto [S_{\mathrm{c}}(T)]^{-\theta/(d-\theta)} \propto [\xi(T)]^{\theta} \propto [N_{\mathrm{corr,4}}(T)]^{\theta/d}, \tag{5.18}$$

where the last equality was obtained by using the assumption $N_{\mathrm{corr,4}} \sim \xi^d$. By plotting $N_{\mathrm{corr,4}}$ as a function of $\sigma_{\mathrm{CRR}}(T)$, they found that the power law predicted by RFOT is well verified, and that θ lies in the range 2.2–2.5.

5.5.2 The non-linear susceptibility: A probe for dynamical correlations and their critical nature

In the previous section, we have seen that the temperature (or density, or relaxation time, etc.) dependence of the average number of dynamically correlated molecules in a supercooled liquid close to T_{g}, could be evaluated by using the dynamical susceptibility χ_{T}. We have also seen that possible limitations of χ_{T} made the comparison with other methods highly desirable. An alternative method relies on the non-linear response of the system to an external excitation (electrical, mechanical, etc.). By using an extended fluctuation–dissipation relation, Bouchaud and Biroli (Bouchaud and Biroli, 2005) showed that N_{corr} should be directly proportional to the non-linear susceptibility χ_3, which relates the response to the excitation at three times the frequency of the excitation. For an electrical excitation for instance:

$$\chi_3(\omega, T) = \frac{\epsilon_0 (\Delta\chi_1)^2 a^3}{k_{\text{B}}T} N_{\text{corr}} H(\omega\tau), \tag{5.19}$$

where ω is the pulsation, χ_1 the linear susceptibility with $\Delta\chi_1 = \chi_1(\omega = 0) - \chi_1(\omega \to \infty)$, a^3 the volume occupied by a molecule, $\tau = \tau_\alpha(T)$ the typical relaxation time at temperature T, and H a scaling function that goes to zero for low and large arguments. H and χ_3 are complex quantities. χ_3 gives the dominant term of the third-harmonics part of the response,

$$P(t)/\epsilon_0 = \left(E_0\chi_1(\omega) + \frac{3}{4}E_0^3\chi_{\bar{3}}(\omega) + \ldots \right) e^{-i\omega t} + \left(\frac{1}{4}E_0^3\chi_3(\omega) + \ldots \right) e^{-3i\omega t} + \ldots, \tag{5.20}$$

where $E_0 e^{-i\omega t}$ is the applied electric field and $P(t)$ the response of the supercooled liquid, i.e. the complex polarization. The response is an infinite sum of terms corresponding to odd frequencies, ω, 3ω, 5ω, $n\omega$, etc. (Thibierge *et al.*, 2008). The even terms are zero because the response to $-E_0 e^{-i\omega t}$ is the opposite of the response to $E_0 e^{-i\omega t}$. Each of those terms n is itself an infinite sum of terms involving the susceptibilities of order n, $n+2$, $n+4$, etc., but in Eq. (5.20), we restricted ourselves to the dominant ones. Equation (5.20) is easy to establish by writing the response P as a function of the excitation E in the time domain:

$$\frac{P(t)}{\epsilon_0} = \int \chi_1(t-t')E(t')dt' + \int \chi_3\left(t-t_1', t-t_2', t-t_3'\right) E\left(t_1'\right) E\left(t_2'\right) E\left(t_3'\right) dt_1' dt_2' dt_3', \tag{5.21}$$

where $\chi_1(t)$ and $\chi_3(t_1, t_2, t_3)$ are, respectively, the linear and cubic non-linear susceptibilities in the time domain. Equation (5.20) is obtained by Fourier transforming Eq. (5.21) in the case of a sinusoidal excitation field $E(t)$. The non-linear susceptibilities that appear in the responses at frequencies ω and 3ω, are, respectively, $\chi_{\bar{3}}(\omega) = \chi_3(-\omega, \omega, \omega)$ and $\chi_3(\omega) = \chi_3(\omega, \omega, \omega)$ (Thibierge *et al.*, 2008), $\chi_3(\omega_1, \omega_2, \omega_3)$ (which is invariant by permutation of its arguments) being the Fourier transform of $\chi_3(t_1, t_2, t_3)$.

The developments leading to Eq. (5.19) were inspired by spin-glass physics (Bouchaud and Biroli, 2005; Maglione *et al.*, 1986; Kirkpatrick and Wolynes, 1987; Zippelius, 1994; Donati *et al.*, 2002). A spin glass is a set of disordered interacting spins S_i with frustration (Binder and Young, 1986; Mydosh, 1993; Vincent, 2007). The transition from the high-temperature paramagnetic state to the low-temperature spin-glass state is a phase transition. However, the order associated with the spin-glass phase is not that of spin-spin correlations as in, e.g., ferromagnetic systems. It is characterized by a *four-point* correlation function $[< S_i S_j >^2]$, where $< >$ denotes a thermal average and $[\,]$ a spatial average. As a consequence, the divergence of the correlation length at the transition is associated through a fluctuation–dissipation (FD) relation with that of the *non-linear* magnetic (static) susceptibility (Miyako *et al.*, 1951; Omari *et al.*, 1986; Lévy and Ogielski, 1986; Vincent and Hammann, 1987; Lévy, 1988), contrary to the ferromagnetic transition where the two-point correlation $[< S_i S_j >]$ and the associated linear susceptibility are the relevant quantities. For the case of structural glasses, Bouchaud and Biroli (Bouchaud and Biroli, 2005) started from the four-point correlation function $G_4(r'-r,t) = <\delta\rho(r,t{=}0)\,\delta\rho(r,t)\,\delta\rho(r',t{=}0)\,\delta\rho(r',t)> - C^2(t)$, with $C(t) = <\delta\rho(r,t{=}0)\,\delta\rho(r,t)>$ which is reminiscent of the spin-glass one $[< S_i S_j >^2]$: G_4 evaluates to what extent the density variation during a time interval [0,t] at point $\mathbf{r}$, is correlated to that at point $\mathbf{r}'$ during the same time interval, while $[< S_i S_j >^2]$ evaluates to what extent the fact that spin i has a given orientation is correlated to the fact that spin j has another given orientation (which can be different from that of spin i). Note that, due to quenched disorder, the signs of the correlations between two spins, separated by a given distance r, are random in space: This is why the average of the two points correlation function, namely $< S_i S_j >$, decreases extremely fast in space, over a distance of the order of the lattice spacing a, much smaller than the length $\xi \gg a$ giving the range of the correlation between one given spin and the rest of spins. This suggests that the use of $< S_i S_j >^2$ is required to capture the amorphous long-range order in spin-glasses. Intuitively, the similarity of the two situations could be given by: "Inside a correlated volume, a change at point $\mathbf{r}$ (or at spin i) implies also a change at point $\mathbf{r}'$ (or spin j)" (A dynamical picture of the spin-glass case is developed in Section II.A of Ref. (Berthier *et al.*, 2007*a*)). A difference between the two situations is that $[< S_i S_j >^2]$ is time independent, contrary to G_4, a point that is related to the fact that the static or ac non-linear susceptibilities are considered in each of the two cases. Extending the FD relation between the correlation function and the non-linear susceptibility from spin-glasses to structural glasses is not straightforward because of the absence of quenched disorder in the second case (Bouchaud and Biroli, 2005; Bouchaud *et al.*, 1997). The calculations of Bouchaud and Biroli have two important consequences. First, a measurable quantity, $\chi_3(\omega)$ is directly related to the size $N_{\rm corr}$ of the dynamical correlations, while the relation between $\chi_{\rm T}$ and $N_{\rm corr,T}$ was hampered by the presence of possible basic limitations (see the introduction of Section 5.5). Secondly, as for spin glasses, a divergence or an increase of $\max_\omega(|\,\chi_3(\omega)\,|)$ when T decreases should be considered as an indication of the possible presence of an underlying phase transition for structural glasses (Bouchaud and Biroli, 2005).

We note here that ac thermal measurements at a frequency three times that of the excitation frequency have already been performed in supercooled liquids (Birge, 1986). However, they should not be confused with the non-linear susceptibility measurements of interest here. Clearly, such ac thermal measurements allow measurement of the frequency-dependent specific heat in the *linear* regime, and the measured third harmonics is related to the electrical and thermal circuits involved in the experiment.

χ_3 is measurable using standard dielectric spectroscopy techniques already used to measure the linear susceptibility χ_1 (Dixon *et al.*, 1990; Kudlik *et al.*, 1999; Lunkenheimer *et al.*, 2000; Weinstein and Richert, 2007*c*), in which the supercooled liquid is the dielectric layer of a capacitor: χ_3 is proportional to the current at pulsation 3ω, $I(3\omega)$ for an applied excitation voltage at pulsation ω, $V(\omega)$ (Thibierge *et al.*, 2008).

However, the magnitude of the current $I(3\omega)$ due to the non-linear response is very low: For glycerol, the ratio of $I_{3\omega} = \mid I(3\omega) \mid$ to the modulus of the current at pulsation ω, $I(\omega) = V(\omega)/Z(\omega)$, where $Z(\omega)$ is the impedance of the capacitor, is of the order of $10^{-5} - 10^{-7}$ for a standard voltage $V(\omega)(\mathrm{rms}) \sim 10\,\mathrm{V}$ and a capacitor thickness of 10–20 μm (Thibierge *et al.*, 2008). This is lower than typical harmonic distortions of standard electronics. This experimental difficulty should be overcome by increasing the magnitude of $V(\omega)$ (Weinstein and Richert, 2007*c*) or by using filtering or bridge techniques (Thibierge *et al.*, 2008). Figure 5.18 shows an example of such a current dependence as a function of the applied voltage: The two different experimental techniques used agree well, thus demonstrating their reliability. A possible spurious contribution to $I(3\omega)$ is that of heating effects: The power dissipated in the sample oscillates at a frequency 2ω, thus leading to a 3ω contribution to $P(t)$. Such an effect must be evaluated and controlled in order to ensure that it remains negligible (Crauste-Thibierge *et al.*, 2010).

The third-harmonics non-linear susceptibility has been measured by Crauste-Thibierge *et al.* for supercooled glycerol at temperatures ranging from $T_g + 4\,\mathrm{K}$ to $T_g + 35$ K ($T_g \approx 190$ K) by using a two-sample bridge technique (Crauste-Thibierge *et al.*, 2010). Figure 5.19 shows the normalized magnitude of χ_3, $X(\omega, T) = \mid \chi_3(\omega) \mid k_B T/[(\Delta\chi_1)^2 a^3 \epsilon_0]$ as a function of frequency for three temperatures. According to Eq. (5.19), X should be $N_{\mathrm{corr}} \mid H(\omega\tau) \mid$. Two remarkable results confirming the

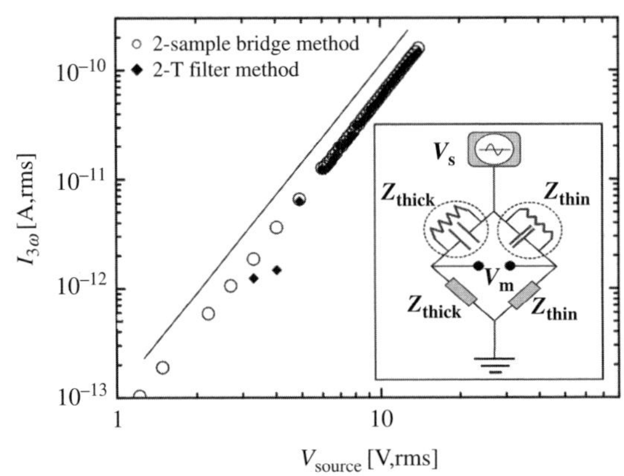

Fig. 5.18 Magnitude $I_{3\omega}$ of the current at frequency 3ω measured through a capacitor in which supercooled glycerol is the dielectric, as a function of the voltage magnitude at frequency ω applied to the capacitor (Thibierge *et al.*, 2008). $I_{3\omega}$ is proportional to the modulus of the cubic non-linear susceptibility. Open circles: The measurement method is that of the two-sample bridge technique shown in the inset. Closed diamonds: Same current measured using a twin-T notch filter method described in (Thibierge *et al.*, 2008). The continuous line is the power-3 dependence. The experimental conditions were $T = 210.2$ K and $\omega/2\pi = 43.76$ Hz. From (Thibierge *et al.*, 2008).

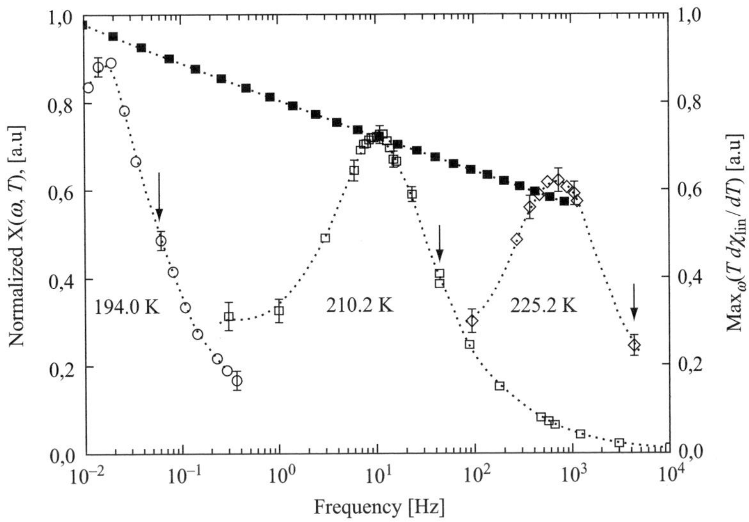

Fig. 5.19 Open symbols: For each of the three temperatures labeling the curves, $X(\omega, T)$, the normalized modulus of the non-linear susceptibility $\chi_3(\omega, T)$ in supercooled glycerol is shown as a function of the frequency $f = \omega/2\pi$. $X(\omega, T)$ is defined in order to be $N_{\text{corr}} \mid H(\omega, T) \mid$ according to Eq. (5.19) (see text). The arrows indicate the three relaxation frequencies (for which χ_1'' is maximum) corresponding to the three temperatures. Closed squares: Number of correlated molecules estimated by using $T\chi_T(T)$ (see Eq. (5.10)) as a function of the relaxation frequency $1/\tau_\alpha(T)$. Arbitrary normalization factors have been applied in order that X remains below 1, and that the $T\chi_T$ line coincides with the maximum of $X(\omega, 210.2\,\text{K})$.

predictions (Bouchaud and Biroli, 2005; Tarzia *et al.*, 2010) appear in Fig. 5.19. First, X is peaked at a frequency f^* of the order of $1/\tau_\alpha(T)$, more precisely $f^* \approx 0.21/\tau_\alpha(T)$; secondly the height of the peak (proportional to N_{corr} according to Eq. (5.19)) increases as the temperature decreases. To fully confirm the agreement with the predictions (Bouchaud and Biroli, 2005), the scaling in Eq. (5.19) , i.e. that H (thus X and the phase of χ_3) depend only on the product $\omega\tau$ must be verified. Thorough investigations of this point show that the scaling is well obeyed (Crauste-Thibierge *et al.*, 2010). A weak departure from scaling at low frequency can be explained by the presence of "trivial" dielectric saturation effects (Böttcher and Bordewijk, 1973; Déjardin and Kalmykov, 2000; Weinstein and Richert, 2007*c*) superimposed on the genuine collective effect.

These results demonstrate experimentally that the cubic non-linear susceptibility is a relevant tool to evaluate the $N_{\text{corr}}(T)$ dependence. Note that, as what is measured is the product $N_{\text{corr}} \mid H \mid$, the real number of correlated molecules cannot be obtained directly, H being unknown. Its peak value is expected to be lower than 1 in most cases, since calculations for independent Brownian molecules give $X(\omega, T) \leq 0.2$ (Crauste-Thibierge *et al.*, 2010; Déjardin and Kalmykov, 2000).

As we have seen above, it is important to compare the two different methods (based on χ_T and χ_3) used to estimate the size N_{corr} of the dynamic heterogeneities. In Fig. 5.19, we can see that the f_α ($= 1/\tau_\alpha$) dependencies of the two methods are close. Thus, the possible limitations of the χ_T method evoked above should not hold at least for glycerol, and the statement that a unique dynamical length scale governs the growth of dynamical susceptibilities (Berthier *et al.*, 2007*a*) is confirmed. It was also found in Ref. (Crauste-Thibierge *et al.*, 2010) that the overall behavior of $\chi_3(\omega)$ and $\chi_T(2\omega)$ were close to each other, as predicted in Ref. (Tarzia *et al.*, 2010). Note that the N_{corr} values obtained through χ_3 or χ_T cannot be compared directly to the sizes obtained by NMR (see Section 5.3.3) since, as evoked above and below, unknown normalization factors are involved in the N_{corr} calculations.

For $\omega\tau > 1$, $\chi_3(\omega)$ exhibits a power-law behavior with an exponent -0.65 ± 0.04 (Crauste-Thibierge *et al.*, 2010). This behavior is predicted by mode-coupling theory (MCT) (Tarzia *et al.*, 2010), and the experimental exponent is compatible with the MCT prediction $-b \approx -0.6$ for glycerol (Lunkenheimer *et al.*, 1996). However, as MCT should be relevant only at much higher temperatures, the authors of Ref. (Crauste-Thibierge *et al.*, 2010) suggest that such a power-law regime for large ω with the same exponents for χ_1 and χ_3 is a generic property, valid outside the MCT regime, which would be related to the incipient criticality of the system.

An important aspect of the agreement between the experimental data (Crauste-Thibierge *et al.*, 2010) and the predictions (Bouchaud and Biroli, 2005; Tarzia *et al.*, 2010) in what concerns the non-linear susceptibility $\chi_3(\omega, T)$, is that this agreement suggests the possible existence of an underlying phase transition. The correlation length that would diverge at the critical temperature should be that of the dynamically correlated heterogeneities, and the associated susceptibility should be χ_3, which as we have seen increases when T decreases. The long-standing conjecture of an underlying phase transition (Gibbs and DiMarzio, 1958; Kirkpatrick and Wolynes, 1987; Kirkpatrick *et al.*, 1989; Sethna *et al.*, 1991; Bennemann *et al.*, 1999; Franz and Parisi, 2000; Donati *et al.*, 2002; Ritort and Sollich, 2003; Bouchaud and Biroli, 2004, 2005; Dyre, 2006; Tarzia *et al.*, 2010) would imply a critical temperature (such as the Kauzmann temperature T_K) well below T_g. Clearly, the experimental investigation of the critical behavior at such temperatures is extremely difficult because of the tremendous relaxation times below T_g; this is why the $\chi_3(\omega, T)$ results are important clues of the criticality.

The χ_3 measurements (Crauste-Thibierge *et al.*, 2010) represent a challenge for models and theories in which dynamical correlations play a central role: They should reproduce not only the increase of N_{corr} as $T \to T_g$ (see Fig. 5.19) but also the shape of $\chi_3(\omega, T)$, given by the H function in eqn (5.19), which carries significant information on the physics of the dynamical correlations. The phenomenological model of Richert *et al.* (Richert and Weinstein, 2006; Weinstein and Richert, 2007*c*; Huang and Richert, 2008*b*) is of particular interest because it already accounts for non-linear susceptibility data at the excitation frequency 1ω. It assumes that the supercooled liquid is a collection of "Debye-like" dynamical heterogeneities (DH), each of them having its own relaxation time τ_{DH}. The electrical power absorbed by each DH raises its temperature, with a dc and an ac component. These components lead to 1ω and 3ω heating contributions to the polarization $P(t)$. The 3ω contribution should be considered as resulting from the dynamical correlations related to the glassy dynamics (Bouchaud and Biroli, 2005; Tarzia *et al.*, 2010), phenomenologically described through the heterogeneous heating of the DHs (Richert and Weinstein, 2006; Weinstein and Richert, 2007*c*; Huang and Richert, 2008*b*). Further work is needed to investigate the 3ω contribution within such a model. Still, further studies of interest would be to look at the non-linear susceptibility $\chi_3(\omega, T)$ predicted by models in which dynamical correlations appear as a consequence of *ab-initio* physical assumptions.

5.6 Other experiments related to heterogeneity

This section outlines a few experimental observations that do not detect heterogeneity directly, but our understanding of these phenomena is bound to remain incomplete if heterogeneous dynamics is not accounted for.

5.6.1 Translation–rotation decoupling

(R. RICHERT)
Most correlation functions associated with the primary structural relaxation of glass-forming liquids display essentially the same temperature dependence for a certain material, irrespective of the technique (Schröter and Donth, 2000; Stickel *et al.*, 1996). Typical examples are viscosity, dielectric relaxation, shear-stress relaxation, photon correlation spectroscopy, molecular reorientation by NMR, etc. While self-diffusivity is one of the most fundamental transport properties of liquids, the diffusion constant D has been observed by NMR techniques to deviate from other measures of structural relaxation such as viscosity or rotational time scales (Chang and Sillescu, 1997; Fujara *et al.*, 1992). It turned out that diffusivity was a factor of about 3 more effective at T_c than what was expected from the behavior at higher temperatures. Here, T_c refers to the critical temperature of the idealized mode-coupling theory (Götze and Sjögren, 1992). Later, probe (Cicerone and Ediger, 1996; Wang and Ediger, 1997; Cicerone *et al.*, 1995*b*) and self-diffusion (Qi *et al.*, 2000; Swallen *et al.*, 2003; Mapes *et al.*, 2006; Swallen *et al.*, 2006, 2009) experiments were extended to near T_g and revealed that the decoupling becomes much more pronounced than what had been observed near T_c. This "enhanced diffusion" phenomenon is seen for molecular liquids, polymers, as well as metallic glass formers (Meyer, 2002; Zöllmer *et al.*, 2003). At the glass transition, the separation can increase to factors of 100 to 1000, as indicated in Fig. 5.20. These significant decoupling effects are observed only for liquids that are fragile ($m > 70$) and if diffusion probes are used that match the size of the liquid constituents (Ngai, 1999; Rajian *et al.*, 2006).

Typical reference curves with which $D(T)$ data would be compared are viscosity $\eta(T)$ or some rotation related relaxation time $\tau_{\text{rot}}(T)$. The expectation for how these quantities should be related stems in part from hydrodynamic equations, the Stokes–

Einstein (SE) and Debye–Stokes–Einstein (DSE) relations

$$D = \frac{k_B T}{6\pi\eta r} \text{ and } \tau_{\text{rot}} = \frac{1}{\ell(\ell+1)D_{\text{rot}}} = \frac{8\pi\eta r^3}{\ell(\ell+1)k_B T}, \tag{5.22}$$

where r is the radius of the spherical particle considered (Stillinger and Hodgdon, 1994). The expectation regarding the product $D\tau_{\text{rot}}$ is simply a value that no longer depends on temperature or viscosity, $D\tau_{\text{rot}} = cr^2$, where the constant c is only a matter of the rank ℓ of the orientational measurement. While the translation–rotation decoupling is often referred to a violation of the SE and DSE laws, these relations hold only for particles of a size large enough so that the structure of the liquid is irrelevant and the liquid behavior is entirely determined by its viscosity η. Therefore, the hydrodynamic equations make no predictions for the dynamics of a neat liquid or for self-diffusion. On the other hand, the relations do appear to apply in the high-temperature regime, $T > T_c$, and the observations often fit the so-called fractional SE law, $D \propto (T/\eta)^c$, with an exponent c that is commonly near 0.7. Because only the very fragile materials show clear decoupling, a fragility dependence such as $c(m)$ has not been established, but the fact that c tends to unity for smaller m values does suggest a systematic fragility dependence of the exponent c. In conclusion, the question remains as to what is changing from high to low temperatures that results in the gradual failure of the DSE relation.

The connection of the above issue to heterogeneity is founded in the notion that a particle can not be envisioned to undergo translational motion across several molecular diameters without altering its orientation. This conclusion would be necessary in the case of homogeneous dynamics, but can be circumvented in a heterogeneous scenario

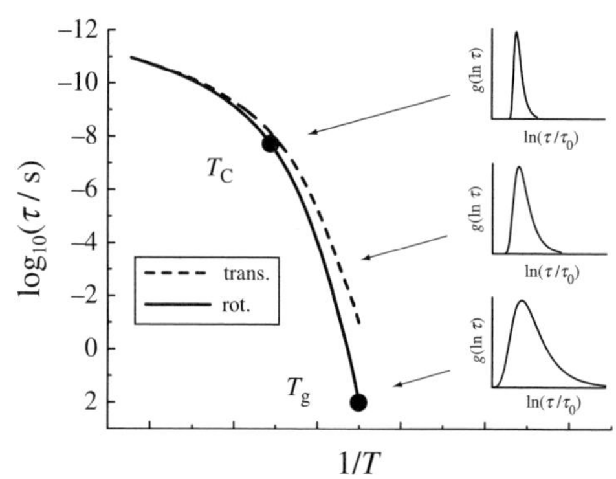

Fig. 5.20 Sketch of a typical rotational time scale τ_{rot} versus reciprocal temperature for a fragile supercooled liquid (solid curve). The dashed curve indicates the decoupling of diffusivity after translating D to some equivalent time constant τ_{trans}. Between T_c and T_g the separation of the curves increases from a third of a decade to two or three orders of magnitude. Diffusivity data would be shifted so that τ_{trans} matches τ_{rot} at the highest temperatures, $T > T_c$. The insets display the change in relaxation time dispersion anticipated by several models aimed at rationalizing this translation–rotation decoupling.

(Stillinger and Hodgdon, 1994; Tarjus and Kivelson, 1995; Cicerone *et al.*, 1997; Ediger, 1998; Xia and Wolynes, 2001). Within this picture of spatially dispersed dynamics, one can argue that the average time scale of rotational motion is governed by the slower contributions within the distribution of relaxation times, $D_{\mathrm{rot}} \propto <\tau>^{-1}$, while the faster time constants are weighted higher in determining the average translation time with D_{trans} being approximated by $<\tau^{-1}>$ (Cicerone *et al.*, 1997). In this picture, the gradual increase of decoupling as the temperature is lowered is seen as a consequence of the change in the extent of relaxation-time dispersion (Hall *et al.*, 1997), as indicated in Fig. 5.20. This view seems entirely consistent with the observation that more fragile materials display higher levels of decoupling at T_g: first, fragile liquids are considered to have broader relaxation-time distributions at T_g, secondly, there has been the expectation that all liquids have lost much of their dispersive relaxation character for temperatures around T_c and beyond.

The above picture has been generally accepted over several years, until it turned out that it is particularly the class of very fragile liquids that obey time– temperature superposition (TTS), that is their width of the relaxation-time dispersion does not change within the $T_g - T_c$ range (Wang and Richert, 2007*a*). Examples for considerable decoupling at T_g combined with TTS are *o*-terphenyl (Richert, 2005), tris-naphthylbenzene (Richert *et al.*, 2003), sucrose benzoate (Rajian *et al.*, 2006), ROY (Sun *et al.*, 2009). As a result, the rationale involving the idea that rotation and translation sense different moments of the distribution is not tenable. Few alternatives have been proposed (Diezemann *et al.*, 1998; Richert and Samwer, 2007; Graessley, 2009).

5.6.2 Exponential probe rotation

(R. RICHERT)

Molecular probes are often used in order to assess the dynamics of a liquid (Richert and Blumen, 1994). The above-mentioned techniques of optical depolarization (Wang and Richert, 2004*b*), photobleaching (Cicerone and Ediger, 1993), and single-molecule studies all rely on probes that are assumed to report the environmental dynamics. How well the guest molecule will reflect the host dynamics is a matter of its size and shape, and only two situations are simple. If the probe molecule is practically identical to the liquid constituents, guest and host dynamics will be indistinguishable. If the probe is a sufficiently large object on the structural length scales of the liquid, its dynamics can be safely predicted by hydrodynamics. The question that is addressed here is what determines the transition from heterogeneous to hydrodynamic probe dynamics.

Using molecular probes of various sizes in *o*-terphenyl (OTP), Ediger and coworkers found that larger probes (e.g. rubrene in OTP) display more exponential rotation correlation functions, whereas probes that match the liquid in terms of size (e.g. tetracene in OTP) are subject to the relaxation-time dispersion of the neat liquid (Cicerone *et al.*, 1995*a*). In this viscous range of *o*-terphenyl, Ediger's results of very slow rate exchange (Wang and Ediger, 1999) led to the conclusion that spatial averaging is at the origin of this Debye character of the large probes (Cicerone *et al.*, 1995*a*).

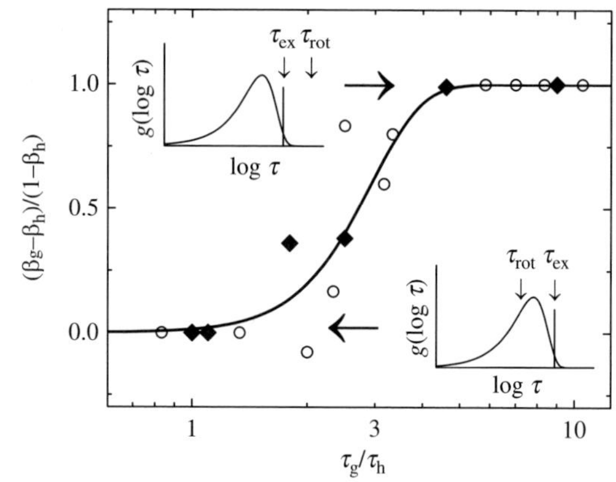

Fig. 5.21 Guest and host dynamics in viscous supercooled liquids derived from optical depolarization experiments (open circles) and from dielectric relaxation measurements (solid diamonds). The graph shows how the probe rotation evolves from dispersive to Debye-type character as the relaxation time of the guest becomes slower than that of the host. The line serves as a guide only. The insets displays the time-averaging interpretation in terms of the transition from $\tau_g < \tau_{ex}$ to $\tau_g > \tau_{ex}$.

Other extensive data is available from phosphorescence depolarization studies of molecular triplet-state probes in organic low molecular weight glass-forming liquids (Wang and Richert, 2004*b*; Yang and Richert, 2002). The rotation correlation functions of rank 2 for the probes were compared with the simultaneously measured Stokes-shift correlation function $C(t)$ of the same probes, where $C(t)$ reports the solvent response in the immediate vicinity of the chromophores. Both guest and host dynamics were analyzed by stretched exponentials yielding the KWW parameters τ_g, β_g and τ_h, β_h for the guest and the host behavior, respectively. In order to generate a value that gauges how much more exponential the guest relaxation is over that of the host, the function $\rho = (\beta_g - \beta_h)/(1 - \beta_h)$ is employed. This quantity is zero if guest and host dispersions are equal, and unity if the guest displays Debye-type dynamics with $\beta_g = 1$. The optical data are shown as open symbols in Fig. 5.21 in terms of ρ versus the time scale ratio τ_g/τ_h (after shifting a factor of 6 to the left on the abscissa scale). The data follow a common trend of undergoing a sharp transition to exponential probe rotation where τ_g begins to exceed $6\tau_h$ by a factor of $2 - 3$ or more (Huang and Richert, 2006*a*). The seemingly arbitrary factor of 6 is meant to account for the typical difference in time scale between the faster solvation dynamics and a single-particle correlation time (see Section 5.4.1).

The above ambiguity inherent in comparing guest and host dynamics on the basis of different correlation functions has been eliminated by a dielectric relaxation approach to probe rotation (Shahriari *et al.*, 2004; Huang *et al.*, 2005). In this case, the non-polar host liquid is doped with a dipolar guest at the 1 wt% level. The mismatch in guest/host dipole moment is designed such that the 1% guest concentration leads to a signal of comparable magnitude with that originating from the 99% host contribution.

In this manner, dynamics of the probe and of the liquid are measured simultaneously and in terms of the same correlation function. The results from these dielectric data are included in Fig. 5.21 as solid symbols and indicate the same transition towards Debye-type probe rotation for $\tau_g/\tau_h > 3$. This transition can be identified for average dielectric relaxation times ranging from 5 s to 50 ns, i.e. there is no indication of a change across most of the T_g to T_c range (Huang and Richert, 2006*b*).

In all cases compiled in Fig. 5.21 the probe size did not exceed that of the liquid molecules by more than a factor of two. Therefore, time averaging is a more likely explanation than spatial averaging over the heterogeneous dynamics of the host liquid, very much in the sense of the environmental fluctuations put forward by Anderson and Ullman (Anderson and Ullman, 1967) and models used in the context of NMR (Sillescu, 1971), solvation dynamics (Richert, 2001*c*), and photochromic probes (Richert, 1988*b*), and other processes (Zwanzig, 1990). As indicated by the insets of Fig. 5.21, the transition towards $\beta_h = 1$ is understood as the rotation time becoming slower than the time scale τ_{ex} of rate exchange, and the probe will eventually sense an averaged environment while reorienting, as in the hydrodynamic limit.

5.7 Conclusions

In the above, heterogeneity has been discussed sorted according to the technical approach or experimental method. Many of the techniques have actually emerged from the interest in solving the nature of the relaxation behavior in supercooled liquids. This concluding section critically summarizes the results with focus on the statement made with regards to heterogeneity *per se*, to the exchange times involved, and to the topology and length scales of dynamically distinct domains.

5.7.1 Heterogeneous versus homogeneous dynamics

(R. RICHERT)
Irrespective of the technique used and the material studied, heterogeneous dynamics has been found at the origin of overall dispersive relaxation behavior in practically all cases. The claim of heterogeneity alone, however, will not specify the dispersion intrinsic in each local response. To assess this intrinsic non-exponentiality, the net correlation function is expressed as a superposition of KWW-type decays with “intrinsic” stretching quantified by β_{intr},

$$\phi(t) = \phi_0 \int_{-\infty}^{\infty} g(\tau) \exp\left[-(t/\tau)^{\beta_{\text{intr}}}\right] d\tau. \tag{5.23}$$

If $\phi(t)$ itself is approximated by a stretched exponential with exponent β_{KWW} as in Eq. (5.1), then β_{intr} could be anywhere in the range $\beta_{\text{KWW}} \leq \beta_{\text{intr}} \leq 1$. The case $\beta_{\text{intr}} = \beta_{\text{KWW}}$ implies that $g(\tau) = \delta(\tau - \tau_{\text{KWW}})$ and the case of homogeneous dynamics. The other extreme is given by $\beta_{\text{intr}} = 1$, so that the intrinsic responses are purely exponential and $g(\tau)$ will account for the entire overall relaxation time dispersion. As a continuous metric for the degree of heterogeneity, the following quantity η has been proposed (Böhmer *et al.*, 1998*a*),

$$\eta = \frac{\beta_{\text{intr}} - \beta_{\text{KWW}}}{1 - \beta_{\text{KWW}}}. \tag{5.24}$$

Most experimental findings point to $\eta = 1$ or purely exponential intrinsic modes ($\beta_{\text{intr}} = 1$). Dipolar solvation dynamics has been found to be particularly sensitive to the value of β_{intr}, and $\beta_{\text{intr}} = 1$ provided a superior description of the experimental data over the $\beta_{\text{intr}} = 0.9$ case (Wendt and Richert, 2000).

As discussed in Section 5.6.2 above, the observation of a Debye-type susceptibility or exponential correlation decay in a supercooled liquid is not necessarily an indicator of dynamic homogeneity. Where dispersive dynamics are expected, Debye-type processes usually arise from a mode that is slower than the structural relaxation itself. Wong and Angell have interpreted the Debye-type behavior of the large ion pair tri-*n*-butyl ammonium picrate as molecular probe in *ortho*-terphenyl observed by Davies, Hains, and Williams (Davies *et al.*, 1973) as a result of the solvent molecules having "themselves undergone so many rearrangements that all local environmental effects on the probe relaxation have been averaged out" (Wong and Angell, 1976). This early example is a special case of the systematic probe-size variation outlined above, where the hydrodynamic limit of an exponential single-particle rotation correlation function has been reached. Another common example for such behavior is the class of monohydroxy alcohols and some other hydrogen-bonding liquids, in which the prominent dielectric process is of Debye character (Hansen *et al.*, 1997; Murthy and Tyagi, 2002; Wang and Richert, 2004*a*). The feature is again explained readily on the basis of this polarization mode being significantly slower than the true α-process (Huth *et al.*, 2007), and the effect of heterogeneous structural dynamics has been eliminated by time averaging (Anderson and Ullman, 1967; Huang and Richert, 2006*a*). Rotational dynamics gradually approach Debye-type behavior as the temperature is increased (Wang and Richert, 2007*a*), and if the α-process itself is no longer dispersive, the question of homogeneous versus heterogeneous dynamics ceases to be sensible.

5.7.2 Fluctuations in time

(R. RICHERT)
If a certain relaxation time constant τ is assigned to the domain with the liquid, e.g. a value slower or faster than the ensemble average, then this τ can not persist indefinitely, as ergodicity is not restored until time and ensemble averages are equal. The time scales involved in the fluctuations of these rate are commonly referred to as rate exchange (or heterogeneity lifetime). With τ_{ex} we denote the average time involved in a domain experiencing a considerable change in its rate or time constant τ. If the ensemble-averaged structural relaxation time τ_α changes, e.g. as a result of a different temperature or pressure, we expect the exchange time τ_{ex} to adjust accordingly. Therefore, it is useful to quantify rate exchange by a memory parameter, $Q_{\text{sel}} = \tau_{\text{ex}}/\tau_{\text{sel}}$ or $Q_\alpha = \tau_{\text{ex}}/\tau_\alpha$ (Heuer, 1997), which should be associated with the

caveat that the ratio can refer to the mean of the spectrally selected ensemble (τ_{sel}) or of the full ensemble (τ_α).

Across the literature, values for Q between 1 and $> 10^6$ can be found, as itemized above. A value of 1 is for Q_{sel}, which translates into a Q_α of around 3 (depending on filter efficiency). In fact, $Q_\alpha \approx 3$ is the prevailing result for rate exchange, and in the majority of cases no significant temperature dependence of Q is found. For many structural relaxations, the probability density of relaxation times displays a rather sharp cutoff at τ values slightly exceeding the mean value τ_α. As an example, for the common Cole–Davidson-type susceptibility, $\chi^\star(\omega) = \chi_\infty + \Delta\chi(1 + i\omega\tau_0)^{-\gamma}$, no modes are found with $\tau > \tau_0$. The results of $Q_\alpha = 3$ are consistent with saying that τ_{ex} coincides with this longest time scale of the system. In fact, this can be turned around to arrive at the view that there is a wider and more symmetric underlying distribution of relaxation times, but the exchange generates the cutoff and is therefore responsible for the asymmetry and KWW-type features. Such approaches, a gaussian probability density of $\ln \tau$ plus exchange, have been used to compute KWW-type relaxations in the presence of exchange (Richert, 2001*c*).

A typical realization of exchange effects is the following: for each mode with relaxation time τ_i, a time t_{ex} determined at random from the exponential characterized by the time constant of the exchange process, τ_{ex}. After the time t_{ex} has elapsed, a new random number from the probability density $g(\tau)$ is picked, irrespective of the previous value (Richert, 2001*c*). However, we do not know that this is a realistic image of exchange. Faster and slower modes could have different effective τ_{ex} values, a τ_i could undergo a more diffusive motion through τ-space. Furthermore, one can assume a coupling between exchange and a nearby relaxation event as suggested recently (Rehwald *et al.*, 2010), or one can use a model in which relaxation itself erases memory of the current τ_i entirely (Diezemann, 2002).

5.7.3 Fluctuations in space

(R. RICHERT)

The independence of fast and slow modes is often associated with the picture of spatial clustering or dynamically distinct domains in real space. The majority of experimental techniques, even those that clearly demonstrate heterogeneity, do not imply the spatial separation of the distinct modes. The only direct experimental approach to the spatial scales of heterogeneous dynamics is the 4D CP NMR technique, leading to length scales between 2 and 4 nm for poly(vinylacetate) and *o*-terphenyl, and a smaller value of 1 nm for glycerol (Reinsberg *et al.*, 2001, 2002).

While not necessarily equal, the above results are very similar to the $1 - 3.5$ nm sizes for ξ typically cited (Hempel *et al.*, 2000) in the context of cooperative rearranging regions (CRRs) introduced by Adam and Gibbs (Adam and Gibbs, 1965). If the cooperativity length scale ξ is understood as the (average) distance required for two molecules to relax independently, one should expect similar values for the domain size of heterogeneous dynamics. Spatial scales within the dynamics of supercooled liquids also emerge from theoretical considerations (Berthier, 2004; Berthier *et al.*, 2005), but the topologies involved are not necessarily simple (Stevenson *et al.*, 2006). Molecular dynamics simulations and colloidal materials are systems for which the spatial aspect of heterogeneity can be observed directly, and those cases are treated in separate chapters of this book.

Acknowledgments

The authors thank R. Böhmer and M.D. Ediger for helpful comments on the manuscript.

References

Adam, G. and Gibbs, J.H. (1965). *J. Chem. Phys.*, **43**, 139.

Adhikari, A.N., Capurso, N.A., and Bingemann, D. (2007). *J. Chem. Phys.*, **127**, 114508.

Anderson, J.E. and Ullman, R. (1967). *J. Chem. Phys.*, **47**, 2178.

Angell, C.A. (1991). *J. Non-Cryst. Solids*, **131-133**, 13.

Angell, C.A., Ngai, K.L., McKenna, G.B., McMillan, P.F., and Martin, S.W. (2000). *J. Appl. Phys.*, **88**, 3113.

Bagchi, B. and Chandra, A. (1991). *Adv. Chem. Phys.*, **80**, 1.

Bennemann, C., Donati, C., Baschnagel, J., and Glotzer, S.C. (1999). *Nature*, **399**, 246.

Berthier, L. (2004). *Phys. Rev. E*, **69**, 020201.

Berthier, L., Biroli, G., Bouchaud, J.-P., Cipelletti, L., El Masri, D., L'Hôte, D., Ladieu, F., and Pierno, M. (2005). *Science*, **310**, 1797.

Berthier, L., Biroli, G., Bouchaud, J.-P., Kob, W., Miyazaki, K., and Reichman, D.R. (2007*a*). *J. Chem. Phys.*, **126**, 184503.

Berthier, L., Biroli, G., Bouchaud, J.-P., Kob, W., Miyazaki, K., and Reichman, D.R. (2007*b*). *J. Chem. Phys.*, **126**, 184504.

Binder, K. and Young, A. P. (1986). *Rev. Mod. Phys.*, **58**, 801.

Birge, N. O. (1986). *Phys. Rev. B*, **34**, 1631.

Biroli, G., Bouchaud, J.-P., Miyazaki, K., and Reichman, D.R. (2006). *Phys. Rev. Lett*, **97**, 195701.

Blochowicz, T. and Rössler, E.A. (2005). *J. Chem. Phys.*, **122**, 224511.

Böhmer, R. (1998*a*). *Phase Transit.*, **65**, 211.

Böhmer, R. (1998*b*). *Curr. Opin. Solid State Mater. Sci.*, **3**, 378.

Böhmer, R., Chamberlin, R.V., Diezemann, G., Geil, B., Heuer, A., Hinze, G., Kuebler, S.C., Richert, R., Schiener, B., Sillescu, H., Spiess, H.-W., Tracht, U., and Wilhelm, M. (1998*a*). *J. Non-Cryst. Solids*, **235-237**, 1.

Böhmer, R., Diezemann, G., Hinze, G., and Rössler, E. (2001). *Prog. NMR Spectrosc.*, **39**, 191.

Böhmer, R., Diezemann, G., Hinze, G., and Sillescu, H. (1998*b*). *J. Chem. Phys.*, **108**, 890.

Böhmer, R., Hinze, G., Diezemann, G., Geil, B., and Sillescu, H. (1996). *Europhys. Lett.*, **36**, 55.

Böhmer, R., Ngai, K.L., Angell, C.A., and Plazek, D.J. (1993). *J. Chem. Phys.*, **99**, 4201.
Böhmer, R., Schiener, B., Hemberger, J., and Chamberlin, R.V. (1995). *Z. Phys. B*, **99**, 91.
Böttcher, J.F. and Bordewijk, P. (1973). *Theory of Electric Polarization.* Elsevier, Amsterdam.
Bouchaud, J.-P. and Biroli, G. (2004). *J. Chem. Phys.*, **121**, 7347.
Bouchaud, J.-P. and Biroli, G. (2005). *Phys. Rev. B*, **72**, 064204.
Bouchaud, J.-P., Cugliandolo, L., Kurchan, J., and Mézard, M. (1997) In *Spin-Glasses, Random Fields* (ed. A. P. Young). World Scientific, Singapore.
Capacioli, S., Ruocco, G., and Zamponi, F. (2008). *J. Phys. Chem.*, **B112**, 10652.
Castillo, H.E., Chamon, C., Cugliandolo, L.F., and Kennett, M.P. (2002). *Phys. Rev. Lett.*, **88**, 237201.
Chamberlin, R.V. (1999). *Phys. Rev. Lett.*, **83**, 5134.
Chamberlin, R.V., Schiener, B., and Böhmer, R. (1997). *Mater. Res. Soc. Symp. Proc.*, **455**, 117.
Chang, I. and Sillescu, H. (1997). *J. Phys. Chem. B*, **101**, 8794.
Cicerone, M.T., Blackburn, F.R., and Ediger, M.D. (1995*a*). *J. Chem. Phys.*, **102**, 471.
Cicerone, M.T., Blackburn, F.R., and Ediger, M.D. (1995*b*). *Macromolecules*, **28**, 8224.
Cicerone, M.T. and Ediger, M.D. (1993). *J. Chem. Phys.*, **97**, 10489.
Cicerone, M.T. and Ediger, M.D. (1995). *J. Chem. Phys.*, **103**, 5684.
Cicerone, M.T. and Ediger, M.D. (1996). *J. Chem. Phys.*, **104**, 7210.
Cicerone, M.T., Wagner, P.A., and Ediger, M.D. (1997). *J. Phys. Chem. B*, **101**, 8727.
Crauste-Thibierge, C., Brun, C., Ladieu, F., L'Hôte, D., Biroli, G., and Bouchaud, J.-P. (2010). *Phys. Rev. Lett.*, **104**, 165703.
Crider, P.S. and Israeloff, N.E. (2006). *Nano Lett.*, **6**, 887.
Crider, P.S., Majewski, M.R., Zhang, J., Oukris, H., and Israeloff, N.E. (2007). *Appl. Phys. Lett.*, **91**, 013102.
Crider, P.S., Majewski, M.R., Zhang, J., Oukris, H., and Israeloff, N.E. (2008). *J. Chem. Phys.*, **128**, 044908.
Dalle-Ferrier, C. (2009). Ph.D. thesis, University Paris Sud 11, France.
Dalle-Ferrier, C., Eibl, S., Pappas, C., and Alba-Simionesco, C. (2008). *J. Phys.: Condens. Matter*, **20**, 494240.
Dalle-Ferrier, C., Thibierge, C., Alba-Simionesco, C., Berthier, L., Biroli, G., Bouchaud, J.-P., Ladieu, F., L'Hôte, D., and Tarjus, G. (2007). *Phys. Rev., E* **76**, 041510.
Davies, M., Hains, P.J., and Williams, G. (1973). *J. Chem. Soc. Faraday II*, **69**, 1785.
Déjardin, J.L. and Kalmykov, Yu.P. (2000). *Phys. Rev. E*, **61**, 1211.
Deschenes, L.A. and Vanden Bout, D.A. (2002). *J. Phys. Chem. B*, **106**, 11438.
Diezemann, G. (2002). *J. Chem. Phys.*, **116**, 1647.
Diezemann, G., Sillescu, H., Hinze, G., and Böhmer, R. (1998). *Phys. Rev. E*, **57**, 4398.

Dixon, P.K., Wu, L., Nagel, S.R., Williams, B.D., and Carini, J.P. (1990). *Phys. Rev. Lett.*, **65**, 1108.

Donati, C., Franz, S., Glotzer, S.C., and Parisi, G. (2002). *J. Non-Cryst. Solids*, **307**, 215.

Duvvuri, K. and Richert, R. (2003). *J. Chem. Phys.*, **118**, 1356.

Duvvuri, K. and Richert, R. (2004). *J. Phys. Chem. B*, **108**, 10451.

Dyre, J.C. (2006). *Rev. Mod. Phys.*, **78**, 953.

Ediger, M.D. (1998). *J. Non-Cryst. Solids*, **235-237**, 10.

Ediger, M.D. (2000). *Ann. Rev. Phys. Chem.*, **51**, 99.

Ediger, M.D., Angell, C.A., and Nagel, S.R. (1996). *J. Phys. Chem.*, **100**, 13200.

Fleming, G.R. and Cho, M. (1996*a*). *Ann. Rev. Phys. Chem.*, **47**, 109.

Fleming, G.R. and Cho, M. (1996*b*). *Ann. Rev. Phys. Chem.*, **47**, 109.

Fourkas, J.T. (2001). *Optics Letters*, **26**, 211.

Fourkas, J.T., Kivelson, D., Mohanty, U., and Nelson, K.A. (ed.) (1997). *Supercooled Liquids: Advances and Novel Applications* ACS Publications, Washington.

Fragiadakis, D., Casalini, R., and Roland, C. M. (2009). *J. Phys. Chem.*, **B113**, 13134.

Franz, S. and Parisi, G. (2000). *J. Phys. Condens. Matter*, **12**, 6335.

Fujara, F., Geil, B., Sillescu, H., and Fleischer, G. (1992). *Z. Phys. B*, **88**, 195.

Garrahan, J.P. and Chandler, D. (2002). *Phys. Rev. Lett.*, **89**, 035704.

Gibbs, J.H. and DiMarzio, E.A. (1958). *J. Chem. Phys.*, **28**, 373.

Glotzer, S.C. (2000). *J. Non-Cryst. Solids*, **274**, 342.

Götze, W. and Sjögren, L. (1992). *Rep. Prog. Phys.*, **55**, 241.

Graessley, W.W. (2009). *J. Chem. Phys.*, **130**, 164502.

Grebenkin, S.Y. (2008). *J. Phys. Chem. B*, **112**, 15369.

Grebenkin, S.Y. and Bol'shakov, B.V. (2007). *J. Phys. Chem. B*, **111**, 5921.

Grebenkin, S.Y. and Syutkin, V.M. (2007). *Phys. Rev. B*, **76**, 054202.

Hall, D.B., Dhinojwala, A., and Torkelson, J.M. (1997). *Phys. Rev. Lett.*, **79**, 103.

Hansen, C., Stickel, F., Berger, T., Richert, R., and Fischer, E.W. (1997). *J. Chem. Phys.*, **107**, 1086.

Hempel, E., Hempel, G., Hensel, A., Schick, C., and Donth, E. (2000). *J. Phys. Chem. B*, **104**, 2460.

Heuer, A. (1997). *Phys. Rev. E*, **56**, 730.

Heuer, A., Tracht, U., Kuebler, S.C., and Spiess, H.W. (1999). *J. Mole. Struct.*, **479**, 251.

Heuer, A., Wilhelm, M., Zimmermann, H., and Spiess, H.W. (1995). *Phys. Rev. Lett.*, **75**, 2851.

Hinze, G. (1998). *Phys. Rev. B*, **57**, 2010.

Hinze, G., Diezemann, G., and Basche, T. (2004). *Phys. Rev. Lett.*, **93**, 203001.

Horng, M.-L., Gardecki, J.A., and Maroncelli, M. (1997). *J. Phys. Chem. A*, **101**, 1030.

Howell, F.S., Bose, R.A., Macedo, P.B., and Moynihan, C.T. (1974). *J. Phys. Chem.*, **78**, 639.

Hu, Z. and Margulis, C.J. (2006*a*). *Proc. Natl. Acad. Sci.*, **103**, 831.

Hu, Z. and Margulis, C.J. (2006*b*). *J. Phys. Chem. B*, **110**, 11025.

Huang, W. and Richert, R. (2006*a*). *J. Non-Cryst. Solids*, **352**, 4704.

Huang, W. and Richert, R. (2006*b*). *J. Chem. Phys.*, **124**, 164510.

Huang, W. and Richert, R. (2008*a*). *J. Phys. Chem. B*, **112**, 9909.

Huang, W. and Richert, R. (2008*b*). *Eur. Phys. J. B*, **66**, 217.

Huang, W. and Richert, R. (2009). *J. Chem. Phys.*, **130**, 194509.

Huang, W., Shahriari, S., and Richert, R. (2005). *J. Chem. Phys.*, **123**, 164504.

Huth, H., Wang, L.-M., Schick, C., and Richert, R. (2007). *J. Chem. Phys.*, **126**, 104503.

Ito, N., Duvvuri, K., Matyushov, D.V., and Richert, R. (2006). *J. Chem. Phys.*, **125**, 024504.

Jäckle, J. (1986). *Rep. Prog. Phys.*, **49**, 171.

Jäckle, J. and Richert, R. (2008). *Phys. Rev. E*, **77**, 031201.

Jeffrey, K.R., Richert, R., and Duvvuri, K. (2003). *J. Chem. Phys.*, **119**, 6150.

Johari, G.P. and Goldstein, M. (1970). *J. Chem. Phys.*, **53**, 2372.

Jurjiu, A., Friedrich, C., and Blumen, A. (2002). *Chem. Phys.*, **284**, 221.

Kircher, O., Diezemann, G., and Böhmer, R. (2001). *Phys. Rev. B*, **64**, 054103.

Kirkpatrick, T.R., Thirumalai, D., and Wolynes, P.G. (1989). *Phys. Rev.*, **A40**, 1045.

Kirkpatrick, T.R. and Wolynes, P.G. (1987). *Phys. Rev.*, **B36**, 8552.

Kohlrausch, R. (1854). *Pogg. Ann. Phys.*, **91**, 179.

Kudlik, A., Benkhof, S., Blochowicz, T., Tschirwitz, C., and Rössler, E. (1999). *J. Molec. Struct.*, **479**, 201.

Kuebler, S.C., Heuer, A., and Spiess, H.W. (1997). *Phys. Rev. E*, **56**, 741.

Kulzer, F. and Orrit, M. (2004). *Ann. Rev. Phys. Chem.*, **55**, 585.

Lačević, N., Starr, F.W., Schrøder, T.B., and Glotzer, S.C. (2003*a*). *J. Chem. Phys.*, **119**, 7372.

Lačević, N., Starr, F. W., Schröder, T. B., and Glotzer, S. C. (2003*b*). *J. Chem. Phys.*, **119**, 7372.

Lévy, L. (1988). *Phys. Rev. B*, **38**, 4963.

Lévy, L.P. and Ogielski, A.T. (1986). *Phys. Rev. Lett.*, **57**, 3288.

Lubchenko, V. and Wolynes, P.G. (2003). *J. Chem. Phys.*, **119**, 9088.

Lubchenko, V. and Wolynes, P.G. (2007). *Annu. Rev. Phys. Chem.*, **58**, 235.

Lunkenheimer, P., Pimenov, A., Dressel, M., Goncharov, Yu. G., Böhmer, R., and Loidl, A. (1996). *Phys. Rev. Lett.*, **77**, 318.

Lunkenheimer, P., Schneider, U., Brand, R., and Loidl, A. (2000). *Contemp. Phys.*, **41**, 15.

Maglione, M., Höchli, U. T., and Joffrin, J. (1986). *Phys. Rev. Lett.*, **57**, 436.

Mapes, M.K., Swallen, S.F., Kearns, K.L., and Ediger, M.D. (2006). *J. Chem. Phys.*, **124**, 054710.

Marcus, A.H., Schofield, J., and Rice, S.A. (1999). *Phys. Rev. E*, **60**, 5725.

Maroncelli, M. (1993). *J. Mol. Liquids*, **57**, 1.

Maroncelli, M., MacInnis, J., and Fleming, G.R. (1989). *Science*, **243**, 1674.

Mason, T.G. and Weitz, D.A. (1995). *Phys. Rev. Lett.*, **74**, 1250.

Meyer, A. (2002). *Phys. Rev. B*, **66**, 134205.

Miyako, Y., Chikazawa, S., T, Saito, and Yuochunas, Y.G. (1951)). *J. Phys. Soc. Jpn.*, **46**, 79.

Moerner, W.E. (1994). *Science*, **265**, 46.

Murthy, S.S.N. and Tyagi, M. (2002). *J. Chem. Phys.*, **117**, 3837.
Mydosh, J. A. (1993). *Spin Glasses: An Experimental Introduction.* London - Washington.
Ngai, K.L. (1999). *J. Phys. Chem. B*, **103**, 10684.
Nickel, B. (1989). *J. Lumin.*, **44**, 1.
Nielsen, A.I., Christensen, T., Jakobsen, B., Niss, K., Olsen, N.B., Richert, R., and Dyre, J.C. (2009). *J. Chem. Phys.*, **130**, 154508.
Omari, R., Prejean, J.J., and Souletie, J. (1986). *J. Phys.*, **44**, 1069.
Oukris, H. and Israeloff, N.E. (2010). *Nature Phys.*, **6** 135.
Palmer, G., Stein, D., Abrahams, E., and, P.W., Anderson (1984). *Phys. Rev. Lett.*, **53**, 958.
Phillips, J.C. (1996). *Rep. Prog. Phys.*, **59**, 1133.
Qi, F., El Goresy, T., Böhmer, R., Döβ, A., Diezemann, G., Hinze, G., Sillescu, H., Blochowicz, T., Gainaru, C., Rössler, E., and Zimmermann, H. (2003). *J. Chem. Phys.*, **118**, 7431.
Qi, F., Schug, K.U., Dupont, S., Döβ, A., Böhmer, R., Sillescu, H., Kolshorn, H., and Zimmermann, H. (2000). *J. Chem. Phys.*, **112**, 9455.
Qiu, X.H. and Ediger, M.D. (2003). *J. Phys. Chem. B*, **107**, 459.
Rajian, J.R., Huang, W., Richert, R., and Quitevis, E.L. (2006). *J. Chem. Phys.*, **124**, 014510.
Rehwald, C., Rubner, O., and A., Heuer (2010). *Phys. Rev. Lett.*, **105**, 11780.
Reinsberg, S.A., Heuer, A., Doliwa, B., Zimmermann, H., and Spiess, H.W. (2002). *J. Non-Cryst. Solids*, **307-310**, 208.
Reinsberg, S.A., Qiu, X.H., Wilhelm, M., Spiess, H.W., and Ediger, M.D. (2001). *J. Chem. Phys.*, **114**, 7299.
Richert, R. (1985). *Chem. Phys. Lett.*, **118**, 534.
Richert, R. (1988*a*). *Macromolecules*, **21**, 923.
Richert, R. (1988*b*). *Chem. Phys.*, **122**, 455.
Richert, R. (1993). *Chem. Phys. Lett.*, **216**, 223.
Richert, R. (1997). *J. Phys. Chem. B*, **101**, 6323.
Richert, R. (2000). *J. Chem. Phys.*, **113**, 8404.
Richert, R. (2001*a*). *Europhys. Lett.*, **54**, 767.
Richert, R. (2001*b*). *J. Chem. Phys.*, **114**, 7471.
Richert, R. (2001*c*). *J. Chem. Phys.*, **115**, 1429.
Richert, R. (2002). *J. Phys.: Condens. Matter*, **14**, R703.
Richert, R. (2005). *J. Chem. Phys.*, **123**, 154502.
Richert, R. and Blumen, A (ed.) (1994). Disorder Effects on Relaxation Processes. Springer, Berlin.
Richert, R. and Böhmer, R. (1999). *Phys. Rev. Lett.*, **83**, 4337.
Richert, R., Duvvuri, K., and Duong, L.-T. (2003). *J. Chem. Phys.*, **118**, 1828.
Richert, R. and Heuer, A. (1997). *Macromol.*, **30**, 4038.
Richert, R. and Richert, M. (1998). *Phys. Rev. E*, **58**, 779.
Richert, R. and Samwer, K. (2007). *New J. Phys.*, **9**, 36.
Richert, R. and Wagener, A. (1991). *J. Phys. Chem.*, **95**, 10115.
Richert, R. and Weinstein, S. (2006). *Phys. Rev. Lett.*, **97**, 095703.

Risken, H. (1989). *The Fokker-Planck Equation.* Springer, Berlin.
Ritort, F. and Sollich, P. (2003). *Adv. Phys.*, **52**, 219.
Schiener, B., Bhmer, R., Loidl, A., and Chamberlin, R.V. (1996). *Science*, **274**, 752.
Schiener, B., Chamberlin, R.V., Diezemann, G., and Böhmer, R. (1997). *J. Phys. Chem.*, **107**, 7746.
Schmidt-Rohr, K. and Spiess, H.W. (1991). *Phys. Rev. Lett.*, **66**, 3020.
Schmidt-Rohr, K. and Spiess, H.W. (1994). *Multidimensional Solid-State NMR, Polymers.* London.
Schob, A., Cichos, F., Schuster, J., and von Borczyskowski, C. (2004). *Eur. Polymer J.*, **40**, 1019.
Schröter, K. and Donth, E. (2000). *J. Chem. Phys.*, **113**, 9101.
Schuster, J., Cichos, F., and von Borczyskowski, C. (2003). *Eur. Phys. J. E*, **12**, S75.
Schuster, J., Cichos, F., and von Borczyskowski, C. (2004). *Eur. Polym. J.*, **40**, 993.
Sethna, J.P., Shore, J.D., and Huang, M. (1991). *Phys. Rev.*, **B44**, 4943.
Shahriari, S., Mandanici, A., Wang, L.-M., and Richert, R. (2004). *J. Chem. Phys.*, **121**, 8960.
Shi, X. and McKenna, G.B. (2005). *Phys. Rev. Lett.*, **94**, 157801.
Sillescu, H. (1971). *J. Chem. Phys.*, **54**, 2110.
Sillescu, H. (1999). *J. Non-Cryst. Solids*, **243**, 81.
Smets, G., (1983). *Adv. Polym. Sci.* **50**, 17.
Sokolov, I.M., Klafter, J., and Blumen, A. (2002). *Phys. Today*, **55**, 48.
Stevenson, J.D., Schmalian, J., and Wolynes, P.G. (2006). *Nature Phys.*, **2**, 268.
Stickel, F., Fischer, E.W., and Richert, R. (1996). *J. Chem. Phys.*, **104**, 2043.
Stillinger, F.H. and Hodgdon, J.A. (1994). *Phys. Rev. E*, **50**, 2064.
Sun, Y., Xi, H., Ediger, M.D., Richert, R., and Yu, L. (2009). *J. Chem. Phys.*, **131**, 074506.
Swallen, S.F., Bonvallet, P.A., McMahon, R.J., and Ediger, M.D. (2003). *Phys. Rev. Lett.*, **90**, 015901.
Swallen, S.F., Mapes, M.K., Kim, Y.S., McMahon, R.J., Ediger, M.D., and Satija, S. (2006). *J. Chem. Phys.*, **124**, 184501.
Swallen, S.F., Traynor, K., McMahon, R.J., Ediger, M.D., and Mates, T.E. (2009). *J. Phys. Chem. B*, **113**, 4600.
Tarjus, G. and Kivelson, D. (1995). *J. Chem. Phys.*, **103**, 3071.
Tarzia, M., Biroli, G., Bouchaud, J.-P., and Lefèvre, A. (2010). *J. Chem. Phys.*, **132**, 054501.
Thibierge, C., L'Hôte, D., Ladieu, F., and Tourbot, R. (2008). *Rev. Sci. Instrum.*, **79**, 103905.
Tracht, U., Wilhelm, M., Heuer, A., Feng, H., Schmidt-Rohr, K., and Spiess, H.W. (1998). *Phys. Rev. Lett.*, **81**, 2727.
Tracht, U., Wilhelm, M., Heuer, A., and Spiess, H.W. (1999). *J. Magn. Reson.*, **140**, 460.
Vidal Russell, E. and Israeloff, N.E. (2000). *Nature*, **408**, 695.
Vidal Russell, E., Israeloff, N.E., Walther, L.E., and Alvarez Gomariz, H. (1998). *Phys. Rev. Lett.*, **81**, 1461.
Vincent, E. (2007). *Lecture Notes in Physics*, **716**, 7.

Vincent, E. and Hammann, J. (1987). *J. Phys. C: Solid State Phys.*, **20**, 2659.
Wagner, H. and Richert, R. (1997). *Polymer*, **38**, 255.
Walther, L.E., Israeloff, N.E., Vidal Russell, E., and Alvarez Gomariz, H. (1998*a*). *Phys. Rev. B*, **57**, R15112.
Walther, L.E., Vidal Russell, E., Israeloff, N.E., and Alvarez Gomariz, H. (1998*b*). *Appl. Phys. Lett.*, **72**, 3223.
Wang, C.-Y. and Ediger, M.D. (1997). *Macromolecules*, **30**, 4770.
Wang, C.-Y. and Ediger, M.D. (1999). *J. Phys. Chem. B*, **103**, 4177.
Wang, C.-Y. and Ediger, M.D. (2000). *J. Chem. Phys.*, **112**, 6933.
Wang, L.-M. and Richert, R. (2004*a*). *J. Chem. Phys.*, **121**, 11170.
Wang, L.-M. and Richert, R. (2004*b*). *J. Chem. Phys.*, **120**, 11082.
Wang, L.-M. and Richert, R. (2007*a*). *Phys. Rev. B*, **76**, 064201.
Wang, L.-M. and Richert, R. (2007*b*). *Phys. Rev. Lett.*, **99**, 185701.
Ware, W.R., Lee, S.K., Brant, G.J., and Chow, P. (1971). *J. Chem. Phys.*, **54**, 4729.
Weeks, E.R., Weitz, D.A., Crocker, J.C., Levitt, A.C., and Schofield, A. (2002). *Phys. Rev. Lett.*, **89**, 095704.
Weinstein, S. and Richert, R. (2005). *J. Chem. Phys.*, **123**, 224506.
Weinstein, S. and Richert, R. (2007*a*). *Phys. Rev. B*, **75**, 064302.
Weinstein, S. and Richert, R. (2007*b*). *J. Phys.: Condens. Matter*, **19**, 205128.
Weinstein, S. and Richert, R. (2007*c*). *Phys. Rev. B*, **75**, 064302.
Wendt, H. and Richert, R. (2000). *Phys. Rev. E*, **61**, 1722.
Williams, G. and Watts, D.C. (1970). *Trans. Faraday Soc.*, **66**, 80.
Wong, J. and Angell, C.A. (1976). *Glass Structure by Spectroscopy.* New York.
Xia, X. and Wolynes, P.G. (2000). *Proc. Natl. Acad. Sci. USA*, **97**, 2990.
Xia, X. and Wolynes, P.G. (2001). *J. Phys. Chem. B*, **105**, 6570.
Yang, M. and Richert, R. (2002). *Chem. Phys.*, **284**, 103.
Zippelius, A. (1994). *Phys. Rev. B*, **29**, 2717.
Zöllmer, V., Rätzke, K., Faupel, F., and Meyer, A. (2003). *Phys. Rev. Lett.*, **90**, 195502.
Zondervan, R., Kulzer, F., Berkhout, G.C.G., and Orrit, M. (2007). *Proc. Natl. Acad. Sci. USA*, **104**, 12628.
Zwanzig, R. (1990). *Acc. Chem. Res.*, **23**, 148.

6
Dynamical heterogeneities in grains and foams

Olivier DAUCHOT, Douglas J. DURIAN
and Martin van HECKE

Abstract

Dynamical heterogeneities have been introduced in the context of the glass transition of molecular liquids and the length scale associated with them has been argued to be at the origin of the observed quasi-universal behavior of glassy systems. Dense amorphous packings of granular media and foams also exhibit slow dynamics, intermittency and heterogeneities. We review a number of recent experimental studies of these systems, where one has direct access to the relevant space-time dynamics, allowing for direct visualizations of the dynamical heterogeneities. On the one hand, these visualizations provide a unique opportunity to access the microscopic mechanisms responsible for the growth of dynamical correlations. On the other hand, focusing on the differences in these heterogeneities in microscopically different systems allows us to discuss the range of the analogies between molecular thermal glasses and athermal glasses such as granular media and foams. Finally, this review is the opportunity to discuss various approaches to actually extract quantitatively the dynamical length scale from experimental data.

6.1 Introduction

Granular media and foams exhibit a wide range of complex flow phenomena, some familiar, some surprising, but often poorly understood (Jaeger *et al.*, 1996; Cates *et al.*, 1998; Liu and Nagel, 2001; Duran, 2000; MiDi, 2004; Aranson and Tsimring, 2006; Kraynik, 1988; Wilson, 1989; Prud'homme and Khan, 1996; Weaire and Hutzler, 1999; Dauchot, 2007). The constituents of these materials, macroscopic grains and gas bubbles, are so large that thermal fluctuations do not cause appreciable agitations, and their interactions are dissipative. Hence, when unperturbed, these materials jam into a metastable, disordered state, and to create dynamics energy needs to be supplied, in the form of, e.g., shearing or vibrations.

The reason why these materials feature in a book on glasses is that their dynamical behavior often is "glassy"—slow dynamics, intermittency and heterogeneities are key phrases in describing their behavior. For example, when repeatedly tapping a loose packing of grains, their density slowly increases, but instead of exhibiting an exponential relaxation to an asymptotic density, the relaxation process is logarithmically slow and, moreover, exhibits memory effects, evidencing aging (see (Dauchot, 2007) for a review). In addition, when granular media flow, they typically do so inhomogeneously. Far away from the main flowing region this results in very slow creeping flows, where fluctuations are relatively large and where the response is sluggish. Foams exhibit similar phenomenology.

Dynamical heterogeneities are a key characteristic of glassy dynamics in thermal systems (Sillescu, 1999; Ediger, 2000; Glotzer, 2000; Lačević *et al.*, 2003; Cipelletti and Ramos, 2005), so a natural question to ask is the nature and organization of the fluctuations in granular media and foams. As we will outline in this chapter, the macroscopic glassy features of these materials are indeed accompanied by heterogeneous fluctuations at the microscopic, i.e. bubble or grain scale. To whet the reader's appetite, in Fig. 6.1 we show a graphic example of dynamical heterogeneity in top views of a system of air-fluidized beads. The coloring in this graph represents the persistent area order parameter, which quantifies how much the area of the Voronoi

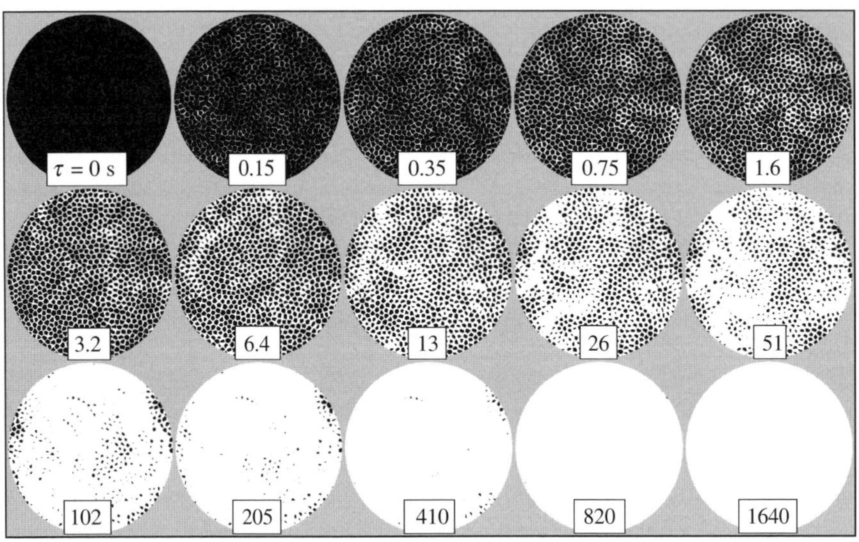

Fig. 6.1 An illustration of dynamical heterogeneities in an air-fluidized monolayer of beads at an area packing fraction of 0.79 (from Ref. (Abate and Durian, 2007)). The Voronoi tessellation is computed at each time step and the white regions represent for each bead the space that has come outside of the initial Voronoi cell after a delay τ as labelled in the figure.

cell surrounding a grain has changed in a given lag time. Note that the dynamics appear homogeneous at early and long times, but are spatially heterogeneous at intermediate times–most noticeably for the images at 13 s $\leq \tau \leq$ 102 s.

We will focus here on examples of heterogeneous dynamics of foams and granulates. Since crowding plays a crucial role for these materials in their dense, glassy phase, one should perhaps not be surprised that grain and bubble motion is inherently heterogeneous, and that fluctuations become spatially correlated—for one grain to move in a dense system, many other grains have to get out of the way. Open questions include how far one can push the analogies between molecular and colloidal glasses on the one hand, and foams and granular media on the other hand, and what one can learn about the differences in the heterogeneities in microscopically different systems.

An important experimental advantage of granular media and foams is that one has direct access to the relevant space-time dynamics, allowing for direct observations of the heterogeneous behavior. Moreover, these materials can be brought close to and often through a jamming transition, in the vicinity of which dynamical properties can be expected to change dramatically with control parameters. Finally, grains and foams have different microscopic interactions, that are quite well understood—grains are essentially undeformable and have inelastic and frictional interactions, whereas foam bubbles are easily deformed and have viscous interactions. Hence, by comparing their behavior, robustness of various glass-forming / jamming / heterogeneity scenarios can be probed. Moreover, the more complex interactions of grains means that heterogeneities can be probed in two physically distinct regimes—a relatively low packing density regime where grain interactions are dominated by collisions, and a higher-density regime where frictional contacts dominate the interactions.

The term "jamming" has evolved to have many meanings. It was originally proposed as an umbrella concept, meant to apply equally to the glass transition in molecular and colloidal liquids as to the cessation of flow in grains and foams (Liu and Nagel, 2001). Two different definitions are offered in Ref. (O'Hern *et al.*, 2003*b*). The first is that jamming is said to occur when a system develops a yield stress, and hence has mechanical rigidity. However, as a practical matter it is not possible to test whether a material truly has a yield stress, or whether the stress-relaxation time is too long to measure. So, alternatively jamming is said to occur when a system develops a relaxation time that exceeds a reasonable experimental timesscale, e.g. 1000 s. This is similar in spirit to defining the glass transition to occur when the viscosity exceeds 10^{13} poise, a large but arbitrary value. These two definitions are perfectly consistent when the relaxation time refers to rigidity, in terms of the decay time of the macroscopic shear-stress-relaxation modulus. However, it's a different notion, not always well distinguished in the literature, to define jamming in terms of the time scale for microscopic reorganization of structural degrees of freedom like the set of topological nearest neighbors. Here, we often use jamming in a rather loose sense, referring to dramatic slowing down of the dynamics and a qualitative change of the behavior from freely flowing to being stuck.

The outline of this chapter is as follows. In Section 6.2 we discuss heterogeneities in agitated granular media. These include air-fluidized granular systems where the grain interactions are dominated by collisions, the grains are driven randomly and isotropically, and the system is quasi-two-dimensional (Section 6.2.1), and dense 2D granular systems where grain interactions are frictional (Sections 6.2.2 and 6.2.3). In Section 6.2.2 the system is driven by slow oscillatory shear, and real-space rearrangements play a role, while in Section 6.2.3 the system is driven by horizontally shaking its support plate, and real-space rearrangements are substantially less than one grain diameter during the duration of the experiment. In Section 6.3 we describe observations of heterogeneities in granular flows in inclined plane, rotating drum and pile-flow geometries—for these flows, the grain interactions are a mix between collisions and enduring contacts, the flow is driven by bulk shear forces, and the system is three-dimensional. Finally, in Section 6.4 we describe observations of heterogeneities and large fluctuations in foams. We close this chapter by a discussion of commonalities and differences between these systems, and in the appendix discuss various approaches for calculating the dynamical susceptibility χ_4 that quantifies the dynamical heterogeneities. The reader who is not familiar with dynamical susceptibilities and how they relate to four-point dynamical correlators should refer to this appendix and the first chapters of the present book.

6.2 Heterogeneities in agitated granular media

6.2.1 Growing dynamical length scale in a monolayer fluidized bed

Quasi-two-dimensional monolayers of shaken, sheared, and fluidized grains have been important as model systems, both because it's difficult to adequately probe an opaque 3D medium but more crucially because their packing density may be controlled and varied free from gravitational compaction. For fluidization, the setup involves a submonolayer of beads that typically roll without slipping upon a fine screen up through which air is uniformly blown. The upward drag need not fully offset gravity to excite motion—rather, the grains are stochastically kicked within the plane by the shedding of turbulent vortices. As reported in Ref. (Abate and Durian, 2006) the air-fluidized grains therefore experience random ballistic motion at short times and random diffusive motion at long times. At intermediate times, and at high enough densities, the grains also exhibit an interval of subdiffusive motion where they collide multiple times with a long-lived set of neighbors. As the packing density is increased, the duration of this "caging" increases until all motion ceases at random close packing. Concurrently, there is little change in packing structure, though the pair correlation function does exhibit a growing first peak and a split second peak.

On approach to jamming by addition of grains to increase the packing density, the dynamics slows down and becomes heterogeneous, as illustrated in Fig. 6.1. This has been quantified by measures of the cluster size and of the chain length for the intermittent fast-moving regions, as well as through use of dynamic four-point susceptibilities based on three similar order parameters, which essentially characterize how much the local structure has evolved (see appendix). Data for the one based on the overlap of the Voronoi tessalation (see Fig. 6.1) are plotted in Fig. 6.2 for a sequence of different packing fractions. The upper plot is a measure of the temporal relaxation averaged over the whole system. It clearly demonstrates the slowing down

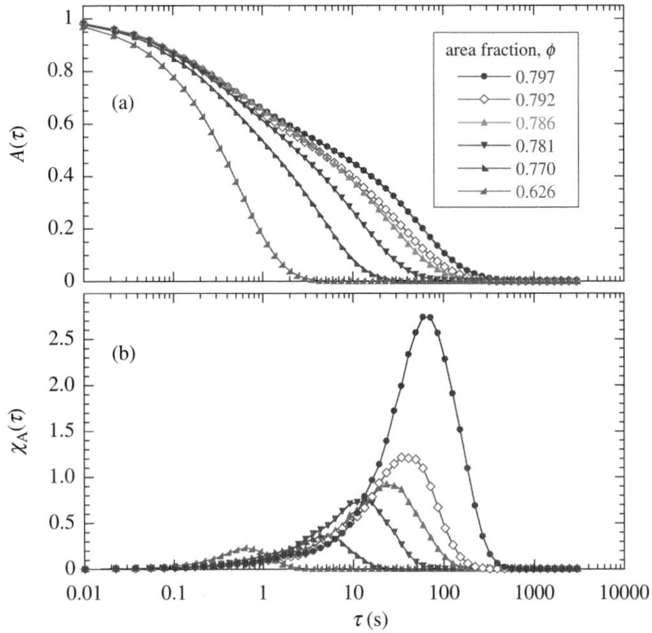

Fig. 6.2 (a) Average relaxation, and (b) associated susceptibility, vs. delay time for a sequence of packing fractions in the monolayer fluidized-bed experiment (from Ref. (Abate and Durian, 2007)).

of the dynamics when the packing fraction increases. The bottom one shows that the peak in the corresponding susceptibility rises on approach to jamming indicating the presence of a growing length scale (Keys *et al.*, 2007).

6.2.2 Building blocks of dynamical heterogeneities

Recently, the dynamics of a dense bidisperse monolayer of disks under cyclic shear has been investigated in Saclay (Marty and Dauchot, 2005; Dauchot *et al.*, 2005; Candelier *et al.*, 2009*a*). The experimental setup is shown in Fig. 6.3-lhs). The dominant feature of the grain trajectories is the so-called cage effect (see Fig. 6.3-rhs): at short times, particles are trapped by their neighbors, while at longer times particles leave their cage and diffuse through the sample by successive cage jumps. In these experiments the packing fraction is large and cage jumps necessarily lead to displacements of neighboring particles—this observation is at the root of the idea of cooperative motion and dynamical heterogeneities.

As is illustrated in Fig. 6.4, cage jumps are organized in clusters that avalanche to build up the long-term dynamical heterogeneities (Candelier *et al.*, 2009*a*). The distribution in space and time of the cage jumps is far from homogeneous. The left panel of Fig. 6.4 illustrated that cage jumps form clusters in space, occurring on a relatively short time scale $\tau_{\text{jump}} \simeq 10$. The cluster-size distribution is well described

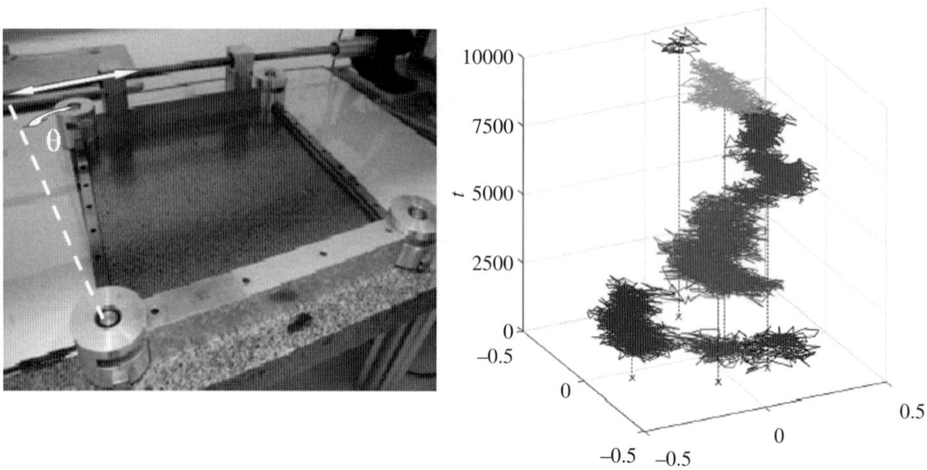

Fig. 6.3 The cyclic shear cell experiment (from Ref. (Marty and Dauchot, 2005)). A bidimensional, bidisperse granular material, composed of about 8000 metallic cylinders of diameter 5 and 6 mm in equal proportions, is sheared quasi-statically in a horizontal deformable parallelogram. The shear is periodic, with an amplitude $\theta_{\rm max} = \pm 5^\circ$. The volume accessible to the grains is maintained constant by imposing the height of the parallelogram, so that the volume fraction is a constant ($\phi \simeq 0.84$). Up to 4000 grains located in the center of the device are tracked by a high-resolution digital camera that takes a picture each time the system is back to its initial position $\theta = 0^\circ$. The unit of time is then one cycle, one experimental run lasting 10 000 cycles. The unit of length is chosen to be the small particle diameter d. Images are taken at each cycle and the resulting stroboscopic trajectories of the grains exhibit typical cages separated by cages jumps (rhs).

by a power law $\rho(N_{\rm c}) \approx N_{\rm c}^{-\alpha}$, where $N_{\rm c}$ is the number of grains within a cluster and $\alpha \in [3/2 - 2]$. The distribution of the lag times separating two adjacent clusters exhibits an excess of small times as well as an excess of large times as compared to a Poissonian uncorrelated process, and can be described by the superposition of two distributions: one for the long times, corresponding to the distribution of the time spent by the particles in each cage, and one for short delays between adjacent clusters that suggest a facilitation mechanism among clusters, the origin of which remains to be found. As a result of these two time scales, the clusters form avalanches well separated in time and space. Finally, selecting a time interval of length $\tau_{\rm DH}$ corresponding to the time scale for which dynamical heterogeneities are maximal, initiated at the beginning of a given avalanche, Fig. 6.4-rhs displays the spatial organization of the clusters in the avalanche. One can see how the clusters spread and build up a region of identical temporal decorrelation and thereby conclude that the avalanches *are* the dynamical heterogeneities.

Theoretical approaches based on dynamic facilitation usually focus on kinetically constrained models (Ritort and Sollich, 2003; Garrahan and Chandler, 2002; Toninelli *et al.*, 2006). They are characterized by a common mechanism leading to slow dynam-

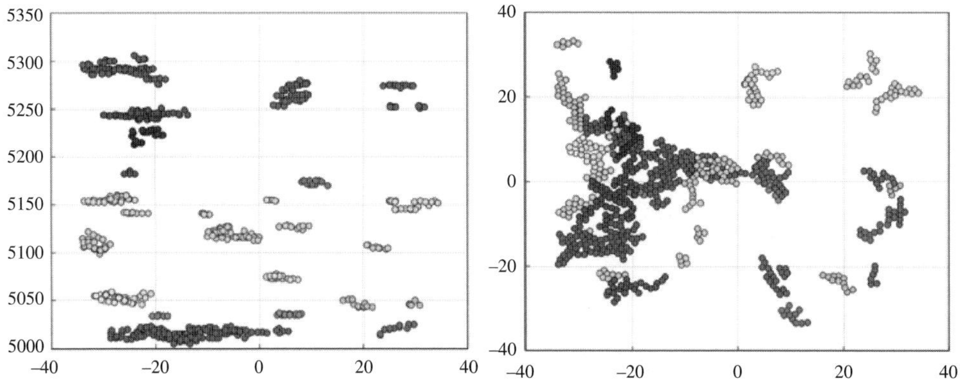

Fig. 6.4 Spatio-temporal organization of the cage jumps (from Ref. (Candelier *et al.*, 2009*a*)). Left: Time of cage jump (vertical axis) vs. its x-coordinate (horizontal axis). Right: Spatial location of cage jumps, showing that cage jumps facilitate each other to form dynamical heterogeneities.

ics: relaxation is due to mobile facilitating regions that are rare and move slowly across the system. Here, we find a dynamics characterized by avalanches inside which clusters are facilitating each other. However, in the present system facilitation is not conserved as in kinetically constrained models since the first cluster of an avalanche is far from any other possible facilitating region. Recent observations (Candelier *et al.*, 2009*b*) in the fluidized-bed experiment described in the previous section confirm that indeed facilitation becomes less and less conserved and a less and less significant mechanism when approaching jamming. Also, it has been shown numerically that the mechanisms described in this section also hold in a repulsive supercooled liquid (Candelier *et al.*, 2009*c*). This is a remarkable fact given the fundamental difference between the athermal granular system and the thermal structural liquid.

6.2.3 Criticality across the jamming transition

Once the system has entered the glass phase, its relaxation time has become much larger than the experimental time scale and it has fallen off equilibrium. However, one can still increase the packing fraction under external vibration, up to some value, where a finite fraction of particles will need to overlap to accommodate the increase of packing fraction. At that point, the pressure feels the hardcore repulsion of the grains and jamming occurs.

Lechenault *et al.* considered the dynamics of a bidisperse monolayer of disks under horizontal vibration (Lechenault *et al.*, 2008*a*)—see Fig. 6.5. The quench protocol produces reproducible, very dense configurations with structural relaxation time τ_α much larger than the experimental time scales. The pressure in the absence of vibration falls to zero at the jamming transition $\phi_J \in [0.8417, 0.8422]$, and in this system the density can be increased beyond this transition. One then observes that long-time correlations, accompanied by the growth of spatial correlations, are maximal at ϕ_J.

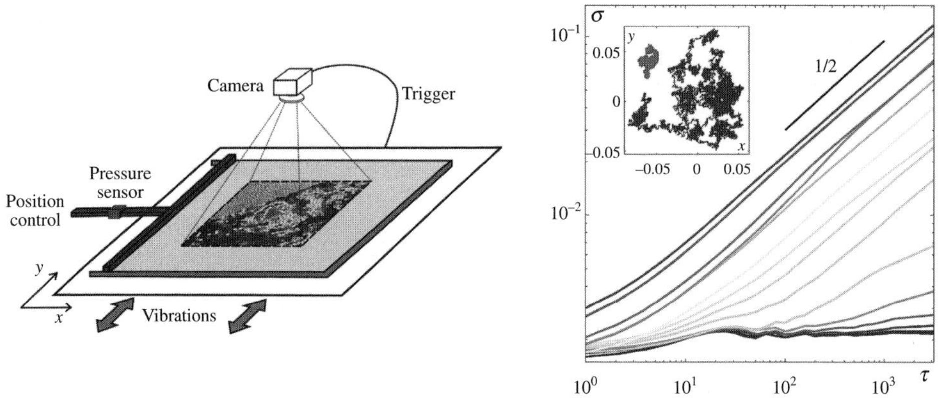

Fig. 6.5 A monolayer of bidisperse grains is driven close to jamming by successive compression steps under horizontal vibration (from Ref. (Lechenault *et al.*, 2008*a*)). Left: Setup. A bidisperse monolayer of 8500 brass cylinders of diameters $d_{\mathrm{small}} = 4 \pm 0.01$ mm and $d_{\mathrm{big}} = 5 \pm 0.01$ mm laid out on a horizontally vibrated glass plate ($frequency : 10\,\mathrm{Hz}, amplitude : 10\,\mathrm{mm}$). A lateral mobile wall allows variation of the packing fraction by tiny amounts ($\delta\phi/\phi \sim 5 \times 10^{-4}$) within an accuracy of 10^{-4}. The pressure exerted on this wall is measured by a force sensor inserted between the wall and the stage. The stroboscopic motion—in phase with the oscillating plate—of a set of 1500 grains in the center of the sample is tracked by a CCD camera. Lengths are measured in d_{small} units and time in cycle units. Right: the root mean square displacements exhibits a strongly subdiffusive behavior at short time before recovering diffusive motion. Note that even for the loosest packing fraction, the total displacement on the duration of the experiment does not exceed 0.1 grain diameter. The inset shows two trajectories, in dark gray for the loosest packing fraction, in light gray for the highest ones.

Here, a snapshot of the displacement field reveals the existence of a superdiffusive motion organized in channel currents meandering between blobs of blocked particles.

Figure 6.5-rhs displays the root mean square displacement as a function of the lag τ for various packing fractions ϕ. The very small values of $\sigma_\phi(\tau)$ at all time scales are consistent with the idea that the packing remains in a given structural arrangement. At low packing fractions $\phi < \phi_{\mathrm{J}}$, and at small τ the mean square displacement displays a subdiffusive behavior before recovering a diffusive regime at longer time scales. As the packing fraction is increased, the typical lag at which this crossover occurs becomes larger and, at first sight, does not seem to exhibit any special feature for $\phi \simeq \phi_{\mathrm{J}}$. Above ϕ_{J}, an intermediate plateau appears before diffusion resumes. A closer inspection of $\sigma^2_\phi(\tau)$ reveals an intriguing behavior, that appears more clearly on the local logarithmic slope $\nu = \partial \log \sigma_\phi(\tau)/\partial \log(\tau)$ (see (Lechenault *et al.*, 2008*a*,*b*)). For packing fractions close to ϕ_{J} and after the subdiffusive regime, the motion becomes *superdiffusive* at intermediate times corresponding to large-scale currents shown in Fig. 6.6a.

To characterize the various diffusion regimes, these authors define three characteristic times: $\tau_1(\phi)$ as the lag at which $\nu(\tau)$ first reaches $1/2$, corresponding to the start of the superdiffusive regime, $\tau_{\mathrm{sD}}(\phi)$ when $\nu(\tau)$ reaches a maximum $\nu^*(\phi)$ (peak of the superdiffusive regime), and $\tau_D(\phi)$ beyond which the system recovers the diffusive regime. These characteristic time scales are plotted as a function of the packing fraction in Fig. 6.6b. Whereas τ_1 does not exhibit any special features across ϕ_J, both τ_{sD} and τ_D are strongly peaked at ϕ_J.

Finally, one can extract the typical size of these currents by computing the dynamical susceptibility $\chi_4(\tau)$, which quantifies the number of particles moving in a correlated manner and exhibits a maximum at τ^*. Interestingly τ^* behaves like τ_{sD}, a further proof that in the present case, superdiffusion and dynamical heterogeneities are related. Recently, a deeper analysis of the same data have revealed that the superdiffusive behaviour must be attributed to the emergence of Levy flights in the dispacement distributions rather than to long-time correlations, suggesting the existence of rapid cracks of all scales rather than the progressive development of soft regions (Lechenault *et al.*, 2010).

Another way to characterize the spatial correlation is to compute the spatial correlator of the displacement field amplitude for a lag τ^* (see appendix). The authors could demonstrate that this spatial correlator also called a four-point correlation function obeys critical scaling $G_4(\vec{r}, \tau^*) \propto \frac{1}{r^\alpha} \mathcal{G}\left(\frac{r}{\xi_4}\right)$, with $\alpha \simeq 0.15$ in the vicinity

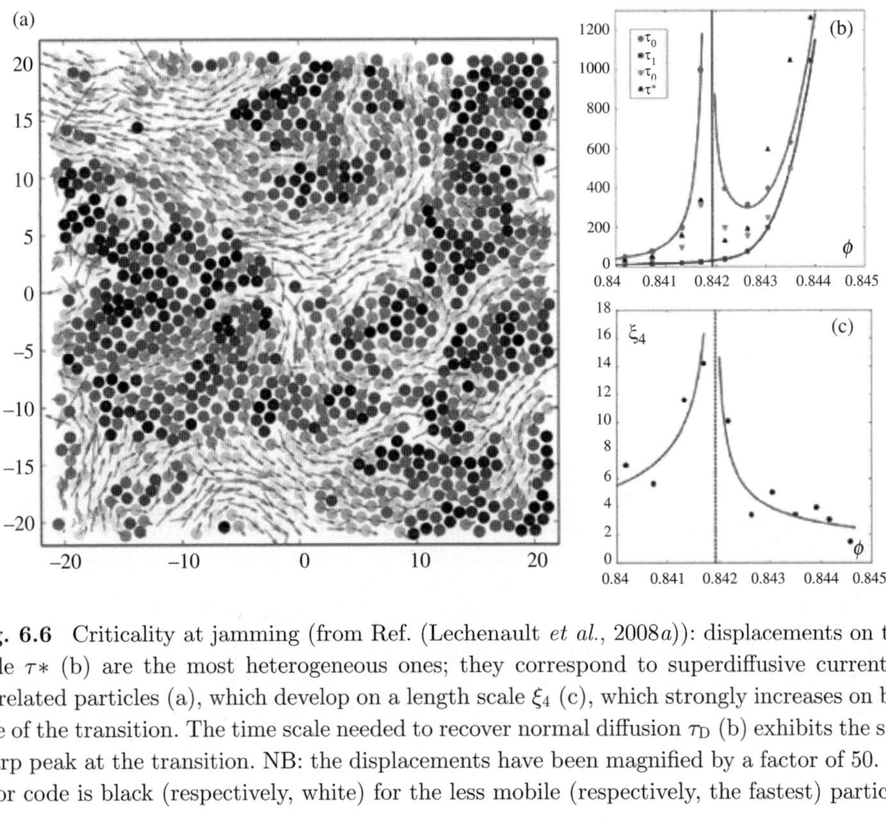

Fig. 6.6 Criticality at jamming (from Ref. (Lechenault *et al.*, 2008*a*)): displacements on time scale $\tau*$ (b) are the most heterogeneous ones; they correspond to superdiffusive currents of correlated particles (a), which develop on a length scale ξ_4 (c), which strongly increases on both side of the transition. The time scale needed to recover normal diffusion τ_D (b) exhibits the same sharp peak at the transition. NB: the displacements have been magnified by a factor of 50. The color code is black (respectively, white) for the less mobile (respectively, the fastest) particle.

of ϕ_J. $\xi_4(\phi)$, plotted on Fig.6.6c, is the length scale over which dynamical correlations develop. This scaling form, together with the strong increase of both $\xi_4(\phi)$ and $\tau_{sD}(\phi)$ over a minute range of ϕ, is the strongest evidence that the jamming fraction ϕ_J is indeed a critical point, where a static pressure appears *and* long-range dynamical correlations develop.

6.3 Heterogeneities in granular flows

6.3.1 Flow rules

Three different granular flow regimes are to be distinguished. *Rapid flows* are fairly dilute. The main grain interactions are through collisions, and this regime is described well within the framework of the kinetic theory (Savage and Jeffrey, 1981; Goldhirsch, 2003). *Slow flows* are dense, and the grain interactions are dominated by the frictional contact forces. This is the regime associated with soil mechanics (Nedderman, 1992), although existing descriptions for such slow flows are rather incomplete and have limited predictive power (Fenistein *et al.*, 2004; Deboeuf *et al.*, 2005). *Liquid-like granular flows* constitute the intermediate regime, where both inertia and friction are important and grain interactions are a mix between enduring contacts and collisions. This last regime has been widely investigated recently (MiDi, 2004; Savage, 1998; Losert *et al.*, 2000; Mills *et al.*, 1999; Aranson and Tsimring, 2002; Cortet *et al.*, 2009; Lemaître, 2002; Deboeuf *et al.*, 2006; Katsuragi *et al.*, 2010).

The crucial progress made recently comes from dimensional analysis (Iordanoff and Khonsari, 2004; MiDi, 2004; da Cruz *et al.*, 2005) that suggests that, in simple incompressible unidirectional uniformly sheared flows, there is only one dimensionless number that governs the flow: the so-called *inertial number* $I = \dot{\gamma} d/\sqrt{P/\rho}$, which is a function of bead diameter d, grain density ρ, global pressure P and global shear rate $\dot{\gamma}$. The rheology is then set by requiring that the ratio of shear to normal stresses is given by an effective friction coefficient that depends on the inertial number only: $\tau/P = \mu(I)$. Such a relation has first been evidenced, numerically, in plane shear (Iordanoff and Khonsari, 2004; da Cruz *et al.*, 2005) and, experimentally, in inclined-plane (MiDi, 2004) configurations. A local tensorial extension of this relation was recently proposed by Jop *et al.* (Jop *et al.*, 2006) as a constitutive law for dense granular flows, and these authors succeeded to fit the surface velocity profile for a steady unidirectional flow down an inclined plane with walls. Microscopically, it has been proposed by Ertaş and Halsey (Ertas and Halsey, 2002) that the motion of grains in dense granular flows occurs through clusters, whose size is controlled by the stress distribution. In the remainder of this section we will focus now on heterogeneities arising in this flow regime.

6.3.2 Granular flows in rotating drums

Pouliquen (Pouliquen *et al.*, 2003) has experimentally studied the velocity fluctuations of grains flowing down a rough inclined plane. He has shown that grains at the free surface exhibit fluctuating motions, which are correlated over a few grain diameters. Surprisingly, the correlation length is not controlled by the thickness of the flowing layer but by the inclination only. The correlation length is maximum at low inclination and decreases at high inclination, in a similar way as the critical thickness below which, for a given inclination, the flow stops (Daerr and Douady, 1999).

Bonamy *et al.* (Bonamy *et al.*, 2002) have also observed clusters of particles in the steady-flow regime in a rotating drum. In the recorded region, located at the center of the drum, the granular surface flow presents the now well-known velocity profile, linear in the flowing surface flow and exponential in the quasi-static bed (see Fig. 6.7b). By tracking the particles Bonamy *et al.* found that the velocity fluctuations of two beads in contact tend to have correlated orientations. Figure 6.7c displays the resulting clusters—here clusters are defined as consisting of particles in contact with velocity fluctuations aligned to within 60°. The authors further showed that the distribution of the number of beads in a cluster is a power law with a cutoff given by the flow thickness, thereby enforcing an earlier scenario proposed by Ertaş and Halsey (Ertas and Halsey, 2002).

What are the time scales governing these spatially correlated clusters? Deboeuf *et al.* (Deboeuf *et al.*, 2003) studied the related question of the typical relaxation times of the granular assembly inside the drum, once the flow is stopped. For that purpose the drum is first rotated in the well-known regime of intermittent avalanches, and then stopped just after an avalanche has occurred. The subsequent relaxation events are then recorded with a standard CCD camera, which takes images of the pile every 15 s. Denoting the fraction of beads that have moved between two acquisitions by $\delta A(t)$, the slow relaxation of the pile can be characterized.

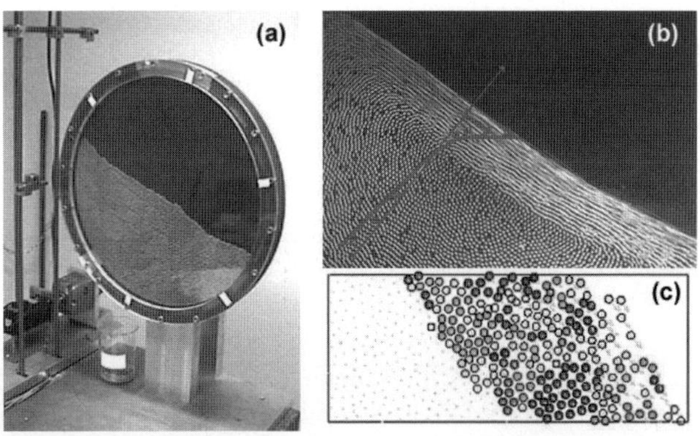

Fig. 6.7 Rotating-drum experiment (from Ref. (Bonamy *et al.*, 2002)). (a) Setup consisting of a rotating drum of diameter 45 cm and gap of 7 mm, half-filled with steel beads of diameter $d = 3 \pm 0.05$ mm. A quasi-2D packing is obtained but with a local 3D microscopic disordered structure. A fast camera allows tracking the 60 per cent of the beads observed through the transparent side wall of the tumbler. The rotating velocity of the drum is varied from 1 rpm to 8 rpm. (b) Linear (respectively, exponential) velocity profile in the upper flowing layer (respectively, the lower static layer). (c) Clusters of beads with correlated velocity fluctuation orientation in quasi-2D flow; typical frame of the clusters for $\Omega = 8\,\text{rpm}$.

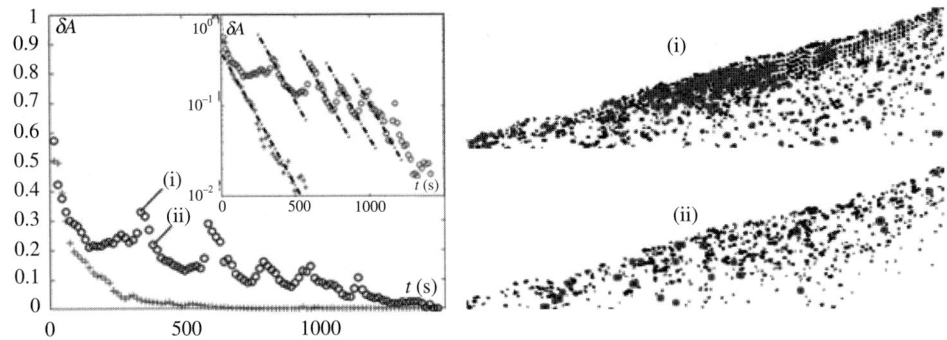

Fig. 6.8 Intermittent relaxations following a sudden flow arrest (from Ref. (Deboeuf *et al.*, 2003): the evolution of the fraction of mobile grains $\delta A(t)$ for two different realizations at similar pile slopes: (•) $\theta = 15$ deg; (+) $\theta = 16.5$ deg. Inset is the log-lin plot of the same data. Notice the exponential decay rate, which is identical in the monotonous case (+) and in the intermittent case (•). On the rhs, displacements in the pile during a burst event (i) and during an exponential decay period (ii). The dark pixels correspond to positions where a displacement has occurred in the 15 s preceding the considered time step. The light overlay indicates the pixels, where displacements have occurred successively during 30 s following the given time step (see text for details).

Two qualitatively different types of behavior can be found, as illustrated in Fig. 6.8. During the very first time steps the relaxation process is identical in both records: the bulk of the pile relaxes rapidly from bottom to top on time scales of the order of 15 s—typical events involve isolated bead displacements on short time- and length scales. The relaxation process then slows down in a subsurface layer of thickness [10–20] bead diameters—this subsurface layer may relax very differently from one realization to another (see Fig. 6.8).

In one case, one observes a simple exponential decay of the subsurface layer activity with a characteristic time scale of the order of 200 s. In the other case, intermittent bursts interrupt periods of exponential decay, with the same time scale as in the first case (see inset of Fig. 6.8-lhs). The competition between the exponential relaxation and the reactivation bursts results in a much slower relaxation. A visual inspection reveals that the reactivation bursts correspond to collective motions of grain clusters, whereas the exponential decay involves individual bead displacements—see Fig. 6.8-rhs. For the case of burst events those displacements persist in time and are spatially correlated, forming grain clusters.

Altogether, the above experiments reveal the existence of correlated clusters, which seemingly control the thickness of the steady flow and the relaxation times of the avalanches in the intermittent regime. Such clusters are purely dynamical in the sense that they involve spatial correlations of the dynamics, not of the local structure inside the pile. A characterization of these dynamical correlations in terms of dynamical heterogeneities, as introduced for the study of glasses, has not been done in the case of the rotating drum, at least not in a systematic manner, but was done for the flow down a pile, as we will discuss now.

6.3.3 Granular flows down a pile

One striking feature of granular flow down a pile is that the flow near the surface can be very smooth and fluid-like, while simultaneously far below the surface the heap appears to be a completely static solid. This is true even at very high mass flux, when the surface flow is steady and independent of time. This situation has now been extensively studied in a simpler geometry where the heap is confined in the narrow gap between two transparent sidewalls through which the grains may be measured optically. Several groups report that the velocity profile along the sidewall decreases nearly exponentially with depth (Lemieux and Durian, 2000; Khakhar *et al.*, 2001; Komatsu *et al.*, 2001; Andreotti and Douady, 2001; Jop *et al.*, 2005; Djaoui and Crassous, 2005; Richard *et al.*, 2008). Similar localized flow behavior is found for grains in a rotating drum (Rajchenbach, 1990; du Pont *et al.*, 2005; Cortet *et al.*, 2009; MiDi, 2004), as well as in Couette (Howell *et al.*, 1999; Mueth *et al.*, 2000; Bocquet *et al.*, 2002) and split-ring (Fenistein *et al.*, 2004) cells. Due to the exponential character of the velocity profiles for continuous heap flow, the shear rate is highest near the top free surface where the velocity is highest, and it decays almost exponentially with depth, too. Thus, the grains experience neighbor changes most frequently and are most unjammed near the top, and they become progressively jammed as a function of depth.

The nature of the jamming transition for continuous heap flow, controlled as a function of depth, was recently studied and compared with jamming transitions for uniform systems controlled as a more usual function of temperature or density or shear (Katsuragi *et al.*, 2010). Measurement of the static structure factor and pair correlation function for grains along the sidewall show that the spatial arrangement of grains is slightly dilated in the first layer or two due to saltation. At greater depths, there is no noticeable change in structure to accompany the dramatic decreases in velocity and shear rate. Such behavior is a hallmark feature: glass and jamming transitions are dynamical, and are not controlled by the growth of a correlation length associated with instantaneous (static) order.

There are two interesting features in the dynamics that both grow with depth on approach to jamming. The first is the ratio of the characteristic grain fluctuations speed, δv, to the flow speed, v_x; the former is measured by speckle-visibility spectroscopy, while the latter is measured by particle-image velocimetry (Katsuragi *et al.*, 2010). While both speeds decrease with depth, the fluctuations do so more slowly in accord with $\delta v \propto \sqrt{v_x}$. The relative rise of δv over v_x for greater depth signifies an increase in jostling and hence in dissipation at decreased driving rates, that ultimately results in jamming; it also means that the flow does not simply slow down without change in character. In particular, the second interesting feature is that the character of the dynamics becomes increasingly heterogeneous on approach to jamming. This is seen by measurement of an overlap order parameter and associated susceptibility, $\chi_4(\tau)$, based on a novel image correlation method that does not rely on

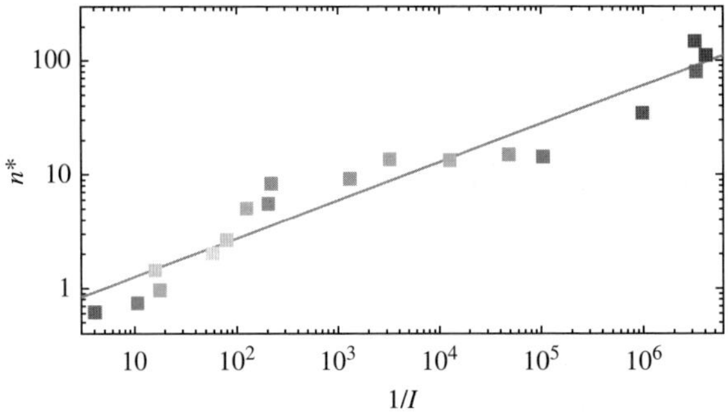

Fig. 6.9 Number n^* of grains in a heterogeneity for heap flow versus inverse inertial number, $I = \dot{\gamma}d/\sqrt{P/\rho}$, where $\dot{\gamma}$ is the shear rate, d is the grain diameter, P is the local pressure, which depends on depth, and ρ is the grain density (from Ref. (Katsuragi *et al.*, 2010)). The line is the best fit to a power law, giving $n^* \sim (1/I)^{0.33\pm0.02}$ in accord with simulation of a system undergoing uniform shear (Hatano, 2008).

particle tracking (Katsuragi *et al.*, 2010). At all depths, $\chi_4(\tau)$ displays a peak vs. τ that is located very close to the grain radius divided by v_x, and hence that, slows to longer times at greater depths. More importantly, the height of the peak, ${\chi_4}^*$, increases nearly exponentially with depth. This means that the dynamics become increasingly heterogeneous on approach to jamming.

To compare with other systems, it is more appropriate to consider the growth in the number n^* of grains in a heterogeneity as a function of shear rate rather than of depth. For this, n^* is computed from ${\chi_4}^*$, as shown in the appendix, and the shear rate is characterized by the inertial number (MiDi, 2004). For the experiment, I is maximum at about 0.2 near the surface and decays nearly exponentially with depth. The scaling displayed in Fig. 6.9 of the size of the heterogeneities with dimensionless shear rate is a power-law relation $n^* \sim I^{-1/3}$.

6.4 Foams, frictionless soft spheres

Foams are dispersions of gas bubbles in a liquid, stabilized by surfactants (Fig. 6.10-left) (Kraynik, 1988; Wilson, 1989; Prud'homme and Khan, 1996; Weaire and Hutzler, 1999). A crucial parameter is the liquid fraction, or wetness of the foam, which specifies the volume fraction of the liquid phase. When the liquid fraction is too large, the individual gas bubbles do not touch and the material is unjammed—one refers to this as a bubbly liquid rather than a foam. Below a critical liquid fraction—around 36% per cent for 3D foams—bubbles can no longer avoid each other and undergo a jamming transition. What is particular for foams is that vanishingly small liquid fractions can easily be reached, where the foam essentially consists of very thin liquid layers meeting in quasi-1D plateaux borders, which themselves meet in vertices—such foams are called dry foams. In systems under gravity, drainage can cause gradients in the wetness from very dry at the top to wet at the bottom.

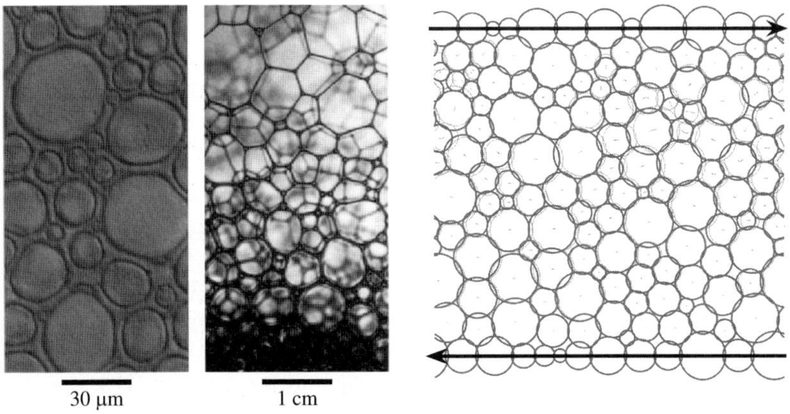

Fig. 6.10 Left: Gas bubbles in a shaving cream and in a vial of soapy water about 30 min after shaking. The former is about 92% gas, while the latter has a vertical gradient in wetness due to gravity. Right: The bubble model as introduced in (Durian, 1995, 1997): bubble positions just before (light gray) and after (dark gray) a shear-induced rearrangement, with trajectory of the centers shown by the small arrows centered in the bubbles.

It is interesting to compare bubbles in foam with grains in a sandpile. In common, both are comprised of large packing units that experience negligible thermal motion and that tend to be jammed. But there are many contrasts: First, grains in a pile are effectively incompressible, and pack at packing fractions *below* random close packing, while bubbles in a foam are readily deformed and squashed together *above* random close packing. Secondly, grains are subject to static and sliding friction, as well as to collisional dissipation, whereas the bubble contacts are through a liquid film that typically does not support static friction.

This has important implications for differences between the jamming of foams and the jamming of grains. Hard grains essentially are always close to jamming, but due to the friction, they are not necessarily critical—the jamming transition for frictional particles is usually not critical, and is not characterized by a unique packing fraction or contact number (van Hecke, 2010). In contrast, the jamming transition for foams (and emulsions) has all the hallmarks of the theoretically well-studied jamming of soft frictionless spheres at point J. Here, the jamming transition corresponds to a precise packing fraction and contact number, and materials near point J exhibit a diverging length scale and non-trivial power-law scaling of their elastic moduli (O'Hern *et al.*, 2003*a*; van Hecke, 2010). Some of these features were, in fact, discovered first in numerical simulations of simple models of foams (Bolton and Weaire, 1990; Durian, 1995, 1997) (see Fig. 6.10).

Flows of granular media and foams also exhibit essential differences. Granular flow requires dilatation, and can be separated into slow and fast flow, depending on the role of inertia. In contrast, foam flows are highly damped due to viscous interactions, and accomplished by bubble deformation and rearrangement with no dilatation—inertia plays essentially no role.

6.4.1 Unjamming of foams

There are at least three ways to unjam foams. The first is simply to allow the foam to coarsen: with time gas will diffuse from high- to low-pressure bubbles, which generally causes smaller bubbles to shrink and larger ones to grow. This is driven by surface tension through Laplace's law, and serves to reduce the total interfacial area. As coarsening proceeds, the bubbles rearrange into different packing configurations and hence can relax macroscopically imposed stress. The time scale for rearrangement can be comparable to that for size change, as in very dry foams. But for fairly wet foams, the rearrangements can be very much faster. Such avalanche-like rearrangements are a kind of dynamical heterogeneity, where a localized region of neighboring bubbles briefly mobilizes and comes to rest in a new configuration. For opaque foams, these events may be captured by diffusing-wave spectroscopy and its variants (Durian *et al.*, 1991; Cohen-Addad and Hohler, 2001; Gittings and Durian, 2006). The measured signal gives the time between successive rearrangements at a scattering site, averaged over both time and the volume of the sample through which the photons diffuse. As the foam coarsens, the time between events is found to grow as a power law of time. The DWS signal also includes subtle contributions from continuous motion of the gas liquid interfaces, due to both thermal fluctuations (Gopal and Durian, 1997) and also the coarsening process (Gittings and Durian, 2008; Sessoms *et al.*, 2010).

Time-resolved versions of DWS allow further information to be extracted (see Fig. 6.11). In Ref. (Mayer *et al.*, 2004), $\chi_4(\tau)$ is measured by fluctuations in the decay rate of the DWS correlation function. As the foam coarsens, the peak location moves to longer times in accord with the growing time between events. Furthermore, the peak height also grows—perhaps because the scattering volume contains a decreasing number of bubbles and perhaps because the system is becoming progressively jammed. In Ref. (Gittings and Durian, 2008), the scattering volume is decreased to the point that the dynamics of individual rearrangements may be followed with speckle-visibility spectroscopy. This allows access to a second important time scale—the *duration* of the

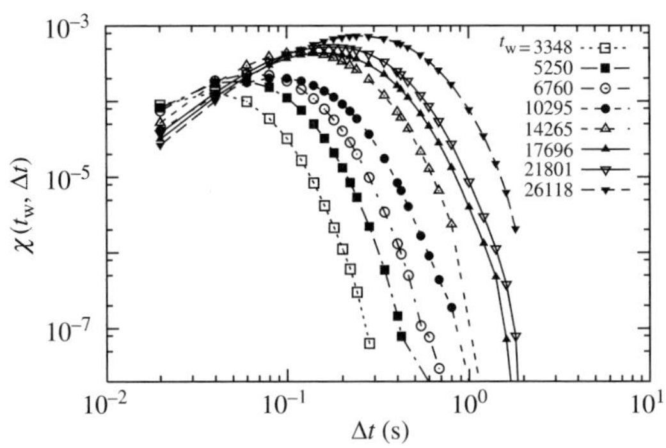

Fig. 6.11 Dynamical susceptibility measured by light scattering for a coarsening foam at different wait times t_w, labeled in seconds (Mayer *et al.*, 2004). Note that the peaks shifts to longer times and grows in height as the foam ages.

rearrangement events. In addition, the spatial distribution of successive events may now be studied with a recently introduced photon correlation imaging technique (Duri *et al.*, 2009; Sessoms *et al.*, 2010).

Foams may also be unjammed by application of shear. As probed by DWS (Earnshaw and Jaafar, 1994; Gopal and Durian, 1995), the shear-induced rearrangements appear similar to the coarsening-induced rearrangements but occur at a frequency proportional to the strain rate. So, coarsening dominates at very low strain rates, small compared to the reciprocal of the time between coarsening-induced events. At very high strain rates, compared to the reciprocal of the *duration* of events, the bubble-scale dynamics are qualitatively different. Rearranging bubbles no longer have enough time to lock into a locally stable configuration before having to rearrange again. Thus, successive events merge into continuous flow, and bubble–bubble interactions are dominated by dissipative forces rather than surface-tension forces. This is evidenced by a change in the functional form of the DWS correlation function (Gopal and Durian, 1999), similar to the change due to diffusive vs. ballistic microscopic motion. The effect of altered microscopic dynamics on the macroscopic rheology may be seen to some extent in the shape of the stress vs. strain rate flow curve; however, it is much more apparent in the transient stress jump and decay observed when a small step-strain is superposed on steady shear (Gopal and Durian, 2003). In particular, the transient shear modulus and stress relaxation time both decrease vs. strain rate at a characteristic scale set by yield strain divided by event duration (Gopal and Durian, 2003).

Finally, the third approach to unjam foams is by increasing the liquid content. In the dry limit, the bubbles are polyhedral and separated from their neighbors by thin curved soap films. The addition of liquid causes inflation not of the films but of the plateau borders and vertices at which the films meet. Thus, progressively wetter foams have progressively rounder bubbles, which unjam when the liquid fraction rises to about 36% and the bubbles are randomly close-packed spheres filling about 64% of space (in 3D). This unjamming has been measured in terms of the vanishing of the shear modulus and the yield strain vs. liquid fraction (Saint-Jalmes and Durian, 1999). While rearrangements play no role in this transition, there are nonetheless interesting changes in dynamics. First, as seen in simulation, there is a growing time scale for stress relaxation (Durian, 1995). This is accompanied by, and in fact may be due to, bubble displacements that become increasingly non-affine as the liquid fraction approaches unjamming (Durian, 1997) as illustrated on Fig. 6.12. Non-affine response has been implicated in a $\sqrt{i\omega}$ contribution to the complex shear modulus (Liu *et al.*, 1996; Gopal and Durian, 2003), and in the non-trivial scaling of the shear modulus with packing fraction (O'Hern *et al.*, 2003*b*; Ellenbroek *et al.*, 2009).

6.4.2 Flow of 2D foams

Experiments on two-dimensional foams under shear have yielded tremendous insight, in part because the full bubble-packing structure can be readily imaged and tracked as a function of time but also because the dry limit may be modeled in terms of idealized topological features (Bolton and Weaire, 1990; Herdtle and Aref, 1992; Okuzono and

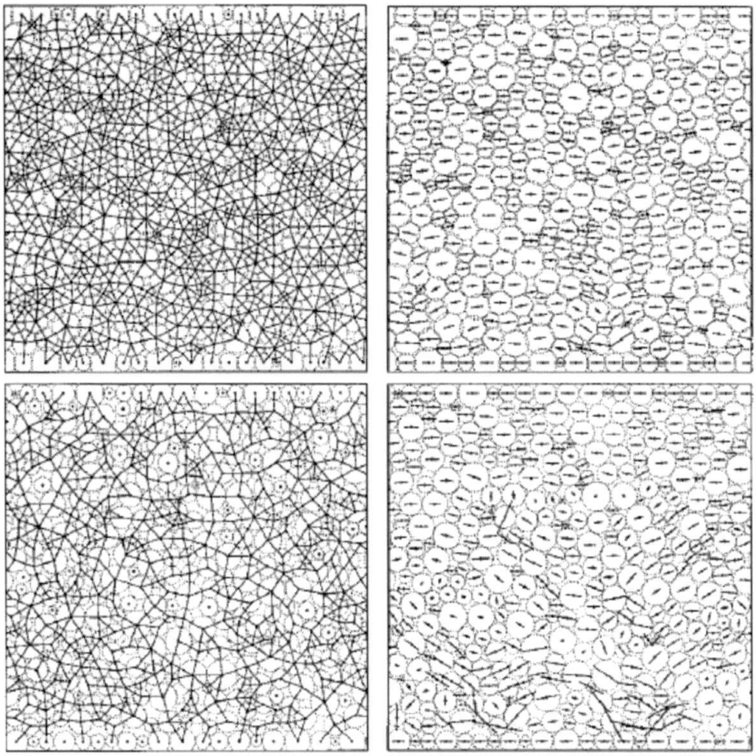

Fig. 6.12 Bubble positions and spring network (left column), for the model of Refs. (Durian, 1995, 1997). The top row is for a packing fraction of 1 and the bottom row is for a packing fraction of 0.84. The right column depicts the motion that occurs for small-amplitude shear strain, showing that it becomes more non-affine on approach to unjamming.

Kawasaki, 1995). Pioneering measurements on shear bubble rafts date back to Argon (Argon and Kuo, 1979), who sought analogy with the flow of metallic glasses.

In recent years, a variety of studies have addressed the flow of quasi-2D foams, which consist of a single layer of macroscopic ($d > 1\,\text{mm}$) bubbles. Such single layers can be made by freely floating bubbles on the surface of a surfactant solution (Wang, 2006), by trapping them between a top glass plate and the surfactant solution (Katgert and van Hecke, 2008; Wang, 2006), or by trapping them between two parallel glass plates (Debregeas and di Meglio, 2001). The confining glass plates enhance the stability of the foam, but also introduce additional drag forces that lead to the formation of shear bands in the foam (Wang, 2006).

While the time-averaged flow profiles in such geometries have received much attention (Debregeas and di Meglio, 2001; Lauridsen *et al.*, 2004; Wang, 2006; Janiaud and Hutzler, 2006; Katgert and van Hecke, 2008) here we will briefly outline recent work on the fluctuations around the average flows. As shown by Debregeas (Debregeas and di Meglio, 2001) and Lauridsen *et al.* (Lauridsen *et al.*, 2004), the instantaneous flow field exhibits swirly, vortex-like motion, commonly observed in other flowing systems near jamming also. Moreover, T1 events (local changes in the contact topology) are readily observed in these systems (Lauridsen *et al.*, 2004; Wang, 2006).

The probability distributions describing the instantaneous bubble velocities exhibit fat tails (Wang *et al.*, 2006). Consistent with this, Möbius *et al.* established that, for a given local strain rate, the probability distributions of bubble displacements exhibit fat tails for short times, develop exponential tails for intermediate times and finally become Gaussian. The occurrence of purely exponential distributions at a sharply defined time defines the relaxation time t_r, which coincides with the crossover time from superdiffusive to diffusive behavior, and also with the Lindeman criterion (Mobius and van Hecke, 2008).

Surprisingly, t_r is not proportional to the inverse of the strain rate that would be the simplest relation consistent with dimensional arguments, but instead exhibits a non-linear relationship with the strain rate. This has a direct consequence for the probability distributions of bubble displacements taken at a fixed strain: the width of this distribution grows as $\dot{\gamma} \to 0$. This so-called sublinear scaling, which has been observed in simulations (Ono and Liu, 2003), implies that these flows are not quasi-static, but rather that the amount of fluctuations increases for slower flows—not dissimilar to what we discussed above for granular pile flows.

For collections of viscous bubbles with known bubble–bubble interactions, the balance of work done on the system and the energy dissipated at the local scale, immediately dictates this sublinear scaling (Ono and Liu, 2003; Tighe and van Hecke, 2010). It has recently been suggested by Tighe *et al.* that the non-trivial scaling of the fluctuations also governs the non-trivial rheology of foams—where the global relation between strain rate and stress does not follow directly from the local relation between relative bubble motion and drag forces (Olsson and Teitel, 2007; Katgert and van Hecke, 2008; Tighe and van Hecke, 2010). This provides an intriguing link between non-trivial behavior at micro- and macroscale—how the spatial organization of the strong bubble fluctuations associated with sublinear scaling connects to dynamical heterogeneities in foams is at present an important open question.

6.5 Discussion

We have discussed the heterogeneities that arise in a variety of weakly driven systems near jamming. These heterogeneities have unveiled the existence of a dynamical length scale and the associated time scale responsible for the slow relaxation of these systems. On the one hand, the existence of such a length scale can be argued to be at the origin of the quasi-universal behavior observed in these glassy systems. On the other hand, the observed length scale is always rather small, say smaller than 10 particles diameters and the effect of the different microscopic mechanisms may still be significant.

For example, for the fluidized grains, interactions are mainly collisional, for the dense grain systems across the jamming transition they are mainly frictional, for the pile flows they are a combination of collisions and enduring contacts, while for foams the interactions are viscous.

Another important difference between these systems is that different sets of coordinates can be expected to characterize their states. For example, the structure of collisional grains is set by their positions only, and structural relaxation will be related to real-space motion and cage breaking, while for dense granular assemblies that do not show substantial motion of the grains, the relevant degrees of freedom may be the contact forces, and relaxations may not be dominated by particle motion but rather by changes in the contact forces.

The microscopic mechanism of dissipation, which differs between these systems, may also play an important role. All energy fed into these systems by shear or agitations needs to be dissipated by relative motion of neighboring particles. Energy balance then may lead to the so-called sublinear scaling of fluctuations (Ono and Liu, 2003)—the details of the energy dissipation then actually set the width of the distributions characterizing the fluctuations. Is it also responsible for the spatial organization of these fluctuations? More work is needed to clarify this.

As a matter of fact, while it is rather obvious that in dense systems, local rearrangements will couple to neighboring particles, it is far from clear what mechanism governs the spatial organization of these relaxation events. On the one hand, it is tempting, following recent work (Wyart, 2005; Brito and Wyart, 2006), to conjecture that the correlated currents observed here are related to the extended *soft modes* that appear when the system loses or acquires rigidity near jamming. Under the action of a mechanical drive the system should fail along these soft modes. On the other hand, recent investigations (Lechenault *et al.*, 2010) suggest that motion of frictional grains in the vicinity of the jamming transition can be interpreted as microcrack events on all scales undermining the usefulness of harmonic modes as a way to rationalize the dynamics.

Finally, the nature of the relation, if any, between the rheological response of the materials and the dynamical heterogeneities is far from being understood. Whether the emergence of a large length scale near jamming controls the rheology is still a matter of debate. It is also possible that the answer to this question is different on both side of the transition. Further studies in this matter, including non-linear rheology and local probe experiments (Habdas *et al.*, 2004; Dollet *et al.*, 2005; Geng and Behringer, 2005; Candelier and Dauchot, 2009) will certainly contribute to uncover new and probably unexpected effects in this exciting field of soft-matter physics.

6.6 Appendix: how to measure χ_4, and the dangers

The dynamic susceptibility $\chi_4(l, \tau)$ has emerged as a powerful statistical tool for characterizing dynamical heterogeneities (Sillescu, 1999; Ediger, 2000; Glotzer, 2000; Lačević *et al.*, 2003; Cipelletti and Ramos, 2005). However, its definition is somewhat involved and there are pitfalls that must be recognized and avoided if physical meaning is to be extracted from its use. We offer the following guide to help in this regard.

The first ingredient is an ensemble-averaged dynamical self-overlap order parameter, $Q_t(l, \tau)$, defined such that the contribution from each particle p $Q_{p,t}(l, \tau)$ is some function that decays vs. delay time τ from one to zero as the particle moves a characteristic distance l away from its location at time t. At very short (respectively, very large) τ all particles have moved a distance much less (respectively, much larger) than the length l and their contribution to $Q_t(l, \tau)$ is very nearly 1 (respectively, nearly 0) with little variance for different start times. By contrast, at intermediate τ when particles in mobile regions have moved more than l and immobile regions have moved less than l, fluctuations in the number of mobile regions cause $Q_t(l, \tau)$ to vary noticeably around its average. In essence, the role of $\chi_4(l, \tau) \equiv NVar(Q_t)$ is thus to capture fluctuations in the number of fast-moving mobile regions. Therefore, as shown in Fig. 6.2, $\chi_4(l, \tau)$ vs. τ is generally expected to rise from zero to a peak at delay time $\tau*$ when the typical displacement is near l and then decay back to zero.

For the purpose of clarification, let us first introduce a simplified picture of a system of N particles with a fluctuating number M_t of mobile regions of size n and assume that the order parameter is Q_0 in all fast regions and Q_1 in slow regions. Then, the order parameter is given by a weighted average of these values over the total number nM_t of particles in the fast regions and the number $N - nM_t$ of particles in the slow regions: $Q_t = [f_t Q_0 + (1 - f_t)Q_1]$, where $f_t = nM_t/N$ is the fraction of mobile particles. From this, one readily obtains the averaged control parameter and χ_4:

$$\bar{Q} = \bar{f}\,\delta Q + Q_1, \tag{6.1}$$

$$\chi_4 = N\,\mathrm{Var}(Q) = n\,\bar{f}\,\delta Q^2\,\mathrm{Var}(M)/M, \tag{6.2}$$

where $\delta Q = Q_1 - Q_0$ is a measure of how different are fast and slow particles. If one assumes that there is a large number of mobile regions and that they are decorrelated, then $\mathrm{Var}(M) \sim M$ and one can in principle measure the size n of these regions from the above relations.

We now use the above expressions to discuss the precise way of implementing the above procedure.

Choice of the order parameter

Most simply, it may be taken as the average over particles of step functions that drops discontinuously from one to zero when a particle moves a distance l (Glotzer, 2000; Lačević *et al.*, 2003; Keys *et al.*, 2007). A smooth Gaussian cutoff may also be used (Marty and Dauchot, 2005; Dauchot *et al.*, 2005); then the order parameter appears like a dynamic structure factor, which motivates calling $\chi_4(l, \tau)$ a dynamical susceptibility. The advantage of these two choices is that l may then be adjusted according to the relevant physics. For example, l approximately equal to particle radius is appropriate for the usual caging, where a totally new configuration is obtained when the particles move about a particle size. Alternative choices have been made in which the cutoff length l is set to probe the topological features of the particle arrangement: the persistent area given by the fraction of space in the same Voronoi cell, and the persistent bond given by the fraction of Voronoi neighbors that remain, after a delay time τ (Abate and Durian, 2007). But for a compressed pack of particles subject to shaking / shearing, a totally new configuration of frictional contacts arises without change in neighbors; then, a much smaller value of l is appropriate (Lechenault *et al.*, 2008*a*). One can also fix l on the basis of the cross-correlation of grayscale images (Katsuragi *et al.*, 2010). When in doubt, it is useful to consider the full behavior of $\chi_4(l, \tau)$ vs. both l and τ (Lechenault *et al.*, 2008*a*).

Dependence on the packing fraction
As soon as one is interested in the dependence of the size of the mobile regions at the peak of χ_4, n^*, on the packing fraction ϕ , one should (i) check that the relevant length scale l does not vary too much with ϕ, which is usually the case, (ii) properly normalize χ_4 by $\bar{f}\delta Q^2$, since the difference in mobility between the fast and the slow particles may well vary with the packing fraction too.

Finite-size effects
As stated above, it is necessary to have a large enough number of independant mobile regions in order to ensure $\text{var}(M) \sim M$. Since one also expects large values of n^* close to the transition of interest, satisfying the above condition requires the use of very large systems, typically of the order of $N = 100\, n^*$. These size effects rapidly become critical since the relative error on the measure of $\text{var}(Q)$ scales like $\sqrt{N/T}$, where T is the duration of the acquisition.

One sees that an educated use of χ_4 requires the knowledge of both δ_Q and $\bar{f}$, or equivalently the knowledge not only of the ensemble-averaged $Q_t(l, \tau)$, but of all individuals $Q_{p,t}(l, \tau)$. Clearly this is not always the case, and one can already have some insights in the dynamical heterogenities following the above procedure, but keeping in mind the caveats we have just listed. However, when one has the possibility of tracking the particles and thereby has a direct access to the local dynamics, the dynamical heterogeneities are more precisely characterized by calculating directly the spatial correlation of the dynamics, namely the four-point correlator $G_4(r, l, \tau)$:

$$G_4(r, l, \tau) = \langle Q_{p,t}(l, \tau) Q_{p',t}(l, \tau) \rangle_{d_{p,p'}=r} - \langle Q_t(l, \tau) \rangle^2 , \qquad (6.3)$$

where $< . >_{d_{p,p'}=r}$ means that the average is computed over all pairs of particles separated by the distance r. The typical length scale of the dynamical heterogeneities ξ_4 is then readily obtained from the spatial dependance of this correlator. Obviously computing $G_4(r, l, \tau)$ is a more intensive task than the computation of χ_4.

References

Abate, A. R. and Durian, D. J. (2006). *Phys. Rev. E*, **74**(3), 031308.
Abate, A. R. and Durian, D. J. (2007). *Phys. Rev. E*, **76**(2), 021306.
Andreotti, B. and Douady, S. (2001). *Phys. Rev. E*, **63**(3), 031305.
Aranson, I.S. and Tsimring, L.S. (2002). *Phys. Rev. E*, **65**(6), 61303.
Aranson, I. S. and Tsimring, L. S. (2006). *Rev. Mod. Phys.*, **78**, 641.
Argon, A. S. and Kuo, H. Y. (1979). *Mater. Sci. Eng.*, **39**(1), 101–109.

Bocquet, L., Losert, W., Schalk, D., Lubensky, T. C., and Gollub, J. P. (2002). *Phys. Rev. E*, **65**(1), 011307.

Bolton, F. and Weaire, D. (1990). *Phys. Rev. Lett.*, **65**(27), 3449–3451.

Bonamy, D., Daviaud, F., and Laurent, L. (2002). *Phys. Rev. Lett.*, **89**(034301-1).

Brito, C. and Wyart, M. (2006). *Europhys. Lett*, **76**(1), 149–155.

Candelier, R. and Dauchot, O. (2009). *Phys. Rev. Lett.*, **103**, 128001.

Candelier, R., Dauchot, O., and Biroli, G. (2009*a*). *Phys. Rev. Lett.*, **102**(8), 088001.

Candelier, R., Dauchot, O., and Biroli, G. (2009*b*). Evolution of dynamical facilitation approaching the granular glass transition. arXiv.org:0912.0472.

Candelier, R., Widmer-Cooper, A., Kummerfeld, J. K., Dauchot, O., Biroli, G., Harrowell, P., and Reichman, D. R. (2009*c*). Avalanches and dynamical correlations in supercooled liquids. arXiv.org:0912.0193.

Cates, M. E., Wittmer, J. P., Bouchaud, J. P., and Claudin, P. (1998). *Phys. Rev. Lett.*, **81**(9), 1841.

Cipelletti, L. and Ramos, L. (2005). *J. Phys. Condens. Matter.*, **17**(6), R253–R285.

Cohen-Addad, S. and Hohler, R. (2001). *Phys. Rev. Lett.*, **86**(20), 4700.

Cortet, P. P., Bonamy, D., Daviaud, F., Dauchot, O., Dubrulle, B., and Renouf, M. (2009). Relevance of visco-plastic theory in a multi-directional inhomogeneous granular flow. *Europhys. Lett.*, **88**, 1400.

da Cruz, F., Emam, S., Prochnow, M., Roux, J.N., and Chevoir, F. (2005). *Phys. Rev. E*, **72**(2), 21309.

Daerr, A. and Douady, S. (1999). *Nature*, **399**(6733), 241–243.

Dauchot, O. (2007). In *Ageing and the Glass Transition* (ed. MPRS), Chapter 4, p. 161. Springer, Berlin.

Dauchot, O., Marty, G., and Biroli, G. (2005). *Phys. Rev. Lett.*, **95**(26), 265701.

Deboeuf, S., Bertin, EM, Lajeunesse, E., and Dauchot, O. (2003). *Eur. Phys. J. B-Condens. Matter Complex Syst.*, **36**(1), 105–113.

Deboeuf, S., Dauchot, O., Staron, L., Mangeney, A., and Vilotte, J.P. (2005). *Phys. Rev. E*, **72**(5), 51305.

Deboeuf, S., Lajeunesse, E., Dauchot, O., and Andreotti, B. (2006). *Phys. Rev. Lett.*, **97**(15), 158303.

Debregeas, G. Tabuteau, H and di Meglio, J.M. (2001). *Phys. Rev. Lett.*, **87**, 178305.

Djaoui, L. and Crassous, J. (2005). *Granular Matter*, **7**(4), 185–190.

Dollet, B., Elias, F., Quilliet, C., Raufaste, C., Aubouy, M., and Graner, F. (2005). *Phys. Rev. E*, **71**(3), 31403.

du Pont, S. C., Fischer, R., Gondret, P., Perrin, B., and Rabaud, M. (2005). *Phys. Rev. Lett.*, **94**, 048003.

Duran, J. (2000). *Sands, Powders, and Grains: An Introduction to the Physics of Granular Materials.* Springer, New York.

Duri, A., Sessoms, D. A., Trappe, V., and Cipelletti, L. (2009). *Phys. Rev. Lett.*, **102**(8), 085702–4. DOI: 10.1103/PhysRevLett.102.085702.

Durian, D. J. (1995). *Phys. Rev. Lett.*, **75**(26), 4780–4783.

Durian, D. J. (1997). *Physical Review E*, **55**(2), 1739–1751.

Durian, D. J., Weitz, D. A., and Pine, D. J. (1991). *Science*, **252**(5006), 686–688.

E. Janiaud, D. Weaire and Hutzler, S. (2006). *Phys. Rev. Lett.*, **97**, 038302.

Earnshaw, J. C. and Jaafar, A. H. (1994). *Physical Review E*, **49**, 5408.
Ediger, M. D. (2000). *Annu. Rev. Phys. Chem.*, **51**, 99–128.
Ellenbroek, W. G., Zeravcic, Z., van Saarloos, W., and van Hecke, M. (2009). *Europhys. Lett.*, **87**(3), 34004.
Ertas, D. and Halsey, T.C. (2002). *Europhys. Lett.*, **60**, 931.
Fenistein, D., van de Meent, J. W., and van Hecke, M. (2004). *Phys. Rev. Lett.*, **92**(9), 094301.
Garrahan, Juan P. and Chandler, David (2002). *Phys. Rev. Lett.*, **89**, 035704.
Geng, Junfei and Behringer, R. P. (2005, Jan). *Phys. Rev. E*, **71**(1), 011302.
Gittings, A. S. and Durian, D. J. (2006). *Appl. Opt.*, **45**, 2199–2204.
Gittings, A. S. and Durian, D. J. (2008). *Physical Review E*, **78**(6), 066313.
Glotzer, S. C. (2000). *J. Non-Cryst. Solids*, **274**(1-3), 342–355.
Goldhirsch, I (2003). *Annu. Rev. Fluid Mech.*, **35**, 267.
Gopal, A. D. and Durian, D. J. (1995). *Phys. Rev. Lett.*, **75**(13), 2610–2613.
Gopal, A. D. and Durian, D. J. (1997). *J. Opt. Soc. Am. A*, **14**(1), 150–155.
Gopal, A. D. and Durian, D. J. (1999). *J. Colloid Interf. Sci.*, **213**(1), 169–178.
Gopal, A. D. and Durian, D. J. (2003). *Phys. Rev. Lett.*, **91**(18), 188303.
Habdas, P., Schaar, D., Levitt, A.C., and Weeks, E.R. (2004). *Europhys. Lett.*, **67**(3), 477–483.
Hatano, T. (2008). *arXiv:0804.0477v2*.
Herdtle, T. and Aref, H. (1992). *J. Fluid Mecha.*, **241**, 233–60.
Howell, D., Behringer, R.P., and Veje, C. (1999). *Phys. Rev. Lett.*, **82**(26), 5241–5244.
Iordanoff, I. and Khonsari, MM (2004). *J. Tribol.*, **126**, 137.
Jaeger, H. M., Nagel, S. R., and Behringer, R. P. (1996). *Rev. Mod. Phys.*, **68**, 1259.
Jop, P., Forterre, Y., and Pouliquen, O. (2005). *J. Fluid Mech.*, **541**, 167–192.
Jop, P., Forterre, Y., and Pouliquen, O. (2006). *Nature*, **441**(7094), 727–730.
Katgert, G. Mobius, M. E. and van Hecke, M. (2008). *Phys. Rev. Lett.*, **101**, 058301.
Katsuragi, H., Abate, A. R., and Durian, D. J. (2010). *Soft Matter*, **6**, 3023.
Keys, A. S., Abate, A. R., Glotzer, S. C., and Durian, D. J. (2007). *Nature Phys.*, **3**(4), 260–264.
Khakhar, D. V., Orpe, A. V., Andresén, P., and Ottino, J. M. (2001). *J. Fluid Mech.*, **441**, 255.
Komatsu, T. S., Inagaki, S., Nakagawa, N., and Nasuno, S. (2001). *Phys. Rev. Lett.*, **86**, 1757.
Kraynik, A.M. (1988). *Ann. Rev. Fluid Mech.*, **20**, 325–357.
Lačević, N., Starr, F. W., Schroder, T. B., and Glotzer, S. C. (2003). *J. Chem. Phys.*, **119**(14), 7372–7387.
Lauridsen, J., Chanan, G., and Dennin, M. (2004). *Phys. Rev. Lett.*, **93**(1), 018303.
Lechenault, F., Candelier, R., Dauchot, O., Bouchaud, J. P., and Biroli, G. (2010). Super-diffusion around the rigidity transition: Levy and the lilliputians. arXiv.org:1001.1765. *Soft Matter*, **6**, 3054.
Lechenault, F., Dauchot, O., Biroli, G., and Bouchaud, JP (2008*a*). *Europhys. Lett.*, **83**, 46003.

Lechenault, F., Dauchot, O., Biroli, G., and Bouchaud, J.-P. (2008*b*). *Europhys. Lett.*, **83**.

Lemaître, A. (2002). *Phys. Rev. Lett.*, **89**(19), 195503.

Lemieux, P.-A. and Durian, D. J. (2000). *Phys. Rev. Lett.*, **85**, 4273.

Liu, A. J. and Nagel, S. R. (ed.) (2001). *Jamming and Rheology: Constrained Dynamics on Microscopic and Macroscopic Scales.* Taylor and Francis, New York.

Liu, A. J., Ramaswamy, S., Mason, T. G., Gang, H., and Weitz, D. A. (1996). *Phys. Rev. Lett.*, **76**(16), 3017–20.

Losert, W., Bocquet, L., Lubensky, TC, and Gollub, JP (2000). *Phys. Rev. Lett.*, **85**(7), 1428–1431.

Marty, G. and Dauchot, O. (2005). *Phys. Rev. Lett.*, **94**(1), 15701.

Mayer, P., Bissig, H., Berthier, L., Cipelletti, L., Garrahan, J. P., Sollich, P., and Trappe, V. (2004). *Phys. Rev. Lett.*, **93**(11), 115701.

MiDi, GDR (2004). *Eur. Phys. J. E: Soft Matter Biol. Phys.*, **14**(4), 341–365.

Mills, P., Loggia, D., and Texier, M. (1999). *Europhys. Lett.*, **45**, 733.

Mobius, M. E. Katgert, G. and van Hecke, M. (2008). *arXiv:cond-mat/0811-0534*.

Mueth, D. M., Debregeas, G. F., Karczmart, G. S., Eng, P. J., Nagel, S. R., and Jaeger, H. M. (2000). *Nature*, **406**(6794), 385–9.

Nedderman, R.M. (1992). *Statics and Kinematics of Granular Materials.* Cambridge University Press, Cambridge.

O'Hern, C.S., Silbert, L.E., Liu, A.J., and Nagel, S.R. (2003*a*). *Phys. Rev. E*, **68**(1), 11306.

O'Hern, C. S., Silbert, L. E., Liu, A. J., and Nagel, S. R. (2003*b*). *Phys. Rev. E*, **68**(1), 011306.

Okuzono, T. and Kawasaki, K. (1995). *Physical Review E*, **51**(2), 1246–53.

Olsson, P. and Teitel, S. (2007). *Phys. Rev. Lett.*, **99**, 178001.

Ono, I. K. Tewari, S. Langer, S.A. and Liu, A. J. (2003). *Phys. Rev. E*, **67**, 061503.

Pouliquen, Olivier, Belzons, M, and Nicolas, M (2003). *Phys. Rev. Lett.*, **91**, 014301.

Prud'homme, Robert K. and Khan, Saad A. (1996). *Foams: Theory, Measurements, and Applications.* Marcel Dekker Inc., New York.

Rajchenbach, J. (1990). *Phys. Rev. Lett.*, **65**, 2221.

Richard, P., Valance, A., Metayer, J. F., Sanchez, P., Crassous, J., Louge, M., and Delannay, R. (2008). *Phys. Rev. Lett.*, **101**(24), 248002.

Ritort, F. and Sollich, P. (2003). *Adv. Phys.*, **52**(4), 219–342.

Saint-Jalmes, A. and Durian, D. J. (1999). *J. Rheol.*, **43**(6), 1411–1422.

Savage, S.B. (1998). *J. Fluid Mech.*, **377**, 1–26.

Savage, S.B. and Jeffrey, D.J (1981). *J. Fluid Mech.*, **110**, 255.

Sessoms, David A., Bissig, Hugo, Duri, Agnes, Cipelletti, Luca, and Trappe, Veronique (2010). *Soft Matter*, **6**(13), 3030–3037.

Sillescu, H. (1999). *J. Non-Cryst. Solids*, **243**, 81–108.

Tighe, B. Woldhuis, E. Remmers, J. van Saarloos, W. and van Hecke, M. (2010). *Preprint*.

Toninelli, C., Biroli, G., and Fisher, D.S. (2006). *Phys. Rev. Lett.*, **96**(3), 35702.

van Hecke, M. (2010). *J. Phys. Condens. Matter*, **22**, 031101.
Wang, Y. H., Krishan, K., and Dennin, M. (2006). *Physical Review E*, **74**(4), 041405.
Weaire, D. and Hutzler, S. (1999). *The Physics of Foams.* Clarendon Press, New York.
Wilson, A.J. (ed.) (1989). *Foams: Physics, Chemistry and Structure.* Springer-Verlag, New York.
Wyart, M. (2005). *Ann. Phys. Fr*, **30**(3), 1–96.
Wang, Y. H. Krishan, K. Dennin, M. (2006). *Phys. Rev. E*, **73**, 031401.

7

The length scales of dynamic heterogeneity: results from molecular dynamics simulations

Peter HARROWELL

Abstract

Over times shorter than that required for relaxation of enthalpy, a liquid can exhibit striking heterogeneities. The picture of these heterogeneities is complex with transient patches of rigidity, irregular yet persistent, intersected by tendrils of mobile particles, flickering intermittently into new spatial patterns of motion and arrest. The study of these dynamic heterogeneities has, over the last 20 years, allowed us to characterize cooperative dynamics, to identify new strategies in controlling kinetics in glass-forming liquids and to begin to systematically explore the relationship between dynamics and structure that underpins the behavior of amorphous materials. Computer simulations of the dynamics in atomic and molecular liquids have played a dominant role in all of this progress. While some may be uneasy about this reliance on modeling, it is unavoidable, given the amount of microscopic detail needed to characterize the dynamic heterogeneities. The complexities revealed by these simulations have called for new conceptual tools. In this chapter, I have tried to provide the reader with a clear and complete account of how these tools have been developed in terms of the literature on kinetic length scales in molecular dynamics simulations. Through the "prism" of these length scales, this chapter addresses the question what have we learnt about dynamic heterogeneities from computer simulations?

7.1 Introduction

What is it that distinguishes a glass from a crystal? Starting with the most casual inspection, the presence of oriented planes, grain boundaries or edges would indicate a crystal since the glass must be isotropic (and, hence, amorphous). Looking closer, the presence of sharp intensity peaks in scattered radiation at large angles indicates the presence of repeated parallel planes of density. The absence of such planes is typically all it takes for us to label the material "disordered". At the level of the constituent particles (atoms or molecules), however, the most striking physical consequence of being in a glass rather than a crystal is the large number of different local environments (i.e. structural heterogeneity) in the former. Daama and Villars (Daama and Villars, 1997) have established that over 90% of the known 17 000 inorganic crystal structures with intermetallic structure types consist of no more than 4 distinct coordination environments. This is in clear contrast to a corresponding glass. Reverse Monte Carlo analysis of EXAFS measurements of the intermetallic glass $Ni_{80}P_{20}$ (Luo and Ma, 2008) have identified over 15 different coordination environments. Collective dynamics serves to amplify these structural variations, translating the often subtle differences in local configurations into dramatic variations in local relaxation kinetics. The result is that the approach to the glass transition is typically marked by striking spatial heterogeneity in dynamics. Many of the landmark insights of material science—dislocation-mediated plasticity, the Peierls barrier, Nabarro–Herring creep, grain-boundary mobility—can be comfortably re-expressed as manifestations of dynamic heterogeneities. In crystals, these objects are identifiable as defects and therefore have explicit structural signatures but, by shifting our focus to the spatial distribution of dynamics rather than order, we can expand our study of localized relaxation processes to materials for which we have no *a priori* notion of the structural origin of the localization. From this perspective, dynamic heterogeneities provide a genuine opportunity to develop a universal description of dynamic localization and collective relaxation in condensed matter—ordered and disordered, in equilibrium or out.

Over the last 20 years, there has been a steady growth in the appreciation of the ubiquity of dynamic heterogeneities in disordered materials and of the value of their study. The subject of dynamic heterogeneities has been considered in a number of reviews of the glass transition (Poole, 1998; Kob, 1999; Glotzer, 2000; Jäckle, 2002; Andersen, 2005; Binder and Kob, 2005; Heuer, 2008). These spatial fluctuations, which take the form of transient kinetic domains, represent an extension of the "traditional" phenomenology of disordered materials ("traditional" referring to thermodynamics and bulk-averaged scattering and dynamic susceptibilities); an extension capable of providing spatial information about the fluctuations associated with the collective dynamics without requiring any insight as to the particle arrangements responsible. The existence of heterogeneities can also provide a link between different aspects of the phenomenology—stretched relaxations, fragility, etc. (Perera and Harrowell, 1996). While computational methods may dominate their study and theoretical goals provide much of the motivation, it is worth emphasizing that dynamic heterogeneities are not a theoretical construction but a physical fact and their description is, in the end, of value in its own right.

To talk about dynamic heterogeneities in anything other than pictures we need some quantities to measure and kinetic lengths represent the most versatile of these, making them something of a *lingua franca* of glassy dynamics. What do we mean by a kinetic length? Prior to the study of dynamic heterogeneities, the kinetic length was attributed, somewhat vaguely, to a size of collective motion or a cooperative rearrangement. In this chapter we shall review what we now know about kinetic length scales associated with dynamic heterogeneities as a result of molecular dynamics (MD) simulations. The restriction to MD simulations, while omitting many important aspects of the research into dynamic heterogeneities, ensures that we shall only consider those aspects of cooperative dynamics that we can explicitly connect to the positions and momenta of particles. We shall not cover the work on lattice models of glass-forming liquids and we shall not cover the various theoretical treatments of the glass transition. Interested readers can find excellent reviews of these important topics in refs. (Ritort and Sollich, 2003) and (Cavagna, 2009), respectively, and in the chapters of this volume.

We shall take the view that MD simulations confront us with all of the essential complexities of real supercooled liquids and glasses without, necessarily, providing a quantitatively accurate model of any one specific glass former. The great boon of these simulations, i.e. the provision of all particle positions and momenta at as many instances as desired, is also the source of the rather stringent obligation that they exert. After all, if you can access all information, then the failure to explain any aspect of the glass transition that can be captured by a simulation can only be attributed to our personal failure in asking the right question. Dynamic heterogeneity owes much of its existence as a recognized phenomenon to MD simulation studies. While we now have mesoscopic (Weeks *et al.*, 2000; Abate and Durian, 2007) and macroscopic (Dauchot *et al.*, 2005) analogs of liquids in which the dynamic heterogeneities can be directly seen, it is fair to say that these discoveries owe their validation as models of microscopic heterogeneities to comparisons with computer modeling.

The idea of a kinetic length scale, which extends back, at least, to the work of Adam and Gibbs (Adam and Gibbs, 1965) in 1965 (if not further back (Jenckel, 1939)), is that the growing degree of dynamic correlation in a supercooled liquid can be measured in terms of a length that grows with cooling. To be accurate, Adam–Gibbs did not talk of a length scale but, rather, a number of particles required for a collective rearrangement. Converting this number into a length requires some sort of ancillary assumption concerning the connectivity of these particles. There is an undeniable wishfulness involved in expecting to be able to replace something so poorly defined as the "extent of cooperative motion" with something as concrete as a length. Computer simulations have provided the opportunity (obligation, really) to put this aspiration to the test and develop explicit expressions (theoretical or algorithmic) for this kinetic length scale.

MD simulations require interparticle potentials. Simulations of supercooled liquids require interactions that are computationally simple in order to get as many time steps out of a computer as possible and yet geometrically complex enough to stave off crystallization long enough to allow characterization of the metastable liquid. In Table 7.1 we provide details of a number of models, all based on binary atomic alloys, that have proved popular and that form the basis of many of the simulation studies of the glass transition.

In this review we shall consider five approaches to calculating a kinetic correlation length from MD simulations: direct measures of the spatial distribution of mobility, 4-point correlation functions and the associated susceptibility χ_4, finite-size effects, kinetic correlations at interfaces and the crossover lengths for kinetic properties.

Table 7.1 Details of a number of model glass-forming liquids. Note that BMLJ1 and BMLJ2 differ in that the former includes a strong preference for AB neighbors, while the latter does not impose any specific chemical ordering. The soft-sphere (SS) model is the same as the BMLJ2 system, except for the absence of the attractions.

Label	Interparticle Potential	Composition	Reference
BMLJ1 *(Kob–Andersen)*	$\varphi_{ij}(r) = 4\epsilon_{ij}([\sigma_{ij}/r]^{12} - [\sigma_{ij}/r]^6)$ $\sigma_{AA} = 1.0, \sigma_{BB} = 0.88,\ \sigma_{AB} = 0.8$ $\epsilon_{AA} = 1.0, \epsilon_{BB} = 0.5, \epsilon_{AB} = 1.5$	A_4B	a
BMLJ2 *(Wahnström)*	$\varphi_{ij}(r) = 4\epsilon_{ij}([\sigma_{ij}/r]^{12} - [\sigma_{ij}/r]^6)$ $\sigma_{AA} = 1.2, \sigma_{BB} = 1.0,\ \sigma_{AB} = 1.1$ $\epsilon_{AA} = \epsilon_{BB} = \epsilon_{AB} = 1.0$	AB	b
SS *(soft sphere)*	$\varphi_{ij}(r) = \epsilon_{ij}(\sigma_{ij}/r)^{12}$ $\sigma_{AA} = 1.2, \sigma_{BB} = 1.0,\ \sigma_{AB} = 1.1$ $\epsilon_{AA} = \epsilon_{BB} = \epsilon_{AB} = 1.0$	AB	c
SD *(soft disk)*	$\varphi_{ij}(r) = \epsilon_{ij}(\sigma_{ij}/r)^{12}$ (in 2D) $\sigma_{AA} = 1.4, \sigma_{BB} = 1.0,\ \sigma_{AB} = 1.2$ $\epsilon_{AA} = \epsilon_{BB} = \epsilon_{AB} = 1.0$	AB	d

References: (a) Kob and Andersen, 1995, (b) Wahnström, 1991, (c) Bernu *et al.*, 1987, (d) Perera and Harrowell, 1998

7.2 Kinetic lengths from displacement distributions

Historically, the first approach (Deng *et al.*, 1989) to analyzing dynamic heterogeneities was to make a map of them. Maps are appealing. They retain a large amount of information. A single kinetic length scale, as we shall see, represents a major (and often uncontrolled) discarding of most of this information. The appeal of a map must be weighed against their principal shortcoming—they provide information (a lot of it, admittedly) about a single instance of heterogeneity only. Considerable care, therefore, is generally required in extracting the statistically significant features from the noise.

The earliest example of displacement maps being used to specifically characterize dynamic heterogeneities can be found in a 1989 paper by Deng, Argon and Yip (Deng *et al.*, 1989). This paper contains a section with the prophetic title *Inhomogeneities and the clustering of atomic motions.* Inspection of maps of the particle displacements over a chosen time interval (see Fig. 7.1 for an example), reveal the generic features of the heterogeneities. Typically, we find connected domains of slow and of fast particles, with the former typically more compact than the latter. Collective strains within the "slow" domains coexist with low-dimensional flows among the more mobile particles. Forced to identify lengths we might choose the average extent of slow domains or the fast domains or we might consider the length scale over which displacement direction is correlated. Alternatively, we might ask about the size of the "core" regions where large displacements appear to be directed randomly or, instead, determine the average separation between such localized reorganizations. Our choice would, of course, be

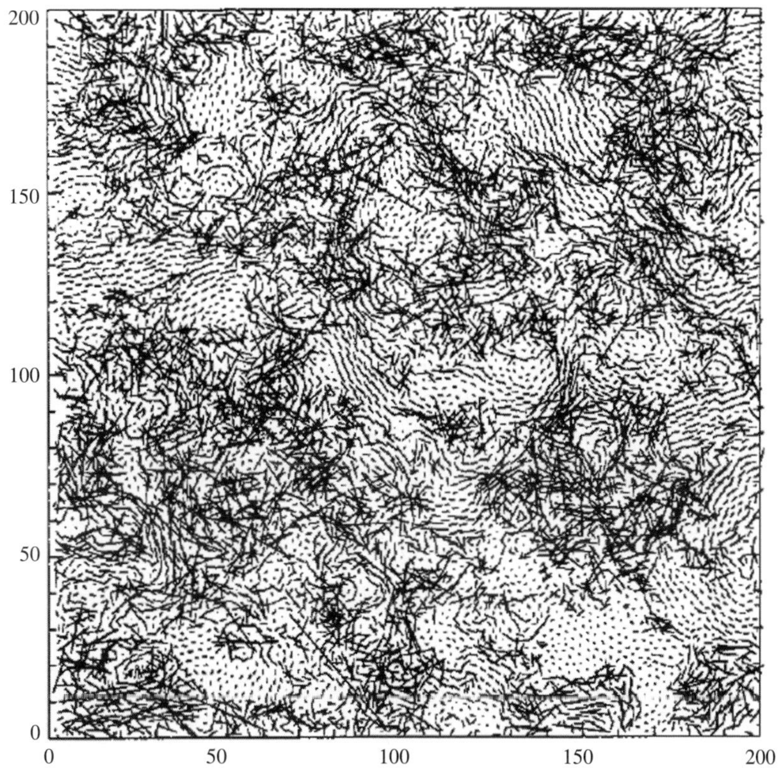

Fig. 7.1 The particle motions $\vec{r}_i(t + 50\tau_\alpha) - \vec{r}_i(t)$, where τ_α is the structural relaxation time for a polydisperse mixture of hard disks. [Reproduced with permission from ref. (Doliwa and Heuer, 2000).]

simplified if an argument existed that established that all of these lengths scaled in a similar fashion. Unfortunately, no such argument exists.

In 1995, Hurley and Harrowell (Hurley and Harrowell, 1995) extracted a kinetic length scale for the one-component soft-disk liquid in terms of the decay in the variance of relaxation time maps as the linear dimension of the scaling volume increased using a box scaling method. (That these calculations were carried out on an equilibrium liquid—and only later on a supercooled mixture (Perera and Harrowell, 1998)—underscores the point that dynamic heterogeneities are not restricted to supercooled liquids.) The relaxation time was defined as the first passage time for a particle's displacement to exceed a threshold distance that was chosen to maximize the kinetic length. The kinetic length in the SD liquid was found to exhibit a super-Arrhenius increase with decreasing temperature (Perera and Harrowell, 1998; Perera, 1998; Perera and Harrowell, 1999).

In 1997 Kob *et al.* (Kob *et al.*, 1997) introduced an analysis of dynamic heterogeneities based on determining the statistics of clusters of various kinetic subpopulations. Mobile particles were defined in such a way that they comprised $\sim$ 5% of the particles with the largest displacements over a time interval corresponding to the maximum in the non-Gaussian parameter (see Section 7.6). A characteristic size was obtained as S, the mass-weighted average cluster size. For the BMLJ1 liquid, the temperature variation of S was fitted as $S = 0.975/(T - 0.431)^{0.687}$ (Donati *et al.*, 1999*b*). Previously, the mode-coupling theory (Götze and Sjögren, 1992) had predicted the divergence of the relaxation time via a similar functional form, i.e. $(T - T_c)^{-\varsigma}$. In the BMLJ1 liquid, $T_c \sim 0.43$ (Kob *et al.*, 1997). The analogous average cluster size for the 5% *slowest* particles exhibited little variation with T. The authors suggested that the apparent divergence of the size of the mobile clusters was significant, possibly linked to the divergence predicted by the mode-coupling theory. While an increase in the kinetic length with cooling is confirmed by all approaches, some care needs to be taken in interpreting an increase in cluster size when one is looking at clusters of some subpopulation of fixed size. A decrease in the total number of such clusters will force an increase in the size of those remaining, simply as an artefact of how the clusters are defined.

The cluster analysis has been applied to a number of systems: polymers (Bennemann *et al.*, 1999; Gebremichael *et al.*, 2001), water (Mazza *et al.*, 2006), SiO_2 (Vogel *et al.*, 2004; Vogel and Glotzer, 2004), the Dzugutov potential (Gebremichael *et al.*, 2004) and polar diatomic molecules (Palomar and Sesé, 2008). Giovambattista et al (Giovambattista *et al.*, 2005) have calculated the fractal dimension of the mobile clusters in supercooled water. They found that large clusters exhibited a fractal dimension of $\sim$ 2, a value similar to that predicted (Lamarcq *et al.*, 2002) for low-energy excitations in a spin-glass. Vollmayr-Lee *et al.* (Vollmayr-Lee *et al.*, 2002) have extended the mobile cluster analysis to a BMLJ1 mixture *below* its glass-transition temperature. Below T_g, mobility is determined using the mean amplitude of a particle's fluctuation about its mean position. The authors report that clusters of mobile particles defined by this means were more compact and shorter lived than the analogous clusters above T_g. As to the question of the connection between the kinetic lengths of mobile particles on either side of the glass transition, strong correlations have been demonstrated (Widmer-Cooper and Harrowell, 2006) above T_g in the SD mixture between the spatial distribution of mobile particles and those particles that, over short times, exhibited large-amplitude fluctuations (analyzed in terms of local Debye–Waller factors (Widmer-Cooper and Harrowell, 2006))—a criterion similar in spirit to that used by Vollmayr-Lee *et al.* (Vollmayr-Lee *et al.*, 2002).

Having introduced the cluster analysis, Donati *et al.* (Donati *et al.*, 1998) augmented it in order to examine the correlation between displacement directions and the positions of mobile neighbors. Starting with the same fraction of mobile particles, they introduced an additional requirement for belonging to a cluster—for two neighboring mobile particles to belong to the same cluster the new position of one particle has to lie within a selected distance of the neighbor's old position. This new overlap condition strengthens the interpretation of the clusters as representing a cooperative motion rather than simply some general aggregation. Striking pictures of string-like arrangements of displacements (i.e. collective movements that are correlated only along the direction of flow) have been presented for the BMLJ1 (Donati *et al.*, 1998) and SS (Kim and Yamamoto, 2000) mixtures. The observation of low-dimensional particle flows raises some interesting questions about the mechanisms of cooperative dynamics and readers are encouraged to read Refs. (Vogel *et al.*, 2004; Gebremichael *et al.*, 2004) closely for details of the complex motions by which displacement strings are formed.

Since first presented in 1998 (Donati *et al.*, 1998), images of strings of particle displacements have proved a popular *leif-motif* for dynamic heterogeneities in general. To what degree is this representation accurate? Are dynamic heterogeneities generally string-like or are extended strings of displacements simply the more eye-catching members of a broad distribution of cluster shapes? There is not a lot of information that addresses this specific question. Focusing on the most mobile $\sim 5\%$ of particles in a BMLJ1 mixture, Donati *et al.* (Donati *et al.*, 1998) demonstrated that the displacement vectors exhibited a local forward alignment and that the clusters defined using the overlap condition were described by an exponential distribution of the size with the average number increasing from ~ 1.4 ($T = 0.55$) to ~ 2.2 ($T = 0.45$). The mass-weighted average size is larger, reaching ~ 15 at $T = 0.45$ (Donati *et al.*, 1999*b*). While the authors of Ref. (Donati *et al.*, 1999*b*, 1998) refer to these clusters as "strings", the overlap condition for cluster membership establishes only a flow and does not specifically establish the dimensionality of that flow. While the overlap condition will certainly include string-like correlations, it will also include some collective flows and strains in general. The existence of such collective flows in liquids at rest has been known for some time. Alder and Wainwright (Alder and Wainwright, 1967) demonstrated in 1967 the local forward alignment between particle velocities in an equilibrium liquid, part of a pattern that strongly resembled the solution of the Navier–Stokes equations about a moving particle. Doliwa and Heuer (Doliwa and Heuer, 2000) have reproduced this pattern in a study of the displacements in a binary hard-disk mixture. To establish string-like correlations one would like to establish the average coordination number of particles in the mobile cluster is ≤ 2. This quantity has not been determined, but Donati *et al.* (Donati *et al.*, 1999*b*) have reported a fractal dimension of 1.75 for mobile clusters in the BMLJ1 model (obtained over a single order of magnitude data set). There are indisputably string-like objects such as the self-avoiding random walk in 3D with a similar fractal dimension (1.66 from Flory's argument (Flory, 1969)). There are, however, other random objects with similar fractal dimensions—on a 3D lattice we have the random cluster below percolation with a fractal dimension of 2 and the backbone of the percolating cluster with a dimension $\sim$ 1.7 (Stauffer and Aharony, 1994)—for which the description "string-like" really doesn't apply. The cumulative data clearly points to mobile clusters that are *not* compact 3D objects but it leaves open the question of what they *are*. There is certainly a need for more systematic studies of the shape (and its dependence on temperature and the time interval used to define mobility) as well as the length scale of dynamic heterogeneities.

In simulations of the SD and SS mixtures, Yamamoto and Onuki (Yamamoto and Onuki, 1997, 1998*a*,*b*, 1999) shifted the focus from particle displacements to changes in the topology of a configuration in the form of "broken bonds". Their analysis involved generating spatial maps of the positions where initially nearest-neighbor pairs first moved far enough apart for the two-particle bond to be considered broken.

These authors showed that the structure factor $S_b(q)$ of the broken bonds in both 2D and 3D, accumulated over a fixed time interval, obeyed the small-q expansion of $S_b^{-1}(q)$, i.e.

$$S_b(q) \approx \frac{S_b(0)}{1+\xi^2 q^2} \tag{7.1}$$

where ξ represents the associated length. (The functional form in Eq. (7.1) corresponds to an exponential decaying correlation between broken bonds in real space.) Observing that $S_b(q)$ appeared independent of temperature (and, hence, ξ) at large q, the authors concluded that $S_b(0) \sim \xi^2$, the Ornstein–Zernicke result, and noted the analogy with critical fluctuations.

Analogous correlation functions have also been developed to describe the spatial correlations between displacements. In 1998, Poole *et al.* (Poole *et al.*, 1998) introduced the displacement–displacement correlation function $G_u(\vec{r}, \Delta t)$, defined as follows. For the displacement amplitude field $u(\vec{r}, t, \Delta t)$ given by

$$u(\vec{r}, t, \Delta t) = \sum_{i=1}^{N} |\vec{r}_i(t+\Delta t) - \vec{r}_i(t)| \delta(\vec{r} - \vec{r}_i(t)) \tag{7.2}$$

we can write

$$G_u(\vec{r}, \Delta t) = \int d\vec{r}' < [u(\vec{r}' + \vec{r}, t, \Delta t) - < u >][u(\vec{r}', t, \Delta t) - < u >] >. \tag{7.3}$$

By analogy with the relationship between the density–density correlation function, the variance of the number of particles N and the compressibility κ, Donati *et al.* (Donati *et al.*, 1999*a*) write

$$\int d\vec{r} G_u(\vec{r}, \Delta t) =< [U - < U >]^2 > \equiv < u >< U > kT\kappa_u, \tag{7.4}$$

where $U = \int d\vec{r} u(\vec{r}, t, \Delta t)$ is the total displacement and κ_u is a time-dependent dynamic susceptibility. For the BMLJ1 model, the maximum value of κ_u with respect to the displacement time interval Δt varies with T as $(T - 0.435)^{-0.84}$. From the decay of the density correlations with separations or, equivalently, working with the analogous structure factor and Eq. (7.1) as in ref. (Yamamoto and Onuki, 1998*a*), a kinetic length can be extracted. This kinetic length corresponds to the correlation length of fluctuations in the displacement amplitude field. A number of models have been analyzed using this approach: the BMLJ1 mixture (Poole *et al.*, 1998; Donati *et al.*, 1999*a*), molecular (Qian *et al.*, 1999) and polymeric (Bennemann *et al.*, 1999; Gebremichael *et al.*, 2001) liquids and polydisperse hard spheres (Doliwa and Heuer, 2000).

The connection between length scales and relaxation times represents the main motivation for looking at the length scales in the first place. Empirical power-law relations between a relaxation time and the length of the form

$$\tau = A\xi^z \tag{7.5}$$

were reported in all of the above cases, often with large exponents. In the two-dimensional SD mixture $z \sim 4$ from both the box scaling (Perera and Harrowell, 1998) and broken-bond structure factor (Yamamoto and Onuki, 1998*a*). An even larger exponent is found in the BMLJ1 mixture (Donati *et al.*, 1999*b*), where $\tau_\alpha \sim S^{4.5}$. Since S is the average (mobile) cluster size, it will be related to a length ξ via $S \sim \xi^\nu$, where $1 < \nu < 3$, implying, when substituted back into the power-law relation with τ_α, a value of $z > 4.5$. The BMLJ2 and SS mixtures, in contrast, have modest exponents: $z \sim 2$ (bond breaking) (Yamamoto and Onuki, 1998*a*) for the SS mixture and $z \sim 2.34$ (4-point correlation) (Lačević *et al.*, 2003) in the BMLJ2 model. We remind the reader that the BMLJ2 and SS models have the same short-range repulsions and associated length scales. The smaller the value of z, the larger the kinetic length required to achieve a given relaxation time. Does this mean that the structures responsible for ξ in these two liquids are mechanically less robust than in the SD and BMLJ1 liquids, since they must be larger to achieve an equivalent stability?

The difference between the value of z for the SS and SD liquids is also noteworthy since both only make use of short-range repulsions. They differ, of course, in their spatial dimension. Yamamoto and Onuki (Yamamoto and Onuki, 1998*a*) comment that the difference in z represents the only significant difference they observed in the features of dynamic heterogeneities between 2D and 3D. Where direct comparisons have been carried out (Doliwa and Heuer, 2000; Yamamoto and Onuki, 1998*a*), there is no evidence that the long-wavelength anomalies that are a well-documented feature of crystals and liquids in 2D have any significant impact on the glass transition in 2D. In this context, it is worth noting that the phenomenology of supercooled liquids in 3D and 4D is also very similar, with the most significant difference being that the breakdown of the scaling between the diffusion constant and the relaxation time (the Stokes–Einstein relation) is somewhat weaker in the higher dimension (Eaves and Reichman, 2009).

The kinetic length scales described in this section—direct measures of the spatial distribution of particle displacements or bonds breaking—have proven valuable descriptive tools in establishing the reality of dynamic heterogeneities and their dependence on temperature. In Table 7.2 we provide a summary of these approaches, identifying the time scales and/or-lengths that must be chosen to resolve the transient heterogeneities in each case and the methods used to assign values to these quantities. In an important development, explicit in box scaling and the displacement susceptibility approaches and implicit in the mobile cluster method, it becomes legitimate to select a quantity, not based on some particular physical justification, but rather on the pragmatic goal to maximize the resolution of the heterogeneity.

Is there a *best* measure of a kinetic length? There does appear to be a consensus concerning the choice of how to obtain a kinetic length. The kinetic length of choice is that associated with χ_4, based on 4-point correlations and closely related to the displacement–displacement correlations introduced by Poole *et al.* (Poole *et al.*, 1998). As we shall see in the following section, χ_4 has a number of appealing features, not least being its close connection with the formalism of spin-glasses. Comforting as such consensus can be, it is worth pointing out that χ_4 provides the same information as that contained in the other measures reviewed in this section. Its popularity

Table 7.2 A summary of the various methods of calculating kinetic lengths and the associated length- and time scales required in each case.

Method	Selected length/ time scales	Manner of assigning values
box scaling[a]	threshold length	maximize kinetic length
mobile cluster[b]	observation time	maximize non-Gaussian parameter
	threshold length	maximum length r^* for which $G(r^*,t) = G_{\text{gaus}}(r^*,t)$
bond breaking[c]	observation time	$0.5\tau_b$ where τ_b is the average bond-breaking time
	maximum bond length	a length lying between the first two peaks of $g_{ij}(r)$
displacement–displacement correlation function[d]	observation time	maximize susceptibility $\kappa_u(t)$

[a](Hurley and Harrowell, 1995), [b](Kob *et al.*, 1997), [c](Yamamoto and Onuki, 1998*a*), [d](Poole *et al.*, 1998)

depends, not on the superiority of its description of dynamic heterogeneities, but on its accessibility, particularly from experiments, and on its potential role in developing theoretical treatments. Whether χ_4 provides a sufficiently complete account of cooperative dynamics is a question we shall return to in Section 7.7.

7.3 Kinetic lengths from 4-point correlations functions

A minimal description of dynamic heterogeneities requires that we measure the statistical correlations between the movement of pairs of particles. This description requires a correlation involving four positions: $\vec{r}_1(t)$, $\vec{r}_1(t+\tau)$, $\vec{r}_2(t)$ and $\vec{r}_2(t+\tau)$. The first calculation of such 4-point correlation functions from MD simulations of a Lennard-Jones mixture (similar to BMLJ2 except the size ratio $\sigma_{11}/\sigma_{22} = 5/8$) was reported by Dasgupta *et al.* (Dasgupta *et al.*, 1991) in 1991. The motivation of these calculations was to test for the presence of a growing length scale associated with fluctuations of the Edwards–Anderson order parameter, $\lim_{t\to\infty} < \delta n(\vec{r},t_o)\delta n(\vec{r},t_o+t) >$, where $n(\vec{r},t)$ is the density field. Such a growth in the length scale of an order parameter fluctuation would have provided evidence of a thermodynamic glass transition. Fixing the spatial separation (i.e. $|\vec{r}_1 - \vec{r}_2|$) at $2\sigma_{11}$, the authors found no evidence of correlated fluctuations at any time. It is likely that this negative result was a consequence of the choice of time interval. As we shall see, fluctuations of the density autocorrelation function exhibit a maximum at around τ_α, while the calculations of Dasgupta *et al.* (Dasgupta *et al.*, 1991) were, by design, carried out in the plateau interval of the relaxation function, well short of this time scale.

In 1999, Franz, Donati, Parisi and Glotzer (Franz *et al.*, 1999; Donati *et al.*, 2002) described how the displacement–displacement correlations introduced to quantify the spatial distribution of particle movement (see previous section) could be reformulated in terms of 4-point correlations.[1] This reformulation involved a conceptual convergence of a method introduced to describe the spatial correlations of displacements with the formal tools developed to treat fluctuations in spin-glasses, such as motivated Dasgupta *et al.* (Dasgupta *et al.*, 1991). Beginning in 2000, Glotzer and coworkers (Lačević *et al.*, 2003; Glotzer *et al.*, 2000; Lačević *et al.*, 2002) presented a detailed account of the 4-point correlation function formalism as applied to supercooled liquids. Here, we shall summarize their expressions (Lačević *et al.*, 2003) for a kinetic length scale using a 4-point correlation.

The structural relaxation is described by the quantity

$$Q(t) = \sum_{i=1}^{N}\sum_{j=1}^{N} w|(\vec{r}_i(0) - \vec{r}_j(t)|) \tag{7.6}$$

a measure of the degree to which a configuration at time t still overlaps the initial arrangement. Overlap of a particle with itself or another particle at an earlier time is established through the introduction of a window function $w(r)$ (where $w(r) = 1$ if $|r| \leq a$ and zero, otherwise). The self-part $Q_s(t)$ of the relation function can be defined as

$$Q_{\mathrm{s}}(t) = \sum_{i=1}^{N} w(|\vec{r}_i(0) - \vec{r}_i(t)|). \tag{7.7}$$

Lačević *et al.* (Lačević *et al.*, 2003) showed that this self-part is the dominant contributor to the relaxation function, the dynamic susceptibility and the kinetic length. Physically, this means that the essential relaxation event corresponds to the departure of a particle from its own initial "site". Equation (7.7) provides an interesting link with the previous approaches (Hurley and Harrowell, 1995; Kob *et al.*, 1997) in which the description of dynamic heterogeneities involved considering each particle moving beyond some threshold distance from its initial position. Similarly, Stein and Andersen (Stein and Andersen, 2008; Stein, 2007) examine the 4-point correlations of a mobility defined as

$$\mu(\vec{r}, t) = \sum_{i=1}^{N_A} \delta(\vec{r} - \vec{r}_i(t))\mu_i(t), \tag{7.8}$$

where $\mu_i(t) = 1$ if $|\vec{r}_i(t + t^*) - \vec{r}(t)| \geq d$ and zero otherwise.

Toninelli *et al.* (Toninelli *et al.*, 2005) and Flenner and Szamel (Flenner and Szamel, 2007), along with others, have used the (microscopic) self-intermediate scattering function $F_{\mathrm{s}}(q, t)$ instead of the self-overlap function $Q_s(t)$. There is no fundamental

[1] To get the chronology straight during this busy period, note that the paper (Donati *et al.*, 2002) first appeared as a preprint cond-mat/9905433 in 1999.

difference between the two relaxation functions. The choice of the scattering wavevector q in $F_s(q,t)$ selects the reference length scale analogous to the choice of the value of a in the overlap function. The authors in ref. (Toninelli *et al.*, 2005) considered how various models of collective behavior (elastic modes, domain-wall fluctuations, etc.) were represented at the level of the 4-point correlations. Flenner and Szamel (Flenner and Szamel, 2007) showed that the correlations of the fluctuations in the microscopic self intermediate function exhibited an anisotropy associated with the direction of particle motion relative to the vector between particle pairs. This latter result expressed, in terms of the 4-point correlation function, the anisotropy that had previously been established by Doliwa and Heuer (Doliwa and Heuer, 2000) who demonstrated that the kinetic length scale (obtained using Eq. (7.3)) along the direction of the particle displacement was roughly twice as long (in a binary hard-sphere mixture) as that along a direction perpendicular to the particle displacement.

A susceptibility $\chi_4(t)$, analogous to $\kappa_u(t)$ defined previously for the displacement correlations, is defined in terms of the fluctuations of the relaxation function $Q(t)$, i.e.

$$\chi_4(t) = \frac{\beta V}{N^2}(< Q^2(t) > - < Q(t) >^2). \tag{7.9}$$

The expression for the susceptibility $\chi_4(t)$ (Eq. (7.9)) in terms of the variance of the structural relaxation function $Q(t)$ is remarkable in that it provides information about the extent of dynamic heterogeneities (see below) without ever requiring that the dynamics be spatially resolved (as the methods in Section 7.2 all did). Qualitative considerations of the fluctuations of $Q(t)$ also provide a useful way of differentiating the order-parameter fluctuations, envisioned by Dasgupta *et al.* (Dasgupta *et al.*, 1991), from the fluctuations in dynamics that are the subject of this review. In Fig. 7.2 we present sketches of two distinct types of fluctuations in Q. The top panel represents fluctuations of the height of the plateau. The Edwards–Anderson order parameter for spin-glasses is this plateau height in an arrested system. An alternative type of fluctuation involves variations of the relaxation time (Fig. 7.2, middle panel). These two types of fluctuations are easily distinguished in their respective susceptibilities (Fig. 7.2, bottom panel). Where the plateau fluctuations result in a low amplitude $\chi_4(t)$, extended over the whole plateau time region, the dynamic fluctuations typically produce a more sharply peaked $\chi_4(t)$, with the maximum occurring at roughly the structural relaxation time. Compare these two (idealized) possibilities with the $\chi_4(t)$ calculated from simulations of the BMLJ2 mixture (Lačević *et al.*, 2003) in Fig. 7.3 and we find that fluctuations in dynamics, sketched as option b) in Fig. 7.2, provide a reasonable correspondence with the simulated liquid. The implication is that it is the dynamic fluctuations, as opposed to those of the plateau/order parameter, that dominate the observed susceptibility in the supercooled liquid. Kirkpatrick and Thirumalai (Kirkpatrick and Thirumalai, 1988) were the first to point out that the fluctuations contributing to $\chi_4(t)$ could arise from these two distinct sources: order-parameter fluctuations and fluctuations in the dynamics themselves.

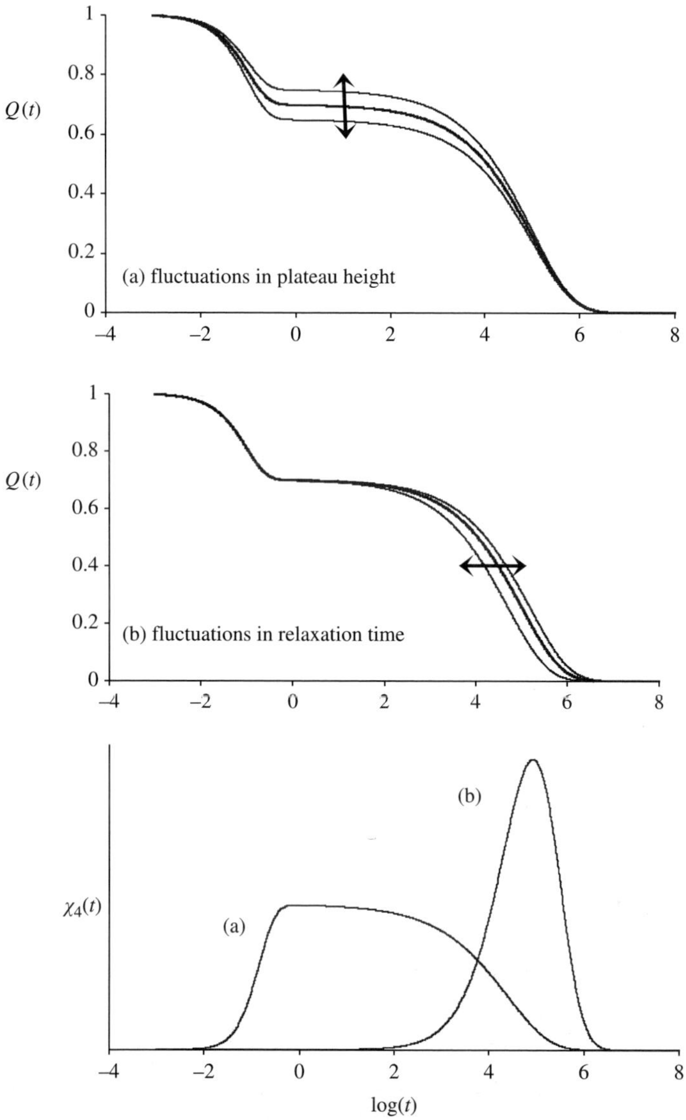

Fig. 7.2 Sketches of the fluctuations in $Q(t)$ due to (a) fluctuations in the plateau height and (b) fluctuations in the magnitude of the structural relaxation time. The time-dependent dynamic susceptibilities $\chi_4(t)$, calculated using Eq. (7.7) for both types of fluctuations, are plotted in the bottom panel.

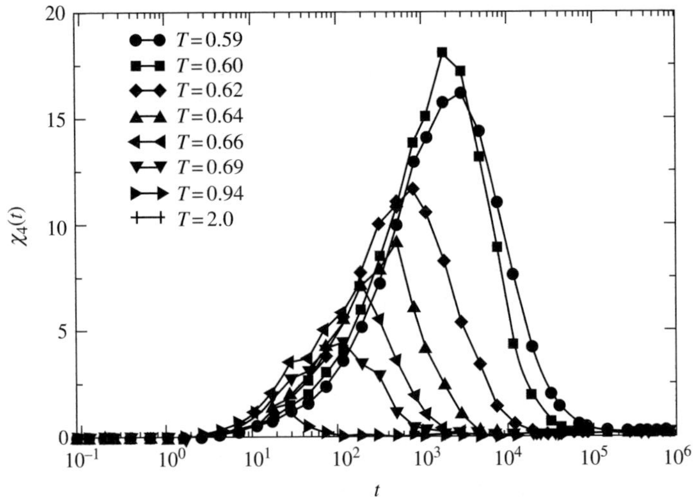

Fig. 7.3 The time and temperature dependence of $\chi_4(t)$ for the BMLJ2 mixture. The time corresponding to the maximum in $\chi_4(t)$ is found to be similar in value to the structural relaxation time τ_α and to exhibit a similar T dependence. [Reprinted with permission from ref. (Lačević *et al.*, 2003) Copyright 2003, American Institute of Physics.]

The susceptibility $\chi_4(t)$ can be directly obtained (Lačević *et al.*, 2003) from the 4-point correlation function via

$$\chi_4(t) = \beta \int d\vec{r} g_4(\vec{r}, t), \tag{7.10}$$

where

$$g_4(\vec{r}, t) = \frac{1}{N\rho} \left\langle \sum_{ijkl} \delta(\vec{r} - \vec{r}_k(0) + \vec{r}_i(0)) w(|\vec{r}_i(0) - \vec{r}_j(t)|) w(|\vec{r}_k(0) - \vec{r}_l(t)|) \right\rangle - \left\langle \frac{Q(t)}{N} \right\rangle^2 \tag{7.11}$$

or

$$g_4(r, t) \equiv g_4^{\mathrm{ol}}(r, t) - \left\langle \frac{Q(t)}{N} \right\rangle^2, \tag{7.12}$$

where we have assumed that the correlations are isotropic and so only depend on the magnitude of the separation. The significance of $g_4^{\mathrm{ol}}(r, t)$ is that it corresponds to the pair correlation function of those particles at $t = 0$ that end up overlapping with a particle at the later time t. It is this quantity, or rather its Fourier transform,

$$S_4^{\mathrm{ol}}(\vec{q}, t) = \int d\vec{r} g_4^{\mathrm{ol}}(r, t) exp(-i\vec{q} \cdot \vec{r}) \tag{7.13}$$

that Lačević *et al.* (Lačević *et al.*, 2003) use to obtain the kinetic length ξ_4 whose dependence on t is plotted in Fig. 7.4. As emphasized in ref. (Lačević *et al.*, 2003), considerable care needs to be taken in ensuring that the 4-point correlator used to extract the kinetic length does not include a weak $O(1/N)$ tail associated with bulk fluctuations. As is generally the case, fluctuations, such as measured by the 4-point correlations, are very dependent on the choice of ensemble. Dalle-Ferrier *et al.* (Dalle-Ferrier *et al.*, 2007) provide a useful discussion of this point.

Toninelli *et al.* (Toninelli *et al.*, 2005) have presented a thorough discussion of the behavior of $\chi_4(t)$ over various time domains. The kinetic length increases in time as the heterogeneous character of the unrelaxed domains is exposed by the dynamics until it reaches a maximum, beyond which it decays to zero as the persistent domains finally succumb to relaxation and homogeneity is re-established. The temperature dependence of the maximum length $\xi_4(t^*)$ in the BMLJ2 liquid is fitted by $\xi_4(t^*) = 0.82(T/T_c - 1)^{-0.82}$ with an apparent divergence at the mode-coupling temperature $T_c - 0.57$ (Lačević *et al.*, 2003). Such apparent divergences generally do not actually *eventuate* and there is a considerable literature discussing the various interpretations of both the power-law temperature dependence and the "avoidance" of the singularity (Cavagna, 2009). As already mentioned, a scaling law, $\tau_\alpha \sim \xi_4(t^*)^z$, was found to hold with $z = 2.34$.

The length scale a imposed by the window function $w(r)$ plays a significant role in selecting the physical character of the fluctuations being measured. As pointed out in Ref. (Lačević *et al.*, 2003), select a too small and the results are dominated by vibrational motions that obscure any heterogeneities, select a too large and the notion of overlap quickly loses any physical significance as each particle can overlap with multiple particles. Dauchot *et al.* (Dauchot *et al.*, 2005) have chosen a value for a the same way that t^* is chosen, i.e. by finding the value that maximizes χ_4. Chandler *et al.* (Chandler *et al.*, 2006) pointed out that the use of a small a

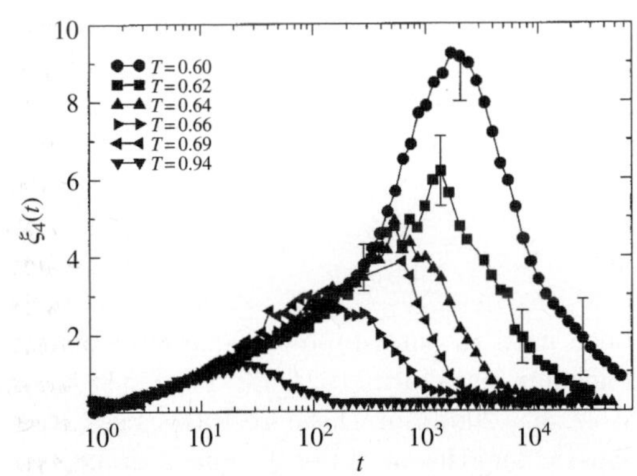

Fig. 7.4 The time and temperature dependence of the kinetic length $\xi_4(t)$ in the BMLJ2 mixture. [Reprinted with permission from ref. (Lačević *et al.*, 2003). Copyright 2003, American Institute of Physics.]

(roughly, $a/\sigma \leq 0.3$) results in the spatial clustering of immobile particles dominating the observed fluctuations, whereas the use of a large a (i.e. $0.5 \leq a/\sigma$) results in a $\chi_4(t)$ reflecting correlations in mobile particles. This length-dependent crossover is related (Chandler *et al.*, 2006) to the non-Fickian-to-Fickian crossover discussed in Section 7.6. Charbonneau and Reichman (Charbonneau and Reichman, 2007) have described how the dependence of χ_4 on the reference length scale differs when comparing liquids whose arrest is dominated by short-range repulsions and those dominated by short-ranged attractions.

The great attraction of the 4-point correlation function approach to the kinetic length is *not* the 4-point correlation functions, $g_4(r, t)$ or $S_4(q, t)$, which are at least as difficult to use to calculate a kinetic length as any of the other methods described in the previous section. The real appeal is in the quantity $\chi_4(t)$ and related susceptibilities. As the space integral of the 4-point correlation function (Eq. (7.10)), $\chi_4(t^*)$, the maximum susceptibility, represents a "volume of correlation" or, alternatively, a number N_{corr} of correlated particles (Dalle-Ferrier *et al.*, 2007). As the variance of the relaxation function $Q(t)$ (Eq. (7.9)), $\chi_4(t)$ is no more difficult to calculate than the relaxation function itself. This latter virtue marks the superiority of $\chi_4(t)$ over the analogous susceptibility, $\kappa_u(t)$, from the displacement correlations (defined in Eq. (7.4)). Comparing Figs. 7.3 and 7.4, it is evident that $\chi_4(t)$ exhibits a qualitatively similar time dependence to that of the kinetic length $\xi_4(t)$ with a peak at some intermediate time, corresponding to a maximum in the differentiation of overlapping and non-overlapping domains.

This theoretical accessibility has been extended towards experimental accessibility in a series of papers (Dalle-Ferrier *et al.*, 2007; Berthier *et al.*, 2005*a*, 2007*a*,*b*; Biroli *et al.*, 2006) starting with Berthier *et al.* (Berthier *et al.*, 2005*a*) in 2005. In Ref. (Berthier *et al.*, 2005*a*), the authors showed how a lower bound on the value of $\chi_4(t^*)$ could be obtained from 3-point correlations defined as the response of the structural relaxation function (i.e. $Q(t)$ or its analog) to a change in a control parameter, such as temperature, pressure or density. This connection between a 3-point correlations and N_{corr} has been explored in some detail (Dalle-Ferrier *et al.*, 2007). Starting with the maximum of the 3-point susceptibility $\chi_{\mathrm{T}}(t^*)$ associated with the dynamic heterogeneities that are correlated to local enthalpy fluctuations, the number of correlated molecules is given (in the NPT ensemble) by

$$N_{\mathrm{corr}} = \sqrt{\frac{k_{\mathrm{B}}T^2}{\Delta c_{\mathrm{P}}}} \max\{|\chi_{\mathrm{T}}(t)|\}. \tag{7.14}$$

Equation (7.14) has been used to determine the size and temperature dependence of $\mathrm{N_{corr}}$ for a range of molecular glass formers (Dalle-Ferrier *et al.*, 2007). The discussion in Ref. (Dalle-Ferrier *et al.*, 2007) of these results raises a number of important questions regarding kinetic lengths in general. The correlation volumes at T_{g} were found to be modest in size. This result offers hope that simulations can contribute to the description of cooperative dynamics closer to T_{g} than previously thought. It also cautions against arguments based on large separation of length scales in the supercooled liquid. Dalle-Ferrier *et al.* (Dalle-Ferrier *et al.*, 2007) cast doubt on any simple connection between N_{corr} and the cooperatively rearranging regions as imagined by Adam and Gibbs (Adam and Gibbs, 1965). There is a quite basic difficulty in trying to translate observed correlations in mobility into the mechanism responsible for those correlations. We shall return to this point at the end of the review.

If both $\chi_4(t^*)$ and $\xi_4(t^*)$ can provide information about the extent of dynamic heterogeneities, what, exactly, is the connection between them? If one assumes that the spatial distribution of the mobile particles is scaled by a single length ($\xi_4(t^*)$ in this case), it follows (Biroli *et al.*, 2006; Biroli and Bouchaud, 2004; Whitelam *et al.*, 2004) that $\chi_4(t^*)$ and $\xi_4(t^*)$ are related by a power law,

$$\chi_4(t^*) = A(T)\xi_4(t^*)^{2-\eta}. \tag{7.15}$$

Stein and Andersen (Stein and Andersen, 2008; Stein, 2007) have confirmed the power-law relation for the BMLJ1 mixture and found $\eta = -2.2$. More recently, Karmakar *et al.* (Karmakar *et al.*, 2010*a*) have carried out similar calculations but using a larger number of particles and reported a significantly smaller exponent, $\eta = -0.4$. Karmakar *et al.* (Karmakar *et al.*, 2010*b*) argue that the extraction of the kinetic correlation length from the 4-point structure factor $S_4(q,t)$ by fitting an Ornstein–Zernicke expression, i.e. $S_4(q,t) = \chi_o/(1 + q^2\xi^2)$, is inaccurate unless large systems (i.e. $N \sim 10^5$) are used.

Robust in definition and simple to apply, the susceptibility $\chi_4(t^*)$ has proved a popular measure of the extent of cooperative motion. As obtained from the fluctuation of the 2-point correlation, $\chi_4(t^*)$s have been reported for colloids (Duri and Cipelletti, 2006; Cipelletti *et al.*, 2003; Brambilla *et al.*, 2009; Ballesta *et al.*, 2008), granular material (Dauchot *et al.*, 2005; Keys *et al.*, 2007) and foams (Mayer *et al.*, 2004). In simulations, $\chi_4(t^*)$ is being used to study cooperative behavior in a wide range of systems—including those in non-Euclidean spaces and out of equilibrium. Some examples of the latter applications: Sausset and Tarjus (Sausset and Tarjus, 2010) have calculated $\chi_4(t^*)$ for a liquid of Lennard-Jones disks on a hyperbolic surface, Furukawa *et al.* (Furukawa *et al.*, 2009) have analyzed the 4-point correlation functions in a liquid under steady shear (finding mobile regions to be elongated), and Parsaeian and Castillo (Parsaeian and Castillo, 2008) have studied the effects of aging on $\chi_4(t^*)$. Abraham and Bagchi (Abraham and Bagchi, 2008) have demonstrated that the low-temperature magnitude of $\chi_4(t^*)$ in a polydisperse Lennard-Jones mixture decreases with increasing width of the particle size distribution.

7.4 Kinetic lengths from finite size-analysis

If heterogeneities are characterized by a length, then it follows that, as the size of a finite system approaches this inherent length, properties of the dynamics should show a dependence on the system size. For this finite-size effect to be linked specifically to a kinetic length scale it is important to also establish that the static correlations in the liquid are independent of the system size over the size range studied. In 1992, Dasgupta and Ramaswamy (Dasgupta and Ramaswamy, 1992) reported the absence of any size dependence in the relaxation time of a density autocorrelation function for a binary Lennard-Jones mixture with a radius ratio of 5/8. This study, however, failed to collect data for temperatures between $0.8T_g$ and $0.99T_g$, despite what appears to be an onset of a finite-size dependence in this temperature interval. The authors, expecting to see the time scale *decrease* with system size, interpreted the (slight) signs of the opposite trend as supporting their conclusion that there was no system-size dependence. In 2000, Kim and Yamamoto (Kim and Yamamoto, 2000), studying the SS mixture for sizes $N = 108$, 1000 and 10 000, found that as the temperature was lowered the relaxation time τ_α exhibited a dependence on the system size, the relaxation time increasing as the system size decreased (see Fig. 7.5). No dependence of the pair distribution function on system size was found for all temperatures studied.

The influence of system size on the relaxation times in the BMLJ1 mixture has been the subject of a number of studies. In 2003 Doliwa and Heuer (Doliwa and Heuer, 2003) calculated the diffusion constants for four system sizes ($N = 65$, 130, 260 and 1000) down to T_c (~ 0.43). At the lowest temperature, they found $D_{65}/D_{130} \sim 1.2$ but with a large error (± 0.2). They concluded that the size dependence was small, significantly less (and in the opposite direction) to that observed in the SS mixture (Kim and Yamamoto, 2000). Stein (Stein, 2007), working with larger systems ($N = 1000$ and 8000), found a slight decrease in the diffusion constant for the smaller system but, again, the difference was within the noise. In 2009, Karmakara *et al.* (Karmakar *et al.*,

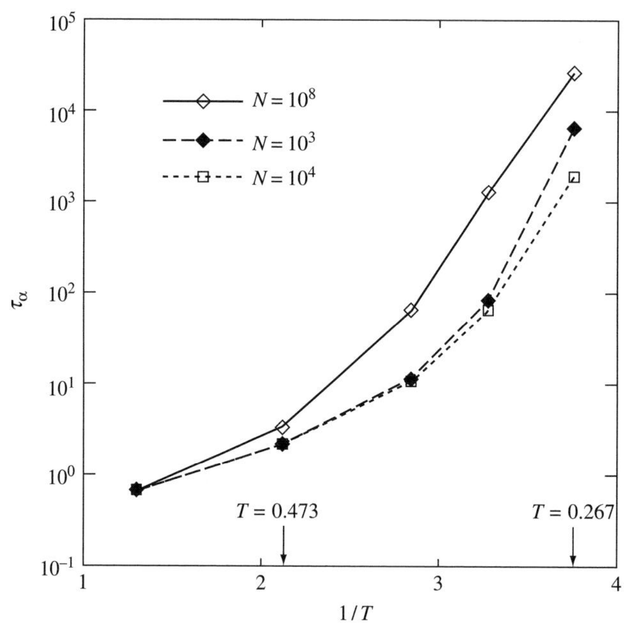

Fig. 7.5 The temperature dependence of τ_α for $N = 108$ (open diamonds), 10^3 (closed diamonds), and 10^4 (open squares) for the SS mixture. [Reproduced with permission from ref. (Kim and Yamamoto, 2000).]

2009) determined the relaxation time in a large number of systems across the size range $50 \leq N \leq 1600$ range of system sizes down to $T = 0.45$. They find a systematic increase in relaxation time with decreasing system sizes for N (see Fig. 7.9). The onset value of N below which this finite-size effect is observed increases from ~ 100 at $T = 0.8$ to ~ 200 at $T = 0.45$. These workers also studied the size dependence of $\chi_4(t^*)$ and, using a method first adapted from the study of critical phenomena to the glass problem by Berthier (Berthier, 2003), determined a length scale from this data that increased on cooling from 2.1 ($T = 0.70$) to 6.2 ($T = 0.45$). An increase in the relaxation time with decreasing system size has also been reported in simulations of silica (Zhang *et al.*, 2004; Teboul, 2006) but, unfortunately, there is no confirmation that the static properties remained unchanged, a non-trivial condition given the long-range character of the potential in this model.

With the exception of ref. (Doliwa and Heuer, 2003), the data from simulations presents a picture of a modest but systematic increase in the volume associated with cooperative motion on cooling. Significantly, relaxation is *slower* in small systems. We shall consider the implications and possible origin of this behavior in Section 7.7. Confinement, whether imposed by walls or arising from inherent fluctuations in the supercooled liquid, appears to represent one generic mechanism for slowing down relaxation.

7.5 Kinetic lengths at amorphous interfaces

If your goal is to clarify a phenomenon in the homogeneous liquid, the inclusion of an interface is not usually a good idea. The problem is that the interface will typically perturb the liquid structure significantly and, thus, so alter the phenomenon from that found in the bulk as to obscure any connection with the homogeneous situation. There is, of course, considerable interest in exactly such perturbed situations in the context of glass transitions actually taking place in confined geometries. This subject has been reviewed by a number of authors (Binder and Kob, 2005; Baschnagel and Varnick, 2005). Within the artificial world of simulations, however, it is possible to contrive a surface that is structurally neutral by simply freezing the motions of some portion of the liquid. Such walls are an example of the imposition of a kinetic constraint. The interesting question is then to determine the length scale over which the influence of such a constraint is propagated into the unconstrained liquid. The idea of using an interface to establish the kinetic length scale was described (Butler and Harrowell, 1991) in 1991 in the context of a lattice model of glassy kinetics.

Scheidler *et al.* (Scheidler *et al.*, 2000, 2004) have reported simulation studies of the BMLJ1 mixture adjacent to a rough wall made up of the frozen liquid. These workers have considered both a cylindrical pore (Scheidler *et al.*, 2000) and a planar wall (Scheidler *et al.*, 2004). While the idea of a frozen liquid wall is simply sketched, its implementation takes some care. In Ref. (Scheidler *et al.*, 2004) the temperature of the liquid used to produce the frozen walls was adjusted to minimize any structural perturbation and an additional repulsive potential was included to prevent particle penetration into the wall. Scheidler *et al.* (Scheidler *et al.*, 2004) fitted the relaxation time $\tau_q(z)$ (associated with the decay of a self-intermediate relaxation time $F_s(q, z, t)$

at a distance z from the wall) for the BMLJ1 mixture to three different functions of the normal distance z from the planar wall. The best fit was obtained with the following expression,

$$\ln\left[\frac{\tau_q(z)}{\tau_o}\right] = A(T)\exp\left[-\frac{z}{\xi_o(T)}\right]. \tag{7.16}$$

The kinetic length ξ_o increased by a factor of 3 as the temperature dropped to $T = 0.5$. The increase in ξ_o with temperature was fitted with a simple Arrhenius form. The temperature dependence of ξ_o appears to be quite different from the $(T - T_c)^{-\gamma}$ dependence reported for the length of the dynamic heterogeneities in the bulk for the same model (Donati *et al.*, 1999*b*). Obtaining values of the surface kinetic length at temperatures closer to T_c has been frustrated by poor statistics.

Where Scheidler *et al.* (Scheidler *et al.*, 2000, 2004) have described how effective a frozen liquid is in imparting its immobility to an adjacent mobile liquid, Cavagna *et al.* (Cavagna *et al.*, 2007; Biroli *et al.*, 2008) have examined how the frozen liquid can actually constrain the configurations available to the mobile liquid. The use of amorphous boundaries to establish an equilibrium length scale associated with structural correlations was described by Bouchaud and Biroli (Bouchaud and Biroli, 2004) and Montanari and Semerjian (Montanari and Semerjian, 2006). Instead of a frozen half-plane, Cavagna *et al.* (Cavagna *et al.*, 2007) have considered a spherical shell of frozen SS liquid, enclosing the mobile liquid in a volume of radius R. We emphasize that this study does not deal with the kinetics of the confined liquid but, instead, uses an amorphous wall to extract a *static* length scale. It is included in this review of kinetic lengths, in part, because of the obvious methodological parallels with the work of Scheidler *et al.* (Scheidler *et al.*, 2004). There is also a general expectation that kinetic length scales derive from underlying static length scales. Mark Ediger (Ediger, 2000) expressed the situation with admirable delicacy, "*At present, it is an article of faith that something in the structure is responsible for dynamics that can vary by orders of magnitude from one region of the sample to another at T_g*". Recent work (Cammarota *et al.*, 101) has suggested that the origin of the growing kinetic length scale might lie in a separation of phases characterized by distinct degrees of the amorphous order. In Ref. (Biroli *et al.*, 2008), the authors determine, using an accelerated Monte Carlo algorithm, the degree of overlap q_c of the ensemble of configuration at the center of the sphere as a function of the radius R. They found that the overlap decayed to the random value q_o as

$$q_c(R) - q_o = \Omega\exp[-(r/\xi)^{\zeta}] \tag{7.17}$$

with both the length ξ and the exponent ζ increasing with supercooling—at $T = 0.482$ the length $\xi = 0.617$ and the exponent $\zeta \sim 1$ and both increase on cooling so that at $T = 0.203$, $\xi = 3.82$ and $\zeta = 4.00$.

An earlier study of a liquid confined within a frozen liquid shell was presented by Sim *et al.* (Sim *et al.*, 1998, 1999). These authors studied a single-component Lennard-Jones liquid in 2D where the crystallization was frustrated by the disorder of the frozen wall. They did not present any systematic results associated with a kinetic length. However, the model is of potential interest due to the simplicity of the confined liquid and the possibility of an explicit counting of allowed configurations. The model represents an extension of the classic problem of packing disks in finite containers (Desmond and Weeks, 2009). Sim *et al.* (Sim *et al.*, 1999) reported some specific examples of collective reorganization events.

The geometry of the pinned particles—planes, tubes or spherical cavities—reflect the particular problem that inspired the authors. In the case of Scheidler *et al.* (Scheidler *et al.*, 2004), this was the influence of confinement on glassy dynamics, while for Cavagna *et al.* (Cavagna *et al.*, 2007) it was to test the idea of droplet excitations as conceived in the mosaic theory (Kirkpatrick and Wolynes, 1987; Kirkpatrick *et al.*, 1989; Xia and Wolynes, 2001; Lubchenko and Wolynes, 2003; Bouchaud and Biroli, 2004). What about a spatial distribution of pinned particles that does not impose any particular correlation? Supercooled liquids subject to random pinnings have been studied by a number of groups (Kim, 2003; Lin *et al.*, 2006). In 2003 Kim (Kim, 2003) reported the effect of pinning the positions of a fixed number N_{d} of randomly selected particles in the SS mixture. He found that the relaxation time τ_α scaled with N_{d} and T as

$$\tau_\alpha(T, N_{\mathrm{d}}) \propto \exp[N_{\mathrm{d}}/T^\nu], \tag{7.18}$$

with $\nu = 3.7$. In trying to extract a kinetic length scale from this calculation, Kim resorted to the following argument. There is one length scale imposed by the defects, i.e.

$$\xi_{\mathrm{d}}(T) \propto (V/N_{\mathrm{d}})^{1/3} \tag{7.19}$$

and there is another, the intrinsic kinetic length $\xi(T)$, the one we are actually interested in. If one assumes that (a) the intrinsic kinetic length is unperturbed by the pinned particles and (b) that at the glass transition (defined as τ_α equalling some big number) $\xi_{\mathrm{d}}(T) = \xi(T)$, then it follows that

$$\xi(T) \propto T^{-\nu/3}. \tag{7.20}$$

The argument is awkward. In particular, assumption (b) above neglects the cross-over to simple unpinned behaviour when $\xi_{\mathrm{d}}(T) > \xi(T)$. The methodology, however, has potential as a general tool for exploring kinetic and static correlations through the imposition of dilute random pinnings.

7.6 Kinetic lengths from crossover behavior

We shall complete our survey of length scales with those that arise most directly from the dynamical processes of interest and, therefore, perhaps represent the most pressing demand for our attention. The length scales described in this section take as their starting point the existence of some sort of length-dependent crossover in a physical property explicitly associated with particle motion. The existence of dynamic heterogeneities leads, not surprisingly, to a range of physical behavior that deviates from that expected of a uniform system. As one probes the behavior over length scales larger than that of the heterogeneity, the "classical" behavior is recovered and, accordingly, a length scale can be associated with this crossover to homogeneity.

Self-(tracer) diffusion is one such phenomenon that exhibits a cross-over that is manifest in changes in the time-dependent displacement probability distribution $G(r,t)$ (the van Hove distribution function). The story around $G(r,t)$ in glass-forming liquids is quite rich and so we shall take a moment to sketch out some of the main points, at least to clarify some terminology. In 1990, Odagaki and Hiwatari (Odagaki and Hiwatari, 1990) noted that, at fixed times, $G(r,t)$ underwent a transition from Gaussian to non-Gaussian as a liquid was supercooled. Hurley and Harrowell (Hurley and Harrowell, 1996) pointed out that this was an expected consequence of the increase in dynamic heterogeneity; specifically, the presence of persistent kinetic subpopulations. Dynamic heterogeneities have also been invoked to resolve another puzzle involving self-diffusion. Sillescu and coworkers (Fujara *et al.*, 1992) had shown experimentally that, on supercooling, fragile liquids exhibited a breakdown in the scaling between the diffusion constant D and both the shear viscosity (the Stokes–Einstein relation) and the rotational diffusion constant (the Debye expression). This phenomenon is actually evident just among the different length scales of the self-intermediate scattering function $F_s(q,t)$, the Fourier transform of $G(r,t)$. The temperature dependence of the relaxation time of $F_s(q,t)$ exhibits, on supercooling, an increasing anomalous dependence on q, with the small-q time scale behaving like D (as it must) and the large-q relaxation exhibiting a temperature dependence similar to that of the shear viscosity (Perera and Harrowell, 1998; Chandler *et al.*, 2006; Chaudhuri *et al.*, 2007). This general loss of a single time scale on supercooling is referred to as "decoupling". A number of groups (Cicerone and Ediger, 1996; Berthier, 2004; Berthier *et al.*, 2005*b*) have reached the following consensus regarding the origin of this decoupling. The idea is that different transport properties correspond to different moments of the distribution of microscopic times, so their decoupling at a particular wavevector is associated with the growth of dynamical heterogeneity (as manifest in the broadening of the distribution of microscopic times) over the corresponding length scale. Chaudhuri *et al.* (Chaudhuri *et al.*, 2007) have pointed out that the presence of an exponential tail in the van Hove function is a signature of the presence of slow and fast particles and can account for the decoupling of diffusion and structural relaxation. The continuous-time random walk model they propose has been further quantified by Hedges *et al.* (Hedges *et al.*, 2007) who demonstrated that the ratio of the persistence time over the exchange time (the times for the first move and between subsequent moves, respectively) grows rapidly in the supercooled liquid.

The explanation of decoupling provided in refs. (Chaudhuri *et al.*, 2007; Hedges *et al.*, 2007) requires that the different transport processes can be related to a common distribution of microscopic times. While this condition is met for the relaxation of $F_s(q,t)$ at different q, it is not clear that it is met by the Stokes–Einstein breakdown itself since diffusion and viscosity correspond to quite different physical processes and, therefore, are associated with quite distinct distributions of microscopic times. Other approaches (Hodgdon and Stillinger, 1993) to the decoupling of diffusion and viscosity avoid this particular criticism by retaining explicit coupling between local mobility fluctuations and stress relaxation.

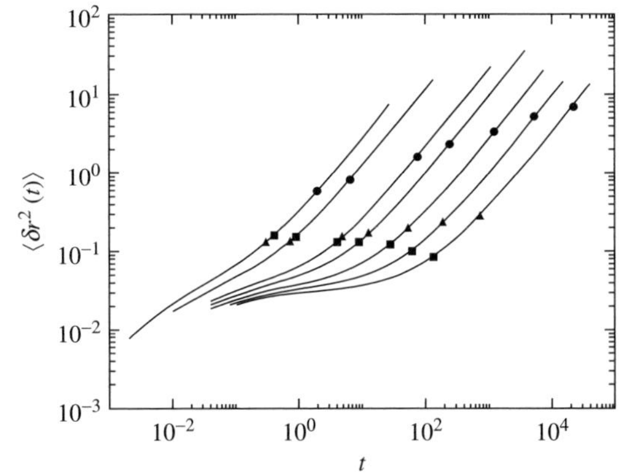

Fig. 7.6 The time dependence of the mean square displacement for the A particles in the BMLJ1 model for $T = 1.0$, 0.8, 0.6, 0.55, 0.50, 0.47 and 0.45 listed from left to right. The symbols are placed at different characteristic times. Squares: the time at which the standard non-Gaussian parameter reaches the maximum value. Triangles: the α relaxation time τ_α. Circles: the onset time for Fickian diffusion. [Reproduced with permission from ref. (Szamel and Flenner, 2006).]

Over a long enough time, particles will sample a sufficient number of kinetic environments so as to recover standard or Fickian diffusion. Szamel and Flenner (Szamel and Flenner, 2006) have determined the time τ_F over which a particle must, on average, move before $G(r,t)$ becomes Gaussian for the BMLJ1 mixture simulated using Brownian dynamics. Their results are shown in Fig. 7.6. They observe that τ_F is larger than the relaxation time τ_α and that this difference increases as the temperature is lowered. A length scale can be obtained from Fig. 7.6 simply by reading off the mean square displacement at $t = \tau_F$. This length l^* ranges up to 2.5 small particle diameters and corresponds to the distance that a particle must on average move before exhibiting Fickian diffusion. The value of l^* from Ref. (Szamel and Flenner, 2006) is roughly half that predicted by the expression for the crossover length due to Berthier *et al.* (Berthier *et al.*, 2005*c*), $l^* \propto \sqrt{D\tau_\alpha}$, but, at large supercoolings, the two lengths exhibit a similar temperature dependence.

Stariolo and Fabricius (Stariolo and Fabricius, 2006), in a study of the BMLJ1 mixture, reported the appearance of a new crossover in the self-intermediate scattering function at around τ_α, in addition to the cross-over to Fickian diffusion at longer times. The authors associated the earlier cross-over with the length scale of the dynamic heterogeneities as measured by χ_4. A link between the length scale of the cross-over to Fickian behavior and the length scale of the dynamic heterogeneities as obtained from the 4-point correlations has been explored by Berthier (Berthier, 2004). Studying the BMLJ1 model, he demonstrated that a suitably normalized product of the q-dependent relaxation time and the diffusion constant from a wide range of temperatures could be collapsed onto a single master curve when the wavevector q was scaled by the length scale of dynamic heterogeneities.

Is the crossover to Fickian behavior governed by a time scale τ_F or a length scale l^*? Given the transient character of dynamic heterogeneities, the answer appears to be a matter of taste. A crossover involving an unequivocally *static* distribution of heterogeneities has been studied by Barrat and coworkers (Leonforte *et al.*, 2005). These authors demonstrated that the continuum elastic description of a disordered polydisperse mixture of Lennard-Jones particles at zero temperature broke down for length scales less than ~ 23 particle diameters. Over length scales smaller than this crossover length, the material exhibited non-affine response to an applied strain and this length gives the lower wavelength bound for the applicability of classical eigenvectors.

The connection between non-affine displacements and the approach to the glass transition has been made explicit in a study by Mosayebi *et al.* (Mosayebi *et al.*, 2010). The local potential-energy minima of the BMLJ1 model have been collected from MD trajectories as a range of temperatures. These inherent structures are subjected to a static shear deformation at $T = 0$ and the spatial distribution of the resulting non-affine displacements calculated. These authors find that a characteristic length scale of the non-affine field grows as the temperature from which the inherent structures are obtained is lowered.

While the complexities of normal modes of disordered materials has not traditionally been associated with dynamic heterogeneities, a growing body of evidence suggests that the two phenomena are correlated (Schober *et al.*, 1993; Oligschleger and Schober, 1999; Brito and Wyart, 2006, 2007; Widmer-Cooper *et al.*, 2008, 2009). It is possible that the large, but finite, length scales identified in Ref. (Leonforte *et al.*, 2005) represent a useful $T = 0$ limit for the length scale of dynamic heterogeneities.

So far in this review we have made no mention of a length scale associated specifically with stress relaxation. Certainly, the overwhelming emphasis of simulation studies of the glass transition has been on single-particle dynamics, in spite of the central role of viscosity in defining the glass transition. There are, however, a number of interesting papers on growing length scales associated with transverse momentum fluctuations, the kernel of the Green–Kubo expression for shear viscosity. In 1995, Mountain (Mountain, 1995) used the transverse momentum autocorrelation function to obtain the dispersion curve for the supercooled SS mixture. He identified a growing length scale as the longest wavelength associated with propagating transverse modes. This wavelength, obtained by extrapolating the dispersion curve for the supercooled liquid to the point where the mode frequency vanished, showed a strong increase at large supercoolings. At the temperature at which Yamamoto and Onuki (Yamamoto and Onuki, 1998*a*) found a kinetic length of ~ 10, Mountain reports a length of ~ 55 (a value roughly three times the length of his own simulation box). Ahluwalia and Das (Ahluwalia and Das, 1998), using a mode-coupling theory, have argued that the length identified by Mountain will diverge as T_g is approached from above. There is a serious difficulty in associating the length described by Mountain and Das with the kinetic length scales that are the subject of this review. As pointed out by Hiwatari and Miyagawa (Hiwatari and Miyagawa, 1990), a straightforward application of viscoelastic theory results in the prediction that the longest wavelength associated with transverse propagation is proportional to η, the shear viscosity. This relation simply reflects the condition that, for a mode to propagate, its frequency must exceed $1/\eta$, the value set by the dissipation. The growing length described by Mountain (Mountain, 1995) is a direct consequence of the growing relaxation time and is quite independent of any microscopic correlation length associated with the physical origin of this growing relaxation time.

An alternate and more informative treatment of the transverse momentum correlation function has been presented by Kim and Keyes (Kim and Keyes, 2005). These workers have calculated the time integral of the k-dependent transverse momentum autocorrelation function. This quantity, essentially the zero-frequency component of the correlation function, is the wavevector-dependent shear viscosity $\eta(k)$. The authors (Kim and Keyes, 2005) found that $\eta(k)$, calculated for the BMLJ1 model, could be fitted with the following functional form,

$$\frac{\eta(k)}{\eta(0)} = 1 + a\tanh(k\xi(T)). \tag{7.21}$$

Consistent with Eq. (7.21), $\eta(k)/\eta(0)$ plotted against $k\xi(T)$ collapses the data from different temperatures onto a single master curve (as shown in Fig. 7.7) that decays with increasing wavevector. The length scale ξ was found to increase from 0.13 to 1.62 σ_{AA} on cooling. Kim and Keyes (Kim and Keyes, 2005) go on to argue that the increase in ξ can be directly linked to the breakdown in the Stokes–Einstein scaling through the use of a mode-coupling expression due to Keyes and Oppenheim (Keyes and Oppenhiem, 1973).

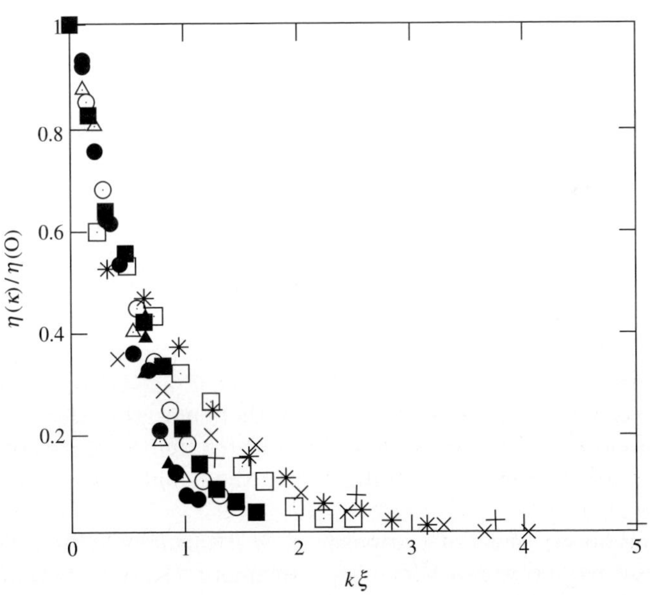

Fig. 7.7 The wavevector-dependent viscosity for the BMLJ1 mixture divided by the $k = 0$ value, vs. the reduced wavevector for 200 K, 160 K, 120 K, 100 K, 80 K and 60 K. The value of ξ, the kinetic length, is obtained from the fitting function in Eq. (7.21). [Reproduced with permission from ref. (Kim and Keyes, 2005).]

Furukawa and Tanaka (Furukawa and Tanaka, 2009), studying the SS model, have extended the Kim–Keyes observations in two ways. First, they have shown that the increasing wavevector dependence of $\eta(k)$ is due entirely to the transverse component of the momentum flux and, secondly, they have established (again, for the SS mixture) that the length extracted from $\eta(k)$ has the same magnitude and temperature dependence as χ_4 obtained, as described in Section 7.3, from fluctuations in the structural relaxation function. Puscasu *et al.* (Puscasu *et al.*, 2010*a*,*b*) have simulated the wavevector dependence of the velocity kernel for a range of liquids including diatomic molecules and short-chain alkanes and, for the latter model, report the growth of a very large length scale as the supercooling is increased.

The decay of $\eta(k)$ with wavevector k reflects the decrease in dissipation as the wavelength shortens. As harmonic solids show no dissipation, it is tempting to associate the growing length identified in Refs. (Kim and Keyes, 2005; Furukawa and Tanaka, 2009) as being associated with the characteristic length scale of such solid-like domains. The (rough) superposition of the normalized $\eta(k)$s from both high and low temperatures, as shown in Fig. 7.7, suggests that, in the supercooled liquid, stress relaxes similarly to that in the high-temperature liquids except that the elementary objects are now rigid clusters with a linear dimension ξ instead of the atomic components. In 1989, Ladd and Alder (Ladd and Alder, 1989) described the stretched tail of the shear stress autocorrelation function in hard-sphere liquids near freezing. (Their evocative label—the "molasses" tail—does not seem to have caught on.) Subsequently, Isobe and Alder (Isobe and Alder, 2009, 2010) have argued that the long-time relaxation of the shear stress is dominated by the lifetime of rigid clusters in the liquid.

7.7 What lengths influence relaxation?

Does a kinetic length provide the unified description of cooperative dynamics for which it was intended? We have seen that the relaxation time of a glass-forming system can be increased by either decreasing the temperature or decreasing the system size below some threshold value. Karmakar *et al.* (Karmakar *et al.*, 2009) have examined the dependence of both the relaxation time τ and the susceptibility $\chi_4(t^*)$ (χ_4^P in their notation) of the BMLJ1 mixture as a function of temperature and number of particles. Their results are shown in Figs. 7.8 and 7.9. Fixing the system size at a large value, say $N = 1000$, we see that χ_4^P and τ both increase as T decreases, similar to behavior already described in Section 7.3. If, however, we hold T fixed, the variation of χ_4^{P} and τ with respect to N have opposite signs. The authors note that this result is contrary to the expectations of finite-scale scaling. It indicates that χ_4^{P} does not contain all of the information about the collective processes in the liquid necessary to establish the relaxation time.

The puzzling observations of Karmakar *et al.* (Karmakar *et al.*, 2009) had been foreshadowed by earlier work. Kim and Yamamoto (Kim and Yamamoto, 2000) demonstrated that the increase in the relaxation time they observed for the small system was accompanied, not unexpectedly, by a truncation of the size of mobile clusters. The point here is that the more extended mobile regions may represent greater mobility for the system, not less. Jäckle and coworkers (Froböse *et al.*, 2000; Jäckle and

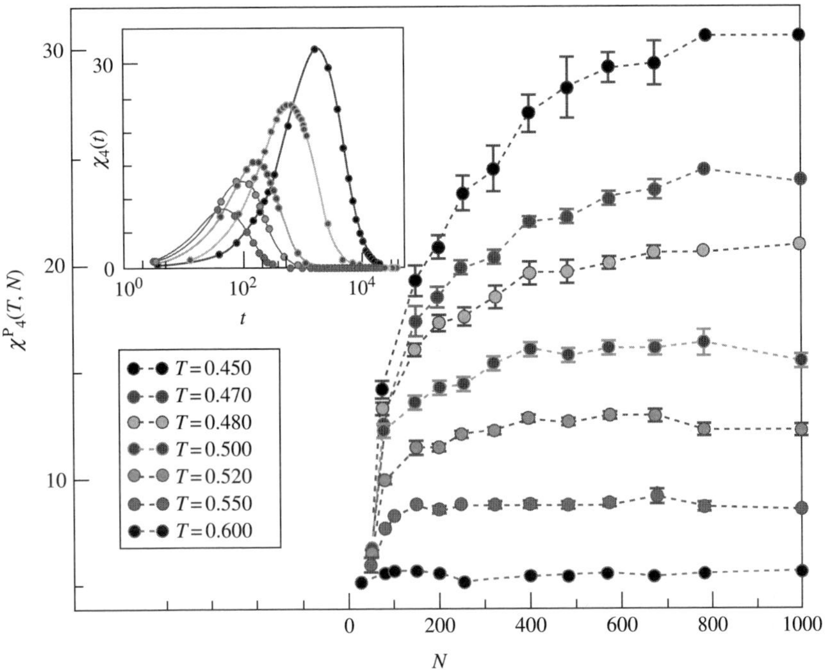

Fig. 7.8 Peak height of the dynamic susceptibility, $\chi_4^P(T,n)$ for the BMLJ1 model plotted as a function of system size N for different temperatures. For each temperature, $\chi_4^P(T,n)$ increases with system size, and saturates for large system sizes. $\chi_4^P(T,n)$ also increases as the temperature is lowered. Insert: $\chi_4^P(T,n)$ plotted as a function of time. In the main plot, temperature increases from the top curve to the bottom. In the insert, temperature decreases moving from the left curve to the right. [Reproduced with permission from Ref. (Karmakar *et al.*, 2009).]

Kawai, 2001) demonstrated that, even in systems exhibiting dynamic heterogeneities, the diffusion of particles still involved significant coupling to extended viscoelastic flows (i.e. collective motions in which particles retain their neighbors) dominated by the longest-wavelength transverse modes in the system. Decreasing the system size removes the longer-wavelength modes and, possibly, stiffens the surrounding medium, leading to a decrease in mobility. Such a scenario would be expected to exhibit a very different system-size dependence to the way size influenced the dynamic heterogeneities themselves. The spatial correlations in small displacements, as opposed to the large ones that typically define "mobile" particles, are an important component of the length that characterizes relaxation, both incoherent (e.g. self-diffusion) and coherent (e.g. stress relaxation). In a recent study of structural relaxation (Widmer-Cooper *et al.*, 2009), it was shown that, of the particle movements that have contributed irreversibly to relaxation, 60% (at the lowest temperature studied) could be categorized as strain-like, meaning that they involved the loss of no more than one of the initial neighbors.

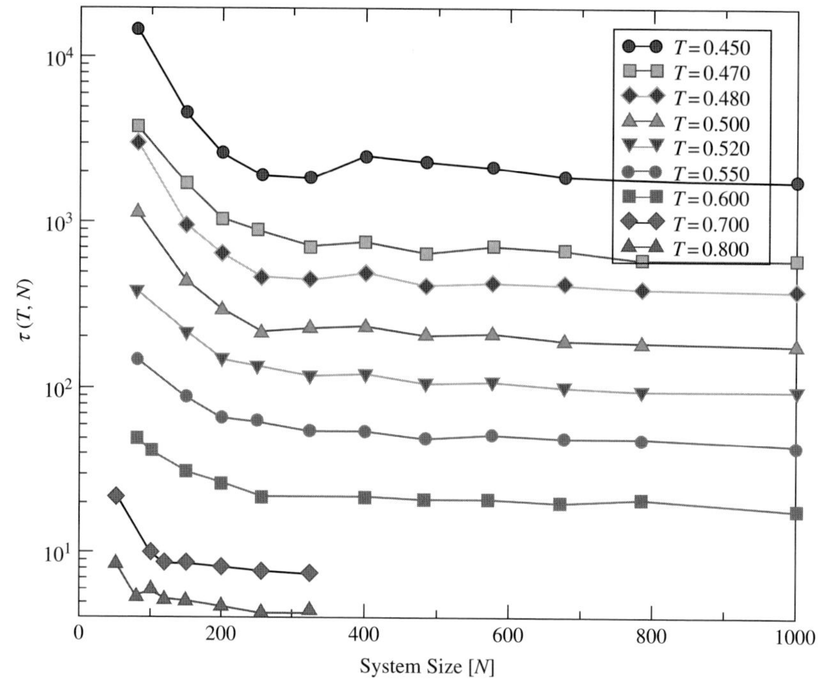

Fig. 7.9 Relaxation times as a function of temperature and system size in the BMLJ1 mixture. For the smallest temperature, $\tau(T, N)$ increases by approximately a decade from the largest to the smallest system size. Temperature increases from the top curve to the bottom. [Reproduced with permission from Ref. (Karmakar *et al.*, 2009).]

7.8 Conclusions

The measures of a kinetic length scale reviewed in this chapter have succeeded in a number of things. They have confirmed, by the act of measurement, the existence of spatial heterogeneity of the kinetics and the coarsening of this distribution on cooling. This descriptive success is an important milestone. It allows for the comparison between different glass formers, between the spatial character of the dynamics and that of the static properties of the supercooled liquid and between various theoretical treatments of the glass transition. The repeated observation of power-law relations between a kinetic length and a relaxation time is significant. Such behavior is suggestive of relaxation processes governed by the fluctuations in domain size. An alternative, in which the length corresponded to the dimension of an object involved in an activated process, would be expected to exhibit a different relation (Cavagna, 2009), i.e.

$$\tau \propto \exp\left[\frac{A\xi^{\zeta}}{k_B T}\right]. \tag{7.22}$$

The power-law exponents have been found to lie, roughly, between 2 and 4. If we think of the glass transition as being the temperature at which a relaxation time as increased by a factor of, let us say, 10^{10}, then these power laws would require the kinetic length at this glass transition to have increased by a factor of between 10^2 and 10^5. The reports of temperature dependences for the kinetic length with singularities at the mode-coupling temperature are generally regarded as a signature, not of a divergence, but of a crossover to some alternative relaxation mechanism. To date, there is no evidence for the large increases in kinetic lengths suggested by the observed power law. The conclusion is then that the time–length relationship extracted over a limited range of temperatures in simulations does not continue to hold at lower temperatures. The slow down in the growth of the kinetic length on cooling has been noted in experiments (Dalle-Ferrier *et al.*, 2007; Brambilla *et al.*, 2009). In simulations of the BMLJ1 mixture, Berthier *et al.* (Berthier *et al.*, 2007*b*) reported that for $T \leq 0.47$, the growth of the dynamic susceptibility with respect to τ_α becomes much slower than that observed at higher temperatures, "perhaps logarithmically slow". This last comment from (Berthier *et al.*, 2007*b*) is a reference to the prediction of a logarithmic relation $\xi \sim (\ln \tau_\alpha)^\varsigma$, such as expressed in Eq. (7.22), from theories that invoke an activated process (Lubchenko and Wolynes, 2003; Bouchaud and Biroli, 2004).

The kinetic lengths reviewed in Sections 7.2 and 7.3 all provide useful and, essentially similar, measures of the dynamic heterogeneities. An important question is, however, how well can they account for the length scales implicit in finite-size effects, interfacial correlations and crossover behavior? Furukawa and Tanaka (Furukawa and Tanaka, 2009) have reported that the characteristic length obtained from the q-dependence of the transverse momentum fluctuations exhibits a similar size and temperature dependence to $\chi_4(t^*)$. Berthier's (Berthier, 2004) demonstration that the Fickian crossover can be scaled by the length scale from $S_4(q,t)$ establishes a similar link. On the other hand, the increase in time scale due to the reduction in system size points to the role of correlations not included in the dynamic susceptibility. There remains an open problem to establish a clear identification of an explicit kinetic length (i.e. analogous to the lengths defined in Sections 7.2 and 7.3) with an implicit length scale such as demonstrated in Sections 7.4, 7.5 and 7.6.

As already touched upon, there is a gap between description and mechanism, one that only becomes truly evident now that the problem of description has been largely solved. To date, it is not clear that dynamic heterogeneities have clarified the mechanisms of relaxation. Part of the challenge in addressing the issue of mechanism certainly lies in refining what it is we actually want explained. The growth of the activation energy on cooling fragile liquids is generally associated with the growth of the number of elementary processes that must occur in series to achieve relaxation. There is certainly clear evidence, through the finite-size results, amorphous interface studies and the observation of the characteristic length in the transverse momentum fluctuations, that the mechanisms responsible for relaxation of particle positions and stress do have characteristic lengths and that these grow as the temperature decreases. What is missing is a description of those elementary process by which the observed length scales are generated.

There remain a number of interesting and open challenges that arise directly from what we have already learnt from MD simulations. We shall conclude by listing a few. First, there is evidence that the number of mobile particles (however one might define them) in a supercooled liquid undergoes substantial fluctuations in time. The democratic particle approach, introduced by Appignanesi and coworkers (Appignanesi *et al.*, 2006*a*,*b*), provides one explicit measure of the intermittent fluctuations in the number of particles involved in large displacements. From experiments on granular systems (Candelier *et al.*, 2009) and, subsequently, MD simulations (Candelier *et al.*, 2010), Dauchot and coworkers have identified the intermittent appearance of spatially correlated bursts of enhanced mobility among particles. Christened "avalanches", this intermittent behavior appears to increase as the glass transition is approached. Clarification of the significance of these fluctuations is important. Is there a new length scale (or a hierarchy of new lengths) describing the volume over which the avalanche occurs? Is relaxation at low temperatures increasingly dominated by intermittent bursts of activity and, if so, what is happening during the quiescent intervals that leads to initiating a mobility event?

A second challenge is to make direct connection between dynamic heterogeneities and coherent processes like shear-stress relaxation. The growing length scale associated with the transverse momentum fluctuations (Kim and Keyes, 2005; Furukawa and Tanaka, 2009; Puscasu *et al.*, 2010*b*) provides a starting point. Along with particle mobility, dissipation is also becoming increasingly heterogeneously distributed as the temperature drops. To understand such phenomena we need to understand how to construct a description of viscoelastic behavior where the dissipation can be localized even as the elastic behavior becomes more extensive with the approach to the glass state.

Finally, it remains a fundamental tenet that the ultimate origin of a kinetic length scale lies in length scales associated with structural correlations. Understanding this link between structure and dynamics is a problem that, currently, only simulations can address. There is no shortage of aspirants for the missing link: the spatial distribution of localized soft modes (Schober *et al.*, 1993; Oligschleger and Schober, 1999; Brito and Wyart, 2006, 2007; Widmer-Cooper *et al.*, 2008, 2009), the mosaic length scale (Cavagna *et al.*, 2007; Biroli *et al.*, 2008), the physical extent of clusters of locally preferred structures (Coslovich and Pastore, 2009; Tanaka *et al.*, 2010; Lerner *et al.*, 2009; Mossa and Tarjus, 2006; Pedersen *et al.*, 2010) to name only some. Should the relevant structural length scales ever be successfully unearthed it is quite possible that the description of dynamic heterogeneities may be rendered irrelevant, replaced by the more convenient and illuminating account provided by the relevant structure. That dynamic heterogeneitiy may prove the agent of its own demise (as a descriptor of cooperativity, that is) is a real possibility. Until then, the spatial distribution of dynamics remains our most general and concrete description of the complex dynamics associated with a liquid's passage to rigidity in the glass state.

Acknowledgments

I gratefully acknowledge valuable conversations with David Reichman, Grzegorz Szamel, Gilles Tarjus and Peter Daivis. I would also like to thank Luca Cipelletti for his generous help in preparing the manuscript. This work has been supported in part by the Australian Research Council through the Discovery Grant program.

References

Abate, A. R. and Durian, D. J. (2007). *Chaos*, **17**, 041107.

Abraham, S. E. and Bagchi, B. (2008). *Phys. Rev. E*, **78**, 051501.

Adam, G. and Gibbs, J. H. (1965). *J. Chem. Phys.*, **43**, 139.

Ahluwalia, R. and Das, S. P. (1998). *Phys. Rev. E*, **57**, 5771.

Alder, B. J. and Wainwright, T. E. (1967). *Phys. Rev. Lett.*, **18**, 988.

Andersen, H. C. (2005). *Proc. Nat. Acad. Sci.*, **102**, 6686.

Appignanesi, G.A., Rodriguez Fris, J.A., and Frechero, M.A. (2006*a*). *Phys. Rev. Lett.*, **96**, 237803.

Appignanesi, G.A., Rodriguez Fris, J.A., Montani, R. A., and Kob, W. (2006*b*). *Phys. Rev. Lett.*, **96**, 057801.

Ballesta, P., Duri, A., and Cipelletti, L. (2008). *Nature Phys.*, **4**, 550.

Baschnagel, J. and Varnick, F. (2005). *J. Phys.: Condens. Matter.*, **17**, R851.

Bennemann, C., Donati, C., Baschnagel, J., and Glotzer, S. C. (1999). *Nature*, **399**, 246.

Bernu, B., Hansen, J. P., Hiwatari, Y., and Pastore, G. (1987). *Phys. Rev. A*, **36**, 4891.

Berthier, L. (2003). *Phys. Rev. Lett.*, **91**, 055701.

Berthier, L. (2004). *Phys. Rev. E*, **69**, 020201(R).

Berthier, L., Biroli, G., Bouchaud, J-P., Cipelletti, L., El Masri, D., L'Hote, D., Ladieu, F., and Pierno, M. (2005*a*). *Science*, **310**, 1797.

Berthier, L., Biroli, G., Bouchaud, J.-P., Kob, W., Miyazaki, K., and Reichman, D. (2007*a*). *J. Chem. Phys.*, **126**, 184503.

Berthier, L., Biroli, G., Bouchaud, J.-P., Kob, W., Miyazaki, K., and Reichman, D. (2007*b*). *J. Chem. Phys.*, **126**, 184504.

Berthier, L., Chandler, D., and Garrahan, J. P. (2005*b*). *Europhys. Lett.*, **69**, 320.

Berthier, L., Garrahan, J. P., and Chandler, D. C. (2005*c*). *Europhys. Lett.*, **69**, 320.

Binder, K. and Kob, W. (2005). *Glassy Materials and Disordered Solids.* World Scientific, Singapore.

Biroli, G. and Bouchaud, J.-P. (2004). *Europhys. Lett.*, **67**, 21.

Biroli, G., Bouchaud, J.-P., Cavagna, A., Grigera, T. S., and Verrocchio, P. (2008). *Nature Phys.*, **4**, 771.

Biroli, G., Bouchaud, J.-P., Miyazaki, K., and Reichman, D. (2006). *Phys. Rev. Lett.*, **97**, 195701.

Bouchaud, J.-P. and Biroli, G. (2004). *J. Chem. Phys.*, **121**, 7347.

Brambilla, G., El Masri, D., Pierno, M., Berthier, L., Cipelletti, L., Petekidis, G., and Schofield, A.B. (2009). *Phys. Rev. Lett.*, **102**, 085703.

Brito, C. and Wyart, M. (2006). *Europhys. Lett.*, **76**, 149.

Brito, C. and Wyart, M. (2007). *J. Stat. Mech.*, L08003.

Butler, S. and Harrowell, P. (1991). *J. Chem. Phys.*, **95**, 4466.

Cammarota, C., Cavgana, A., Giardina, I., Gradenigo, G., Grigera, T. S., Parisi, G., and Verrocchio, P. (2010). **cond-mat**, arXiv:1001.2539. *Phys. Rev. Lett.*, **105**, 055703.

Candelier, R., Dauchot, O., and Biroli, G. (2009). *Phys. Rev. Lett.*, **102**, 088001.

Candelier, R., Widmer-Cooper, A., Kumerfeld, J. K., Dauchot, O., Biroli, G., Harrowell, P, and Reichman, D. R. (2010). *preprint, arXiv:0912.0193*.

Cavagna, A. (2009). *Phys. Rep.*, **476**, 51.

Cavagna, A., Grigera, T., and Verrocchio, P. (2007). *Phys. Rev. Lett.*, **98**, 187801.

Chandler, D., Garrahan, J. P., Jack, R. L., Maibaum, L., and Pan, A. C. (2006). *Phys. Rev. E*, **74**, 051501.

Charbonneau, P. and Reichman, D. R. (2007). *Phys. Rev. Lett.*, **99**, 135701.

Chaudhuri, P., Berthier, L., and Kob, W. (2007). *Phys. Rev. Lett.*, **99**, 060604.

Cicerone, M.T. and Ediger, M.D. (1996). *J. Chem. Phys.*, **104**, 7210.

Cipelletti, L., Bissig, H., Trappe, V., Ballesta, P., and Mazoyer, S. (2003). *J. Phys.: Condens. Matt.*, **15**, S257.

Coslovich, D. and Pastore, G. (2009). *J. Phys.: Condens. Matter.*, **21**, 285107.

Daama, J. L. C. and Villars, P. (1997). *J. Alloys Compd.*, **252**, 110.

Dalle-Ferrier, C., Thibierge, C., Alba-Simionesco, C., Berthier, L., Biroli, G., Bouchaud, J-P., Ladieu, F., L'Hote, D., and Tarjus, G. (2007). *Phys. Rev. E*, **76**, 041510.

Dasgupta, C., Indrani, A.V., Ramaswamy, S., and Phani, M. K. (1991). *Europhys. Lett.*, **15**, 307.

Dasgupta, C. and Ramaswamy, S. (1992). *Physica A*, **186**, 314.

Dauchot, O., Marty, G., and Biroli, G. (2005). *Phys. Rev. Lett.*, **95**, 265701.

Deng, D., Argon, A. S., and Yip, S. (1989). *Philo. Trans. R. Soc. Lond. A*, **329**, 595.

Desmond, K. W. and Weeks, E. R. (2009). *Phys. Rev. E*, **80**, 1305.

Doliwa, B. and Heuer, A. (2000). *Phys. Rev. E*, **61**, 6898.

Doliwa, B. and Heuer, A. (2003). *J. Phys: Condens. Matter.*, **15**, S849.

Donati, C., Douglas, J. F., Kob, W., Plimpton, S. J., Poole, P. H., and Glotzer, S. C. (1998). *Phys. Rev. Lett.*, **80**, 2338.

Donati, C., Franz, F., Parisi, G., and Glotzer, S. (2002). *J. Non-Cryst. Solids*, **307-310**, 215.

Donati, C., Glotzer, S. C., and Poole, P. H. (1999*a*). *Phys. Rev. Lett.*, **82**, 5064.

Donati, C., Glotzer, S. C., Poole, P. H., Kob, W., and Plimpton, J. S. (1999*b*). *Phys. Rev. E*, **60**, 3107.

Duri, A. and Cipelletti, L. (2006). *Europhys. Lett.*, **76**, 972.

Eaves, J. D. and Reichman, D. R. (2009). *Proc. Nat. Acad. Sci.*, **106**, 15171.

Ediger, M. D. (2000). *Ann. Rev. Phys. Chem.*, **51**, 99.

Flenner, E. and Szamel, G. (2007). *J. Phys.: Condens. Matt.*, **19**, 205125.

Flory, P. J. (1969). *Statistical Mechanics of Chain Molecules.* Interscience Publishers, Ithaca, New York.

Franz, S., Donati, C., Parisis, G., and Glotzer, S. C. (1999). *Philo. Mag. B*, **79**, 1827.

Froböse, K., Kolbe, F., and Jäckle, J. (2000). *J. Phys: Condens. Matter.*, **12**, 6563.

Fujara, F., Geil, B., Sillescu, H., and Fleischer, G. (1992). *Z. Phys. B*, **88**, 195.

Furukawa, A., Kim, K., Saito, A., and Tanaka, H. (2009). *Phys. Rev. Lett.*, **102**, 016001.

Furukawa, A. and Tanaka, H. (2009). *Phys. Rev. Lett.*, **103**, 135703.

Gebremichael, Y., Schrøder, T. B., Starr, F. W., and Glotzer, S. C. (2001). *Phys. Rev. E*, **64**, 051503.

Gebremichael, Y., Vogel, M., and Glotzer, S. C. (2004). *J. Chem. Phys.*, **120**, 4415.

Giovambattista, N., Buldyrev, S. V., Stanley, H. E., and Starr, F. W. (2005). *Phys. Rev. E*, **72**, 011202.

Glotzer, S. C. (2000). *J. Non-Cryst. Solids*, **274**, 342.

Glotzer, S. C., Novikov, N., and Schrøder, T. B. (2000). *J. Chem. Phys.*, **112**, 509.

Götze, W. and Sjögren, L. (1992). *Rep. Prog. Phys.*, **55**, 241.

Hedges, L. O., Maibaum, L., Chandler, D., and Garrahan, J. P. (2007). *J. Chem. Phys.*, **127**, 211101.

Heuer, A. (2008). *J. Phys.: Condens. Matter.*, **20**, 373101.

Hiwatari, Y. and Miyagawa, H. (1990). *J. Non-Cryst. Solids*, **117/118**, 862.

Hodgdon, J. A. and Stillinger, F. H. (1993). *Phys. Rev. E*, **48**, 207.

Hurley, M.M. and Harrowell, P. (1995). *Phys. Rev. E*, **52**, 1694.

Hurley, M. M. and Harrowell, P. (1996). *J. Chem. Phys.*, **105**, 10521.

Isobe, M. and Alder, B. J. (2009). *Mol. Phys.*, **107**, 609.

Isobe, M. and Alder, B. J. (2010). *Prog. Theor. Phys. Supp.*, **184**, 439.

Jäckle, J. (2002). *J. Phys.: Condens. Matter*, **14**, 1432.

Jäckle, J. and Kawai, H. (2001). *Physica A*, **291**, 184.

Jenckel, E. (1939). *Z. Physik. Chem.*, **A184**, 309.

Karmakar, S., Dasgupta, C., and Sastry, S. (2009). *Proc. Nat. Acad. Sci.*, **106**, 3675.

Karmakar, S., Dasgupta, C., and Sastry, S. (2010*a*). *Phys. Rev. Lett.*, **105**, 015701.

Karmakar, S., Dasgupta, C., and Sastry, S. (2010*b*). *Phys. Rev. Lett.*, **105**, 019801.

Keyes, T. and Oppenhiem, I. (1973). *Phys. Rev. A*, **8**, 937.

Keys, A. S., Abate, A. R., Glotzer, S. C., and Durian, D. J. (2007). *Nature Phys.*, **3**, 260.

Kim, J. and Keyes, T. (2005). *J. Phys. Chem. B*, **109**, 21445.

Kim, K. (2003). *Europhys. Lett.*, **61**, 790.

Kim, K. and Yamamoto, R. (2000). *Phys. Rev. E*, **61**, R41.

Kirkpatrick, T. and Wolynes, P. G. (1987). *Phys. Rev. B*, **36**, 8552.

Kirkpatrick, T. R. and Thirumalai, D. (1988). *Phys. Rev. A*, **37**, 4439.

Kirkpatrick, T. R., Thirumalai, D., and Wolynes, P. G. (1989). *Phys. Rev. A*, **40**, 1045.

Kob, W. (1999). *J. Phys: Cond. Mat.*, **11**, R85.

Kob, W. and Andersen, H. C. (1995). *Phys. Rev. E*, **51**, 4626.

Kob, W., Donati, C., Plimpton, S. J., Poole, P. H., and Glotzer, S. C. (1997). *Phys. Rev. Lett.*, **79**, 2827.

Lačević, N., Starr, F. W., Schrøder, T. B., Novikov, V. N., and Glotzer, S. C. (2002). *Phys. Rev. E*, **66**, 030101.

Lačević, N., Starr, F. W., Schrøder, T. B., and Glotzer, S. C. (2003). *J. Chem. Phys.*, **119**, 7372.

Ladd, A.J.C. and Alder, B. J. (1989). *J. Stat. Phys.*, **57**, 473.

Lamarcq, J., Bouchaud, J.-P., Martin, O. C., and Mézard, M. (2002). *Europhys. Lett.*, **58**, 321.
Leonforte, F., Boissière, R., Tanguy, A., Wittmer, J. P., and Barrat, J. L. (2005). *Phys. Rev. B*, **72**, 224206.
Lerner, E., Procaccia, I., and Zylberg, J. (2009). *Phys. Rev. Lett.*, **102**, 125701.
Lin, J.-X., Reichahrdt, C., Nussinov, Z., Pryadko, L. P., and Reichhardt, C.J.O. (2006). *Phys. Rev. E*, **73**, 061401.
Lubchenko, V. and Wolynes, P. G. (2003). *J. Chem. Phys.*, **119**, 9088.
Luo, W. K. and Ma, E. (2008). *J. Non-Cryst. Solids*, **354**, 945.
Mayer, P., Bissig, H., Berthier, L., Cipelletti, L., Garrahan, J. P., Sollich, P., and Trappe, V. (2004). *Phys. Rev. Lett.*, **93**, 115701.
Mazza, M. G., Giovambattista, N., Starr, F. W., and Stanley, H. E. (2006). *Phys. Rev. Lett.*, **96**, 057803.
Montanari, A. and Semerjian, G. (2006). *J. Stat. Phys.*, **125**, 23.
Mosayebi, M., Del Gado, E., Ilg, P., and Öttinger H. C. (2010). *Phys. Rev. Lett.*, **104**, 205704.
Mossa, S. and Tarjus, G. (2006). *J. Non-Cryst. Solids*, **352**, 4847.
Mountain, R. D. (1995). *J. Chem. Phys.*, **102**, 5408.
Odagaki, T. and Hiwatari, Y. (1990). *Phys. Rev. A*, **41**, 929.
Oligschleger, C. and Schober, H. R. (1999). *Phys. Rev. B*, **59**, 811.
Palomar, R. and Sesé, G. (2008). *J. Chem. Phys.*, **129**, 064505.
Parsaeian, A. and Castillo, H. E. (2008). *Phys. Rev. E*, **78**, 060105(R).
Pedersen, U. R., Schrøder, T. B., Dyre, J. C., and Harrowell, P. (2010). *Phys. Rev. Lett.*, **104**, 105701.
Perera, D. N. (1998). *J. Phys.: Condens. Matter*, **10**, 10115.
Perera, D. N. and Harrowell, P. (1996). *Phys. Rev.*, **E**, 54 1652.
Perera, D. N. and Harrowell, P. (1998). *Phys. Rev. Lett.*, **81**, 120.
Perera, D. N. and Harrowell, P. (1999). *J. Chem. Phys.*, **111**, 5441.
Poole, P. H. (1998). *Curr. Opin. Solid State Mater. Sci.*, **3**, 391.
Poole, P. H., Donati, C., and Glotzer, S. (1998). *Physics A*, **261**, 51.
Puscasu, R. M., Todd, B. D., Daivis, P. J., and Hansen, J. S. (2010*a*). *J. Phys.: Condens. Mater.*, **22**, 195105.
Puscasu, R. M., Todd, B. D., Daivis, P. J., and Hansen, J. S. (2010*b*). *J. Chem. Phys.*, **133**, 144907.
Qian, J., Hentschke, R., and Heuer, A. (1999). *J. Chem. Phys.*, **111**, 10177.
Ritort, F. and Sollich, P. (2003). *Adv. Phys.*, **52**, 219.
Sausset, F. and Tarjus, G. (2010). *Phys. Rev. Lett.*, **104**, 065701.
Scheidler, P., Kob, W., and Binder, K. (2000). *Europhys. Lett.*, **52**, 277.
Scheidler, P., Kob, W., and Binder, K. (2004). *J. Phys. Chem. B*, **108**, 6673.
Schober, H. R., Oligschleger, C., and Laird, B. B. (1993). *J. Non-Cryst. Solids*, **156**, 965.
Sim, E., Patashinski, A. Z., and Ratner, M. A. (1998). *J. Chem. Phys.*, **109**, 7901.
Sim, E., Patashinski, A. Z., and Ratner, M. A. (1999). *J. Chem. Phys.*, **110**, 6519.
Stariolo, D. A. and Fabricius, G. (2006). *J. Chem. Phys.*, **125**, 064505.

Stauffer, D. and Aharony, A. (1994). *Introduction to Percolation Theory.* Taylor & Francis, London.

Stein, R. S. L. (2007). Ph.D. thesis, Stanford University.

Stein, R. S. L. and Andersen, H. C. (2008). *Phys. Rev. Lett.*, **101**, 267802.

Szamel, G. and Flenner, E. (2006). *Phys. Rev. E*, **73**, 011504.

Tanaka, H., Kawasaki, T., Shintani, H., and Watanabe, K. (2010). *Nature Mater.*, **9**, 324.

Teboul, V. (2006). *Euro. Phys. J. B*, **51**, 111.

Toninelli, C., Wyart, M., Berthier, L., Biroli, G., and Bouchaud, J. P. (2005). *Phys. Rev. E*, **71**, 041505.

Vogel, M., Doliwa, B., Heuer, A., and Glotzer, S. C. (2004). *J. Chem. Phys.*, **120**, 4404.

Vogel, M. and Glotzer, S. C. (2004). *Phys. Rev. E*, **70**, 061504.

Vollmayr-Lee, K., Kob, W., Binder, K., and Zippelius, A. (2002). *J. Chem. Phys.*, **116**, 5158.

Wahnström, G. (1991). *Phys. Rev. A*, **44**, 3752.

Weeks, E. R., Crocker, J. C., Levitt, A. C., Schofield, A., and Weitz, D. A. (2000). *Science*, **287**, 627.

Whitelam, S., Berthier, L., and Garrahan, J. P. (2004). *Phys. Rev. Lett.*, **92**, 185705.

Widmer-Cooper, A. and Harrowell, P. (2006). *Phys. Rev. Lett.*, **96**, 185701.

Widmer-Cooper, A., Perry, H., Harrowell, P., and Reichman, D. R. (2008). *Nature Phys.*, **4**, 711.

Widmer-Cooper, A., Perry, H., Harrowell, P., and Reichman, D. R. (2009). *J. Chem. Phys.*, **131**, 194508.

Xia, X. and Wolynes, P. G. (2001). *Phys. Rev. Lett.*, **86**, 5526.

Yamamoto, R. and Onuki, A. (1997). *J. Phys. Soc. Jpn.*, **66**, 2545.

Yamamoto, R. and Onuki, A. (1998*a*). *Phys. Rev. E*, **58**, 3515.

Yamamoto, R. and Onuki, A. (1998*b*). *J. Non-Cryst. Solids*, **235-237**, 34.

Yamamoto, R. and Onuki, A. (1999). *Int. J. Mod. Phys. C*, **10**, 1553.

Zhang, Y. G., Guo, G., Refson, K., and Zhao, Y. (2004). *J. Phys:Condens. Mat.*, **16**, 9127.

8
Heterogeneities in amorphous systems under shear

Jean-Louis BARRAT and Anaël LEMAÎTRE

Abstract

The last decade has seen major progresses in studies of elementary mechanisms of deformation in amorphous materials. Here, we start with a review of physically based theories of plasticity, going back to the identification of "shear transformations" as early as the 1970s. We show how constructive criticism of the theoretical models permits to formulate questions concerning the role of structural disorder, mechanical noise, and long-ranged elastic interactions. These questions provide the necessary context to understand what has motivated recent numerical studies. We then summarize their results, show why they had to focus on athermal systems, and point out the outstanding questions.

8.1 Introduction

Rheology and plasticity, although they both investigate the flow of solid materials, are generally considered as two separate fields of materials science. Plasticity deals with the deformation of "hard" solids, characterized by large elastic moduli (typically in the GPa range). Rheology, on the other hand, deals with much softer materials, such as colloidal pastes, foams, or other "complex fluids" (Larson, 1999) with moduli that can vary from a few Pa to kPa. In view of these differences, the experimental tools used to investigate the flow of hard and soft materials differ widely, whether they are mechanical or involve more indirect microscopic characterizations.

Still, if one temporarily forgets about the differences in the scale of stress levels, striking similarities appear in the behavior of these different materials, as illustrated schematically in Fig. 8.1. The differences in stress scales are indeed easily understood in terms of the interactions. The scale for elastic moduli is an energy per unit volume. In hard materials, typical energies will be in the range 0.1–1 eV, and the typical length scales are of order of nanometers, or even smaller. In softer materials, the energy scale

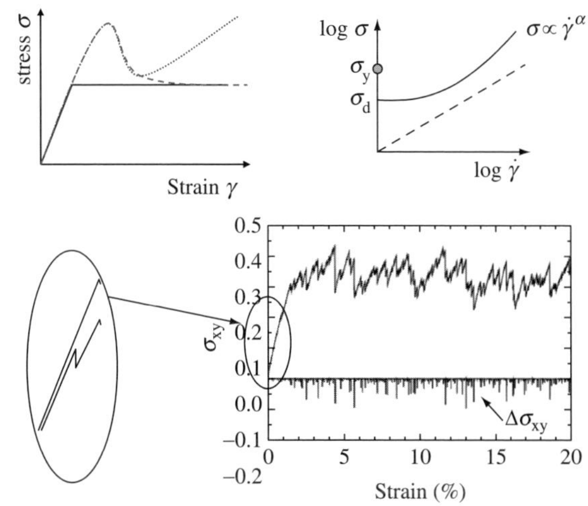

Fig. 8.1 Top left: schematic illustration of a stress–strain curve for the plastic deformation of a solid material. The three curves correspond, respectively, to ideal elastoplastic behavior (a), a metallic (b) and a polymeric system at finite strain rate (c). The strain-hardening part at large strain is specific to polymer systems. Top right: schematic illustration of a flow curve for a soft material; the dotted line is the Newtonian fluid case, while the low strain rate limit of the full curve corresponds to the yield stress. Bottom: actual response of a simulated strained glass at low temperature. Note the large stress fluctuations, which are associated with the finite size of the sample. The stress variations are also shown. The zoom on the elastic part at low strain shows that small plastic events are also observed in this part. Adapted from (Tanguy *et al.*, 2006).

is often comparable to $k_B T$, and length scales of the order of a tenth of a micrometer. Finally, the case of foams corresponds to a stress scale set by the surface tension γ divided by the typical bubble size. It is then not impossible that common physical properties can be found in such widely different systems if the proper elementary units are considered and the appropriate rescalings are made.

In terms of "reduced" parameters, however, the experimental conditions may correspond to very different ranges for the various systems investigated. For example, a foam or a two-dimensional "bubble raft" is essentially always athermal (the thermal fluctuations are irrelevant compared to the energies involved at the scale of individual bubbles) so that it should be compared to metallic or polymer glasses at low temperatures. On the other hand, a not too dense colloidal system at room temperature, in which thermal fluctuations are significant, could be compared to systems close to their glass-transition temperature. The same type of considerations apply to time scales and their comparison with the applied deformation rates, which have a strong influence on the stress–strain curve, as sketched in Fig. 8.1. Typically, the stress peak $\sigma_{\max}$ shown in this figure tends to display a logarithmic increase with the deformation rate $\dot{\epsilon}$ and, when aging is observed, with the age, t_w of the system.

In this chapter, we will limit our considerations to non-crystalline, amorphous materials. In crystalline materials, flow can be described in terms of dislocation motion, and although the interaction between these extended defects may lead to a macroscopic behavior similar to that of amorphous materials (Miguel and Zapperi, 2006), the underlying microscopic physics is different. For amorphous materials it will be shown that flow defects, if they exist, are localized rather than extended. We will also exclude from our considerations the case of real granular materials, which raise several complications such as the importance of gravity and that of friction.

Why is a chapter on sheared materials included in a book on dynamical heterogeneities? The standard plasticity or rheology approach is based on macroscopic constitutive equations, established using symmetry arguments (Lubliner, 2008). These equations relate the stress and strain (or strain history) in the system, within a fully homogeneous, continuous medium description. However, the notion of dynamical heterogeneities naturally emerges when one attempts to reach an understanding of the microscopic mechanisms that underly the macroscopic behavior. Is it possible to identify microscopic heterogeneities, that would in some respects play the role assigned to dislocations in the flow of crystalline materials? What governs the dynamical activity of such heterogeneities, what are their interactions and correlations, and do they organize on larger scales? From the pioneering experiments of Argon (Argon and Kuo, 1979) emerged the notion of "shear transformations", localized (in space and time) rearrangements that govern the plastic activity (Argon, 1982). Such local yield events have been very clearly identified in experiments on bubble rafts, in colloidal systems (Schall *et al.*, 2007), as well as in various atomistic simulations of low-temperature deformation (see, e.g., (Falk and Langer, 1998; Maloney and Lemaitre, 2006; Tanguy *et al.*, 2006)). They are now believed to constitute the elementary constituent of plastic deformation in amorphous solids at low temperature. However, their cooperative organization is far from being understood, although a number of models based of this notion of elementary events have been developed and studied analytically at the mean-field level or numerically. The discussion of these microscopic, dynamical heterogeneities will be the core of this chapter.

Another, very different aspect that escapes the purely macroscopic description of flow is the frequent experimental observation that strain in solid materials can take place in a very heterogeneous manner at the macroscopic level. This phenomenon, described as the existence of "shear bands" or "strain localization", takes place both in hard and soft materials, under various conditions of deformation. Instead of being evenly distributed and uniform through the system (affine deformation), the deformation is concentrated inside a localized region of space, typically a two-dimensional "shear band" with a finite thickness. Within such shear bands where the entire macroscopic deformation is concentrated, the local strain becomes very large in a short time, eventually leading to material failure for hard materials. In soft materials, which, in contrast to hard materials, can sustain steady flow when strained in a pure shear (Couette) geometry, shear bands can become permanent features of the steady-state flow and coexist with immobile parts of the same material. In such cases the flowing part, which is also described as a shear band, occupies a finite fraction of the sample thickness. While shear bands are clearly macroscopic features of the flow, understanding the mechanism of their formation and stability necessitates the introduction of auxiliary state variables and associated length scales, which in turn could be related to the existence of flow heterogeneities at smaller scales. These aspects will be discussed further in Section 8.2.7.

We mention at the end of this introduction that the focus of our chapter will be on simple shear deformation. In most materials, it is expected that the study of such deformations will allow one to unveil the mechanisms of plasticity in more complex situations, as the plastic deformation proceeds with little or no density increase. This is true in metallic glasses or soft condensed system, where plastic deformation often takes place at almost constant volume, but less in network materials with a lower Poisson ratio, where density increase and shear deformation are observed simultaneously (Rouxel *et al.*, 2008).

We would like to point out that the field of plasticity and rheology, even limited to amorphous materials, is a very broad one, and that the subject has been tackled by many groups with widely different backgrounds, from metallurgy to soft matter and granular materials. Such a brief review may in many cases give a very incomplete account of important theoretical developments. We hope, however, that the reader will find in the bibliography the necessary references. In all cases, we tried to point out critically what we see as the limitations of theoretical approaches, in the hope that the reader's interest will be stimulated to test and possibly improve the current models. Our bias in this chapter, consistently with the general topic of the book, will be to put more emphasis on models that involve dynamically heterogeneous, collective behavior, and could be analyzed using tools presented in other chapters. This naturally leads us to insist on results obtained on systems in which thermal noise is not dominant (such as many soft matter systems, or glasses at low temperature) and such heterogeneous behavior in the flow response is expected to be most important.

8.2 Theoretical background

8.2.1 Macroscale

Plasticity at the macroscale. Many tests have been developed to characterize plastic behavior. They involve deforming a piece of material under various conditions: constant stress, constant or oscillatory strain rate, step stress or strain, ... while measuring the response using appropriate observables. Typical stress responses, for a material loaded at constant strain rate, are depicted in Fig. 8.1. When the strain is vanishingly small, the material responds elastically. As strain increases, plasticity progressively sets in and the stress smoothly rounds off. This transient response obviously depends on the evolution of the internal state of the system during loading, hence on sample preparation and loading rate. Further increase of strain may be accompanied, in some circumstances, with the onset of an instability, as deformation localizes along shear bands and leads the material to failure. When this instability can be avoided, however, stress reaches a steady-state plateau value.

In general, it is difficult experimentally to achieve situations in which the flow is truly uniform: external driving must always be applied at boundaries, so that material flow must be inhomogeneous to adapt to these conditions (one exception is the flow of a Newtonian fluid in plane Couette geometry). So, in the flow of complex fluids one often observes a strong non-uniformity in the flow rate that is due to the combination of a non-uniform stress field with a strongly non-linear flow curve, and one often speaks of "localization" in this context. This use of the term "localization" is most often found in rheology, as typical experimental setups, such as the Couette cell necessarily lead to flow profiles that are inhomogeneous (Coussot, 2005). "Localization" in also used in another context, for example, in experiments on "hard" glasses (Schuh *et al.*, 2007), when it refers to an instability that is not directly associated with the macroscopic stress inhomogeneity and can only be observed during transients as it leads to failure. Whether these two forms of localization are related remains an open question.

Constitutive equations. One goal of theories of plasticity is to provide a macroscopic description of the deforming medium, analogous to the Navier–Stokes equations for Newtonian fluids. As illustrated in Fig. 8.1, plastic materials exhibit significant, preparation or age-dependent, transients in which stress does not match its steady-state value. Proper account of either transient response or of instabilities of the steady inhomogeneous flow must rely on a description of how the internal state of the system depends on loading history. This entails identifying variables to properly characterize the material state and providing constitutive equations to complement the relevant conservation equations (momentum, but also possibly density or energy). The search for constitutive equations often rests on the idea that it is possible to provide a local and instantaneous representation of the material state, that is to describe its response in terms of relations between local quantities (such as stresses, strains, energy, density, ...) and their derivatives.

The introduction of internal state variables is also a necessity if one wants to study localization, or instabilities in the material response. For example, it has been shown that constitutive equations that are approximated (grossly) by their steady-state stress/strain-rate relation lead to instability criteria that are rudimentary, and that internal variables have to be introduced to capture non-stationary material response (Rice and Ruina, 1983). Situations when the plastic response is unsteady play a critical role in the identification of relevant state variables. Localization thus attracts a considerable amount of interest not only for its practical importance, but also for the theoretical implications of identifying the proper variables governing its development.

State variables for localization. One of the most obvious state variable is temperature. Clearly, it is expected to increase due to plastic activity when a material is strained under adiabatic conditions. Its role in localization has thus been the subject of a long-standing debate in the metallic-glass community (Schuh *et al.*, 2007). Some insist that localization could be due to local heating and should be treated by introducing thermal dependence in the stress/strain-rate relationship, plus energy-conservation equations. The consensus, however, now is that this argument fails because the dissipated energy due to plasticity is evacuated too fast (Schuh *et al.*, 2007). Other features of material response, such as localization or the peak stress, depend sensitively on age at time scales that are completely separated from those of thermal exchanges. These observations rule out the role of thermal inhomogeneities in most cases. In soft materials, the solvent acts in general as an external bath, so that deformation can be considered to take place under isothermal conditions.

A model proposed recently to account for shear banding or fracture focuses on compressibility and density fluctuations (Furukawa and Tanaka, 2006; Furukawa *et al.*, 2009). Like the thermal theories of localization, it is based on a purely macroscopic description. It assumes that (i) the flow is Newtonian and (ii) stresses assume their steady-state values. The values of shear stress and pressure are thus provided by simple steady-state equations, function of strain rate and density, and localization is in a sense a kinematic effect. The proposed scenario involves a strong density dependence of the viscosity on density or pressure. As a result, a density fluctuation (which is usually neglected in the description of incompressible flows) results in a local decrease of the viscosity, which in turn increases the local shear rate and leads to an amplification of the fluctuation. A linear stability analysis of the Navier–Stokes equation under these conditions shows that for shear rates larger than $(\partial\eta/\partial P)^{-1}$, large wavelength density fluctuations at 45 degrees from the flow direction become linearly unstable. This leads to growth of density fluctuations and potentially to material failure.

Similar arguments have been used to describe the coupling between density fluctuations and rheology in polymer solutions (Helfand and Fredrickson, 1989). This type of scenario predicts that strain localization only operates *above* a critical strain rate, and it was argued in (Furukawa *et al.*, 2009) that it captured a transition to localisation that is seen around $T_{\rm g}$. Experiments on metallic glasses also show, at low temperatures, an enhancement of localization at low strain rates, and a form of localization that now occurs when the strain rate lies below a critical value (Schuh *et al.*, 2007). It thus seems that the localization predicted in (Furukawa *et al.*, 2009) around $T_{\rm g}$ is different from the phenomenon that is discussed in metallic glasses at low temperatures.

Discussion. These difficulties illustrate the problems encountered when formulating theories of plasticity. It seems unlikely that a description involving only the usual thermomechanical observables (stress, pressure, temperature) can be sufficient. We know indeed that these variables do not provide a complete representation of the internal structure of glasses. The state of a glassy system is, by definition, out of equilibrium, and evolves constantly with time in a relaxation process that involves hopping in a potential-energy landscape (PEL) (Stillinger, 1995; Debenedetti and Stillinger, 2001), over distributed energy barriers (Doliwa and Heuer, 2003*a*,*b*,*c*). Deformation modifies the picture as it competes with this relaxation (Utz *et al.*, 2000) and constantly rejuvenates the glassy structure, i.e. allows the system to stay in rather high energy states, and in some cases to reach a non-equilibrium steady state. The question is thus, using a few dynamical equations involving a limited set of state variables, to be able to describe the response of a glass driven out-of-equilibrium by external deformation. To describe the transient stress response upon loading, as exemplified in Fig. 8.1, we would also have to be able to describe the state of the glass that is produced by annealing in terms of variables that can be introduced in a description of plastic response.

The description of flow in amorphous solids that will be presented in the following is essentially based on ideas from elasticity theory, and therefore starts from the low-temperature, solid side of the glass or jamming transition. We mention briefly here another alternative approach, the mode-coupling approach of Fuchs and Cates (Fuchs and Cates, 2002). Let us recall that the mode-coupling approach to dense liquids involves a non-linear feedback mechanism, in which the relaxation dynamics of a given density fluctuation at some **k** is coupled to that of all other fluctuations at different wavevectors. This feedback leads to structural arrest—i.e. absence of relaxation and appearance of a frozen structure—at some finite density and temperature. The effect of shear described by Fuchs and Cates is the advection of density fluctuations, in such a way that the coupling that leads to this non-linear feedback weakens with time, and a relaxation eventually results. A dynamical yield stress is obtained as the zero shear rate limit of the stress under steady shear, and turns out to be non-zero for the high-density, strongly coupled systems that the theory predicts to be in a non-ergodic state at zero shear. The theory was developed further by Brader and Fuchs (Brader *et al.*, 2008), and predicts rheological behavior that is in very reasonable agreement with experimental observations on colloidal glasses, together with a prediction of the deviation from the equilibrium fluctuation–dissipation theorem. However, this mode-coupling description has not been extended, up to now, to the description of dynamical heterogeneities.

8.2.2 Local inelastic transformations

In crystals, the mechanisms of plastic deformation involve the motion and generation of dislocations, which are a class of topological defects of the periodic structure. The possibility to identify and precisely define the objects responsible for irreversible deformation has played and continues to play a key role in formulating phenomenologies and constructing governing equations for crystalline plasticity. In contrast, the state of knowledge regarding amorphous systems lags far behind, because the elementary objects governing the plastic response remain quite difficult to pinpoint.

Most modern theories of plasticity are based on the idea, proposed by Ali Argon in the late 1970s (Argon, 1979), that macroscopic plastic deformation is the net accumulation of local, collective, rearrangements of small volume elements—typically 5–10 particles in diameter. Initially corroborated by the observation of flow in a bubble raft (Argon and Kuo, 1979), this idea is now firmly supported by numerical simulations (Argon *et al.*, 1995; Falk and Langer, 1998; Maloney and Lemaitre, 2004*a*,*b*, 2006; Tanguy *et al.*, 2006), and by a recent experiment (Schall *et al.*, 2007).

In contrast with crystals, however, the notion of rearrangements (also called "local inelastic transformations", "shear transformations", or simply "flips") refers to a process, not to a specific type of defect. Means have been devised to identify "zones" before they flip, via measurements of particle displacement fields, (Lemaitre and Caroli, 2007) or of local elastic moduli (Yoshimoto *et al.*, 2004; Tsamados *et al.*, 2009),

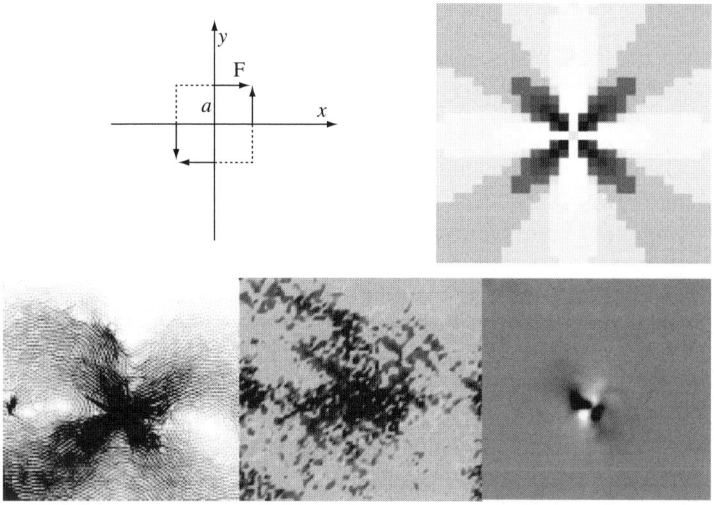

Fig. 8.2 Top: schematic illustration of a force quadrupole that corresponds to the theoretical description of shear transformation zones (left), and its Eshelby stress field. Bottom, from left to right: displacement field, stress response, and energy change associated with a localized plastic event, as observed in the quasi-static simulations described in Section 8.3. Figures adapted from Refs (Picard *et al.*, 2004; Maloney and Lemaitre, 2004*b*; Tanguy *et al.*, 2006).

but these observations can be unequivocally correlated with rearrangements only rather close to yielding, and there is no universally accepted prescription for identifying *a priori* the locations (the "zones") where flips occur. Of course, it seems reasonable to infer that there should be some specific features of the local packing, which make these transformations possible—and are probably related to the fluctuations of local thermodynamic quantities such as energy, stress, or density (free-volume). Many terms have been coined to reflect this notion—"flow defects", (Spaepen, 1977) "τ-defects", or more recently "shear transformation zones" (STZ) (Falk and Langer, 1998)—but the questions of what precisely a zone is, or how zones could be identified before-the-fact, remain largely open.

This idea does not imply that only *shear* transformations exist, or that each rearrangement is a pure shear event. Rearrangements may involve some local changes of volume too, and some theories—such as free-volume approaches—may attempt to take this into account. Occurring at a very local scale, rearrangements are also inevitably broadly distributed. But in view of constructing a theory of stress relaxation, it is the net effect of local contributions to shear strain that clearly matter.

8.2.3 Activation theories

Activation theories attempt to describe plasticity as the net result of independent shear transformations, which are supposed to be rare, thermally activated events. It is thus assumed that the system spends most of its time near local equilibria: clearly, this picture belongs to a low-temperature regime, when the dynamics in a PEL would be described by infrequent hops between local minima. With these postulates, constitutive equations require a specification of transition rates (as a function of energy, density, stress) governing the rearrangements.

Eyring's theory. The simplest form of such an activation theory of material flow dates back to the work of Eyring, and was initially intended to account for the viscous behavior of liquids (above T_g). In Eyring's view, flow proceeds by the motion of single molecules into holes left open by neighboring ones. This obviously misses the possibility that elementary events are in fact collective, but Eyring introduces several assumptions that are still quite fundamental to the field and useful to keep in mind as a reference.

He first assigns a typical value E to the energy barrier that must be overcome to allow such hops—it is, in his view, the energy needed to create a hole—and observes that various types of hops are possible, which can either increase or reduce the macroscopic stress. He then restricts his description to two types of opposite moves, contributing elementary strains of opposite signs $\pm\Delta\epsilon_0$, so that the macroscopic strain rate takes the form:

$$\dot{\gamma} = \Delta\epsilon_0 \left(\mathcal{R}_+ - \mathcal{R}_-\right), \tag{8.1}$$

with $\mathcal{R}_\pm$ the rates of forward and backward moves. These rates are taken to follow an Arrhenius activation law describing $\pm$ hops over activations barriers:

$$\mathcal{R}_\pm = \omega_0 \exp(-E_\pm/kT), \tag{8.2}$$

with ω_0 a microscopic frequency. Eyring assumes that stress induces a linear bias between energy levels, so that the barrier depends linearly on σ: $E_\pm = E_0 \mp \sigma\,\Omega_0$, with Ω_0 an "activation" volume. This leads to writing the stress/strain-rate relation as:

$$\dot{\gamma} = 2\,\omega_0\,\Delta\epsilon_0 \exp\left(-\frac{E_0}{kT}\right) \sinh\left(\frac{\Omega_0\sigma}{kT}\right). \tag{8.3}$$

Of course, the linearization of this relationship for small stresses leads to Newtonian behavior, with an Arrhenius viscosity. At large σ, it is customary to keep only the dominant exponential term and write the stress as:

$$\sigma = \frac{kT}{\Omega_0} \ln\left(\tau_0(T)\dot{\gamma}\right),$$

with $\tau_0(T) = \exp\left(\frac{E_0}{kT}\right)/\omega_0\Delta\epsilon_0$.

Argon's theory. Eyring's formulation relied on a representation of deformation in terms of hops of single atoms or molecules, instead of collective motions. Early representations of plasticity addressed this issue while borrowing from the general framework of Eyring's approach, and in particular the idea that the linear viscous behavior arises from a balance between forward (stress-releasing) flips and a "back-flux". In order to take into account the collective character of elementary transitions, Argon (Argon, 1979) argues that the shear zones can be viewed as inclusions that are elastically coupled to the surrounding medium, and postulates that a flip occurs when a zone elastically deforms up to some critical strain, in the range of $\sim 2-4\%$, at which it becomes unstable. The calculation borrows from Eshelby's work on martensites (Eshelby, 1957), which offers an analytical framework to estimate the total change in elastic energy due to a change in the internal strain state of an inclusion.

The question is how to take into account the fact that the average stress level biases the elastic energy associated with a zone. At high temperatures, Argon computes the stress bias at linear order, like Eyring, leading to an expression similar to Eq. (8.3). At low temperatures, he performs an imposing treatment of the elastic problem, to arrive at a perturbative, second-order, estimate of the effect of stress on the minimum where the system resides. It leads to an expression of the form (Argon, 1979):

$$E_{+} \propto \left(1-\frac{\sigma}{\sigma_{\mathrm{c}}}\right)^{2}, \tag{8.4}$$

where $\sigma_{\mathrm{c}} = \mu(T)\epsilon_{\mathrm{c}}$ is a typical scale of the shear stress needed to reach the strain ϵ_{c} at the yield point and $\mu(T)$ is the shear modulus. Using Eqs. (8.1) and (8.2) and neglecting the backflux, Eq. (8.4) leads to a stress/strain-rate relation of the form: $\sigma - \sigma_{\mathrm{c}} \propto -\left(T \ln\left(\Delta\epsilon_0\, \omega_0/\dot{\gamma}\right)\right)^{1/2}$. The main interest of this theory is that it captures important features of experimental data on metallic glasses, in particular, (i) an apparent singular behavior in the low-T limit, and (ii) the weak $\dot{\gamma}$ dependence over a broader range of temperatures (Schuh *et al.*, 2007).

One main limitation of this approach is that it treats perturbatively—at second order—the effect of stress on the elastic potential of a shear transformation zone. This approach must break down if the stress level is sufficiently large to bring a zone close to instability, as the system then approaches a catastrophe. Indeed, there is evidence from numerical simulations that deformation-induced instabilities correspond to a saddle-node bifurcation (Malandro and Lacks, 1997, 1999; Maloney and Lemaitre, 2004*a*). The energy barrier near instability is thus of the form $E \propto (1-\sigma/\sigma_{\mathrm{c}})^{3/2}$, which, after insertion in Eqs. (8.1) and (8.2) would lead to a stress/strain-rate relation of the form: $\sigma - \sigma_{\mathrm{c}} \propto -\left(T \ln\left(\Delta\epsilon_0\, \omega_0/\dot{\gamma}\right)\right)^{2/3}$. Already formulated by Caroli and Nozières (Caroli and Nozières, 1996) in the context of dry friction, this argument was brought up recently by Johnson and Samwer (Johnson and Samwer, 2005) for metallic glasses, and successfully compared with experimental data.

Discussion. A first objection that should be made against these activation theories is that they treat flips as uncorrelated events. As Bulatov and Argon first noted (Bulatov and Argon, 1994*b*,*c*,*d*), however, each rearrangement creates a long-range elastic field (Eshelby (Eshelby, 1957)), and hence alters the stress in the rest of the system. These stress changes can be viewed as mechanical signals that are emitted by flips and may trigger secondary events (Lemaitre and Caroli, 2009). This mechanism shows up strikingly in numerical simulation of athermal systems, via the emergence of avalanche behavior (Maloney and Lemaitre, 2004*b*; Demkowicz and Argon, 2005; Maloney and Lemaitre, 2006; Bailey *et al.*, 2007; Lerner and Procaccia, 2009; Lemaitre and Caroli, 2009). When these elastic effects are present, it becomes essential to understand the role of the self-generated stress noise in the activation of plastic events themselves.

A second objection is that the above theories obviously ignore the fact that the energy barriers that limit plastic transformation are broadly distributed (Doliwa and Heuer, 2003*a*,*b*,*c*). Moreover, as the material is strained, shear transformation zones are driven towards their instability threshold (Maloney and Lemaitre, 2004*a*; Lemaitre and Caroli, 2007). This is quite different from the situation of a glass undergoing thermal relaxation: here, the distribution of barriers is set by the competition between elastic loading and plastic yielding (Rodney and Schuh, 2009*a*), and some regions of space could, depending on the parameters, present temporarily, very low energy barriers. Rather than being set *a priori*, the relevant barrier distribution must thus result from a complex dynamical process.

Thirdly, activation theories of plasticity ignore the role of fluctuations of local quantities such as free-volume, elastic constants, stresses, etc. As we will see in Section 8.3, however, the idea that yielding can easily be associated with a unique critical value of the local stress or energy is strongly challenged by numerical observations (Tsamados *et al.*, 2008). Moreover, the activation process should depend significantly on details of the atomic packing, in particular via the fluctuations of local elastic moduli (Mayr, 2009; Tsamados *et al.*, 2009), local density or pressure levels (Demkowicz and Argon, 2004, 2005; Argon and Demkowicz, 2006).

8.2.4 Dynamics of the local stress field

A number of models (Bulatov and Argon, 1994*b*,*c*,*d*; Baret *et al.*, 2002; Picard *et al.*, 2002, 2004, 2005) focus on the local stress field and its dynamics, in order to take into account two important features that were identified in the previous discussion: (i) the fact that loading drives the system towards local instability thresholds, hence that barriers result from a dynamical process, and (ii) the fact, that rearrangements may interact via their Eshelby stress fields.

The first of these models was proposed by Argon and Bulatov (Bulatov and Argon, 1994*a*,*b*,*c*,*d*) in a study involving various aspects of plasticity but also of glassy relaxation (Bulatov and Argon, 1994*c*). They showed that even in the absence of deformation, taking into account long-range interactions between shear transformations could lead to net behavior that ressembles that of glassy models incorporating e.g. distributions of time scales. Their model involves a collection of weak zones that are distributed on a triangular lattice. Transformation probabilities are determined from activation theory, with stress-dependent free-energy barriers $\Delta G^*(\sigma) = \Delta F_0 - \Omega\sigma_{\alpha\beta}\Delta\epsilon_{\alpha\beta}$, where $\Delta\epsilon_{\alpha\beta}$ is the strain increment of a transforming zone during transition. Only pure shear transformations are allowed for convenience, but this does not exclude dilation that is introduced via a dilation parameter meant to account for how activation energy depends on pressure. Finally, each transformation alters the stress levels in the whole system via a Green tensor, corresponding to the solution of the Eshelby problem. The zone flips are thermally activated, but the barriers depend on the stress sustained by the zones, so that the introduction of this mechanism allows for correlations between flips.

The gist of Bulatov and Argon's viewpoint is that initially, after a quench, many of these weak zones should be far from their threshold as a consequence of annealing. As stress increases, energy barriers decrease, and some flips might occur. For small values of external stresses, only a few zones, sufficiently near their instability thresholds at the end of annealing, will respond. Moreover, as most zones are far from instability, the stress released by these rare flip events is insufficient to trigger secondary events. Initial loading is thus accompanied by a small number of isolated rearrangements. As stress increases, however, an increasing number of zones come close to their instability thresholds. Mechanical noise then starts to be able to trigger events elsewhere, leading in some cases to localization.

Two more recent models, strongly inspired by the Argon–Bulatov model, have been proposed by Baret and coworkers (Baret *et al.*, 2002) and Picard and coworkers (Picard *et al.*, 2002, 2004, 2005). The model of (Baret *et al.*, 2002) has the interesting feature of incorporating a distribution of threshold values for the local yield stress, but uses extremal dynamics to describe the evolution of the system at vanishingly small strain rate. The model of Picard, on the other hand, does not include disorder, but has the advantage of relative simplicity and of easily incorporating finite strain-rate effects. The model describes the evolution of a scalar stress σ_i on a lattice site via an equation of the form

$$\partial_t \sigma_i = \mu \dot{\gamma} + \sum_j G_{ij} \dot{\epsilon}_{j,\mathrm{plast}}, \tag{8.5}$$

where μ is a shear elastic modulus. The first term describes an elastic loading proportional to an average external strain rate. The second term is the supplementary loading that arises from the plastic activity at all other sites in the system, which is assumed to be transferred instantaneously through an elastic propagator G_{ij}. This plastic activity $\dot{\epsilon}_{j,\mathrm{plast}}$ is computed in turn by assuming that any site that reaches a stress beyond some local critical yield σ_Y releases its stress with a time constant τ. Numerical studies of the model show that, at low strain rates, zones of persistent plastic activities can be observed, with a typical size that tends to diverge as the strain rate vanishes.

8.2.5 Dynamics of distributions

Based on the ideas mentioned above, several approaches have been proposed that describe the state of the system by the distribution of a scalar variable—corresponding to either stress levels (Hebraud and Lequeux, 1998) or energy barriers (Sollich *et al.*, 1997; Sollich, 1998). They are based on some empirical assumptions that are inspired by the physical picture of interacting flow defects. In most of these approaches, an assumption of homogeneous deformation is made, so that the description of macroscopic strain localization is not permitted. Moreover, a scalar description of stress and strain is retained, with the implicit assumption that the scalar stress corresponds to a local shear stress. Despite this apparent simplicity, these models are far richer that simple local constitutive equations, as they introduce into the picture some auxiliary quantity describing the internal state of the system, in the spirit of the "rate and state" models of solid friction (Rice and Ruina, 1983).[1]

Hébraud–Lequeux fluidity model. The simplest of these models is probably the one introduced by Hébraud and Lequeux (Hebraud and Lequeux, 1998). In this model, one deals with an ensemble of sites that can each sustain a stress σ. The central quantity is the probability distribution function (pdf) $P(\sigma, t)$ of the local stress, which is assumed to evolve according to the equation

$$\begin{aligned}\partial_t P(\sigma,t) = &-G_0\dot{\gamma}\partial_\sigma P(\sigma,t) - \tfrac{1}{\tau}H(|\sigma|-\sigma_c)P(\sigma,t)\\ &+\tfrac{1}{\tau}\delta(\sigma)\int_{|\sigma'|>\sigma_c} P(\sigma',t)d\sigma' + D\partial^2_{\sigma^2}P(\sigma,t),\end{aligned} \tag{8.6}$$

where H is the Heaviside step function, and the "stress diffusion constant" D is given self-consistently by

$$D = \frac{\alpha}{\tau}\int_{|\sigma'|>\sigma_c} P(\sigma',t)d\sigma'. \tag{8.7}$$

Equation (8.6) is a simple evolution equation for the pdf: the first term corresponds to elastic loading at constant strain rate, with an elastic modulus G_0. The second and third terms correspond to a description of plastic events that take place with a rate $1/\tau$ for sites that exceed the critical yield stress σ_c; according to the third term, each plastic event corresponds to a complete release of the stress, which is set equal to zero. The last term describes a "diffusion" along the stress scale, that is the result of the average activity (stress redistribution after loading or unloading) of all other sites. This could also be described as a "stress noise", and the intensity of this noise is taken, according to Eq. (8.7), to be proportional to the total plastic activity present in the third term of Eq. (8.6). The coupling parameter α in Eq. (8.7) is the control parameter of the model. It could be interpreted as corresponding to the intensity of the elastic coupling between sites. For small values of α ($\alpha < \alpha_c$) the system is jammed, with a vanishing activity $D = 0$ in the absence of strain, and multiple solutions for $P(\sigma)$. In this jammed situation, the model exhibits a non-zero yield stress (the limit of $\langle\sigma\rangle$ when $\dot{\gamma}$ goes to zero is non-zero) and a complex rheological behavior. Despite its apparent simplicity, this model illustrates how the introduction of couplings between simple elastoplastic elements, even when treated at the mean-field level, can give rise to a complex collective behavior. Such models can serve as a basis for more complex, non-local fluidity models, as discussed below.

Soft glassy rheology. The first, very successful example of a model using the dynamics of a distribution function is probably the "soft glassy rheology" approach of Sollich and coworkers (Sollich *et al.*, 1997; Sollich, 1998). A number of reviews of this approach, which has been very successfully applied to many features of the rheology of soft glasses, including complex strain histories, are available (Cates, 2002). We will

[1]Simpler rate and state models involving a single "fluidity" internal parameter have also been proposed in Refs. (Picard *et al.*, 2002; Coussot *et al.*, 2002).

therefore limit ourselves to a very brief description of the assumptions underlying this model. The system is a collection of independent elastoplastic elements, each of which is trapped in an energy minimum of depth $E < 0$ (relative to some zero-energy level). Each element is also assigned a strain ℓ that varies with time as the global strain γ, and the energy barrier changes with strain as $E \to E(\ell) = E + k\ell^2/2$, where k is a local modulus. The escape from a potential well (corresponding to the local yield of an element) is governed by an Arrhenius-like factor, $\tau_{\text{yield}} \sim \exp(-E(\ell)/X)$. The two key ingredients in the model are (i) a distribution of trap depths that implies, in the absence of external strain and for small values of X, that the system has a very broad distribution of relaxation times, and is effectively in a glassy state, with a "weak ergodicity breaking" described by the trap model of Bouchaud (Bouchaud, 1992) ; (ii) the "effective temperature parameter" X, which activates the dynamics of any given element, and is intended to represent the mechanical noise arising from the yield of all other elements in the sample. The introduction of X is a recognition that, in many systems, thermal motion alone is not enough to trigger local yield events. The system has to cross energy barriers that are very large compared to typical thermal energies. However, the model suffers from the fact that this parameter is not determined self-consistently, and therefore should be considered as adjustable. The Hébraud–Lequeux equations described above can be considered as a first, simplified attempt to obtain self-consistency within this type of framework.

8.2.6 Classical rate-and-state formulations

The models we have described so far rely on the notion of zone flips in a weak sense: they assume macroscopic plastic behavior to be the net effect of sudden local rearrangements that can occur anywhere within the material depending primarily on the local or macroscopic stress level; but their focus is on the processes—the flips—but not on their loci—the zones. Underlying all of these theories is of course the notion that some regions of space present specific traits—low density? modulus?—which facilitate flips. Rate-and-state formulations attempt to incorporate dynamical equations governing the density of the "flow defects" (Spaepen, 1977) or "shear transformation zones" (Falk and Langer, 1998), which are supposed to control plastic activity.

Free-volume. The introduction of free-volume dynamics in rate-and-state formulation of plasticity was pioneered by Spaepen as early as the 1970s. It borrows from Cohen and Turnbull's free-volume theory (Turnbull and Cohen, 1970), which was proposed to explain departures from Arrhenius behavior by introducing a variable that would "replace" temperature in activation factors. A material is seen as decomposed into many subsystems of local free volume v_i. These variables are then assumed to be exponentially distributed (with little justification) and relaxation events are assumed to occur in regions of high free volume $v_i > v_0$, with density $\propto e^{-v_0/v_{\text{f}}}$ (where $v_{\text{f}} = \langle v_i \rangle$ is the average free volume). Spaepen then argues, that the strain rate must then take the form (Spaepen, 1977; Heggen *et al.*, 2005):

$$\dot{\gamma} \propto \Delta\epsilon_0\, \omega_0\, e^{-v_0/v_f}\, \sinh\left(\frac{\Delta\epsilon_0 v_0 \sigma}{2kT}\right). \tag{8.8}$$

To couple the plastic response with changes of density, he proposes to write an evolution equation for the concentration $c_f = e^{-v_0/v_f}$ of high free-volume defects, in the form:

$$\dot{c}_f = -k_r c_f (c_f - c_{eq}), \tag{8.9}$$

with constants k_r and c_{eq}, which depend on temperature and stress. This equation captures the idea that the evolution of c_f is controlled by two antagonistic effets, (i) deformation introduces dilatancy at a rate supposedly proportional to $\dot{\gamma}$ and (ii) high free-volume defects occasionally collapse via a "bimolecular" process—whence the form c_f^2. The competition between these two effects determines an equilibrium value c_{eq} in steady state.

In this argument, the postulate that density fluctuation are exponentially distributed, or the precise form of free-volume creation and destruction terms should of course be questioned—but the same remark can be made regarding the creation/destruction term $\mathcal{X}$ in STZ theory (see further, Eq. (8.11)). A particular oddity in Spaepen's argument is the reference to a "bimolecular" process to explain the disappearance of free volume, as if high-density defects where actual objects, which were also mobile and likely to collapse whenever they meet. Different free-volume equations were proposed later (Lemaître, 2002; Lemaître and Carlson, 2004), with a different interpretation, in particular, of the destruction process. It is based on the simple remark that if v_f controls activation mechanisms and evolves in time, then its decay rate should be of the form e^{-v_1/v_f}, with v_1 an activation volume for the decay process that has no reason to be equal to v_0. The rate of dilatancy is, moreover, assumed to be proportional to $\sigma\dot{\gamma}$. Introducing the notation $\chi = v_0/v_f$, this leads to coupled equations of the form (Lemaître, 2002; Lemaître and Carlson, 2004; Lemaitre, 2006),

$$\dot{\gamma} = \Delta\epsilon_0\, \omega_0\, e^{-1/\chi}\, \sinh\left(\frac{\sigma}{\mu}\right)$$

$$\dot{\chi} = -A_0\, e^{-\kappa/\chi} + \sigma\,\dot{\gamma},$$

which can reproduce various forms of the transient and unsteady response of amorphous materials under shear.

The free-volume paradigm associates (via e^{-v_1/v_f} factors) large changes in relaxation time scales to minute changes of density. Precise analysis of experimental data for different conditions of density and pressure have shown, however, that is was unlikely that the relaxation time scales could be thus governed by the amount of free-volume (Alba-Simionesco *et al.*, 2002; Khonik *et al.*, 2008). The existence of free-volume changes in plastic flow is also in question, as there is experimental evidence of homogeneous flows in which plasticity does not produce significant free volume (Heggen *et al.*, 2005), while dilatancy effects have been observed during the formation of shear bands (Chen and Chuang, 1975; Donovan and Stobbs, 1981; Li *et al.*, 2002). This observation, of course, does not mean that dilatancy actually controls the formation of shear bands, as it may only be a passive marker. We finally note that if the free-volume paradigm is not supported by experimental data for metallic or molecular glasses, it still could be relevant to other systems such as granular materials or colloidals glasses.

Dynamics of STZ densities. The STZ model proposed by Falk and Langer (Falk and Langer, 1998), departs from prior works in its stricter interpretation of the concept of zones. They are seen as sufficiently well-defined objects that pre-exist the occurrence of flips, so that it actually makes sense to speak of their density, which becomes a dynamical state variable, somewhat analogous to a defect density in models of crystalline plasticity. Hence, a constitutive law for amorphous solids must include equations of motion for the density and internal state of these zones. Although recent papers have introduced tensorial formulations (Langer, 2004), the gist of STZ theory is captured by assuming that zones are two-level ($\pm$) systems, which can produce strain changes $\pm\Delta\epsilon_0$ only when they undergo $\pm \to \mp$ internal transitions. This leads to writing an equation à la Eyring (see Eq. (8.1)) for strain rate as:

$$\dot{\gamma} = \Delta\epsilon_0 \left(\mathcal{R}_+ n_+ - \mathcal{R}_- n_- \right), \tag{8.10}$$

where $n_\pm$ stand for the number densities of $\pm$ zones. The dynamical equations for the number densities $n_\pm$ then take the form:

$$\dot{n}_\pm = -\mathcal{R}_\pm n_\pm + \mathcal{R}_\mp n_\mp + \mathcal{X}, \tag{8.11}$$

where the last term accounts for creation/destruction mechanisms that are meant to describe how the mean flow affects zone populations—in practice, it is assumed to depend, e.g., on the macroscopic energy dissipation rate. The rate factors, of course, must then be expressed in terms of parameters such as stress, or density. In particular, a linear dependence in terms of stress will commonly be assumed.

Without going into any further details, it is already possible to see from Eq. (8.10) that there are two ways jamming can be explained within the context of STZ theory. Indeed, the strain rate vanishes when $\mathcal{R}_+ n_+ = \mathcal{R}_- n_-$, which may occur either as the result of a balance between $\pm$ flips (in which case $\mathcal{R}_+ n_+ \neq 0$), or because the effective rate of flip events $\mathcal{R}_+ n_+ = \mathcal{R}_- n_- = 0$.

Early STZ papers (Falk and Langer, 1998; Langer, 2001; Eastgate *et al.*, 2003; Langer and Pechenik, 2003) proposed that jamming occurred primarily via a "polarization" of the medium, the jammed state being when $\mathcal{R}_+ n_+ = \mathcal{R}_- n_- \neq 0$, which occurs for a specific value of the ratio $n_+/n_- = \mathcal{R}_-/\mathcal{R}_+$. The possibility to model jamming in this manner is a direct consequence of the representation of zones as two-level systems, which is also the particularity of STZ theory. Under various assumptions for the transition probabilities and for the creation/destruction term $\mathcal{X}$ in Eq. (8.11), the STZ equations were shown in (Falk and Langer, 1998) to present a steady plastic flow and a transition from jammed to flow at a limit external stress, in a way that is consistent with expectations.

Later works on STZ theory (Lemaître, 2002; Lemaître and Carlson, 2004; Falk *et al.*, 2004; Bouchbinder *et al.*, 2007; Bouchbinder and Langer, 2009*a*,*b*) introduce additional kinetic equations for a state variable χ, called free-volume in (Lemaître, 2002; Lemaître and Carlson, 2004) and "effective temperature" in (Falk *et al.*, 2004; Lemaitre, 2006; Bouchbinder *et al.*, 2007; Bouchbinder and Langer, 2009*a*,*b*), such that $\exp(-1/\chi)$ accounts for a density of "flow defects". The effective temperature is thought as a measure of the degree of disorder in the system, and its dynamics is supposed to result from the competition between plastic work—which drives the system more out-of-equilibrium—and the spontaneous relaxation of the system towards equilibrium. The resulting coupled equations were shown to be able to reproduce various aspects of the out-of-equilibrium response of glassy systems such as the Kovacs effect (Lemaitre, 2006; Bouchbinder and Langer, 2009*a*) or the formation of shear bands (Shi *et al.*, 2007; Manning *et al.*, 2007).

As mentioned above, an important originality of STZ theory—in particular compared to free volume theories that have a very similar mathematical structure—lies in the notion that zones have an internal structure (modelled by the $\pm$ states). The physical motivation behind this idea was very clear in early works (Falk and Langer, 1998; Langer, 2001; Eastgate *et al.*, 2003; Langer and Pechenik, 2003): it permits explanation of jamming by a balance between forward and backward rearrangements ($\mathcal{R}_+ n_+ = \mathcal{R}_- n_-$). This, however, implies that in a jammed system, the plastic activity $\mathcal{R}_+ n_+ + \mathcal{R}_- n_-$ remains non-zero. To solve this problem for athermal systems, Falk and Langer have introduced rates $\mathcal{R}_+$ and $\mathcal{R}_-$- that are exactly zero when the stress is applied opposite to the preferred direction of the STZ. In that case, jamming could come about from an exhaustion of plastic activity, as expected in the low-T range. The introduction of the effective temperature also smooths out the jamming mechanism with a crossover between the backward–forward balance and exhaustion, while the two-level aspect remains needed to account for some specific features such as the Bauschinger effect (Falk and Langer, 2010).

Despite its success in accounting for the macroscopic mechanical behavior of metallic glasses, questions remain about some basic assumptions of the STZ model. The very definition of STZs as "ephemeral, noise-activated, configurational fluctuations that happen to be susceptible to stress-driven shear transformation", as presented in a recent review that covers the subject extensively (Falk and Langer, 2010), makes a direct identification of these zones (e.g. using numerical simulations) quite difficult. The same can be said of the definition of the effective temperature, which, as its free-volume counterpart, does not appear to be directly observable (although it could be related to potential-energy density in the simulations of Refs. (Shi and Falk, 2005*a*; Falk and Langer, 2010)). Consequently, like in all the above-mentioned theories, assumptions have to be introduced concerning the form of activation rates, and how they depend on stress, temperature, or effective temperature. Progress and establishing more precise and microscopic definitions of zones or effective temperature, and understanding these assumptions from a microscopic standpoint, will certainly stimulate future work and may lead to convergence with some of the alternative viewpoints such as SGR or fluctuation–dissipation approaches (Haxton and Liu, 2007).

8.2.7 Non-local rheology approach

All the approaches discussed in the previous sections take for granted that some kind of microscopic local heterogeneity governs the stress–strain relationship in amorphous materials. As we will see below, this aspect has, in essence, been confirmed by particle-based simulations, although the precise characterization of the flow events is still the object of numerous studies. The next question is therefore the description of the spatial organization of these events, both in terms of large-scale fluctuations and of permanent strain localization. The study of large-scale fluctuations is a subject of current interest within particle-based or lattice models (Yamamoto and Onuki, 1998; Maloney and Lemaitre, 2004*b*; Tanguy *et al.*, 2006; Maloney and Lemaitre, 2006; Bailey *et al.*, 2007; Lerner and Procaccia, 2009; Lemaitre and Caroli, 2009), and some aspects will be discussed in Section 8.3. Here, we would like to discuss recent extensions of the mean-field theories described above, that attempt a description of flow heterogeneities based on deterministic partial differential equations. The common point in all the recently proposed approaches is to extend the mean-field rheological description by allowing for spatial variation of one of the parameters, based on a diffusive kernel (Manning *et al.*, 2007; Goyon *et al.*, 2008; Fielding *et al.*, 2009). For example, in (Manning *et al.*, 2007), the effective temperature parameter χ is assumed to obey a diffusion equation, with a diffusion constant that is proportional to the local rate of plastic deformation. In (Fielding *et al.*, 2009), the effective temperature parameter X of the SGR model obeys a relaxation–diffusion equation, with a source term that can be taken to be either proportional to the rate of local plastic activity or to the rate of energy dissipation. Finally, in (Goyon *et al.*, 2008), it is the fluidity—that is the local value of $\dot{\gamma}/\sigma$—that appears inside the diffusive term. All these models have been shown to reproduce various forms of localization, either aiming to model the formation of shear bands in metallic glasses (Manning *et al.*, 2007), or near-wall localization in microfluidics flows of colloidal suspensions (Goyon *et al.*, 2008), or even faults (Daub and Carlson, 2008).

This very important issue currently attracts considerable interest, but it seems from the diversity of the study, and their various claims to actually account for localization, that many forms of diffusive kernels would capture some of the underlying physics, irrespective of their actual microscopic assumptions. We must hence worry about the severity of tests provided by fitting such length scales. Clearly, it seems important to assess how sensitive the resulting behavior is to particular forms of the diffusive terms, in other words, to break down the frontiers between different groups where these various equations are developed, and put them alongside each other for close comparison.

At the microscopic level, the specific mechanisms that justify diffusive terms still have to be specified. At present, the only theoretical attempt to do this on the basis of a more microscopic approach can be found in Ref. (Bocquet *et al.*, 2009). This reference introduces a systematic coarse-graining approach of the lattice elastoplastic model of Ref. (Picard *et al.*, 2004), which eventually results in a coupling between a local rheology described by the Hébraud–Lequeux model with Eq. (8.7) being replaced by a non-local relation between stress diffusivity and the rate of plastic activity. In other words, the stress diffusivity—in stress-space—is caused by the plastic activity in the vicinity of the zone in question—in real space. The Laplacian form of this contribution can be easily understood, as a purely linear gradient in plastic activity leads to a local compensation between neighboring zones with higher and lower activities.

These approaches are particularly attractive and will be amenable to a direct comparison with numerical data obtained from elastoplastic lattice models or from particle-based simulations, and probably with experiments. Such comparisons will be, in a first stage, based on the predictions for strain localization and the characteristic length/times over which this phenomenon is predicted to take place. Further refinements, e.g. including tensorial aspects for the stress tensor, will be necessary to actually predict mechanical response under various perturbations.

8.2.8 Conclusion

What effective temperature? It is clear from the preceding discussion that rate-and-state formulations of plasticity should be seen as various forms of empiricism, rather than actual microscopic theories. This does not, however, diminish the fact that they do reproduce many features of experimental data. This is particularly striking in view of the simplicity of most of these formulations that do not involve complex, distribution-like state variables—like Hébraud–Lequeux or SGR. And of course, rate-and-state equations are potentially quite important as they can easily be incorporated in Navier–Stokes equations, and used to study flows with complex geometries.

It is striking that, taken as an empirical attempt to replace temperature in activation factors by another arbitrary quantity, the introduction of free-volume parallels—and anticipates (Spaepen, 1977)—the recourse to the notion of "effective temperature" (Cugliandolo *et al.*, 1997; Berthier *et al.*, 2000; Barrat and Berthier, 2001; Berthier and Barrat, 2002; Ono *et al.*, 2002; Haxton and Liu, 2007) which seems more acceptable in today's language. This emphasizes the necessity to understand low-temperature activation mechanisms. Clarifying whether they involve stress noise, local fluctuations of pressure, density—like the free-volume theories would argue—or even moduli, however, remain open issues.

Free-volume models have the merit to make a specific, testable (Spaepen, 1977; Heggen *et al.*, 2005), assumption about the relation of activation factors with density. Sollich argues (Sollich *et al.*, 1997; Sollich, 1998), that the effective temperature should represent the noise produced by ongoing rearrangements. Yet, as long as the link between "noise temperature" and defect densities is not established, it seems that these two different families of theories introduce the concept to take into account quite different forms of disorder. In fact, both mechanical noise and local fluctuations in moduli are present (Yoshimoto *et al.*, 2004), and both are related to the underlying distribution of local barrier height that, as we have discussed previously is determined by a dynamic interplay between elastic loading and plastic activity itself. In this regard, it is significant that the STZ and free-volume models require the introduction of dynamical equations for their effective temperature.

What zones? The notion of a shear transformation is inevitably rather fuzzy as, in practice, rearrangements are hard to identify and isolate in space or time. However, it offers a useful framework for rationalizing observations, and articulating theories.

Phenomenologies of plasticity based on this premise must still specify a number of assumptions about flips, mainly in order to answer two questions: what triggers a flip? how much does a flip contribute to stress relaxation and energy dissipation?

These questions, at the core of the construction of theories, have motivated a number of numerical and experimental studies, and some results will be discussed in Section 8.3. The aim is to attempt to check the basic assumptions contained in the theoretical description, and to identify the parameters that should be incorporated into theories. This is of course difficult because zones are not easily identified within the disordered structure, and because material properties, such as elastic moduli strongly fluctuate. Basically, it can be said that a consensus is now established on the fact that elementary flips can be identified at low temperature, and produce a stress release characterized by a quadrupolar symmetry in two dimensions. Many questions, regarding in particular the conditions that trigger a flip, or the distribution of stress drops, remain to be explored accurately. The situation in three dimensions is even more complex, and even the form of the stress field released after one flip has not been studied in any detail. The interaction between flips, the way a local rearrangement changes the probability for a subsequent flip to take place nearby, and the associated time scales, are also subjects of current interest that should be clarified and used as inputs to a theoretical description.

8.3 Particle-based simulations

8.3.1 Introduction

Particle-based simulations, and in particular molecular dynamics (MD) simulations of systems under shear have a long history, as they have been used very early to determine the viscosity of simple fluids. A number of tools have been developed in this context that allow one to integrate the equations of motion for an ensemble of several thousands of interacting particles under conditions of constant (or possibly oscillatory) strain rate, or sometimes to conditions of constant stress. Possible local heating associated with the shear is taken care of using thermostats that do not perturb the shear flow (except possibly at very high shear rates that will not be discussed in this chapter). Another "trick" that allows one to mimic shear in large systems is the use of Lees–Edwards boundary conditions, in which the usual square (cubic) box periodicity is replaced by that of a tilted Bravais lattice, with a tilt angle corresponding to a deformation that increases linearly with time. As a result, the system undergoes a shear deformation driven by the boundaries of the simulation cell. In some cases these Lees–Edwards boundary conditions will be supplemented by a procedure that implements a homogeneous local deformation in parallel with the boundary-driven deformation. In quasi-static deformations (see below), this is achieved by rescaling the particle coordinates at each strain step $\delta\gamma$, in an affine manner, e.g. $X' = X + Y \times \delta\gamma$. In MD simulations an algorithm called the SLLOD algorithm (Allen and Tildesley, 1989) is used to modify in a similar way the evolution of the velocity along the shear direction x, $dv_x/dt = F_x/m + \dot{\gamma} v_y$, so that a linear velocity profile is immediately obtained after application of shear.

In the context of glassy systems, the shear rate is of course an essential quantity that provides natural time scales in systems in which the relaxation time is essentially infinite. Particle-based simulations are rather limited in terms of the time scale, with simulation times that are at best of the order of 10^6 microscopic vibrational periods. If within times of this order one wants to simulate total strains γ of order unity or higher, the resulting shear rates may seem very high. For a metallic system, they would be of the order of $10^8\,\mathrm{s}^{-1}$. Still, the important feature is that a reasonable scale separation between the microscopic, vibrational time scale and the time scale of the deformation is achieved. A common assumption, confirmed to some extent by results described below, is that this time scale separation is sufficient to make such simulations representative of the qualitative behavior in deformed glassy systems, in spite of the very high rates used.

An alternative to these finite shear rate calculations is to use zero temperature, quasi-static simulations. In these simulations the system is always in a local energy minimum. After each elementary deformation step, which is described by an affine transformation of the particle coordinates, the energy is minimized to the "nearest" minimum using a conjugate gradient or steepest-descent algorithm. Note that this "nearest minimum" may, strictly speaking, depend on the minimization algorithm. Here, the notion of time step and duration of the simulation is totally absent, and is replaced by that of an elementary strain step and of total deformation at the end of the simulation. The elementary deformation step is limited by the need to avoid an artificial "tunneling" through energy barriers during the elementary deformation. While this cannot be, strictly speaking, avoided, experience shows that elementary step strains of the order of 10^{-5} are small enough to limit such problems and to produce reproducible trajectories. As each step strain requires a careful minimization, achieving large deformations with the quasi-static method is computationally costly. However, the method has the advantage of providing a well-defined, "zero shear rate, zero temperature" limit for the flow behavior of the system.

A critical issue in the study of glassy systems using simulations is that of equilibration. The properties of a glass are known to depend strongly on the preparation route and on the quenching rate, which is unphysically high in simulations. The mechanical properties under small deformations are expected to be affected by the preparation history; however, if a large and homogeneous deformation can be achieved, the memory of the initial configuration will be erased, and the results will not depend on the preparation route. Note that the hypothesis of a deformation that remains homogeneous at large scales is important here, and that different preparation routes may lead to samples that display a stronger tendency to strain localization, without ever undergoing the homogeneous deformation that would erase its memory. Such effects were described in a model glass with partial quasicrystalline order in Ref. (Shi and Falk, 2006).

Many different systems have been explored under shear using either quasi-static or finite shear rate simulations, periodic (Lees Edwards) cells or driving by external walls. The studies published in the literature vary considerably in their choices for interaction potentials and also in the type of quantities that are characterized in the flowing systems. Broadly speaking, one may distinguish three types of interaction potentials: mixtures of particles interacting by Lennard-Jones-type potentials, that are generally used with the intention of modeling either metallic glasses (Srolovitz *et al.*, 1981), or colloidal suspensions (Stevens and Robbins, 1993). Interactions with strong directional bonding are suitable for systems such as amorphous silicon or silica (Argon and Demkowicz, 2006), and contact interactions are used to model granular systems (Combe and Roux, 2000) or foams (Durian, 1995; Langer and Liu, 1997). Polymers have also been extensively simulated (Argon *et al.*, 1995), in general using Lennard-Jones-type interactions with extra intramolecular bonding that gives rise to strain hardening at large strain under traction.

Particle-based simulations offer the possibility to obtain, within the time- and length scales permitted by simulation, all microscopic information that can be obtained from particle coordinates. As often, the main difficulty is to find the appropriate tools to analyze the large amount of data that is available in order to extract the information relevant to the flow process and its heterogeneity. In the following, we describe some of the generic aspects that emerge from these studies, without attempting an exhaustive literature review. We will distinguish results obtained at relatively high T, where the thermal fluctuations are important compared to those associated with the deformation, and results for low temperature, where microscopic motion originates essentially from the external driving.

8.3.2 Finite-temperature MD

At finite temperatures, the local motion of particles is a complex superposition of thermal and of deformation induced movements. The identification of specific plastic events associated with the deformation is not possible. In practice, this situation is encountered if the temperature is close to the glass-transition temperature T_g, so that the systems at rest usually display significant aging. The appropriate tools for analyzing this regime are largely inherited from studies of the liquid–glass transition. A number of studies have focused on the global (macroscopic) stress–strain relation, and the occurrence of macroscopic, shear-banding instabilities. In spite of the short time scales explored by simulations, stress–strain curves exhibit a behavior that is remarkably similar to the one observed in experiments. The peak stress in the curves shown in Fig. 8.1 depends on shear rate and on the "age" of the system, i.e. on the preparation history (Varnik *et al.*, 2004). This dependence was rationalized by Röttler and Robbins (Rottler and Robbins, 2005) on the basis of "rate and state" ideas borrowed from solid friction. The peak stress varies as a logarithmic function of the strain rate and of the age of the system

$$\sigma_{\max} = \sigma_0 + s_0 \ln \theta + s_1 \ln \dot{\gamma}, \tag{8.12}$$

where θ is an effective age of the system, that depends on the waiting time after the quench and before the strain, and on the time spent under strain. This type of behavior is somewhat similar to what could be expected in a simple Eyring description, with the yield stress being associated with activated events of well-defined energy, which could be a function of the waiting time (Rottler and Robbins, 2005). However, the coefficients do not behave as expected in such a description, and in particular the coefficient s_1 tends to be constant rather than proportional to temperature, which implies that somehow a different description of activation has to be found.

MD simulations have developed in several directions: attempts to get a better understanding of activation mechanisms (Ilg and Barrat, 2007; Haxton and Liu, 2007), studies of strain localization at intermediate scales, and studies of correlation functions that would allow the detection of microscopic dynamical heterogeneities (Onuki and Yamamoto, 1998; Yamamoto and Kim, 2000). Studies of activation tend to support the notion of an effective temperature that would determine the rate at which the system overcomes local energy barrier, and that depends on both temperature and shear rate. Haxton and Liu (Haxton and Liu, 2007) claim that a data collapse for the flow curve can be achieved on this basis, with a single parameter $T_{\text{eff}}(T, \dot{\gamma})$. They further argue that this effective temperature is itself given by the fluctuation–dissipation ratio in the system, determined for various observables (Berthier and Barrat, 2002), (O'Hern *et al.*, 2002). A slightly different approach was chosen in Ref. (Ilg and Barrat, 2007), in which an artificial "reaction coordinate" of a bistable system was coupled to a system under shear, and an Arrhenius behavior with a shear-rate-dependent effective temperature was reported. While this points to the existence of a mechanical noise that would complement—or even replace—the thermal noise, no progress has been made that would allow one to relate the intensity of the noise to the shear rate. A clear justification of the SGR "effective temperature" approach from microscopic simulations is therefore still missing.

Persistent strain localization in the form of "shear bands" has been observed in a relatively small number of MD simulation studies, in 2 as well as 3 dimensions. In general, such a persistent localization is observed under strain conditions that are not fully periodic (see, however, ref. (Shi *et al.*, 2007), where Lees–Edwards, fully periodic, boundary conditions are used), but rather induced by boundaries (either in pure shear or uniaxial conditions). In (Varnik *et al.*, 2003), this localization was observed at the walls of the simulation cell in isothermal simulations. Other examples of strain localization are obtained under uniaxial loading with free boundaries (Shi and Falk, 2005*a*, 2006), multiaxial loading (Bailey *et al.*, 2004), or nanoindentation (Shi and Falk, 2005*b*) sometimes using a notch as initiator. These studies have not resulted in a clear understanding of the microscopic behavior of shear bands, as mentioned in (Bailey *et al.*, 2004), "there is much that is not understood about shear bands. This not only includes why they form in the first place, but also what determines their width, what distinguishes them structurally from the surrounding material . . . ". Simulation studies seem to indicate that some kind of initial heterogeneity (e.g. boundaries) is important in the formation mechanism. Structural differences concerning the local bond-order environment or local potential energy inside the shear band have been reported in some systems (Shi and Falk, 2005*a*; Li and Li, 2006), but they are not observed in all simulations (Varnik *et al.*, 2003). In the system studied by Shi and Falk (Shi and Falk, 2005*a*), the quench procedure was also shown to modify the ability to form shear bands. In rapidly quenched samples higher strain rates lead to increased localization, while the more gradually quenched samples exhibit the opposite strain-rate dependence.

Finally, we mention that indications of large-scale dynamical heterogeneities have been found in early simulation work by Yamamoto and Onuki (Yamamoto and Onuki,

1998) on sheared systems of soft spheres. These authors studied the spatial correlations between bond-breakage events defined over a small, fixed time window. Indications of critical behavior in the quasi-static limit are found in this work, which can be seen as a precursor of more detailed studies of 4-point correlations. Such correlations are studied in some more detail in (Furukawa *et al.*, 2009) which shows for example anisotropy of dynamical heterogeneity in the non-Newtonian regime. There is certainly, however, much room for a detailed study of dynamical heterogeneities in sheared systems at finite temperature as a function of temperature and strain rate.

8.3.3 The zero-temperature solid

Molecular dynamics methods can thus mimic various features of material response that are commonly seen in experiments. But they are of course limited by the time scales they can access, which are orders of magnitude smaller than physical ones in numerical models of metallic glasses—the situation is not as dramatic if we seek to model, e.g., colloidal glasses. Moreover, although simulations permit access to detailed information about the structure and dynamics, the microscopic motions are very blurred at finite temperature, so that it turns out to be quite difficult in practice to extract relevant information.

This motivated interest for low-temperature systems: at finite but low temperature—lower than T_g—a glass would typically spend most of the time vibrating around a single local minimum in the potential-energy landscape. One thus expects that this vibrational component makes a trivial contribution, while the most interesting aspects of dissipation in material response are controlled by hops between local minima—which may in some cases be thermally activated. Therefore, following many studies on glassy relaxation (Stillinger, 1995; Debenedetti and Stillinger, 2001; Doliwa and Heuer, 2003*a*,*b*,*c*), it makes sense to try to separate vibrations from hopping in the PEL.

A lot of information about the mechanical response is thus captured by focusing on the $T = 0$ limit. In the absence of an external drive, a material would just relax to a local minimum and rest. This limit, however, is very useful when studying the response to macroscopic deformation, as is helps focus on deformation-induced changes of local minima (Malandro and Lacks, 1997, 1999).

Deformation-induced changes in the PEL. Let us thus consider a numerical model of a glass at rest in a local minimum. This configuration is produced by energy minimization, starting from a configuration that may have had any thermomechanical history (slow annealing from a high-temperature state, or any form of plastic deformation).

We consider here the case when this minimum is stable and the applied deformation small enough to preserve stability. At zero temperature, the positions of the particles in the local minimum—that is as they adapt to the imposed deformation to preserve mechanical equilibrium—are smooth functions $\{\vec{r}_i(\gamma)\}$ of strain γ. In particular, these strain-induced changes are *exactly* reversible; the system behaves as a perfectly elastic material.

Studies of elasticity in this regime date back to the works of Born and Huang (Huang, 1950; Born and Huang, 1954), who proposed to compute elastic constants by assuming that the relative displacements of all particles are affine, i.e. match the macroscopic strain. With this assumption, it is immediately possible to predict elastic moduli from the pair correlation function. The problem is that the particle trajectories $\{\vec{r}_i(\gamma)\}$, which trace strain-induced changes of a local minimum, are not simply dictated by macroscopic deformation. This approximation is only valid for very simple crystals, but not for glasses, and the displacement $\{\vec{r}_i(\gamma)\}$ contains, in general, some non-affine contribution.

The existence of a non-affine component to the displacement field has been known for a long time, and its existence was hinted at in various papers as a trace of disorder (Alexander, 1998). But it was really brought to the fore when it became clear (Leonforte *et al.*, 2005) that it affects significantly the values of the macroscopic elastic moduli. The existence of a non-affine field reflects the disorder in the local elastic moduli, which results in a heterogeneous elastic response exhibiting long-ranged correlations (DiDonna and Lubensky, 2005; Maloney, 2006). This heterogeneity has been characterized in detail using a systematic coarse-graining approach (Tsamados *et al.*, 2009), with the results that below a scale of a few tens of particle sizes the elastic constants (and especially the shear moduli) differ significantly from their macroscopic values.

Analytical expressions are available (Maloney and Lemaitre, 2004*a*; Lemaître and Maloney, 2006) for the non-affine field and for the corrected moduli. They correspond to the zero-temperature limit of those proposed by Lutsko at finite temperature (Lutsko, 1988). The important point is that the $T = 0$ non-affine field takes the form:

$$\frac{d\vec{r}_i}{d\gamma} = -\mathcal{H}_{ij}^{-1}.\vec{\Xi}_j, \tag{8.13}$$

where $\mathcal{H}$ is the Hessian matrix, and $\vec{\Xi}$ a vector field corresponding to infinitesimal, strain-induced changes of forces on each particle. As $\vec{\Xi}$ can be constructed from the derivatives of the potential function, it is easy to show that it does not vanish or present any singular behavior. Consequently, the non-affine displacements will acquire any singular contribution only from the inversion of the Hessian matrix. Since glasses present many low-lying modes, the non-affine field will pick up information about the existence of soft regions in space (Papakonstantopoulos *et al.*, 2008; Mayr, 2009; Tsamados *et al.*, 2009). These soft regions are precisely those that control plasticity, when a material is driven by strain towards instabilities (Lemaitre and Caroli, 2007).

Although it is, in principle, outside the scope of this review, we briefly mention the special case of granular systems interacting through repulsive contact forces. At zero temperature, these systems lose their rigidity abruptly as the density is decreased, which defines the (un)jamming transition. In the vicinity of the jamming density, the non-affine, heterogeneous response becomes the dominant feature in the elastic deformation of such systems (van Hecke, 2010), and can be associated with a diverging "isostatic" length scale, below which the stability is governed by boundary conditions (Wyart, 2005).

Plastic events in AQS shear. The AQS (athermal quasi-static) protocol consists in applying quasi-static deformation to the zero-temperature solid described previously.

It is implemented by a two-step protocol (Malandro and Lacks, 1997, 1999; Maloney and Lemaitre, 2004*b*, 2006). Starting from an equilibrium configuration: (i) the system is deformed homogeneously by a very small increment; (ii) energy is minimized. If the increments are sufficiently small, the system will be able to track continuously the deformation-induced changes of the occupied minimum. As the procedure is iterated, the local minimum will eventually become unstable. Minimization will then let the system relax to a new configuration that is disconnected from the previous one. On a flow curve, as illustrated in Fig. 8.1, the AQS response shows up as a series of continuous branches—corresponding to the reversible tracking of single minima—and discontinuous jumps—corresponding to "plastic events".

Given that the elastic response is perfectly reversible, the plastic events account exactly for all the dissipation, and they can be identified unambiguously from the discontinuities of the stress curve. This has made possible detailed studies of their organization in space and of their size distribution (Maloney and Lemaitre, 2004*b*; Tanguy *et al.*, 2006; Maloney and Lemaitre, 2006; Bailey *et al.*, 2007; Lerner and Procaccia, 2009).

In some cases, single shear transformations can be observed. This can be done by looking at events of small sizes, which are present in the steady-state flow (Maloney and Lemaitre, 2004*b*, 2006), but more easily found during the early loading phase from an annealed, isotropic state (Tanguy *et al.*, 2006). Another way to access them is to look at the onset of plastic events, which were found to involve a single eigenvalue going to zero (Malandro and Lacks, 1997, 1999; Doye and Wales, 2002). Close to instability, the non-affine field (see Eq. (8.13)) aligns with the vanishing mode and the singular behavior of energy, stresses and moduli can even be predicted (Maloney and Lemaitre, 2004*a*). This has allowed one to study the spatial structure of this mode (Maloney and Lemaitre, 2006; Tanguy *et al.*, 2006), showing that it generically presents the quadrupolar structure and a decay away from its center, both of which are predicted by the Eshelby inclusion model (Eshelby, 1957; Picard *et al.*, 2004).

But in steady flow—that is past some preparation-dependent initial transient (Tanguy *et al.*, 2006)—plastic events are not in general composed of single zone flips but typically involve many of them collectively organized as avalanches (Maloney and Lemaitre, 2004*b*, 2006; Bailey *et al.*, 2007; Lerner and Procaccia, 2009). This last claim is supported, in particular, by measurements of the average size of stress (respectively energy) drops, which scale as L^β (respectively L^α) with system size L, and $\alpha, \beta < 1$. Despite some differences in the reported exponents, this power-law scaling is now accepted as a fact, and proves that the local rearrangements are strongly correlated, in notable contradiction with the mean-field assumption.

Avalanches at finite strain rates. An immediate question is whether avalanches are only a feature of the quasi-static limit, or whether they exist for realistic values of the external parameters. In particular, the unfolding of each avalanche should take some time, determined by the duration of elementary flips and by the propagation times of elastic signals between flip events (Lemaitre and Caroli, 2009). Even if we stick with athermal systems, as soon as a finite strain rate is introduced, the avalanches may start to overlap because of their finite duration. Hence, they can no longer be properly identified as in the quasi-static limit.

So, a first problem when going away from the AQS limit is to define valid observables that allow one to characterize the existence of an underlying avalanche process, i.e. of correlations between local rearrangement events. A protocol has thus been designed to characterize avalanches in 2D athermal systems, from measurements of transverse diffusion (Lemaitre and Caroli, 2009). The principle starts from the observations that in AQS simulations, the transverse diffusion constant exhibits a strong size dependence (Lemaitre and Caroli, 2007; Maloney and Robbins, 2008). This dependence can then be attributed to the organization of Eshelby flips along roughly linear patterns (Lemaitre and Caroli, 2009), so that the diffusion constant can be expressed as a function of a typical avalanche size.

This observation has led to the proposal that the avalanche size should depend on the strain rate as $\ell \propto 1/\dot{\gamma}^{1/d}$ in dimension d (Lemaitre and Caroli, 2009). As $\dot{\gamma}$ decreases, the avalanche size should saturate at a length scale $\propto L$ below some critical strain rate $\gamma_c \propto 1/L^d$, as in any usual crossover. This suggestion is based on an interpretation of the avalanche size as being limited by the screening of Eshelby elastic signals by the background noise due to all the flips in the system.

These scalings are consistent with the rough scaling of the average stress drop as $\langle\sigma\rangle \propto 1/L$ found in (Maloney and Lemaitre, 2004*b*, 2006), but of course, these estimates remain rough, and it is not ruled out that more precise measurements would provide slightly different exponents consistent with the values of β found in (Lerner and Procaccia, 2009). What should remain, however, is that avalanches are present at all physically accessible strain rates, even when they overlap in time, and even when they cannot be accessed via the identification of separated events.

Steps towards a phenomenology of plasticity. The observation of avalanches, and their properties in the quasi-static limit and at finite strain rates provides clear benchmarks for future theories of plasticity. Yet, as usual in studies of amorphous systems, we can observe and characterize the avalanche process; we can conclude that some form of correlation exist between flip events; but the underlying mechanisms that promote these correlations remain quite difficult to identify. In fact, several processes must occur together to make the avalanche behavior possible.

First, the flip–flip interaction must be mediated in some way. Here, the medium is elastic and this is known to produce long-ranged effects. The Eshelby mechanism (Eshelby, 1957; Argon, 1979; Bulatov and Argon, 1994*b*,*c*,*d*) must play a central role, namely, each flip alters the stress in its surroundings, which may push a nearby region past its instability threshold, and hence trigger a secondary instability. The existence of this mechanism is supported by the observation of single flips in AQS simulations (Maloney and Lemaitre, 2004*a*,*b*; Tanguy *et al.*, 2006), by the measurement of the stress decay in space (Maloney and Lemaitre, 2006), and by direct visualization of flip events at finite strain rates (Lemaitre and Caroli, 2009). There are also now direct observations that a primary zone flip can push a nearby one closer to instability (Lemaitre and Caroli, 2007, 2009), thus showing that the Eshelby mechanism is fully at work.

But we must note also that when a system is sheared from carefully annealed, isotropic, state zone flips tend to be isolated instead of organizing as avalanches (Tanguy *et al.*, 2006). There is no reason why there should be any fundamental difference between the basic mechanisms that are at work during the loading phase or in steady state. Therefore, the difference between the early-stage response and the steady-state flow must indicate that the state of the material evolves under loading, in a way that increases the density of near-threshold, soft, zones, consistent with the idea that strain results in progressive advection of the zones towards their instability thresholds (Lemaitre and Caroli, 2007). In steady state, the density of near-threshold regions is high enough, and the Eshelby stress redistribution operates efficiently. In early loading, the density of near-threshold regions would be lower if the system is carefully annealed, so that isolated events can be more easily identified.

Like in a game of dominoes, the Eshelby mechanism makes the avalanche process possible. But it can occur only if the density of near-threshold regions is high enough, which must be a property of the material structure. The question thus re-emerges of how to characterize the regions or "zones" where elementary shear transformation may occur. Could we identify them *a priori*? What would be their density? Do they correlate with some property of the local structure—stress, density, moduli? Up to now, the particle-based studies that have addressed these questions have produced rather disappointing results. First, it appears that the regions in which the localized plastic events take place are not, in general, under particularly high stress. More precisely, the probability of observing a yield event in a region under high stress is indeed higher, but this is balanced by the fact that the number of such regions is small. A good correlation, on the other hand, has been established between yield events and regions with low values of elastic moduli (Mayr, 2009; Tsamados *et al.*, 2009). This suggests that the heterogeneity of local elastic constants should be taken into account in more coarse-grained models. Unfortunately, the local elasticity is already a rather complex property, and attempts to directly relate the probability of yielding to the local atomic structure have not been very successful, although a correlation with the shape of the Voronoi volume was observed in polymer glasses (Papakonstantopoulos *et al.*, 2008). In systems with strongly directional bonding such as amorphous silicon, a correlation could also be established between the density of bonding defects and the local plastic activity (Talati *et al.*, 2009). One must acknowledge that a link is still missing that would allow a better control of plastic properties directly from the design of the microscopic structure.

8.4 Perspectives

We close this review by emphasizing a few key issues that, in our opinion, have to be addressed in order to build a consistent theory of amorphous solids under strain, including the heterogeneous, fluctuating aspects. Most of these points have been discussed in detail in the previous sections.

The present consensus on the existence and importance of "zones" and "flips" as the essential building blocks of the plastic activity makes it strongly desirable to have a better understanding of these zones from the standpoint of the local microstructure.

Many more atomistic simulations, involving efficient sampling techniques allowing for longer simulation runs (Rodney and Schuh, 2009*b*), and using various types of interatomic potentials, will be needed to achieve such an understanding. It might even be the case that a predictive search for microstructural characteristics of flipping zones is illusory, and that the local plastic activity is a result from so many factors that it is essentially unpredictable.

Even if the flips are not associated with well-defined zones at the structural level, the essential features of the current models of elastoplastic behavior are in fact statistical in nature. It is therefore essential to develop tools that allow one to quantify in an unbiased, statistical manner the plastic activity, so that a comparison between models, numerical simulations and experiments is possible. Such a strategy has proven very successful in the field of glassy systems and supercooled liquids at rest, and should be extended to the case of low-temperature, driven amorphous systems. In particular, a statistical description of dynamical heterogeneities in strained systems, a quantification of avalanche distributions (in energy and size), and of the relevant correlation lengths, is still missing. The influence of temperature and strain rate on these quantities should also be a subject of interest.

In the introduction, we insisted on the similarities between "soft" systems probed by rheological experiments, and "hard" systems" such as metallic glasses. The similarities are useful and important in terms of theoretical modelling, still the practical applications are very different. In soft matter, the focus will be on steady state or low-frequency rheological behavior, for which a permanent avalanching regime can be established, and the memory of the initial state is wiped out after a few cycles. In contrast, hard materials undergo irreversible failure after a few per cent of strain, so that a permanent regime cannot be established, and the thermomechanical history of the initial state becomes of crucial importance. Adding the system history as an additional "variable" extends considerably the complexity of the problem, so that with the exceptions of a few studies (Utz *et al.*, 2000; Shi and Falk, 2005*a*; Rottler and Robbins, 2005) most simulation works have focused on steady-state properties. Therefore, the influence of thermomechanical history, and the relevance of the correlated avalanches mechanisms in term of material failure, remain outstanding questions in which simulations studies could be compared to experimental results and theoretical approaches (Falk *et al.*, 2004).

We also mentioned that "hard" and "soft" glasses may differ broadly in terms of the relevant "reduced" parameters, that is after mass, length, and energy scales are made dimensionless. Colloidal glasses, for example, involve low reduced temperatures and high reduced strain rates, while metallic glasses involve converse conditions. If recent numerical works have focused on athermal systems, it is because the absence of thermal fluctuations facilitates the observation of elementary mechanisms of deformation. Indeed, it takes only quite small temperatures—compared to T_g—to induce (high-frequency) fluctuations that are significantly larger than the changes associated with plastic deformation and energy dissipation (Hentschel *et al.*, 2010): this of course limits our capacities of investigation. Understanding whether and how the mechanisms identified at zero temperature carry over to finite-T systems will inevitably become one of the major themes in this field. As often in the field of amorphous systems, rather different approaches can be developed, depending on whether one starts from a "low-temperature" or from a "high-temperature" viewpoint. Examples could be elastoplastic models on the one hand, and the mode-coupling theory (which we mentioned only briefly) on the other. Trying to understand the range of applicability of such different approaches, and possibly to get a consistent (if not unified) view of their relevance for various experimental systems and conditions remains a real challenge.

Finally, we made in the introduction a distinction between the heterogeneities associated with statistical fluctuations of a globally homogeneous strain, and the "macroscopic", long-lived heterogeneities described as strain localization. The description of the latter situation has made some progress recently, with the realization that a mechanism involving the diffusion of some auxiliary state variable (fluidity, effective temperature, free volume) was in general needed to produce such heterogeneities. This auxiliary variable could even be related to some of the statistical aspects mentioned above. However, all these approaches are still oversimplified (scalar stress variable, absence of convection ...) so that direct comparison with realistic experimental geometry is difficult.

Acknowledgments

We acknowledge many useful discussions and collaborations with Lydéric Bocquet, Christiane Caroli, Peter Sollich, Anne Tanguy, Michel Tsamados, and critical comments on the manuscript by Micahel Falk and Jim Langer.

References

Alba-Simionesco, C, Kivelson, D, and Tarjus, G (2002). *J. Chem. Phys.*, **116**(12), 5033.

Alexander, S (1998). *Phys. Rep.*, **296**, 65.

Allen, M.A. and Tildesley, D.J. (1989). *Computer Simulation of Liquids.* Oxford University Press, Oxford.

Argon, A. S. (1979). *Acta Met*, **27**, 47.

Argon, A. S. (1982). *J. Phys. Chem. Solids*, **43**, 945.

Argon, A. S., Bulatov, V. V., Mott, P. H., and Suter, U. W. (1995). *J. Rheol.*, **39**, 377.

Argon, A. S. and Demkowicz, M. J. (2006). *Philo. Mag.*, **86**(25–26), 4153.

Argon, A. S. and Kuo, H. Y. (1979). *Mater. Sci. Eng.*, **39**, 101.

Bailey, N.P., Schiotz, J., and Jacobsen, K.W. (2004). *Phys. Rev. B*, **69**, 144205.

Bailey, N. P., Schiotz, J., Lemaitre, A., and Jacobsen, K. W. (2007). *Phys. Rev. Lett.*, **98**, 095501.

Baret, J. C., Vandembroucq, D., and Roux, S. (2002). *Phys. Rev. Lett.*, **89**, 195506.

Barrat, JL and Berthier, L (2001). *Phys. Rev. E*, **63**, 012503.

Berthier, L, Barrat, JL, and Kurchan, J (2000). *Phys. Rev. E*, **61**, 5464.

Berthier, L. and Barrat, J.-L. (2002). *Phys. Rev. Lett.*, **89**, 95702.
Bocquet, L., Colin, A., and Ajdari, A. (2009). *Phys. Rev. Lett.*, **103**, 036001.
Born, M. and Huang, K. (1954). *Dynamical Theory of Crystal Lattices.* Clarendon Press, Oxford.
Bouchaud, J.-P. (1992). *J. Physique I*, **2**, 1705.
Bouchbinder, E. and Langer, J. S. (2009*a*). *Phy. Rev. E*, **80**, 031132.
Bouchbinder, E. and Langer, J. S. (2009*b*). *Phys Rev. E*, **80**, 031133.
Bouchbinder, E., Langer, J. S., and Procaccia, I. (2007). *Phys. Rev. E*, **75**(3), 036107.
Brader, J. M., Cates, M. E., and Fuchs, M. (2008). *Phys. Rev. Lett.*, **101**, 138301.
Bulatov, V. V. and Argon, A. S. (1994*a*). *J. Non-Cryst. Solids*, **172**, 876.
Bulatov, V. V. and Argon, A. S. (1994*b*). *Model. Simul. Mater. Sci. Eng.*, **2**, 167.
Bulatov, V. V. and Argon, A. S. (1994*c*). *Model. Simul. Mater. Sci. Eng.*, **2**, 185.
Bulatov, V. V. and Argon, A. S. (1994*d*). *Model. Simul. Mater. Sci. Eng.*, **2**, 203.
Caroli, C. and Nozières, P. (1996). In *Physics of Sliding Friction* (ed. B. Persson and E. Tosatti), Volume 311, Series E: Applied Sciences, Dordrecht. NATO ASI Series: Kluwer Academic Publishers.
Cates, M.E. (2002). In *Nonequilibrium Dynamics and Slow Relaxations in Condensed Matter* (ed. J.-L. Barrat, M. Feigelman, J. Kurchan, and J. Dalibard), Proceedings of the 57th Les Houches Summer School. Springer, Berlin.
Chen, H.S. and Chuang, S.Y. (1975). *Appl. Phys. Lett.*, **27**, 317.
Combe, G. and Roux, J.-N. (2000). *Phys. Rev. Lett.*, **85**, 3628.
Coussot, P. (2005). *Rheometry of Pastes, Suspensions, and Granular Materials: Applications in Industry and Environment.* Wiley, Hoboken.
Coussot, P., Nguyen, Q.D.and Huynh, H.T., and Bonn, D. (2002). *Phys. Rev. Lett.*, **88**, 175501.
Cugliandolo, LF, Kurchan, J, and Peliti, L (1997). *Phys. Rev. E*, **55**, 3898.
Daub, E.G. and Carlson, J.M. (2008). *J. Geophys. Res.-Solid Earth*, **113**, B12309.
Debenedetti, P.G. and Stillinger, F.H. (2001). *Nature*, **400**, 259.
Demkowicz, M. J. and Argon, A. S. (2004). *Phys. Rev. Lett.*, **93**, 025505.
Demkowicz, M. J. and Argon, A. S. (2005). *Phys. Rev. B*, **72**, 245206.
DiDonna, BA and Lubensky, TC (2005). *Phys. Rev. E*, **72**, 066619.
Doliwa, B. and Heuer, A. (2003*a*). *Phys. Rev. E*, **67**, 031506.
Doliwa, B. and Heuer, A. (2003*b*). *Phys. Rev. E*, **67**, 030501(R).
Doliwa, B. and Heuer, A. (2003*c*). *Phys. Rev. Lett.*, **91**, 235501.
Donovan, PE and Stobbs, WM (1981). *Acta. Met.*, **29**, 1419.
Doye, J. P. K. and Wales, D. J. (2002). *J. Chem. Phys.*, **116**, 3777.
Durian, D. J. (1995). *Phys. Rev. Lett.*, **75**, 4780.
Eastgate, L. O., Langer, J. S., and Pechenik, L. (2003). *Phys. Rev. Lett.*, **90**, 45506.
Eshelby, J. D. (1957). *Proc. Roy. Soc. London A*, **241**, 376.
Falk, M.L. and Langer, J.S. (2010). *arXiv:1004.4684v1, submitted to Annual Reviews of Condensed Matter Physics.*
Falk, M. L., Langer, J.S., and Pechenik, L. (2004). *Phys. Rev. E*, **70**, 011507.
Falk, M. L. and Langer, J. S. (1998). *Phys. Rev. E*, **57**, 7192.
Fielding, S. M., Cates, M. E., and Sollich, P. (2009). *Soft Matter*, **5**, 2378.
Fuchs, M. and Cates, M. E. (2002). *Phys. Rev. Lett.*, **89**, 098301.

Furukawa, A., Kim, K., Saito, S., and Tanaka, H. (2009). *Phys. Rev. Lett.*, **102**, 016001.

Furukawa, A. and Tanaka, H. (2006). *Nature*, **443**, 434.

Goyon, J., Colin, A., Ovarlez, G., Ajdari, A., and Bocquet, L. (2008). *Nature*, **454**, 84.

Haxton, T. K. and Liu, A. J. (2007). *Phys. Rev. Lett.*, **99**, 195701.

Hebraud, P and Lequeux, F (1998). *Phys. Rev. Lett.*, **81**, 2934.

Heggen, M, Spaepen, F, and Feuerbacher, M (2005). *J. Appl. Phys.*, **97**, 033506.

Helfand, E and Fredrickson, GH (1989). *Phys. Rev. Lett.*, **62**, 2468.

Hentschel, H. G. E., Karmakar, S., Lerner, E., and Procaccia, I. (2010). *Phys. Rev. Lett.*, **104**(2), 025501.

Huang, Kun (1950). *Proc. Roy. Soc. London A*, **203**, 178.

Ilg, P. and Barrat, J. L. (2007). *Europhys. Lett.*, **79**, 26001.

Johnson, WL and Samwer, K (2005). *Phys. Rev. Lett.*, **95**, 195501.

Khonik, S. V., Kaverin, L. D., Kobelev, N. P., Nguyen, N. T. N., Lysenko, A. V., Yazvitsky, M. Y., and Khonik, V. A. (2008, August). *J. Non-Cryst. Solids*, **354**(32), 3896–3902.

Langer, J.S. (2004). *Phys. Rev. E*, **70**, 041502.

Langer, J. S. (2001). *Phys. Rev. E*, **6401**(1), 011504.

Langer, J. S. and Pechenik, L. (2003). *Phys. Rev. E*, **68**, 061507.

Langer, S. A. and Liu, A. J. (1997). *J. Phys. Chem. B*, **101**, 866.

Larson, RG (1999). *The Structure and Rheology of Complex Fluids.* Oxford University Press, Oxford.

Lemaître, A. (2002). *Phys. Rev. Lett.*, **89**, 195503.

Lemaitre, A (2006). In *Jamming, Yielding, and Irreversible Deformation in Condensed Matter* (ed. Miguel, M.C. and Rubi, J.M.), Volume 688, Lecture Notes in Physics, p. 129.

Lemaître, A and Carlson, J (2004). *Phys. Rev. E*, **69**, 061611.

Lemaitre, A. and Caroli, C. (2007). *Phys. Rev. E*, **76**, 036104.

Lemaitre, A. and Caroli, C. (2009). *Phys. Rev. Lett.*, **103**, 065501.

Lemaître, A. and Maloney, C.E. (2006). *J. Stat. Phys.*, **123**, 415.

Leonforte, F., Boissière, R., Tanguy, A., Wittmer, J. P., and Barrat, J.-L. (2005). *Phys. Rev. B*, **72**, 224206.

Lerner, E. and Procaccia, I. (2009). *Phys. Rev. E*, **79**, 066109.

Li, J, Spaepen, F, and Hufnagel, TC (2002). *Philos Mag. A*, **82**, 2623.

Li, QK and Li, M (2006). *Appl. Phys. Lett.*, **88**, 241903.

Lubliner, J (2008). *Plasticity Theory.* Dover Publications, New York.

Lutsko, J. F. (1988). *J. Appl. Phys.*, **64**, 1152–1154.

Malandro, Dennis L. and Lacks, Daniel J. (1997). *J. Chem. Phys.*, **107**, 5804.

Malandro, D. L. and Lacks, D. J. (1999). *J. Chem. Phys.*, **110**, 4593.

Maloney, C. and Lemaitre, A. (2004*a*). *Phys. Rev. Lett.*, **93**, 016001.

Maloney, C. and Lemaitre, A. (2004*b*). *Phys. Rev. Lett.*, **93**, 195501.

Maloney, C. E. (2006). *Phys. Rev. Lett.*, **97**, 035503.

Maloney, Craig E. and Lemaitre, A. (2006). *Phys. Rev. E*, **74**, 016118.

Maloney, C. E. and Robbins, M. O. (2008). *J. Phys. Condens. Matter*, **20**, 244128.

Manning, M. L., Langer, J. S., and Carlson, J. M. (2007). *Phys. Rev. E*, **76**, 056106.

Mayr, S. G. (2009, February). *Phys. Rev. B*, **79**(6), 060201.

Miguel, MC and Zapperi, S (2006). *Science*, **312**, 1151.

O'Hern, C. S., Langer, S. A., Liu, A. J., and Nagel, S. R. (2002). *Phys. Rev. Lett.*, **88**, 075507.

Ono, I.K., O'Hern, C.S., Durian, D.J., Langer, S.A., Liu, A.J., and Nagel, S.R. (2002). *Phys. Rev. Lett.*, **89**, 095703.

Onuki, A. and Yamamoto, R. (1998). *J. Non-Cryst. Solids*, **235**, 34.

Papakonstantopoulos, G.J., Riggleman, R.A., Barrat, J-L, and de Pablo, J. J. (2008). *Phys. Rev. E*, **77**, 041502.

Picard, G., Ajdari, A., Bocquet, L., and Lequeux, F. (2002). *Phys. Rev. E*, **66**, 051501.

Picard, G., Ajdari, A., Lequeux, F., and Bocquet, L. (2004). *Eur. Phys. J. E*, **15**, 371.

Picard, G., Ajdari, A., Lequeux, F., and Bocquet, L. (2005). *Phy. Rev. E*, **71**, 010501(R).

Rice, J.R. and Ruina, A.L. (1983). *J. Appl. Mech.*, **105**, 343.

Rodney, D. and Schuh, C.A. (2009*a*). *Phys. Rev. Lett.*, **102**, 235503.

Rodney, D. and Schuh, C.A. (2009*b*). *Phys. Rev. B*, **80**, 184203.

Rottler, J. and Robbins, M. O. (2005). *Phys. Rev. Lett.*, **95**, 225504.

Rouxel, T., Ji, H., Hammouda, T., and Moreac, A. (2008). *Phys. Rev. Lett.*, **100**, 225501.

Schall, P., Weitz, D. A., and Spaepen, F (2007). *Science*, **318**, 1895.

Schuh, Christopher A., Hufnagel, Todd C., and Ramamurty, Upadrasta (2007). *Acta Mater*, **55**, 4067.

Shi, Y. F. and Falk, M. L. (2005*a*). *Phys. Rev. Lett.*, **95**, 095502.

Shi, Y. F. and Falk, M. L. (2005*b*). *Appl. Phys. Lett.*, **86**, 011914.

Shi, Y. F. and Falk, M. L. (2006). *Phys. Rev. B*, **73**, 214201.

Shi, Y. F., Katz, M. B., Li, H., and Falk, M. L. (2007). *Phys. Rev. Lett.*, **98**(18), 185505.

Sollich, P (1998). *Phys. Rev. E*, **58**, 738.

Sollich, P, Lequeux, F, Hebraud, P, and Cates, ME (1997). *Phys. Rev. Lett.*, **78**, 2020.

Spaepen, F. (1977). *Acta Met.*, **25**, 407.

Srolovitz, D., Maeda, K., and Egami, T. (1981). *Philos. Mag. A*, **44**, 847.

Stevens, Mark J. and Robbins, Mark O. (1993). *Phys. Rev. E*, **48**, 3778.

Stillinger, F. H. (1995). *Science*, **267**, 1935.

Talati, M., Albaret, T., and Tanguy, A. (2009). *Europhys. Lett.*, **86**, 66005.

Tanguy, A., Leonforte, F., and Barrat, J. L. (2006). *Eur. Phys. J. E*, **20**, 355.

Tsamados, M., Tanguy, A., Goldenberg, C., and Barrat, J.-L. (2009). *Phys. Rev. E*, **80**, 026112.

Tsamados, M., Tanguy, A., Leonforte, F., and Barrat, J. L. (2008). *Eur. Phys. J. E*, **26**, 283.

Turnbull, D and Cohen, MH (1970). *J. Chem. Phys.*, **52**, 3038.

Utz, M., Debenedetti, P. G., and Stillinger, F. H. (2000). *Phys. Rev. Lett.*, **84**, 1471.

van Hecke, M. (2010). *J. Phys. Cond. Matt.*, **22**, 033101.

Varnik, F., Bocquet, L., and Barrat, J. L. (2004). *J. Chem. Phys.*, **120**, 2788.

Varnik, F., Bocquet, L., Barrat, J. L., and Berthier, L. (2003). *Phys. Rev. Lett.*, **90**, 095702.
Wyart, M (2005). *Ann. de Physique*, **30**, 1.
Yamamoto, R. and Kim, K. (2000). *J. Phys. IV*, **10**, 15.
Yamamoto, R and Onuki, A (1998). *Phys. Rev. Lett.*, **81**, 4915.
Yoshimoto, K, Jain, TS, Workum, KV, Nealey, PF, and de Pablo, JJ (2004). *Phys. Rev. Lett.*, **93**, 175501.

9

The jamming scenario—an introduction and outlook

Andrea J. LIU, Sidney R. NAGEL,
Wim van SAARLOOS and Matthieu WYART

Abstract

The jamming scenario of disordered media, formulated about 10 years ago, has in recent years been advanced by analyzing model systems of granular media. This has led to various new concepts that are increasingly being explored a variety of systems. This chapter contains an introductory review of these recent developments and provides an outlook on their applicability to different physical systems and on future directions. The first part of the chapter is devoted to an overview of the findings for model systems of frictionless spheres, focusing on the excess of low-frequency modes as the jamming point is approached. Particular attention is paid to a discussion of the crossover frequency and length scales that govern this approach. We then discuss the effects of particle asphericity and static friction, the applicability to bubble models for wet foams in which the friction is dynamic, the dynamical arrest in colloids, and the implications for molecular glasses.

9.1 Introduction

Just over ten years ago it was proposed (Liu and Nagel, 1998) to approach disordered condensed-matter systems in a more unified way than is usually done, by starting from the observation that many systems—not just molecular glasses, but also many soft-matter systems like granular media, colloids, pastes, emulsions, and foams—exhibit a stiff solid phase at high density, provided the temperature and the external forces or stresses are small enough. This proposal led to the idea of a *jamming phase diagram* for particulate systems, redrawn in Fig. 9.1.

The very idea that there would be a generic jamming phase diagram is less trivial than may appear at first sight, since granular media, emulsions and foams are in fact *athermal* systems—they are represented by points in the ground plane of Fig. 9.1, as

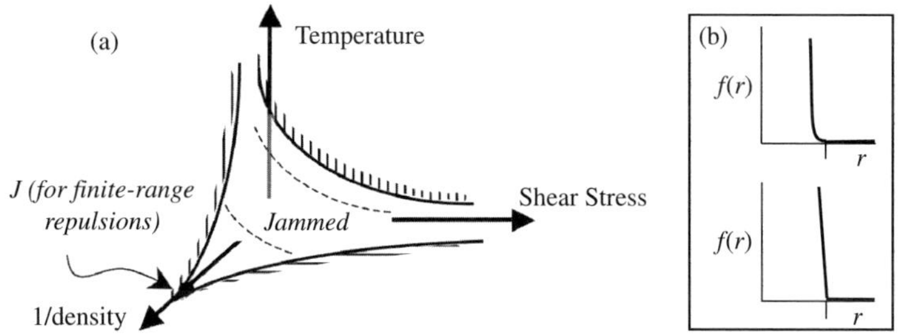

Fig. 9.1 (a) The jamming phase diagram; inside the shaded region, for high density, low temperature and small shear stress, particulate systems are *jammed*, i.e. form a disordered solid phase with a finite resistance to shear. The point J is the so-called "jamming point" for spherical particles with finite-range repulsive forces. It acts like a critical point and organizes the behavior in its neighborhood. (b) The type of forces considered in models of frictionless spheres: the force $f(r)$ is of finite range, i.e. vanishes when the distance r between the centers of the spheres exceeds some well-defined value. The lower case, in which the force increases linearly with the compression δ, is often referred to as one-sided harmonic springs. The Hertzian force of 3D elastic spheres increases as $\delta^{3/2}$ and has a behavior as in the upper figure.

their relevant interaction energies are orders of magnitude larger than the thermal energy scale $k_B T$. While thermodynamics generally guarantees the existence of a unique phase for molecular systems at equilibrium, athermal systems strictly speaking lack the averaging needed to be able to define a unique phase: their disordered jammed states are often history dependent, a phenomenon they share with glasses.

Nevertheless, it has become clear that it is extremely useful to approach these systems with a unifying common focus in mind. For particulate systems like granular media, the constituent particles essentially have strong repulsive interactions of *finite* range: particles that are in contact do interact and repel each other, those that are not in contact, don't. A detailed analysis of the simplest model of this type, that of N repulsive frictionless spheres (or discs in two dimensions) in the large-N limit, has revealed that the point at which they just jam into a rigid phase does indeed have properties reminiscent of an asymptotic or critical point that organizes the behavior in its neighborhood. This is the jamming point, marked J in Fig. 9.1. It is the aim of this chapter to provide an introductory review of these recent findings, and to provide an outlook on the relevance of our present understanding for other systems. We aim our review of jamming on those aspects that are of most interest from the point of view of the main theme of this book, glasses. Hence, we focus in particular on the various length scales that emerge near the jamming point, on the vibrational properties of the marginally jammed state above the jamming transition, and on the connection of these findings with the dynamics.

This approach bears close similarities with investigations of the rigidity of network glasses developed in the 1980s. Instead of density and shear stress, Phillips argued that the key control variable is the glass composition (Phillips, 1979). Based on a model of rigidity percolation proposed by Thorpe, where springs are randomly deposited on a lattice (He and Thorpe, 1985; Boolchand *et al.*, 2005), it was argued that glass properties are controlled by a critical point at zero temperature. The jamming transition can be viewed as a special case of the generic rigidity percolation problem in which the system self-organized to avoid large fluctuations in its structure. Thus, the jamming of particles is qualitatively different from springs randomly placed on a lattice. In some respects, the jamming transition is simpler, allowing the recent conceptual progress reported here.

Unfortunately, because of the focus of this book and space limitations, this chapter cannot do justice to the wonderful recent experimental developments: many new systems have become available or have been analyzed in new ways, and are increasingly driving new directions in the field of jamming. We refer for an overview of recent experiments on colloids to the chapter by Cipelletti and Weeks and for a review of experiments on grains and foams to the chapter by Dauchot, Durian and van Hecke. We will point to relevant experiments where appropriate, without going into details.

9.2 Overview of recent result on jamming of frictionless sphere packings

Close to the jamming point, packings of compressible frictionless spheres exhibit various anomalies including a strongly enhanced density of states at low frequencies. Of course, such packings are a very idealized model system for granular media. We postpone to Section 9.3 a discussion of generalizations of this model such as the role of particle anisotropy and friction.

In computer models of compressible spheres, one usually uses force laws of the type sketched in Fig. 9.1(b): two spheres of radii R_i and R_j experience zero force if the distance r_{ij} between their centers is larger than $R_i + R_j$, and have a force that rapidly increases with the overlap $\delta_{ij} = R_i + R_j - r_{ij}$ if $r_{ij} < R_i + R_j$. A particularly convenient and common choice for computer models is the one-sided harmonic spring model, for which the repulsive force f_{ij} for particles in contact increases linearly with the overlap δ_{ij}. Another choice that one often encounters in the literature for compression of elastic balls, is the Hertzian force law $f_{ij} \sim \delta_{ij}^{3/2}$. In studies of three-dimensional random packings, it usually suffices to take spheres whose radii are all the same, as one easily ends up with a random packing, but in two dimensions one needs to take a polydisperse or bidisperse distribution of discs in order to avoid crystallization. There are various ways to prepare static packings of spheres in mechanical equilibrium. One method is to place particles randomly in a box at a fixed density and quench the system to its closest potential-energy minimum via conjugate-gradient or steepest-descent algorithms (O'Hern *et al.*, 2003). Other methods employ a slow inflation of all the radii[1] to first generate a packing at the jamming threshold where the majority of the particles experience minute forces that balance on each particle. This jamming threshold marks the onset of a non-zero pressure and potential energy. After this initial

[1]This is somewhat like the Lubachevsky–Stillinger method to generate random close packings of hard spheres (Lubachevsky and Stillinger, 1990).

preparation the packing is compressed (or dilated) while continuously minimizing the energy (O'Hern *et al.*, 2003). Other methods slowly adjust the radii so as to steer the pressure in the packing to a prescribed value (Shundyak *et al.*, 2007). We refer to the literature for details.

We note that the onset packing fraction of the jamming transition for an ensemble of states depends on the ensemble. For systems equilibrated at infinite temperature and quenched to $T = 0$, the onset packing fraction corresponds to that of an ensemble in which each state, or local energy minimum, is weighted by the volume of its energy basin. In that case, $\phi_c \approx 0.64$ for monodisperse spheres in three dimensions (O'Hern *et al.*, 2003). However, if the ensemble is prepared by a quench from a system equilibrated at a low temperature, the onset packing fraction is higher (Berthier and Witten, 2009*a*; Chaudhuri *et al.*, 2010). It is clear from the distribution of jamming onsets that this must be the case (O'Hern *et al.*, 2003). In the infinite-system-size limit, the distribution approaches a delta function at $\phi \approx 0.64$ but with non-vanishing tails both on the high-density and low-density side (O'Hern *et al.*, 2003). In systems equilibrated at lower temperatures, lower energy states are weighted more heavily. Such states have higher values of ϕ_c, since the potential energy increases with $\phi - \phi_c$. Thus, the average value of ϕ_c for the ensemble must increase as the system is quenched from lower temperatures, in accordance with the results of Chaudhuri, *et al.* (Chaudhuri *et al.*, 2010), and the system can be described as having a line of jamming transitions extending upwards from $\phi \approx 0.64$, as in Ref. (Mari *et al.*, 2009). In all cases, however, the onset pressure is zero, so the onset is sharp in terms of pressure but not packing fraction, and the vibrational properties, etc. are identical to those described below.

9.2.1 The vibrational density of states of packings

An important concept in condensed-matter systems is the density of states, $D(\omega)$. $D(\omega)\mathrm{d}\omega$ is proportional to the number of states with frequency between ω and $\omega + \mathrm{d}\omega$. Here, ω refers to the frequency of the *vibrational normal modes* of the constituent particles. The density of states concept is also often used for electronic states, but here we focus on the vibrational states of condensed-matter systems. For a crystal, or in fact for any elastic medium, $D(\omega)$ increases at low frequencies as ω^{d-1}, where d is the dimension of space. This generic behavior arises from the fact that sound modes have a dispersion relation $\omega(k)$ that is linear in the wave number k, together with phase-space arguments for the number of modes with wave number between k and $k + \mathrm{d}k$. The Debye scaling law $D(\omega) \sim \omega^2$ for $d = 3$ underlies the ubiquitous T^3 low-temperature specific heat of three-dimensional solids. It serves as an important reference for identifying anomalous behavior—e.g., the well-known enhancement of the specific heat over the Debye law is an indication in glasses for an excess density of excitations at low frequencies.

To analyze the vibrational modes, one obtains the so-called dynamical matrix familiar from solid-state physics. This dynamical matrix is essentially the second derivative of the interparticle potential, and hence has only non-zero elements for particles that are in contact in the packing (for the one-sided harmonic forces these

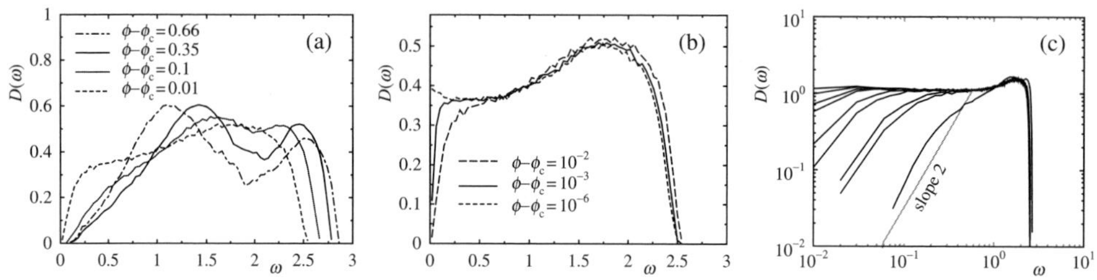

Fig. 9.2 The density of states of vibrational modes of three-dimensional soft-sphere packings with one-sided harmonic forces at varous densities: (a) significantly compressed samples; (b) close to the jamming density ϕ_c (O'Hern *et al.*, 2003); (c) on logarithmic scale (Silbert *et al.*, 2005).

terms are especially simple as the potential is quadratic in the separation). From the diagonalization of the dynamical matrix one then obtains the eigenmodes and their eigenfrequencies ω, and hence $D(\omega)$.

Figure 9.2 shows one of the early results for a three-dimensional packing with one-sided harmonic forces (O'Hern *et al.*, 2003; Silbert *et al.*, 2005). Panel (a) shows that sufficiently far above jamming, $D(\omega)$ vanishes at small frequencies, in qualitative agreement with the Debye scenario, but as the density ϕ is decreased towards the jamming density ϕ_c, the weight at small frequencies increases. Indeed, closer to the jamming density $D(\omega)$ develops a plateau at small frequencies, see panel (b). When plotted on a log-log scale, as in panel (c), the crossover and the emergence of a plateau is even clearer: in these data one observes the ω^2 Debye scaling only for the largest densities, while as the packings are decompressed towards the jamming density, the plateau extends to lower and lower frequencies. Clearly, the closer the packings are to the jamming threshold, the more $D(\omega)$ is enhanced at low frequencies and the larger are the deviations from the usual Debye behavior. We will discuss in Section 9.2.3 the crossover frequency ω^*, that separates the plateau from the lower-frequency downturn.

9.2.2 Isostaticity and marginally connected solids

How do the excess low-frequency modes arise? The answer is related to the fact that at the jamming threshold, packings of frictionless discs and spheres are *isostatic*, i.e. they are marginal solids that can just maintain their stability (Alexander, 1998; Moukarzel, 1998; Head *et al.*, 2001; Tkachenko and Witten, 2000; O'Hern *et al.*, 2003; Roux, 2000; Wyart, 2005). The origin of this is the following. Consider a disordered packing with N_c spheres or discs that have non-trivial contacts,[2] with Z their average contact number. Force balance on each particle implies that the vector sum of the forces adds up to zero; therefore the requirement of force balance on all particles with contacts gives

[2]There is a subtlety here: in a typical packing there is generally a small fraction of "rattlers" or "floaters", particles in a large enough cage of other particles that, in the absence of gravity, can float freely without any contact. These should be left out from the counting below, and from the determination of the average contact numer.

dN_c conditions. If we view the contact forces between the particles as the degrees of freedom that we have available to satisfy these requirements, then there are $ZN_c/2$ such force degrees of freedom. Clearly, assuming no special degeneracies, we arrive at the condition

$$\text{for a stable packing:} \quad Z \geq 2d \tag{9.1}$$

in order that force balance can be maintained.

There is a second constraint in the limit that one approaches the jamming point. As all forces are purely repulsive, in this limit almost all the individual contact forces must approach zero. For force laws like those sketched in Fig. 9.1(b), this mean that as one approaches the jamming point, all non-trivial $ZN_c/2$ contacts must obey the "just-touching" conditions $r_{ij} = R_i + R_j$. The number of degrees of freedom associated with the positions of the centers of the particles with contact is dN_c, so if we think of putting the particles with contacts in the right place to allow them to obey the just-touching conditions, we need to have

$$\text{just-touching conditions at jamming:} \quad Z \leq 2d. \tag{9.2}$$

Clearly, the value $Z_{\text{iso}} = 2d$, the isostatic value, has a special significance. Indeed, the above two conditions imply that as packings approach the jamming point, point J of Fig. 9.1(a), from the jammed side, e.g., by decompressing them, one will have

$$\text{upon approaching point J:} \quad Z \downarrow Z_{\text{iso}} = 2d. \tag{9.3}$$

Note that these results are independent of the presence of polydispersity and the details of the repulsive force law, provided it is of finite range and continuous.[3]

The above argument will be further justified in Section 9.2.4 for frictionless spheres. Figure 9.3(a) shows numerical simulation results for $\Delta Z = Z - Z_{\text{iso}}$, plotted on a log-log scale as a function of the distance from jamming, $\Delta\phi = \phi - \phi_c$, in both 2 and 3 dimensions. In both cases ΔZ goes to zero at jamming: $\Delta Z \sim \sqrt{\Delta\phi}$ (Durian, 1995; O'Hern *et al.*, 2002, 2003). A recent experiment aimed at testing this scaling, albeit in a system with friction, can be found in (Majmudar *et al.*, 2007). The isostaticity concept will be re-examinated in Section 9.3.1 for ellipses.

Note that according to Eq. (9.1) the packings at isostaticity have just enough contacts to maintain stability. In this sense, they are marginal packings, packings at the edge of stability. Moreover, if one imagines a packing with M fewer contacts than dictated by the isostaticity condition, one can according to Eq. (9.2) deform the packing in M different directions in the space of coordinates, while respecting the just-touching constraints (so that particles are not pressed into each other). These therefore correspond to M zero-energy deformation modes in which particles slide past

[3] In passing, we note that ϕ_c, the density of a packing of monodisperse spheres (all the same radii), approaches (O'Hern *et al.*, 2003) the random close packing density of hard spheres. There is an active line of research aimed at analyzing and relating the concepts of jamming and the maximally random jammed state, which has been proposed to replace the concept of the random close-packed state (Torquato *et al.*, 2000; Torquato and Stillinger, 2001).

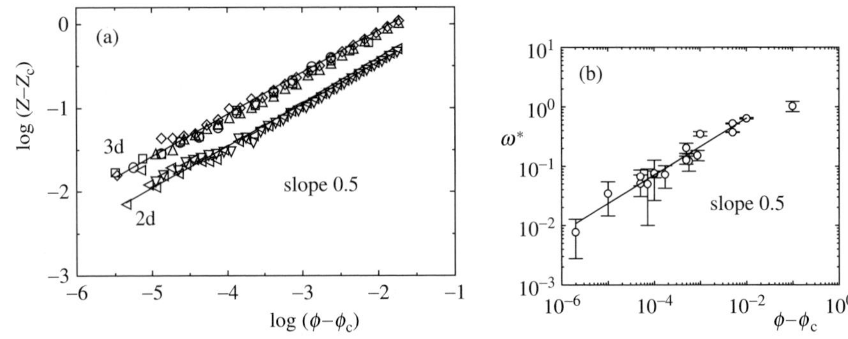

Fig. 9.3 (a) Z scales as $|\Delta\phi|^{0.5}$, from (O'Hern *et al.*, 2003). These results extend those of an earlier numerical study (Durian, 1995). (b) ω^* scales as $|\Delta\phi|^{0.5}$. From (Silbert *et al.*, 2005). Together these two sets of data are consistent with $\omega^* \sim \Delta Z$.

one another. These modes, which are in general global modes, have been called floppy modes (Thorpe, 1983; Alexander, 1998).

9.2.3 The plateau in the density of states and the crossover frequency, ω^*, and length scale, ℓ^*

The development of a plateau in the density of states is intimately connected with the approach to the isostatic jamming point. To see this, we argue as follows (Wyart *et al.*, 2005*a*,*b*). Let us start from an isostatic packing at jamming, sketched in Fig. 9.4(a). Now imagine we disregard ("cut") for a moment the bonds across a square or cube of linear size ℓ, as sketched in panel (b). There are of order ℓ^{d-1} of these surface bonds, and since the orginal packing was isostatic, by cutting the bonds at the surface we create of order ℓ^{d-1} floppy zero-energy modes within the cube. Each of these modes will be very complicated and disordered, but that does not matter here. Next, as

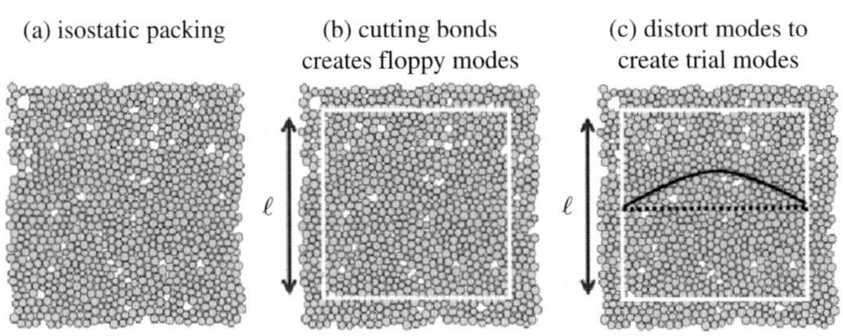

Fig. 9.4 Three stages of creating a low-energy Ansatz for a vibrational eigenmode of the system, used in the argument to derive the flatness of the $D(\omega)$ close to point J, as inspired by (Wyart *et al.*, 2005*a*,*b*). In (c), the solid line illustrates the behavior of the smooth amplitude with which the underlying floppy mode is distorted, and that vanishes at the boundaries of the box. See text for details.

indicated in panel (c), we use each of these floppy modes to create a variational Ansatz for a vibrational eigenmode as follows: we take each of the zero-energy floppy modes created by cutting bonds, and distort it by a smoothly varying sine-like amplitude that vanishes at the boundaries of the box, right at the point where we have artifically cut the bonds. This long-wavelength distorted mode is a good Ansatz for a low-energy eigenmode: when we analyze the potential energy associated with this Ansatz mode, the underlying floppy mode does not contribute to the energy—only the fact that the mode has been made imperfect by the elastic distortion on the scale ℓ contributes at every bond a distortion energy of order $1/\ell^2$ (nearby displacements differ from that in the underlying floppy modes by an amount of order the gradient of the amplitude, so the energy, which involves terms of order the average compression squared, is of order $1/\ell^2$). Hence, this Ansatz mode will have a frequency $\omega_\ell = \mathcal{O}(1/\ell)$.

Of course, in a variational calculation, each of these Ansatz modes will acquire a lower energy upon relaxation, but the number of them will not change (rigorously speaking this argument yields a lower bound on the density of states). In what follows we shall call the modes obtained from the distortion of floppy modes "anomalous", as their nature is very different from that of plane waves. Assuming that the energy does not shift dramatically, we have created $N_\ell \simeq \ell^{d-1}$ modes with frequency up to $\omega_\ell \sim \ell^{-1}$ in a box with of order $V_\ell \sim \ell^d$ particles. Thus, we have

$$\int_0^{\omega_\ell} \mathrm{d}\omega D(\omega) \simeq \frac{N_\ell}{V_\ell} \sim \frac{1}{\ell}. \tag{9.4}$$

If we assume that $D(\omega)$ scales as ω^q for small frequencies, we get

$$(\omega_\ell)^{q+1} \sim \frac{1}{\ell^{a+1}} \sim \frac{1}{\ell}, \tag{9.5}$$

which immediately yields $a = 0$: $D(\omega)$ is flat at low frequencies.

When the packings are compressed so that they have an excess number of bonds $\Delta Z = Z - Z_c$, the above line of reasoning can also be followed to obtain the crossover frequency ω^* and length scale ℓ^* that separate the range dominated by isostaticity effects from the usual elastic behavior. Instead of starting from an isostatic packing, we retrace the above construction for a packing with given excess contact number ΔZ per particle. The total number of excess bonds within the box of Fig. 9.4(b) then scales like $\Delta Z\, \ell^d$. If the number of degrees of freedom $N_\ell \simeq \ell^{d-1}$ created by cutting bonds at the boundary of the box is less than the excess number of bonds in the bulk, no zero-energy modes will be created. This fact was already noticed by Tkachenko and collaborators in inspiring works (Head *et al.*, 2001; Tkachenko and Witten, 2000). Hence, we expect a crossover scale ℓ^* when the two terms balance, i.e. when

$$N_{\ell^*} \simeq (\ell^*)^{d-1} \simeq \Delta Z\, (\ell^*)^d \quad \Longrightarrow \quad \ell^* \sim 1/\Delta Z. \tag{9.6}$$

Likewise, according to the argument above, one expects the crossover frequency ω^* in the density of states above jamming will scale as

$$\omega^* \sim 1/\ell^* \sim \Delta Z. \tag{9.7}$$

The scaling of the crossover frequency, $\omega^* \sim \Delta Z$, has been well documented from a detailed analysis of the density of states, as shown in Fig. 9.2. At finite compression, ω^* is the frequency above which there is a plateau in $D(\omega)$, and below which $D(\omega)$ decreases with decreasing frequency. Early data obtained as a function of excess density $\Delta\phi$ are shown in Fig. 9.3(b), and are found to scale as $\omega^* \sim |\Delta\phi|^{0.5}$. Since $\Delta Z \sim |\Delta\phi|^{0.5}$, see Fig. 9.3(a), this is consistent with the scaling $\omega^* \sim \Delta Z$. An explicit plot of ω^* versus ΔZ can be found in (Wyart *et al.*, 2005*a*).

We stress that in the above argument, it is implicity assumed that the bond strength k remains unchanged. This is true for one-sided harmonic forces, but not for Hertzian interactions that weaken upon approaching the jamming point: $f \sim \delta^{3/2}$ so that $k \sim \delta^{1/2} \sim \sqrt{\Delta\phi}$. Since frequencies scale as $\sqrt{k}$, the above arguments generalize to non-harmonic forces if formulated in terms of scaled frequencies $\tilde{\omega} = \omega/\sqrt{k}$.

While the crossover frequency is relatively easy to extract from the vibrational density of states, the crossover length ℓ^* is more difficult to extract from a mode or from response data. Physically, ℓ^* is the length scale on which the lowest-frequency anomalous modes probe the microscopic structure of the solid, and are therefore sensitive to its fluctuations. Thus, one may expect to observe ℓ^* in the fluctuations of the linear response of the solid, rather than in its mean behavior. Another difficulty lies in the fact that the floppy modes that form the basis for the behavior up to scale ℓ^* are very disordered so that a weak, elastic-like distortion of such modes is difficult to detect. Nevertheless, it has been discovered that the crossover length is easily discernible by eye in the response to a point force or to inflation of a local particle (Ellenbroek *et al.*, 2006, 2009*a*), as Fig. 9.5 illustrates. A detailed analysis of these data has indeed shown that the response is governed by a crossover scale that grows as $1/\Delta Z$, in accord with Eq. (9.6).

9.2.4 Microscopic criterion for stability under compression

So far, our analysis of the density of states has neglected the effect of pressure on the vibrational spectrum. Mundane observations such as the buckling of a straw pushed at its tips show that compression affects vibrational modes, and can even make them unstable. In general, this effect is important for thin objects, but is irrelevant at small strain for bulk solids where plane waves dominate the vibrational spectrum. However, compression plays an important role if floppy modes are present in the solid, as noticed by Alexander in the context of gels (Alexander, 1998). As we now show, this is also true for the anomalous modes introduced above (Wyart *et al.*, 2005*b*).

The analysis is based on the well-known observation, see Fig. 9.6, that if k is the stiffness associated with the longitudinal relative displacements of two particles in contact, then for transverse displacements the stiffness is $k_1 \propto -ke$, where e is the contact strain. For a pure plane wave propagating in an amorphous solid, the transverse and longitudinal components of the relative displacements in the contacts are of the same order. The correction induced on the mode energy by the presence of a finite contact force or strain is thus, in relative terms, of order of the ratio of the two stiffnesses: $-e$. Therefore, the effect of compression on plane waves is negligible at small strain. For anomalous modes the situation is different, because they are

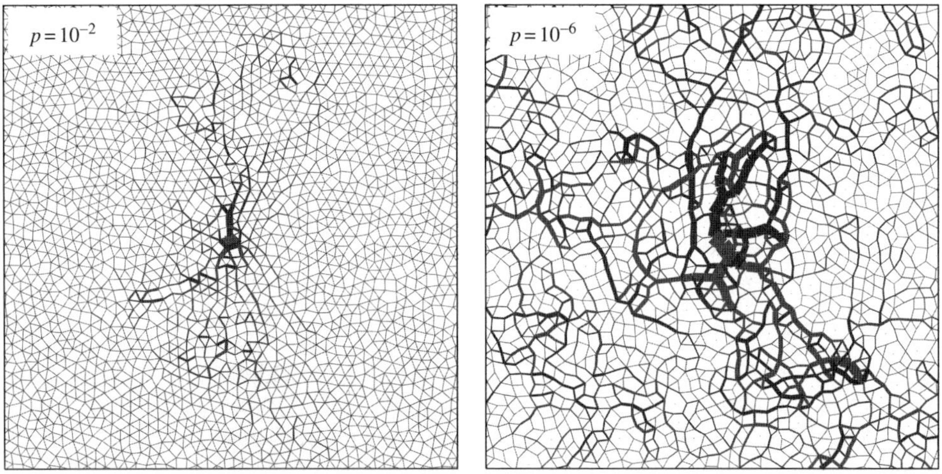

Fig. 9.5 Illustration of the fact that the length scale ℓ^* increases as the jamming threshold is approached. This figure illustrates the response to a force loading in the center, with two different grey levels (due to blue and red in the original paper cited below) indicating an increase or decrease of the force at a contact; the thickness of the lines is proportional to the size of the change in force. The left panel is for a pressure $p = 10^{-2}$, the right one is close to the jamming point, i.e. at pressure $p = 10^{-6}$. In the latter case, the fluctuations are larger and extend over a larger region. A detailed analysis shows that this range grows proportionally to the length scale ℓ^*. From Ellenbroek *et al.* (Ellenbroek *et al.*, 2006, 2009*a*).

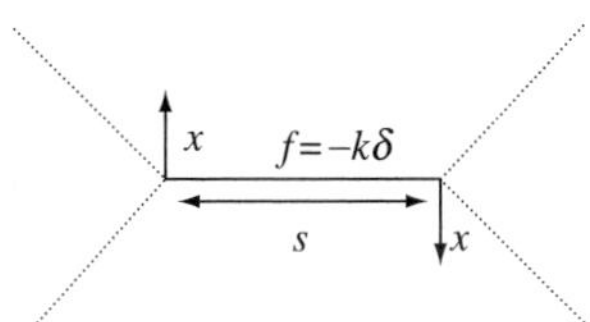

Fig. 9.6 Consider for concreteness a harmonic contact between two particles of length s, of stiffness k, carrying a force $f = -k\delta$, where δ is the contact elongation. If particles move transversally to the contact by an amplitude x as shown with vertical arrows, the contact length increases by an amount proportional to x^2/s following Pythagoras' theorem. The work produced by the contact force is then proportional to $x^2 k\delta/s$, corresponding to a stiffness $k_1 \sim k\delta/s \equiv -ke$, where $e \equiv -\delta/s$ is the contact strain, chosen to be positive for compressed contacts.

built by deforming floppy modes that have no longitudinal relative displacements, but only transverse ones. Once floppy modes are deformed to generate anomalous modes, they gain a longitudinal component of order ΔZ as follows from the variational argument, whereas the transverse component remains of order one (Wyart *et al.*, 2005*b*; Wyart, 2005). Relative corrections in the anomalous mode energies is thus of order $-e/\Delta Z^2$. When these relative corrections reach a constant of order one, anomalous modes become unstable, and the system yields. For particles near jamming, one has $e \propto \Delta\phi$, leading to the condition for stability (Wyart *et al.*, 2005*b*):

$$\Delta Z > \Delta\phi^{1/2} \qquad \text{for all subsystems of size } L > l^*, \tag{9.8}$$

where a numerical pre-factor is omitted. Inequality (9.8) extends Maxwell's criterion to the case of finite compression and is also not a local criterion: it must be satisfied on all subsystems of size larger than l^*. On smaller scales, fluctuations of coordination violating Eq. (9.8) are permitted, as stability can be insured by the boundaries. Configurations where the bound (9.8) is saturated are marginally stable and anomalous modes exist down to zero frequency. Such packings should exhibit the scaling $\Delta Z \sim \Delta\phi^{1/2}$. Configurations obtained by decompression in the vicinity of ϕ_c appear to lie very close to saturation (Wyart *et al.*, 2005*b*). The scaling of Eq. (9.8) was proposed independently by Head (Head, 2005) following a mean-field analysis; there, however, the threshold coordination found was half its correct value.

An argument for why one might expect the system to be marginally stable is provided in Section 9.6.1, where it will be argued that the realization of the bound (9.8) critically affects the dynamics. The criterion (9.8) applies to spheres, but not to ellipses where rotational degrees of freedom matter. It can be shown that in the latter case, compression can have a *stabilizing* effect if the ellipses make on average more contacts where their surface curvature is low. In that case, even an infinitesimal pressure can stabilize zero modes, so that stable packings can be generated although the Maxwell bound is not fulfilled, as shown in Section 9.3.1. The same is true for gels of crosslinked polymers (Alexander, 1998).

9.2.5 Behavior or elastic constants near jamming

The fact that a packing of frictionless spheres or discs is a marginally connected solid, also has its effect on the elastic constants. The behavior of the bulk compression modulus B and shear modulus G always depends on the specific force law, and hence is not universal: for distributions of bonds that are peaked around some average non-zero value, both of them are proportional to the average bond strength k, hence in discussing the effects due to jamming it is useful to divide out this common effect. The individual bond strengths k_{ij} are essentially the second derivatives of the interaction potential evaluated at each contact—for the one-sided harmonic springs these are all the same, but for the Hertzian $f_{ij} \sim \delta_{ij}^{3/2}$ force law one finds $k_{ij} \sim \delta_{ij}^{1/2} \sim f_{ij}^{1/3}$, so that the average bond strength scales with the pressure p as $p^{1/3}$.

An example of data for the elastic moduli is shown in Fig. 9.7, which summarizes data for B and G for packings of two-dimensional discs interacting with Hertzian forces (Ellenbroek *et al.*, 2006, 2009*a*). The bulk modulus B is clearly seen to scale as $p^{1/3}$ (O'Hern *et al.*, 2003), which as we argued above is the scaling of the average bond strength for this force law. However, G decreases much faster, as $p^{2/3}$ (O'Hern *et al.*, 2003): packings close to jamming are much easier to distort by shear than by compression. Results in both two and three dimensions for harmonic (Durian, 1995; O'Hern *et al.*, 2003) and Hertzian potentials (O'Hern *et al.*, 2003) have been determined. From these results, it is generally seen that G/B and ΔZ scale in the

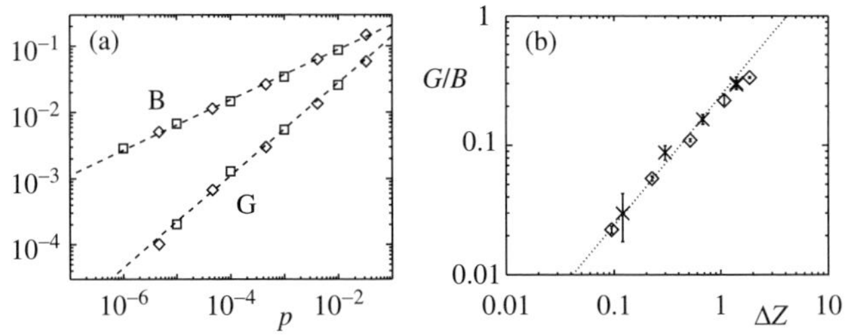

Fig. 9.7 (a) As the jamming point is approached, in this case by lowering the pressure, the shear modulus G becomes much smaller than the bulk compression modulus B (Durian, 1995; O'Hern *et al.*, 2003). These results are for packings with Hertzian forces, for which the bond strength of individual forces scales as $p^{1/3}$. Due to this weakening of the bonds, the bulk modulus B has an overall $p^{1/3}$ scaling, while the shear modulus G has an overall $p^{2/3}$ scaling. (b) The ratio G/B is linear in Δz (Durian, 1995; O'Hern *et al.*, 2003); this scaling is independent of the force law. Squares refer to data obtained from response to a local point force, diamonds to data obtained from a global deformation. After (Ellenbroek *et al.*, 2006, 2009*a*).

same way with compression. The data in Fig. 9.7 are consistent with this, as for a Hertzian interaction, $G/B \sim p^{1/3}$ and $\Delta Z \sim |\Delta\phi|^{0.5} \sim p^{1/3}$. The result that the shear modulus generally scales as $G \sim k\,\Delta Z$, can be understood (Wyart, 2005): independent of the force law, the ratio G/B should scale as $G/B \sim \Delta Z$. Empirical support for this behavior in grains is reviewed in (Agnolin and Roux, 2007).

It is interesting to note that at first sight the fact that G/B goes to zero at the jamming point as ΔZ, may seem to be the anomalous behavior; however, from the point of view of rigidity percolation (Jacobs and Thorpe, 1995), B/k behaves anomalously for packings, as in rigidity percolation both G/k and B/k vanish as ΔZ, upon approaching the isostatic point so that G/B remains constant at the percolation threshold. The difference between the jamming transition and generic rigidity percolation is that in the former case configurations must be such that all the contact forces must be repulsive. This imposes a geometric constraint on the network that is not enforced in standard rigidity percolation models (Wyart, 2005). We refer to (Ellenbroek *et al.*, 2009*b*) for further details.

To compute static properties such as G/B theoretically, one may use effective-medium theory. This approach has been used to study a square lattice with randomly placed next-nearest-neighbor springs (Mao *et al.*, 2009) and random isotropic off-lattice spring networks (Garboczi and Thorpe, 1985; Wyart, 2010). In the isotropic case (rigidity percolation), G/B is independent of ΔZ at the transition, as found in Ref. (Ellenbroek *et al.*, 2009*b*), while in the randomly decorated square lattice $G/B \sim \Delta Z^2$. Both results differ from the scaling observed for jammed packings, $G/B \sim \Delta Z$.

An important topic in granular research has been the issue of the coarse-graining scale needed for elastic behavior to emerge in granular packings (Goldhirsch and Goldenberg, 2002; Goldenberg and Goldhirsch, 2008). In line with the arguments given above, it has been found that near the jamming point, one needs to coarse grain over the scale ℓ^* to see the elastic response emerge (Ellenbroek *et al.*, 2006, 2009*a*): as one approaches the jamming point, one has to coarse grain over larger and larger lengths, diverging as $\ell^* \sim 1/\Delta Z$. That elasticity emerges in granular media on long enough scales is also implicitly confirmed by the data of Fig. 9.7, where squares give the values of the elastic constants obtained from local point response calculations, and diamonds those obtained from global deformations. The two sets of data agree very well (Ellenbroek *et al.*, 2006, 2009*a*).

9.2.6 Diffusivity, quasi-localized modes and anharmonicity

9.2.6.1 Diffusivity

The excess low-frequency vibrational modes associated with the isostatic jamming transition have important consequences for the nature of energy transport in the system. Their contribution to energy transport can be quantified by the energy diffusivity, defined as follows (Allen and Feldman, 1989*a*,*b*). Consider a wavepacket narrowly peaked at frequency ω and localized at $\vec{r}$. This wavepacket spreads out in time, such that the square of the width of the wavepacket at time t, divided by t, is independent of t at large t; this constant defines the diffusivity, $d(\omega)$. The diffusivity was calculated using the Kubo approach within the harmonic approximation for repulsive sphere packings as a function of compression in (Xu *et al.*, 2009*b*; Vitelli *et al.*, 2009). At sufficiently low frequencies, the modes should be plane waves and the diffusivity should follow Rayleigh scattering behavior so that $d(\omega) \sim \omega^{-4}$. Refs. (Xu *et al.*, 2009*b*; Vitelli *et al.*, 2009) show that the diffusivity is nearly flat for frequencies from ω^* up to the Debye frequency, where it drops to zero. Thus, there is a crossover from weakly scattered plane waves to strong scattering, diffusive behavior at ω^*; this can be understood from the heterogeneous nature of the anomalous modes (Wyart, 2005; Vitelli *et al.*, 2009), or by assuming the continuity of the length scale characterizing plane waves and anomalous modes at the crossover frequency, in conjunction with the scaling laws for the shear and bulk moduli discussed in Section 9.2.5 (Vitelli *et al.*, 2009; Xu *et al.*, 2009*b*).

One consequence of the link between the transport crossover and the excess modes (Xu *et al.*, 2009*b*; Vitelli *et al.*, 2009) is that the diffusivity plateau extends all the way down to the lowest frequencies studied as the jamming transition is approached and $\omega^* \to 0$, as shown in Fig. 9.8(a). This suggests that the origin of the diffusivity plateau can be traced to properties of the jamming transition.

Dynamical vibrational properties such as the energy diffusivity can be computed theoretically using dynamical effective-medium theory, a resummation scheme of a perturbation expansion in the disorder also known as the coherent potential approximation (CPA) (Soven, 1969). This approach has been used on the square lattice with randomly placed next-nearest-neighbor springs (Mao *et al.*, 2009) and isotropic disordered spring networks near the isostatic limit (Wyart, 2010). Both calculations yield a crossover to a flat density of states above $\omega^\star \sim \Delta Z$ in good agreement with results for jammed packings. In the isotropic case, Rayleigh scattering is found for $\omega < \omega^\star$ with a scattering length $\ell_s \sim \Delta Z^3/\omega^4$. This calculation predicts a violation of the Ioffe–Regel criterion (according to which the crossover to strong scattering occurs

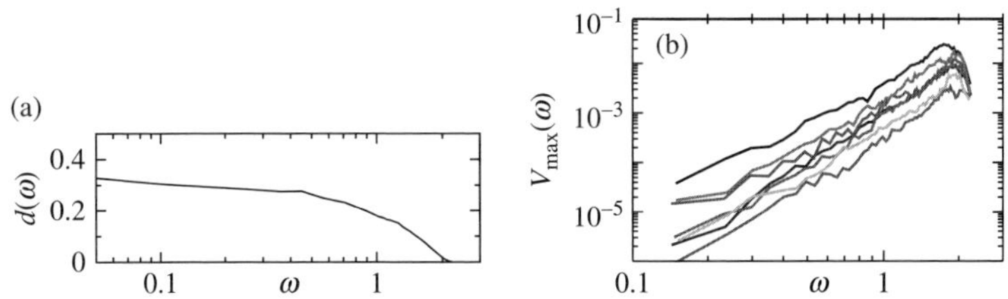

Fig. 9.8 (a) Diffusivity, $d(\omega)$, for a system of spheres interacting via harmonic repulsions very close to the jamming transition. $d(\omega)$ is essentially flat down to zero frequency. From Xu *et al.* (Xu *et al.*, 2009*c*). (b) Energy barrier $V_{\max}(\omega)$ along each mode, before falling into another basin of attraction for a system at $\Delta\phi = 0.1$. The energy barriers along each mode direction provide an upper bound on the energy barriers of the system in a given frequency range, since it is possible that lower-energy barriers could be encountered in directions in phase space corresponding to linear combinations of the modes. From (Xu *et al.*, 2009*a*).

when the scattering length becomes of order of the wavelength) at the crossover frequency $\omega^\star$ since the scattering length $\ell_s \sim 1/\Delta Z$ is much larger than the wavelength $\lambda \sim 1/\sqrt{\Delta Z}$ there. Above $\omega^\star$, the anomalous modes are characterized by a displacement–displacement correlation length that scales as $1/\sqrt{\omega}$, which scales as $1/\sqrt{\Delta z}$ at $\omega^\star$, and a frequency-dependent speed of sound, $c(\omega) \sim \sqrt{\omega}$. This leads to a plateau in the diffusivity consistent with the results of Fig. 9.8. This analysis predicts a drop in the speed of sound near ω^*, a subtle effect recently observed numerically at the Boson peak frequency in model glasses (Monaco and Mossa, 2009). In the case where the isostatic state is a square lattice (Mao *et al.*, 2009), the Ioffe–Regel criterion is satisfied at the crossover frequency ω^* and ℓ_s is of order the lattice spacing. Numerical results for jammed packings are consistent with $\ell_s \sim 1/\sqrt{\Delta z}$ at the crossover.

9.2.6.2 Quasi-localized modes

The nature of the excess low-frequency modes shows a systematic variation with frequency. As the frequency is lowered towards ω^*, numerical studies of three-dimensional sphere packings show that the modes become progressively more heterogeneous with a lower-than-average mode coordination number, and develop high displacement amplitudes in small regions of space (Xu *et al.*, 2009*b*). These quasi-localized modes shift downwards with decreasing compression. These low-frequency, quasi-localized modes have also been observed in two-dimensional packings of soft disks (Zeravcic *et al.*, 2009), but here the quasi-localized modes are mixed in with plane-wave-like modes over the same low-frequency range.

In three dimensions, the frequency of the quasi-localized modes decreases with compression, but the degree of quasi-localization does not appear to depend on compression. This suggests that as the system is decompressed towards the unjamming transition, quasi-localized modes always exist but shift to lower and lower frequency. Quasi-localized modes have been observed in other disordered models (Biswas *et al.*,

1988; Fabian and Allen, 1997; Taraskin and Elliott, 1999; Schober and Ruocco, 2004) and in experiments on glassy polymers (Buchenau *et al.*, 1992, 1996; Vainer *et al.*, 2006). In the soft-potential model, for example, such modes have been shown to lead to strong scattering of vibrations at higher frequencies, so that their frequency corresponds to the Ioffe–Regel crossover frequency in the transport properties (Schober and Oligschleger, 1996), while in model silica, these modes have been identified near the boson peak frequency (Taraskin and Elliott, 1999).

Numerical studies on hard spheres (Brito and Wyart, 2007; Brito and Wyart, 2009) have shown that irreversible particle rearrangements at non-zero temperature occur along the low-frequency modes, observed to be extended in the systems of limited size studied, implying that the lowest activation barriers are in those directions of phase space (Section 9.6.1). This view has been explored in model glasses (Widmer-Cooper *et al.*, 2009*a*), which have also shown that the structural rearrangements occurring at finite temperatures were along low-frequency normal modes, which in that case were quasi-localized.

Figure 9.8(b) shows that for sphere packings, the low-frequency quasi-localized modes have the lowest energy barriers and that the energy barriers decreases systematically with frequency and participation ratio. This result, together with previous observations (Brito and Wyart, 2007; Brito and Wyart, 2009; Widmer-Cooper *et al.*, 2009*a*), should have important consequences for the dynamics and instabilities that occur as the temperature is raised from zero, the system is compressed, or a shear stress is applied. One would expect these modes to shift downwards to zero frequency under the last two circumstances (Maloney and Lemaitre, 2006) and to give rise to highly heterogeneous deformations in the sample. These could be the underlying mechanism for the localized deformations postulated in the picture of shear transformation zones (Falk and Langer, 1998; Langer, 2008).

As the system is decompressed towards the unjamming transition, the barriers between nearby configurations shrink to zero as the amount of interparticle overlap decreases to zero. Thus, in this limit the modes become progressively more anharmonic. This implies that smaller excitation energies or temperatures should suffice to force the system into new energy minima. Thus, the range of temperatures and stresses over which harmonic theory applies should shrink to zero and anharmonic effects should dominate as one approaches the transition. This effect has been computed in networks of springs below or above the isostatic threshold (Wyart *et al.*, 2008), where non-linearities dominate the response to shear when the strain γ is larger than some γ^*, with $\gamma^* \sim |\Delta z|$.

9.2.7 Structural features of the jamming transition

At point J, the pair distribution function has many singular properties (O'Hern *et al.*, 2003; Silbert *et al.*, 2006*a*). In particular, it has been shown that the first peak in $g(r)$ is a δ-function at the nearest-neighbor distance, σ. The area under this δ-function is just the isostatic coordination number: $Z = 6$ in three dimensions. On the high side of this δ-function, $g(r)$ has a power law decay: $g(r) \propto (r - \sigma)^{-0.5}$. It has been proposed that this power law is the vestige of the marginal stability of the configurations visited before reaching ϕ_c (Wyart, 2005). Also, at the jamming transition, the second peak of $g(r)$ is split into what appear to be two singular subpeaks one at $r = \sqrt{3}\sigma$ and the other at $r = 2\sigma$. There is a divergent slope just below each subpeak and a step-function cutoff on its high side (Silbert *et al.*, 2006*a*).

The pair correlation function near the zero-temperature jamming transition has been studied in two experiments. In experiments by Abate and Durian (Abate and Durian, 2006; Keys *et al.*, 2007), a layer of ball bearings was placed on a wire mesh and excited by the upflow of gas through the mesh. The turbulent flow of the gas leads to interactions between the balls and to random motion of the balls. Thus, the effective interaction potential between the balls can be characterized by a hard core determined by the ball diameter and a longer-ranged repulsion due to the gas. In such systems, the kinetic energy of the balls decreases monotonically with increasing density of the balls, such that it reaches zero when the balls are packed to the density of the jamming transition, $\phi_c \simeq 0.84$, measured in terms of the hard-core diameter. The structure of the system, as characterized by the pair correlation function, $g(r)$, was studied along this trajectory to the zero-temperature jamming transition of the hard cores (see the chapter in this book by Dauchot, Durian and van Hecke for a description of the dynamical properties of this system). On approaching ϕ_c, the structure shows an increase of the height of the first peak, g_1. This is consistent with numerical results on soft spheres, which find a divergence in g_1 at the zero-temperature jamming transition. There is also a local maximum in g_1 at $\phi \approx 0.74$, above which no changes in the coordination and geometrical features of Voronoi cells are observed. This secondary maximum at non-zero kinetic energy due to airflow-induced interactions, may correspond to the finite-temperature vestige (Zhang *et al.*, 2009*a*) of the divergence in the height of the first peak of $g(r)$ at the zero-temperature transition discussed below.

Experiments on a two-dimensional bidisperse system of NIPA particles by Zhang, *et al.* (Zhang *et al.*, 2009*a*), measured the pair correlation function as a function of density at essentially fixed, non-zero temperature. The system exhibits a maximum in the height of the first peak of the pair correlation, g_1—see Fig. 9.9. This maximum is interpreted as a thermal vestige of the zero-temperature jamming transition. Corroborating simulations show (Zhang *et al.*, 2009*a*) that the divergence in $g_1 \sim 1/(\phi - \phi_c)$ for soft repulsive spheres at the zero-temperature jamming transition is softened to a maximum at non-zero temperature. The maximum in g_1 decreases in height and shifts to higher density with increasing temperature. The simulations also show that the maximum in g_1 is a structural feature that is apparent as a function of increasing density at fixed temperature, but not as a function of temperature at fixed density or pressure. Recent simulations on the same model by Jacquin, *et al.* (Jacquin and Berthier, 2009) show that the maximum in g_1 appears even in systems at higher temperature or lower pressure that are in thermal equilibrium and can be understood from liquid-state theory.

Note that the divergence in g_1 corresponds to a vanishing of the overlap distance between neighboring particles. Thus, tracking g_1 is equivalent to tracking the overlap distance. The evolution of the overlap distance with temperature is not correlated with the existence of a dynamical glass transition.

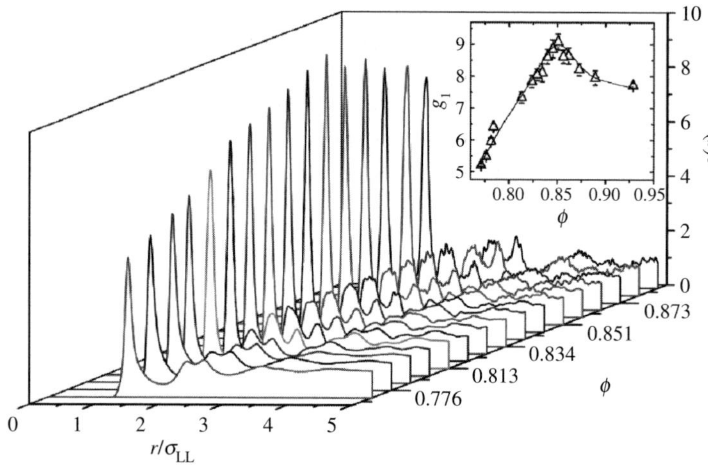

Fig. 9.9 Pair correlation function $g(r)$ for the large particles at different packing fractions ϕ for a bidisperse system of colloids in two dimensions. The inset shows g_1, the height of the first peak of $g(r)$, versus packing fraction. The experimental data shows that g_1 is non-monotonic with a maximum at a packing fraction slightly above ϕ_c, the value at the $T = 0$ jamming transition. The peak is a vestige of that $T = 0$ transition where the first peak of $g(r)$ is a δ-function. Figure reproduced from (Zhang *et al.*, 2009*b*).

9.2.8 Singular length scales at Point J

Several different singular length scales have been identified near the jamming transition. The cutting length, ℓ^*, discussed in Section 9.2.3, has been observed by studying the fluctuations in the change of the force on a contact as a function of the distance of the contact from a small local perturbation (see Fig. 9.5). This length scale diverges as $(\phi - \phi_c)^{-1/2}$, independent of the potential as long as it is repulsive and finite in range. In addition, this length scale is expected to correspond to the wavelength of longitudinal, weakly scattered plane waves just below the crossover frequency ω^* (Silbert *et al.*, 2005), although it has not been seen numerically, due to the small system sizes that can be studied.

A second diverging length scale has been identified from the wavelength of transverse, weakly scattered plane waves just below the cross-over frequency ω^* (Silbert *et al.*, 2005; Vitelli *et al.*, 2009). This length scale diverges as $(\phi - \phi_c)^{-1/4}$. Finally, this length scale emerges within the effective-medium approximation as the decay length for spatial correlations in modes at the frequency $\omega \to \omega^\star+$ (Wyart, 2010).

As noted in Section 9.7, there is a vanishing length scale corresponding to the overlap distance δ between neighboring particles. This distance, which measures the left-hand width of the first peak of the pair correlation function, vanishes as $(\phi - \phi_c)^1$ (O'Hern *et al.*, 2003; Silbert *et al.*, 2006*a*).

Finally, there are other diverging length scales whose origins are not understood. The shift of the position of the jamming transition, ϕ_c, with the linear size of the system, L, yields a length scale that apparently diverges as $|\phi - \phi_c|^{-0.7}$ in both 2 and 3 dimensions (O'Hern *et al.*, 2003). This scaling shows up in simulations in which a hard disk is pushed through a packing below ϕ_c; the transverse distance over which the packing adjusts as the disk passes by diverges with a power law of 0.7 (Drocco *et al.*, 2005). Finally, this power law has been observed for correlations of the transverse velocity on athermal, slowly sheared sphere packings near the jamming transition (Olsson and Teitel, 2007). However, this length scale is only observed for certain models of the dynamics (Tighe *et al.*, 2010).

9.3 Extensions of the results for frictionless spheres

9.3.1 Anisotropic particles: ellipsoids

The counting of degrees of freedom that underlies the above analysis of the soft modes near the jamming point, relies on having perfect spherical and frictionless particles. We briefly review here what happens to the jamming scenario if the particles are frictionless but non-spherical.

The model for which this question has been studied in detail is that of two-dimensional frictionless ellipses in two dimensions Mailman *et al.* (2009) or ellipsoids with one axis of symmetry in three dimensions (Zeravcic *et al.*, 2009). In the latter case, each ellipsoid is essentially characterized by 5 non-trivial degrees of freedom, so a naive counting of the degrees of freedom shows that the isotropic values for such ellipsoids is $Z_{\text{iso}}^{\text{ellipsoid}} = 2 \times 5 = 10$.

Clearly, while the isotropic values jump depending on which degrees of freedom are, and which ones are not, included in the counting, one would expect the physical behavior of such packings to evolve continuously, when the asphericity of the particles is turned on—what is happening?

Let us take the c-axis of our ellipsoids along their symmetry axis; the other axes are then a and $b = a$. We define the ellipticity as the aspect ratio $\varepsilon = c/a$, so prolate ellipsoids (like cigars) correspond to $\varepsilon > 1$ while oblate ones (like M&Ms (Donev *et al.*, 2004)) to $\varepsilon < 1$. Figure 9.10(b) recovers the finding of (Donev *et al.*, 2004; Donev *et al.*, 2007; Sacanna *et al.*, 2007; Wouterse *et al.*, 2007) that as the ellipticity is tuned away from the spherical case $\varepsilon = 1$, the average contact number indeed increases continuously from the spherical value $Z_{\text{iso}} = 6$, as expected.

How to resolve this apparent paradox between the jump in the counting and the continuity of the physical system? How come one can reach values of the average contact number below the isostatic value $Z_{\text{iso}}^{\text{ellipsoid}} = 10$? To answer these questions, it is good to realize that when we reached the conclusion that the isostatic value $Z_{\text{iso}}^{\text{sphere}} = 2d = 6$, we left out the rotational degrees of freedom of each sphere, as these are trivial zero-frequency modes (three per particle) for a system of frictionless spheres. In order to understand the continuity, it is better to include these rotational degrees of freedom from the start, even for the spherical case, and see how they evolve when the spheres are deformed into ellipsoids. Since in the work (Zeravcic *et al.*, 2009) that we will summarize below, the ellipsoids still have one symmetry axis, there are only two non-trivial angular degrees of freedom associated with the anisotropy of these

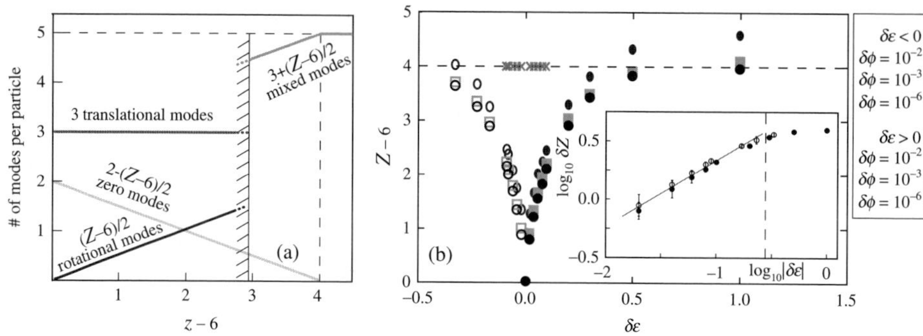

Fig. 9.10 (a) Illustration of how the number of different modes per ellipsoid (excluding rattlers) in the packings at jamming varies as a function of the average contact number Z. As the ellipticity is increased, the average contact number increases. For $Z <\approx 9$, there are two well-defined bands: a rotational band with $(Z-6)/2$ modes per particle and a translational band with three modes per interacting particle, as is the case for spheres (see Fig. 9.11 for details). Upon increasing Z, the number of zero modes decreases as zero modes are converted into finite-frequency rotational modes. Above $Z \approx 9$, there is only one band. (b) Z as a function of the ellipticity $\delta\varepsilon \equiv \varepsilon - 1$ and distance from jamming $\delta\phi$ for 216-particle packings. The sharp decrease around the spherical case $\delta\varepsilon = 0$ is consistent with earlier results (Donev *et al.*, 2004; Sacanna *et al.*, 2007; Wouterse *et al.*, 2007) for hard ellipsoids and spherically capped rods. The log-log plot of δZ versus $|\delta\varepsilon|$ in the inset shows that the rise of Z at jamming is consistent with a $\delta Z \sim \sqrt{|\delta\varepsilon|}$ scaling. The crosses in the main plot for small values of $\delta\varepsilon$ show that twice the measured number of zero-frequency eigenmodes per particle plus $Z-6$ add up precisely to 4, in accordance with panel (a). From Zeravcic *et al.* (Zeravcic *et al.*, 2009).

particles. The relevant isostatic value for these ellipsoids of revolutation is therefore $Z_{\rm iso}^{\rm ellipsoid} = 2 \times 5 = 10$, as already asserted above.

When analyzed more carefully, the counting arguments summarized in Section 9.2.2 imply that whenever $Z < Z_{\rm iso}$, there are $N_{\rm c}(Z_{\rm iso} - Z)/2$ directions in the phase space of the relevant coordinates of the particles along which the packing coordinates can be changed without changing the interaction energy of the particles. This means that each stable configuration must have precisely this number of zero-frequency modes. That this is precisely what happens in the ellipsoid packings, is illustrated by the gray crosses in Fig. 9.10(b); numerically it is found that in the calculation of the density of states of ellipsoid packings one finds precisely the right number of zero modes as expected according to this argument.

Figure 9.10(a) illustrates how, as Z increases with $\delta\varepsilon = \varepsilon - 1$ of the ellipsoids, the number of zero modes per particle with contacts goes down. We thus have more and more non-zero modes as Z increases. What is their nature? Figure 9.10(a) shows that for $|\delta\varepsilon| <$ 0.17, the density of states exhibits two clearly separated bands (Zeravcic *et al.*, 2009). The modes in the upper band of the density of states turn out to be pre-dominantly translational—they are very similar to those found in the sperical packings, and turn out to be characterized by an onset frequency ω^* that scales as

$\Delta Z = Z - 6 \sim |\delta\varepsilon|^{1/2}$: even though contacts at jamming are created by deforming the particles and not by compression, the mechanism that determines the onset of anomalous translational modes at a frequency ω^* is the same! Furthermore, the lower-frequency band that exisits for small ellipticity consists of pre-dominantly rotational modes; it is characterized by an upper frequency that scales linearly in $|\delta\varepsilon|$ (Zeravcic *et al.*, 2009). For excentricities larger than about 0.17, the two bands merge into one mixed band, as Fig. 9.10(b) illustrates.

The scenario that emerges for the number of modes in the various bands is summarized in Fig. 9.11(a)—the upshot is that the jamming scenario as formulated for frictionless spheres, as exemplified by the sharp onset of anomalous modes, withstands the change in shape of the particles in the system.

9.3.2 Packings of frictional spheres

Another relevant extension is to include static friction in the force law. The usual friction law is to take the Mohr–Coulomb law (Johnson, 1985), in which the particles can experience a tangential force f_t that obeys $f_t \leq \mu f_n$, where f_n is the normal force at the contact considered, and μ the Coulomb constant. In the Hertz force-law model, the tangential elastic force builds up continuously, in a well-described way, when a contact gets loaded with a tangential force (Johnson, 1985). This buildup continues up to μf_n, beyond that f_t remains constant and fixed at the value μf_n. In the presence of static friction, therefore, the properties of a packing are very history dependent, as the tangential forces are not only dependent on the configuration, but also on how the configuration was made—see e.g. (Silbert *et al.*, 2002; Zhang and Makse, 2005; Somfai *et al.*, 2005, 2007).

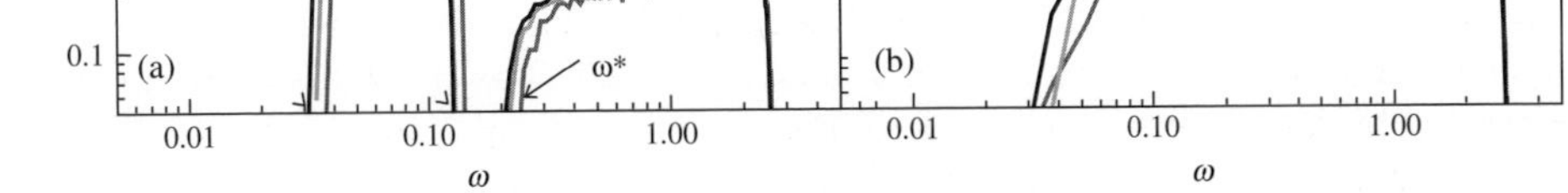

Fig. 9.11 (a) The density of states for slightly oblate ellipsoids, $\delta\varepsilon = -0.04$, for packings close to jamming and at two compressions. The existence of two bands separated by a gap as well as a gap at zero frequency is clearly visible. (b) For larger ellipticities the two bands merge, as is illustrated here for $\delta\varepsilon = -0.33$. The inset shows the data for $\delta\varepsilon = 1$, where $Z \approx Z_c = 10$ at jamming (see panel a). In accordance with this, the gap near zero frequency increases with increasing compression (and therefore increasing Z). From Zeravcic *et al.* (Zeravcic *et al.*, 2009).

If we return to spherical particles, but include the Mohr–Coulomb friction law, again the counting of the number of degrees of freedom is different: the coordinates of the centers of the particles are still the degrees of freedom with which one can satisfy the "just-touching conditions" (9.2) upon approaching the jamming point. On the other hand, the counting of the degrees of freedom in the force and torque balance conditions is different—e.g., in two dimensions the frictional forces add one additional tangential frictional force component for each contact, but torque balance only adds one extra constraint per particle. A straightforward generalization (see, e.g., (Shundyak *et al.*, 2007)) shows that in general with frictional forces stable packings can be obtained provided that $d+1 \leq Z \leq 2d$. Unlike for the frictionless case, where this argument dictates Z completely in the limit that the jamming point is approached, the inclusion of a static frictional force allows a range of possible values of Z. In simulations of frictional systems, it is indeed found that packings at the jamming threshold with different values of Z and density ϕ can be made, by varying in particular the quench rate; moreover, there is a clear trend for the average contact number Z to approach the lower lower limit $d+1$—the isostatic value in the presence of friction—in the limit $\mu \to \infty$ (Shundyak *et al.*, 2007; Song *et al.*, 2008). In line with this, one finds that if one studies the linear vibration–rotation modes of these packings, the density of states only develops a plateau in this large-friction limit (Somfai *et al.*, 2007).

However, it was later discovered that the situation is more subtle. If packings are prepared gently, then one finds that a finite fraction of the contacts in the simulations are at the mobility or Coulomb threshold, i.e., have $f_{\rm t} = \mu f_{\rm n}$ (Shundyak *et al.*, 2007; Henkes *et al.*, 2009). If such a contact is pushed in the direction of increasing friction force, they slip! In other words, a finite fraction of the tangential forces is *not* arbitrary, but constrained to be at the Coulomb threshold. This of course changes the counting of the number of degrees of freedom. Indeed, if one introduces $n_{\rm m}$, the number of contacts per particle that is found at the mobility threshold, then one may introduce a generalized criterion (Shundyak *et al.*, 2007; Henkes *et al.*, 2009)

$$Z \geq (d+1) + \frac{2n_{\rm m}}{d} \equiv Z^{\rm m}_{\rm iso}. \tag{9.9}$$

for the stability of packings. In this scenario, packings that are close to the line $Z = Z^{\rm m}_{\rm iso} = d+1+2n_{\rm m}/d$ are again marginal, i.e. develop many low-lying excitations because the number of contacts that slip under tangential motion is just large enough to make almost zero-energy deformations possible. The idea that gently prepared packings might end up being marginal packings in this way, traces back to an earlier suggestion by Bouchaud (Bouchaud, 2004).

With this generalization, packings with static friction that are gently prepared, do again follow in essence the scenario described above. As illustrated in Fig. 9.12(a), as the pressure on packings is reduced, the points representing these packings in a $Z - n_{\rm m}$ diagram approach the generalized isostaticity line (Henkes *et al.*, 2009). Moreover, if the contacts at the mobility threshold are allowed to slip in analyzing the rotation-vibration dynamics, the density of states of packings close to the jamming threshold is strongly enhanced at low frequencies for *all* values of the friction coefficient μ—see Fig. 9.12(b). Moreover, as the pressure is increased to tune the packings away

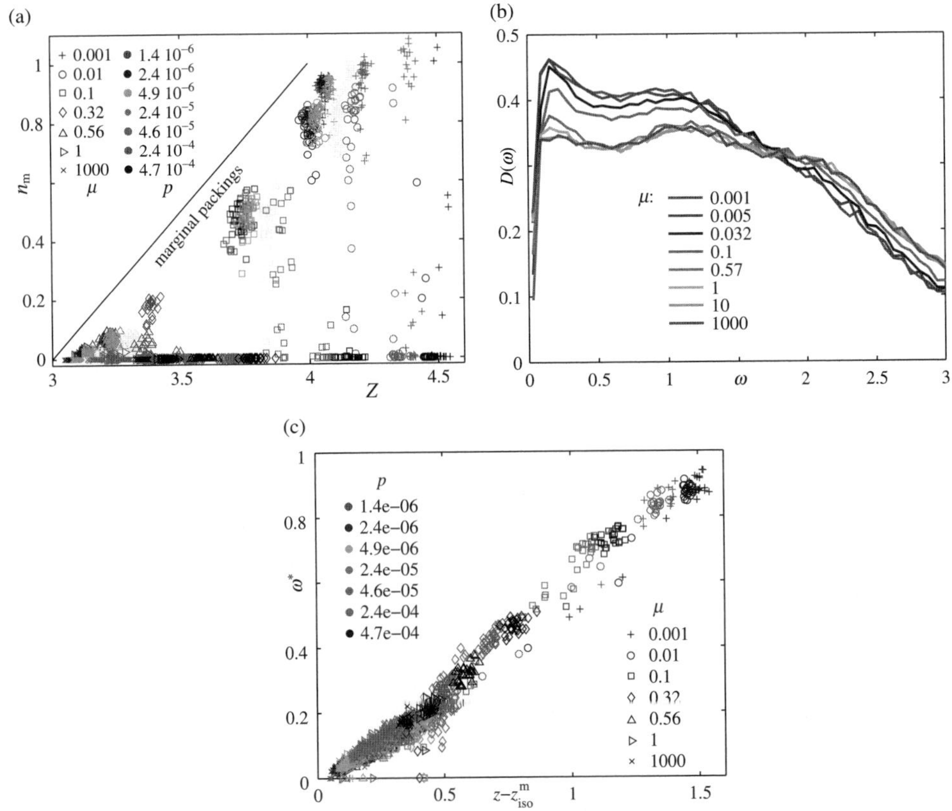

Fig. 9.12 Data from Henkes *et al.* (Henkes *et al.*, 2009) on frictional packings. (a) In gently prepared frictional packings, there a finite fraction of the contacts is found to lie at the mobilization threshold (are "fully mobilized"), i.e. has $f_t = \mu f_n$. In this diagram a large number of packing configurations are plotted in a diagram of Z and n_m, the average number of fully mobilized contacts per particle. For lower pressure, the dots in this diagram representing the packings approach the generalized isostaticity line, defined by Eq. (9.9), where packings are marginal. (b) Density of states for the lowest pressure data for a large range of values of the friction μ. In all cases the density of states is strongly enhanced at low frequencies. The frequency plotted here is actually the scaled frequency $\tilde{\omega}$ mentioned in Section 9.2.3. (c) The crossover frequency ω^* scales as $\Delta Z = Z - Z^{\mathrm{m}}_{\mathrm{iso}}$.

from the generalized isostaticity line, the crossover frequency ω^* in the density of states scales linearly with $\Delta Z = Z - Z^{\mathrm{m}}_{\mathrm{iso}}$, i.e. linearly with the distance from the generalized isostaticity line. This is the first confirmation of the robustness of the concepts reviewed above to frictional particles.

There are still several unresolved subtleties associated with the somewhat singular nature of the Coulomb threshold condition, but we refer for a discussion of these to (Henkes *et al.*, 2009). Another interesting recent development (Henkes *et al.*, 2008) is

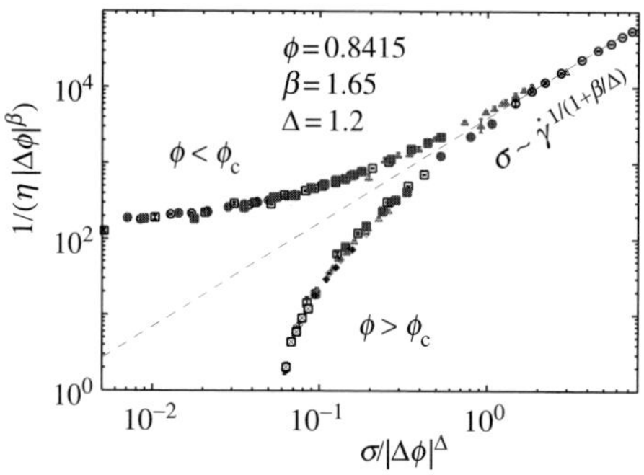

Fig. 9.13 Scaling plot of Olsson and Teitel (Olsson and Teitel, 2007) for simulations of the bubble model under shear. Plotted is the viscosity $\eta = \sigma/\dot{\gamma}$, rescaled by the density, as a function of the applied shear stress σ, with $\dot{\gamma}$ the shear rate. The scaling collapse illustrates that close to the jamming point, non-trivial rheology occurs. Close to the jamming point, we have $\sigma \sim \dot{\gamma}^{1/(1+\beta/\Delta)} \simeq \dot{\gamma}^{0.4}$.

that precursors to avalanches in simulations of tilting of piles of frictional spheres are found in the correlation properties of balls with low contact number.

9.3.3 Finite shear forces: non-trivial rheology in the sheared bubble model

So far we have only discussed static packings and their static and dynamic linear response. An important development of the last two years is that simulations of bubble models under finite shear stress or at a constant shear rate are probing the jamming phase diagram along the stress direction. The bubble model (Durian, 1995, 1997) is essentially a soft-sphere model, enriched with viscous friction, i.e., with dynamical terms proportional to the difference between the velocity of a bubble and that of each one it makes contact with. A particular attractive feature of such bubble-model simulations with dynamic friction is the fact that the relevant reference frame for the static limit is the frictionless sphere or disc model, about which so much is known.

Olsson and Teitel (Olsson and Teitel, 2007) were the first to demonstrate that the rheological data for shear flows in such systems show very good data collapse close to the jamming point of the frictionless case. Figure 9.13 shows a result from their simulations. In this particular plot, the effective viscosity $\eta = \sigma/\dot{\gamma}$ is plotted as a function of the shear stress σ imposed on the system; here $\dot{\gamma}$ is the shear rate. By scaling both quantities properly with the distance $\Delta\phi$ from the jamming density, there is very good data collapse both above and below the critical density. Moreover, at the jamming density, and close to the jamming density for large enough shear rates $\dot{\gamma}$, the data is consistent with a scaling $\sigma \sim \dot{\gamma}^{0.4}$. The excitement in the field is due to the fact that this is one of the first examples of how a simple microscopic model leads to non-trivial rheological behavior at mesoscopic or macroscopic scales, and the fact that this is intimately related to the jamming scenario.

The more customary way to plot rheological data is to plot the stress σ as a function of the imposed shear rate $\dot{\gamma}$; this is how data like that of Fig. 9.13 are usually presented. At the time of writing this review, there is a lot of debate about the precise exponents (Olsson and Teitel, 2007; Hatano, 2008*b*,*a*,*c*; Olsson and Teitel, 2009; Tighe *et al.*, 2010; Haxton and *et al.*, 2009), the existence of a quasi-static limit, the influence of the mean-field approximation employed in the simulations of (Olsson and Teitel, 2007), the question of whether there is a finite velocity correlation length that diverges at the jamming point, or one that is always of order of the system size, etc. From our point of view, the essential question, however, is whether, just like in the static case, the exponents should be rational values that can be understood in terms of simple arguments based on our insight in the response of a marginal solid. To date, the answer to this question is unknown.

9.4 Real physical systems

As stated in the introduction, space limitations do not allow us to pay tribute appropriately to the recent experimental developments that provide new impetus to the field. Some of these are reviewed in the chapters by Cipelletti and Weeks on colloids and by Dauchot, Durian and van Hecke on grains and foams. We wish to draw attention to a few additional points here.

9.4.1 Granular materials

Excited granular matter turns out to be an excellent model system to probe dynamical heterogeneities and the pair correlation function $g(r)$—think of the experiments with discs shaken on a table (Lechenault *et al.*, 2008*a*,*b*) or the ball-bearing experiments (Abate and Durian, 2006; Keys *et al.*, 2007) discussed in Section 9.2.7. Unfortunately, it is extremely difficult to probe the vibrational density of states with grains. The scaling of static properties like the contact number is also difficult to study experimentally: in experiments one will always have friction, whereas most of the clean theoretical scaling predictions are for frictionless spheres—as we have seen in Section 9.3.2, with friction the behavior depends stongly on the way a packing is made, so it is not even clear whether clear scaling behavior should be expected in the experiments on photoelastic discs (Majmudar *et al.*, 2007) aimed at measuring Z as a function of packing density. (However, experiments using photoelastic discs have been very powerful in studying force-chain networks and correlations (Majmudar and Behringer, 2005).) A powerful way to probe elasticity is to use ultrasound measurements, which, in agreement with what we have argued above, support a small G/B ratio in sand (Bonneau *et al.*, 2008; Jacob *et al.*, 2008) and a sharp transition toward strong scattering (Jia *et al.*, 1999) as the frequency increases. Another possibility that is being explored to probe the contact statistics and elastic moduli by using particles that swell when the humidity of the environment is changed (Cheng, 2009). Finally, experiments on avalanches are convenient for probing the length scales involved in failure and flows (Forterre and Pouliquen, 2008). In particular, the angle at which an avalanche starts depends strongly on the thickness of the granular layer. A model based on the stabilization of anomalous modes by the rough fixed boundary at the bottom of the granular layer is consistent with the observed dependence (Wyart, 2009*a*).

9.4.2 Emulsions and foams

It is important to stress here that some of the observations associated with the jamming approach—in particular, the scaling $\Delta Z \sim \sqrt{\Delta\phi}$—already appeared in some form in the early work on bubbles and the bubble model (Durian, 1995, 1997). Moreover, it has become increasingly clear that wet foams are a very good experimental model system for jamming studies, since the friction is dynamic, not static. As a result, models of frictionless spheres or discs provide a reference frame to study the effect of shear. Indeed, recent experiments on sheared foams (Katgert *et al.*, 2009*b*,*a*) hold the promise for testing some of the predictions of the sheared bubble model that were discussed in Section 9.3.3.

Another promising route is to use emulsions or foams. Here too, experiments from over a decade ago (Mason *et al.*, 1995; Saint-Jalmes and Durian, 1999) already gave indications for deviations from the behavior expected for affine deformations. New, more-refined experiments, are called for. Emulsions have recently also been used to probe the structure of static packings (Clusel *et al.*, 2009).

9.4.3 Colloids

Soft colloids, such as foams, emulsions and particles made from poly(N-isopropyl acrylamide) microgels (NIPA particles) (Saunders and Vincent, 1999; Pelton, 2000), can be used to study densities above the jamming transition. In aqueous solution, NIPA colloids swell substantially as the temperature is reduced only slightly. As a result, the packing fraction of the sample can be varied over a wide range with only a minimal change of temperature. The systems are thermal in that the particles are small enough (micròmeter-sized) to exhibit significant Brownian motion. Thus, such systems are well suited for studying jamming at a fixed non-zero temperature.

9.5 Connection with glasses

In this section, we will describe some of the potential connections between the physics that appears at the jamming transition at point J and what happens in real molecular glass formers. There are several phenomena that appear ubiquitously in amorphous solids and glasses yet are quite different from what is found in ordinary crystals.

9.5.1 Properties of glasses to be considered

Density of low-temperature excitations and the Boson peak: There are many more low-frequency excitations in a glass than in a crystal. One of the hallmarks of glasses are the so-called anomalous low-temperature excitations. Since 1971 (Zeller and Pohl, 1971), it has been known that the specific heat, c_V of glasses varies approximately linearly with temperature rather than as T^3 as would be predicted by the Debye calculation where the low-frequency modes are long-wavelength plane waves (Ashcroft and Mermin, 1976). At a somewhat higher frequency, above a frequency ω_{BP}, there is a dramatic excess of vibrational modes in glasses, known as the Boson peak (Sokolov *et al.*, 1995). That the Boson peak should be a ubiquitous property of amorphous solids is surprising since the underlying structure, which would normally determine the phonon spectrum in a crystal, is so different in different glasses. The linear specific heat is surprising since even disordered systems should have well-defined elastic moduli when averaged over large enough wavelengths. Therefore, at low enough frequencies, the modes should remain plane waves with a Debye spectrum. This linear specific heat has been ascribed to the quantum tunneling of groups of atoms between roughly equivalent structural configurations (Anderson *et al.*, 1972; Phillips, 1972) although these configurations have not been unambiguously identified.

Thermal conductivity: In a related vein, the heat transport in glasses is decidedly different from what it is in crystals (Anderson, 1981). At very low temperatures, the thermal conductivity, $\kappa \sim T^2$ rather than $\kappa \sim T^3$ as in crystals. At higher temperatures there is then a plateau region. Above the plateau there is a third regime where κ increases gently, $\kappa \sim T$. This behavior is surprisingly universal between different glasses yet there is no generally accepted explanation of either of the higher-temperature regimes. The lowest-temperature region, where $\kappa \sim T^2$, has been modeled as the scattering of plane waves off the localized tunneling systems used to explain the low-temperature linear term in the specific heat.

Failure under applied stress: When an amorphous solid is compressed or sheared sufficiently, it will start to fail. However, the failure often occurs in the form of rather localized rearrangement events rather than via a long-range catastrophic collapse (Falk and Langer, 1998; Langer, 2008; Widmer-Cooper *et al.*, 2009*a*). Since the system goes unstable when the frequency of a mode reaches zero, it is possible to identify a rearrangement with the lowest-frequency mode just before the rearrangement (Maloney and Lemaître, 2006). However, ideally one would like to predict well in advance where failure will occur, even in large systems with many plane-wave-like modes at low frequencies.

Structure: In metallic and colloidal glasses, the structure has often been described in terms of sphere packings (Finney, 1977; Cahn, 1980; Zallen, 1998). While this appears to be quite natural, there are certain features that have been related to the glassy structure that are not well understood. For example, the first peak in the pair distribution function, $g(r)$ is tall and sharp and there is a split second peak. Moreover, as the temperature is lowered in a supercooled liquid, there is no sign of the onset of rigidity in the static structure factor, $S(q)$; the first peak in $S(q)$ simply grows smoothly in height and decreases in width as the temperature is lowered through the glass-transition temperature, T_g (Busse and Nagel, 1981).

9.5.2 Relating the jamming paradigm to glasses

Low-temperature excitations and the boson peak in covalent glasses

Various theoretical approaches can reproduce a boson peak and some of the features of the corresponding normal modes. They are based on disorder and generally connect the peak to an elastic instability, as in effective medium models (Schirmacher *et al.*, 2007; Schirmacher, 2006), Euclidian random matrix theory (Grigera *et al.*, 2001) or mode coupling (Götze and Mayr, 2000). However, the structural parameters controlling the peak are imposed by hand in these theories (amount of disorder in (Schirmacher, 2006), density in (Grigera *et al.*, 2001) or structure factor in (Götze and Mayr, 2000)). What aspects of the structure are the most relevant for the peak is unclear. Moreover, disorder cannot be a generic explanation for the boson peak: silica has one of the largest peaks, but so do the corresponding crystals, the cristobalites, as shown in Fig. 9.14. It is sometimes argued that silica is a special case. The variational argument of Section 9.2.3 shows that is not so: generically amorphous and crystalline elastic networks must have qualitatively similar low-frequency spectra if their coordination is identical. For example, a cubic lattice has a flat spectrum, as does a random close packing of elastic particles. Thus, coordination rather than positional disorder matters (but disorder in the spring stiffnesses obviously matters and can be incorporated in an improved variational argument (Wyart, 2005)).

The variational argument of (Wyart *et al.*, 2005*a*) has broad applications, as illustrated for silica and other tetrahedral networks following the analysis of (Wyart, 2005, 2009*b*). A particularity of silica or germanium oxide is the weak force constant associated with the rotation of two adjacent tetrahedra, in comparison with deforming the tetrahedra themselves. This suggests modeling silica as an assembly of stiff tetrahedra connected by flexible joints, the RUM model (Tkachenko *et al.*, 2000), as sketched in Fig. 9.14. When this approximation is used on configurations obtained at various pressures to compute the vibrational spectrum, one finds a remarkable similarity between silica and particles near jamming as illustrated in Fig. 9.14. The cause is identical: at reasonable pressures, silica is made of tetrahedra, and 5-fold defects are extremely rare. If joints are flexible, a tetrahedral network is isostatic (there are 6 degrees of freedom per tetrahedron, and 3 constraints per joint, and twice as many joints than tetrahedra) and must have a flat density of states, as observed. For real silica, the rotation of the joints has a finite stiffness, which must shift the anomalous modes forming the plateau by some frequency scale Δ. Using *ab initio* computation of the joint-bending force constant, and the molecular weight of silica one estimates $\Delta = 1.4\,\text{Thz}$. Empirically the density of states of silica indeed has a plateau above approximatively 1 Thz, as expected. This is shown in Fig. 9.14 for numerical simulations. This argument also applies to germanium oxide. For amorphous silicon, the bending force constant of the joints is comparable to the other interactions, since they correspond to the same covalent bond, and one therefore does not expect a boson peak at low frequency; indeed in this material the boson peak is nearly non-existent. Wyart *et al.* (Wyart *et al.*, 2008) extend these arguments to other covalent networks.

Low-temperature excitations and the boson peak in molecular glasses

We can make progress in understanding the excess number of low-frequency excitations in molecular glasses by recalling some of the properties of solids near the jamming transition. At point J, the jamming transition for frictionless repulsive spheres, the density of states has a plateau all the way down to zero frequency. Upon compression, the plateau persists only down to a lower cutoff frequency, ω^*(Silbert *et al.*, 2005; O'Hern *et al.*, 2003) [Figs. 9.2 and 9.3(b)] as explained in (Wyart *et al.*, 2005*a*). Below that frequency, the normal modes are weakly scattered plane waves, while above it

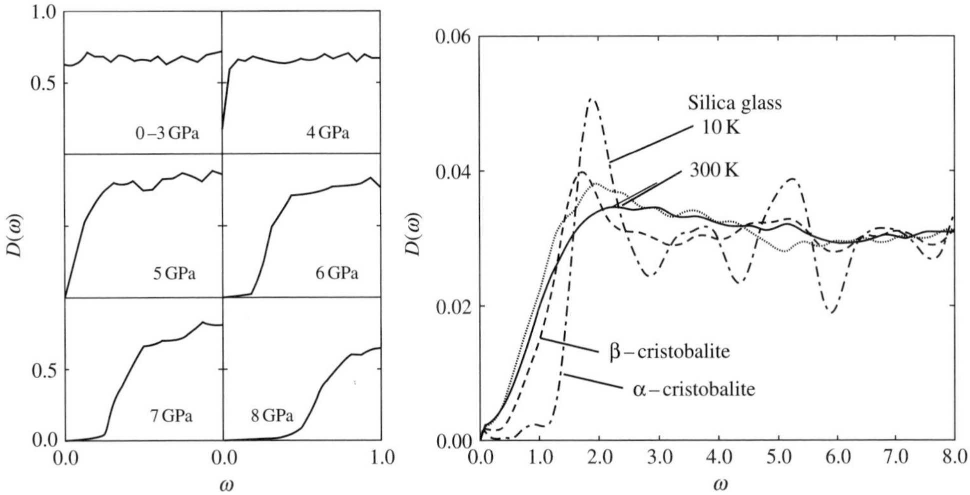

Fig. 9.14 Left: Configuration of amorphous silica as modeled by RUM: SiO_4 tetrahedra are connected by springs of finite stiffness at the joints, which can freely rotate (Tkachenko *et al.*, 2000). Center: Density of states $D(\omega)$ for amorphous silica configurations obtained at various pressures within the RUM approximation. At low pressure, there are only tetrahedra, and $D(\omega)$ is essentially flat. As the pressure increases 5-fold coordinated silicon atoms become more and more numerous, and $D(\omega)$ erodes at low frequency due to increased coordination, very much like in sphere packings (Tkachenko *et al.*, 2004). Right: Computations of $D(\omega)$ for atomic models of silica. Silica glass, β-cristobalite and α-cristobalite all present a plateau above 1 Thz, the boson-peak frequency (Tkachenko *et al.*, 2000).

they are strongly scattered (Silbert *et al.*, 2009; Xu *et al.*, 2009*b*; Vitelli *et al.*, 2009; Wyart, 2010). This jump in $D(\omega)$ appears to be the counterpart to the dramatic increase in the number of vibrational modes that appear at the boson peak in glasses (Silbert *et al.*, 2005; Xu *et al.*, 2007; Wyart, 2005; Wyart *et al.*, 2005*a*).

In order to assert this correspondence more generally, we have ascertained that the physics dominating point J for repulsive spheres is also operative for other systems such as those with particles of non-spherical shape or with particles mediated by long-range and/or attractive interactions. To include such effects, we have studied several different models of glasses that can be analyzed in terms of their proximity to the jamming transition. We refer to Section 9.3.1 on ellipsoids for a brief discussion of what happens when the particles are non-spherical—in essence, the results are consistent with the jamming scenario based on the analysis of frictionless spheres, though in a surprising and non-trivial way.

We have also studied models that were based on variants of the Lennard-Jones interactions between spheres. Such potentials allow attractive as well as long-range interactions so that the average coordination number can be arbitrarily large. Lennard-Jones potentials are more realistic models of molecular systems than the simple harmonic force law described above because the potential decays rapidly with interparticle separation. Particles separated by even a slightly larger distance, interact much more weakly than those closer together.

We have shown (Wyart, 2005; Xu *et al.*, 2007) that the onset of excess modes in these systems derives from the same variational-argument considerations that arise at point J for systems with finite-ranged repulsions. It can again be estimated by analyzing the vibrational energy originating from the excess contacts per particle over the minimum number needed for mechanical stability: $Z - Z_c$. The extra contacts can be divided into Z_1 strong contacts and the remaining $Z - Z_1$ weak ones. Strong contacts shift the energy cost of a mode that is initially zero in the isostatic limit to a non-zero value according to the variational argument outlined above, while weak contacts shift the energy simply by increasing the restoring force for displacements. The energy is then minimized to obtain the fraction of strong contacts. Even though these glasses have a high coordination number, most of the additional contacts can be considered to be weak.

These results are shown in Fig. 9.15 for two models. ω^* is the theoretical prediction for the onset frequency of excess modes based on the variational argument. The onset frequency determined from simulation is given by $\omega^\dagger$. The agreement between ω^* and $\omega^\dagger$ is very good. On the right-hand axis $\delta Z \equiv (Z_1 - Z_c)/Z_c$ is shown. Note that even

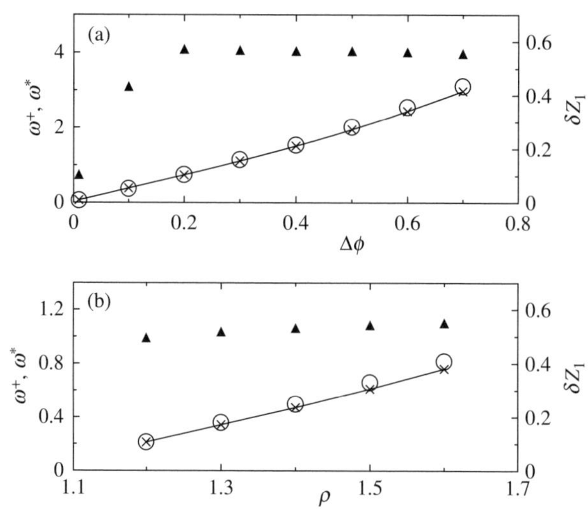

Fig. 9.15 Characteristic frequencies and fractional deviation from isostaticity for two three-dimensional systems based on Lennard-Jones potentials. Left panel shows data for repulsive Lennard-Jones bidisperse mixtures where only the repulsive part of the potential is included versus $\Delta\phi$, distance from the unjamming transition. Right panel shows data versus density, ρ, for a full Lennard-Jones potential with cutoff at $r = 2.5\sigma_{i,j}$ where $\sigma_{i,j}$ is the average diameter of particles i and j. On the left axis, $\omega^\dagger$ (open circles) is the onset frequency of the anomalous modes calculated from numerical simulations and the corresponding predictions for ω^* (crosses) based on a variational calculation. The agreement is very good. Lines are to guide the eye. On the right axis is the fractional deviation from isostaticity: $\delta Z \equiv (Z_1 - Z_c)/Z_c$ (solid triangles). Figure taken from (Xu *et al.*, 2007)

though Z itself can be arbitrarily large, δZ remains small and does not get larger that 0.6. This is why these systems with high coordination can still be understood in terms of the physics of point J and isostaticity. Even though it may be impossible to reach the jamming transition itself, for example because there are long-range interactions or attraction (as in the Lennard-Jones system) the effects of the long-ranged part of the potential can be treated as a correction to the pre-dominant effects of jamming.

Thermal conductivity: The thermal conductivity, κ, of a system near the jamming threshold has been calculated and captures well the thermal conductivity of silica glass above its plateau (Xu *et al.*, 2009*b*; Vitelli *et al.*, 2009). Because the phonons are scattered so strongly by the disorder, phonon–phonon scattering is relatively unimportant so the thermal transport can be calculated within the harmonic approximation without recourse to the anharmonic properties of the modes. This is in strong contrast to the situation for crystals, where anharmonicity must be included to obtain a thermal conductivity that is not infinite. At the jamming threshold, the long-wavelength phonons that give rise to a diverging thermal conductivity in the harmonic limit are completely suppressed by the excess modes, since the density of vibrational states remains flat down to zero frequency. As a result, the thermal conductivity is finite even within the harmonic approximation at this special point. The thermal conductivity can be expressed in terms of the diffusivity, $d(\omega)$ (defined in Section 9.2.6), and heat capacity, $C(\omega)$, of each mode:

$$\kappa(T) = \int_0^\infty D(\omega)\ C(\omega) d(\omega)\ d\omega, \tag{9.10}$$

In a weakly scattering system such as a crystal, $d(\omega) = c\ell(\omega)/3$, where c is the sound speed and $\ell(\omega)$ is the phonon mean-free path. As discussed in Section 9.2.6, Fig. 9.8 shows that at the jamming transition $d(\omega)$ has a very simple shape: it is a plateau that extends from our lowest frequency up to the onset of the high-frequency localized states (Xu *et al.*, 2009*b*; Vitelli *et al.*, 2009), where $d(\omega)$ falls rapidly to zero. The magnitude of the diffusivity in the plateau region, d_0, is small and scales as: $d_0 \sim \sigma^2 \sqrt{\frac{k_{\text{eff}}}{M}}$, where $k_{\text{eff}} = \frac{\partial^2 V(r_{ij})}{\partial r_{ij}^2}$ is the effective spring constant of the system that scales as the bulk modulus, B. Effective-medium theory suggests that these results hold more generally for weakly coordinated disordered elastic networks (Wyart, 2010).

Upon compression, the plateau region in $d(\omega)$ no longer extends down to zero frequency but only to a frequency that is proportional to ω^*. Thus, all the modes above ω^* (excluding the truly localized modes at high frequency) have a nearly constant diffusivity. As we emphasized in the section on low-temperature excitations, ω^* obtained from the density of states corresponds to the frequency of the boson peak. Here, we find that this same frequency also is proportional to the onset of a region of flat diffusivity for thermal transport (Xu *et al.*, 2009*b*; Vitelli *et al.*, 2009). It is just such a region of constant diffusivity that was postulated by Kittel (Kittel, 1949) to explain why the thermal conductivity of glasses has a weak, nearly linear, temperature dependence above the thermal-conductivity plateau. This is consistent with other evidence in glasses that the end of the plateau in $\kappa(T)$ corresponds to the boson peak (Zeller and Pohl, 1971).

Failure under applied stress: As summarized in Section 9.2.6, the vibrations of soft-sphere packings are quasi-localized near the onset of anomalous modes, ω^*: they have large displacements of relatively few particles, while the rest of the mode has a very small amplitude. As the sample is compressed or sheared, it will eventually go unstable and rearrange into a new configuration. At zero temperature, this instability is governed by some quasi-localized mode in the system that goes "soft" so that its frequency is pushed down until it reaches zero (Maloney and Lemaître, 2006). At this point, the sample moves into a new configuration. One cannot predict too much from the purely harmonic properties of a mode about what it will do when it is pushed to such a large amplitude that it goes unstable. However, at the very least one can say that at the point of instigation, the instability will initially have a very low participation ratio. This is consistent with the observation that when a glass fails due to pressure or shear stress, the failure tends to be highly localized in shear-transformation zones (Falk and Langer, 1998; Langer, 2008). So far, however, it has not been possible to identify shear-transformation zones unless they flip (unless a rearrangement occurs). The existence of many low-frequency quasi-localized modes with low energy barriers suggests that it may be possible to identify multiple shear-transformation zones from structural properties of the quasi-localized modes, and to predict which ones will flip *a priori*, even in large systems where there are many plane-wave-like modes at low frequencies.

Structure: At point J, the pair distribution function has many singular properties. In particular, we have shown that the first peak in $g(r)$ is a δ-function at the nearest-neighbor distance, σ. The area under this δ-function is just the isostatic coordination number: $Z = 6$ in three dimensions. On the high side of this δ-function, $g(r)$ has a power-law decay: $g(r) \propto (r - \sigma)^{-0.5}$. Also, at the jamming transition, the second peak of $g(r)$ is split into what appear to be two singular subpeaks one at $r = \sqrt{3}\sigma$ and the other at $r = 2\sigma$. There is a divergent slope just below each subpeak and a step-function cutoff on its high side. The existence of a split second peak in $g(r)$ is a feature that is seen in many glasses, such as metallic and colloidal glasses. One can thus trace the origin of these subpeaks to the geometry at the jamming threshold. One can also ask why, as a supercooled liquid is cooled into the glassy state, there is no signature in the static structure factor, $S(q)$, which is how the structure of molecular glasses is determined from scattering experiments. At the jamming threshold, there are indeed clear signatures in the structure of a sample: these are the divergences just mentioned in $g(r)$. However, because $S(q)$ is the Fourier transform of $g(r)$, the divergences simply appear as slightly more pronounced oscillations in $S(q)$ out to very large wavevector.

9.6 Connection with supercooled liquids

The understanding of the glass transition remains one of the enduring mysteries of condensed-matter physics. There is no doubt that the time scale for relaxation increases dramatically (faster than an Arrhenius law for what are called "fragile" glasses) as the temperature is lowered from the melt towards the glass-transition temperature, T_g, but it is much less clear whether there are any static growing length scales that can be identified that accompany the slowing down (Ediger *et al.*, 1996;

Debenedetti and Stillinger, 2001). However, as is well documented in this book, there is a growing length scale associated with heterogeneities in the dynamics, which become increasingly collective as the viscosity increases. The origin of the super-Arrhenius increase of relaxation time with decreasing temperature (i.e. fragility) is controversial. Is it due to free-volume effects, thermal activation over energy barriers that grow with decreasing temperature, or some other type of cooperative motion?

9.6.1 Glass transition and soft modes in hard-sphere liquids

Goldstein (Goldstein, 1969) has proposed that the slowing down of the dynamics is associated with the emergence of metastable states as a liquid is cooled. Such a roughening of the energy landscape is predicted by mean-field models of liquids and spins (Mezard and Parisi, 1996; Kurchan and Laloux, 1996; Castellani and Cavagna, 2005; Biroli and Bouchaud, 2009) at the so-called mode-coupling temperature T_{MCT}. However, these theoretical approaches lead to a (non-observed) power-law divergence of the viscosity at T_{MCT}, and their interpretation in real space is unclear and actively studied (Franz and Montanari, 2007; Biroli *et al.*, 2006). Empirically, a direct validation of Goldstein scenario in liquids is difficult. Several numerical analysis (Angelani *et al.*, 2000; Broderix *et al.*, 2000; Grigera *et al.*, 2002) have claimed to observe a transition in the energy landscape at T_{MCT}, but in our opinion the interpretation of these results is disputable.[1] Another intriguing aspect of Goldstein proposal is its application to hard spheres, where free volume rather than energy matters.

The connection between microscopic structure and vibrational spectrum established above directly supports the view that the emergence of metastable states slows down the dynamics, and yields a geometrical interpretation of this phenomenon (Brito and Wyart, 2009). In order to see that, the theoretical description of anomalous modes and their associated length scales must be extended to finite temperature. An analogy between hard-sphere liquids and athermal elastic systems can be built (Brito and Wyart, 2006; Brito and Wyart, 2009) in two steps: (i) on short time scales particles rattle rapidly but no large rearrangements occur. As illustrated in Fig. 9.16, a contact force network can be defined, following earlier work on granular matter (Ferguson *et al.*, 2004), which represents the average force exchanged between particles. From such networks a coordination number is computed by counting the pairs of particles with a non-zero contact force. (ii) All the possible hard-sphere configurations associated with a given network can be summed up, leading to a computation of the free energy that can be expressed in terms of the mean positions

[1]In these numerics, equilibrium configurations close to the configurations visited by the dynamics are analyzed. Their number of saddles (or unstable modes) is computed as a function of the temperature, and it is argued that this quantity can be fitted by a curve that vanishes at a temperature T_{MCT}, defined here by fitting the dynamics as $\tau \sim (T - T_{\text{MCT}})^{-a}$, where τ is the α-relaxation time scale and a some exponent. From this observation it is stated that metastable states appear at T_{MCT}. However, this observation is precisely the behavior expected if metastable states had already appeared at larger temperature, implying that the dynamics is activated in the neighborhood of T_{MCT}. Indeed assuming activated dynamics, the number of saddles n_s reflects the probability that a region is in the process of rearranging (crossing a barrier), which occurs with a probability of order $1/\tau$. Thus, one expects $n_s \sim 1/\tau$, which can be fitted with a vanishing quantity at T_{MCT} simply due to the definition of that temperature.

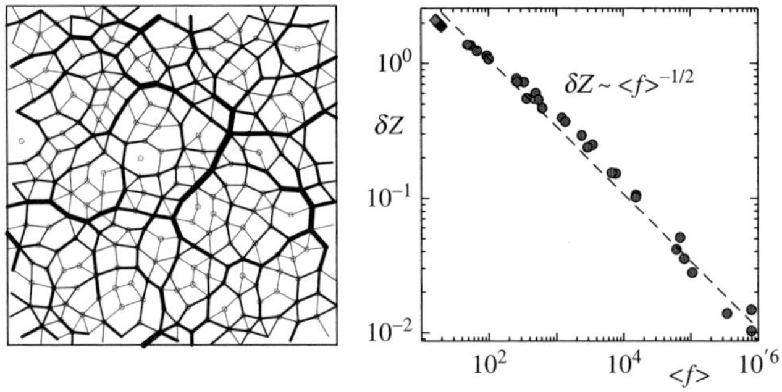

Fig. 9.16 Left: Illustration of the fact that the problem of hard elastic discs at finite temperature and density in the supercooled liquid or glass phase (in this case $\phi_c - \phi = 2 \times 10^{-4}\phi_c$), can, in between rearrangement events, be mapped onto the problem of a static packing with a logarithmic interaction potential (Brito and Wyart, 2006, 2007). Essentially, particles experience a reduced free volume due to collisions with close neighbors, leading to an effective repulsive interaction. Thus, particles exert a force on each other. The plot shows the average contact forces for 256 particles averaged over 10^5 collisions. The points represent the centers of the particles, and the thickness of the line is proportional to the force strength. Since in between rearrangement events the structure is stable, forces are balanced on each particle (Brito and Wyart, 2006, 2007). Right: ΔZ inferred from the coordination of the force network vs. average contact force $\langle f \rangle$, a measure of pressure. Measurements are made in the supercooled liquid (diamonds) and in the glass (circles). The observed coordination is well captured by its minimal value $\Delta Z \sim \langle f \rangle^{-1/2}$ allowing mechanical stability, supporting the view that the glass is marginally stable (Brito and Wyart, 2009).

of the particles. The effective interaction is found to be logarithmic. Expanding the free energy defines a dynamical matrix and vibrational modes that characterize the linear response of the mean particle positions to any imposed external forces within a metastable state.

The results on the vibrational spectra of athermal elastic networks apply to this case as well, and the stability criterion of Eq. (9.8) must be satisfied in any metastable configuration. For hard spheres, it can be written $\Delta Z > p^{-1/2} \sim h^{1/2}$ as the pressure p satisfies $p \sim 1/h \sim 1/\Delta\phi$, where h is the typical average gap among particles in contact. As illustrated in Fig. 9.16, the bound is approximatively saturated and the glass appears to have just enough contacts to maintain its stability. Marginal stability is a fundamental feature of the glass, generically foreign to crystals. It also occurs in some mean-field spin models (Kurchan and Laloux, 1996). In hard spheres it implies the presence of anomalous modes near zero frequency and a curious behavior of the short-time dynamics, in particular a dependence on the mean-square particle displacements scaling as $\langle \delta R^2 \rangle \sim \Delta\phi^{3/2}$ rather than the $\Delta\phi^2$ dependence expected in a crystal (Brito and Wyart, 2009). Those predictions have been confirmed numerically (Brito and Wyart, 2007; Brito and Wyart, 2009; Mari *et al.*, 2009).

Why is a hard-sphere glass marginal? It has been argued (Brito and Wyart, 2009) that this must be so if the viscosity increases very rapidly when metastable states appear in the free-energy landscape. In the $(\Delta Z, h)$ plane sketched in Fig. 9.17,

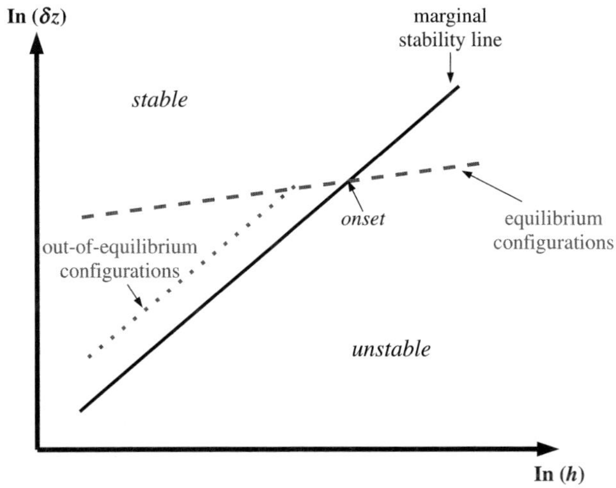

Fig. 9.17 Phase diagram for the stability of hard-sphere configurations, in the coordination δz vs. average gap h plane. The marginal stability line delimits stable and unstable configurations. The dashed line correspond to equilibrium configurations for different ϕ. As ϕ increases, h decreases and the two lines eventually meet. This occurs at the onset packing fraction ϕ_{onset}, where the dynamics become activated and thus intermittent. At larger ϕ, viscosity increases sharply as configurations visited become more stable. For a finite quench rate the system eventually falls out of equilibrium at the glass-transition packing fraction ϕ_0. More stable and more coordinated regions cannot be reached dynamically, and as ϕ is increased further, the system lives close to the marginal stability region, as indicated in the dotted line. The location of the out-of-equilibrium trajectory depends on the quench rate. In the limit of very rapid quench, the out-of-equilibrium line approaches the marginal stability line (Brito and Wyart, 2009).

Eq. (9.8) draws a line separating stable and unstable configurations. At equilibrium the mean value of ΔZ and h are well-defined functions of ϕ, and the corresponding curve is represented by the dashed line in Fig. 9.17. At low ϕ, gaps between particles are large and the configurations visited are unstable. As ϕ increases, the gaps decrease and configurations eventually become stable. This occurs at some ϕ_{onset} when the equilibrium line crosses the marginal-stability line. At larger ϕ, the dynamics becomes activated and intermittent, and the viscosity increases rapidly according to our hypothesis, so that on accessible time scales equilibrium cannot be reached deep in the regions where metastable states are present. As a consequence, the system falls out of equilibrium at some ϕ_0 larger than ϕ_{onset}. Configurations visited must therefore lie close to the marginal stability line, as represented by the dotted line in Fig. 9.17, since more stable, better-coordinated configurations cannot be reached dynamically.

Further support for the view that the viscosity is governed by the presence of an elastic instability near ϕ_{onset} comes from the observation that the sudden rearrangements that relax the structure occur along its softest modes (both in the

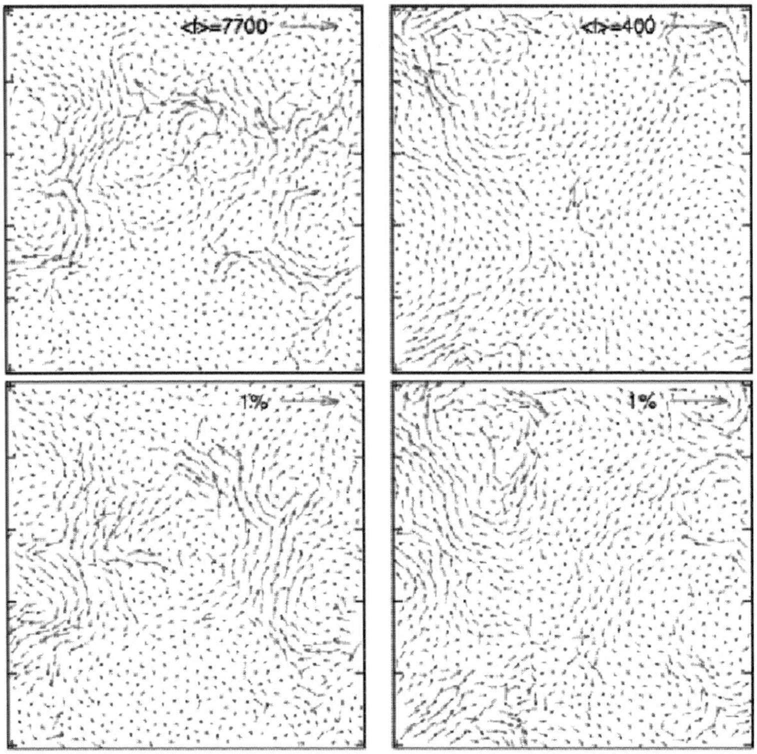

Fig. 9.18 Above: two examples of sudden rearrangements in the glass phase for different average contact force $\langle f \rangle$ for $N = 1024$ particles. Below: projection of these rearrangements on the 1% of the normal modes that contribute most to them, which systematically lie at the lowest frequencies (Brito and Wyart, 2009). This projection captures most of the displacement, indicating that a limited number of (collective) degrees of freedom are active during rearrangements.

liquid and in the glass phase) (Brito and Wyart, 2007), indicating that the lowest activation barriers lie in these directions, as illustrated in Fig. 9.18. Experiments in granular materials (Brito *et al.*, 2010) and numerical observations of normal modes of the energy of inherent structures (Schober *et al.*, 1993; Widmer-Cooper *et al.*, 2009*a*,*b*) reach similar conclusions. The normal modes of the free energy considered here are expected to correlate with the dynamics much better than those of the energy, and must be considered in particular when the interaction potential is strongly non-linear. The collective rearrangements involving a few tens of particles commonly seen in supercooled liquids correspond mostly to a few modes (Brito and Wyart, 2007), an effect even stronger in the glass phase. This analysis supports the view that the anomalous modes are the elementary objects to consider to describe activation, and that the collective aspect of rapid rearrangements is already present at the linear level in the structure of the modes.

This approach permits a quantitative study of activation, in particular to compute the fraction of the spectrum contributing to rearrangements. This quantity decreases rapidly with compression near the glass transition (Brito and Wyart, 2007), implying that fewer and fewer degrees of freedom are relevant for the dynamics. This observed reduction of active degrees of freedom presumably mirrors the slowing down and increasingly collective aspect of the rapid rearrangements that relax the structure (Candelier *et al.*, 2009), but this has yet to be clarified.

Overall, the present approach furnishes a simple spatial picture for the slowing down of the dynamics at intermediate viscosities: metastable states appear when the contact force network becomes sufficiently coordinated to resist the destabilizing effect of pressure. The localization for long times of an individual particle is thus not due to some specific properties of the cage formed by its neighbors, but has to do with the structure of the packing on a length l^* that can be large. Incidentally, the concept of caging, commonly used to picture dynamical arrest, is misleading.

9.6.2 Models of soft finite-range repulsions at finite temperature

By analyzing several models near the jamming threshold, some universal aspects of the slow-relaxation dynamics that appear in one asymptotic limit become apparent (Xu *et al.*, 2009). In particular, when the pressure, p, is small all the relaxation time data can be suitably plotted on a single master curve as a function of a single variable.

The models considered here all have finite-range repulsions. For this study, thermal energy was included in the simulations, and the relaxation time τ of the system was measured by studying when the intermediate scattering function, $f(q,t)$ drops to e^{-1} of its initial value.

Normally in simulations of this type, one measures time in units of $\sigma\sqrt{m/\epsilon}$, where σ is the particle diameter, m is the mass of the particle, and ϵ is the energy scale of the potential. However, in order to produce a collapse onto a master curve, τ is non-dimensionalized in another way by using $\tau\sqrt{p\sigma/m}$. Plotting this quantity versus $T/(p\sigma^3)$ produces an excellent collapse of the data for several different potentials of interaction at all values of $T/(p\sigma^3)$ as long as the non-dimensionalized pressure, $p\sigma^3/\epsilon$ is sufficiently small (Xu *et al.*, 2009). This is shown in Fig. 9.19. This limit is satisfied at low pressures in soft-sphere systems and at any pressure for a hard-sphere system where $\epsilon \to \infty$. Figure 9.19 shows that low-pressure soft-sphere data do indeed collapse with the hard-sphere result. In addition, Ref. (Xu *et al.*, 2009) found that limiting hard-sphere behavior remains a good approximation up to higher values of $p\sigma^3/\epsilon$ as $T/(p\sigma^3)$ increases.

This data collapse has two important implications. First, it shows that at sufficiently low pressures the relaxation at the glass transition is a function of only a single variable. The molecular glass transition is, in this limit, equivalent to the hard-sphere colloidal glass transition. Secondly, Ref. (Xu *et al.*, 2009) shows that there are systematic deviations from this universal behavior at larger values of the second control parameter: $p\sigma^3/\epsilon$. This demonstrates that there are at least two distinct physical processes that enter to produce glassy dynamics. What is important is that they can be cleanly separated, so that for sufficiently small $p\sigma^3/\epsilon$, the rescaled relaxation

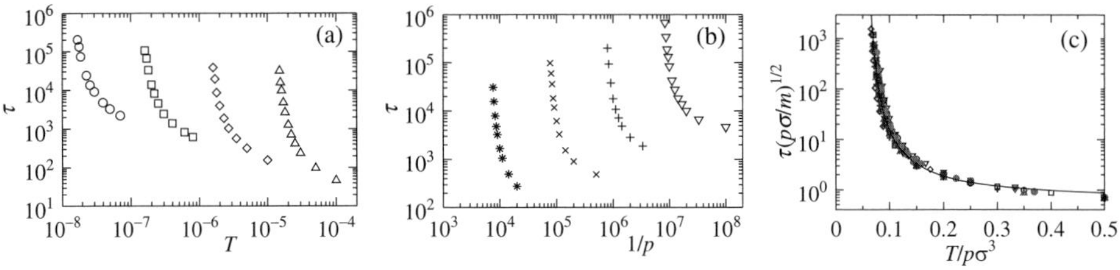

Fig. 9.19 Relaxation time, τ, versus temperature, T, (a) and versus inverse pressure, $1/p$, (b) for a system of spheres interacting via repulsive harmonic potentials. Panel (c) shows that all the relaxation time data for different potentials (harmonic, Hertzian and hard sphere) can be collapsed onto a single master curve. To get this data collapse, the relaxation time is non-dimensionalized by pressure and plotted versus $T/p\sigma^3$. Figure taken from (Xu *et al.*, 2009).

time is only a function of a single variable, $T/(p\sigma^3)$. Note that the existence of such a scaling function still does not tell us whether the rescaled relaxation time diverges at $T/(p\sigma^3) > 0$, implying that a thermodynamic glass transition exists, or at $T/(p\sigma^3) = 0$, implying that the glass transition is an isostatic jamming transition.

In order to understand the consequences of this result for molecular liquids, we must first understand the corrections to the leading hard-sphere behavior when we slowly increase $p\sigma^3/\epsilon$. These studies, have been done for repulsive spheres near the jamming threshold. However, molecular liquids typically have attractions and are at much higher densities than those considered here. To make the correspondence more general, we would have to push these systems farther from the jamming threshold. Nevertheless, this result shows how certain aspects of jamming, inherent in the hard-sphere liquid, are important for glass-forming liquids generally. As we have shown, at least one other distinct contribution to the relaxation must also be considered as the pressure is increased.

As one increases $p\sigma^3/\epsilon$, there are corrections to the leading hard-sphere behavior. It has been argued that these corrections can be collapsed onto a single curve by rescaling the pressure by some factor that varies with packing fraction (Berthier and Witten, 2009*b*).

9.7 Outlook to the future—a unifying concept

The vibrational spectrum of crystals consists of plane waves and leads to a description in terms of linear elasticity. In those systems, one can compute the non-linearities of the excitations and how they scatter from defects. Such a systematic description is lacking in amorphous solids. Various properties, such as thermal transport, force propagation, the vibrational spectrum and resistance to flow present phenomena that are not satisfactorily understood. An inherent difficulty in understanding these phenomena is that they are often governed by processes that occur on small length scales, of the order of a few particle sizes.

The discovery that the jamming threshold of athermal idealized spheres exhibits many properties of a critical point (O'Hern *et al.*, 2003) changes this state of affairs, as it enables one to separate the length scale on which disorder matters from the molecular scale. This has led to the realization that the spectrum of amorphous solids is characterized by a crossover, distinct from Anderson localization, above which modes are extended but not plane-wave-like, and that the characteristic length at which this crossover occurs is decoupled from the molecular length. The corresponding excitations, the anomalous modes, allow one to understand several seemingly disparate anomalies in glasses and granular materials in a coherent fashion. The picture advanced in this review suggests that (1) the length scale l^* corresponding to the cross-over characterizes force propagation in granular packings (Ellenbroek *et al.*, 2006, 2009*a*), (2) the frequency scale of the crossover corresponds to that of the boson peak in glasses (Silbert *et al.*, 2005; Wyart *et al.*, 2005*a*; Xu *et al.*, 2007; Wyart, 2005), (3) the low-energy diffusivity of the modes above the crossover is responsible for the mild linear increase of the thermal conductivity with temperature above the plateau in glasses (Xu *et al.*, 2009*b*; Vitelli *et al.*, 2009; Wyart, 2010), (4) at least for colloidal glasses, the microscopic structure is marginally stable toward anomalous modes (Brito and Wyart, 2006; Brito and Wyart, 2009), (5) the modes appear to govern structural rearrangements in liquids at least in the viscosity range that can be probed numerically (Brito and Wyart, 2007; Brito and Wyart, 2009; Widmer-Cooper *et al.*, 2009*a*)and (6) the structure of the modes corresponds to that of structural arrangements in amorphous solids under mechanical load (Xu *et al.*, 2009*a*; Manning and Liu, 2009). At the linear level those excitations and their consequences are now becoming rather well understood theoretically for sphere packings, with the notable exception of (i) the observed quasi-localization of the lowest-frequency anomalous modes (Xu *et al.*, 2009*a*; Zeravcic *et al.*, 2009) and (ii) the apparently fractal statistics of force propagation below the cutoff length ℓ^* above which continuum elasticity applies.

The jamming community is now beginning to study the non-linearities coupling these excitations. Much would be gained from such knowledge. On the one hand, it is possible that the lowest-frequency anomalous modes are related to two-level systems, which have been proposed on phenomenological grounds to explain transport and specific heat anomalies at sub-Kelvin temperatures. Despite decades of research, the spatial nature of two-level systems has remained elusive. If there is indeed a connection between the highly anharmonic, low-frequency, quasi-localized modes and two-level systems, then the energy barrier separating the two levels should vanish as the isostatic jamming transition is approached. On the other hand, since the low-frequency modes are involved in irreversible rearrangements, understanding their non-linearity is presumably essential for a microscopic description of the flow of supercooled liquids, granular matter and foams.

An exciting recent development is the advent of experiments capable of studying the vibrational properties of systems near the jamming transition (Abate and Durian, 2006; Ghosh *et al.*, 2009; Chen *et al.*, 2010; Gardel, 2009). In laboratory systems, particle motion is typically damped and the interparticle potentials are not necessarily known; this has made it difficult to make direct comparisons between experimental results and numerical and theoretical predictions until now. These complications have recently been overcome (Ghosh *et al.*, 2009; Chen *et al.*, 2010), so experiments should soon be able to test the extent to which predictions for ideal spheres apply to real systems.

Acknowledgments

We would like to thank Carolina Brito, Giulio Biroli, Jean-Philippe Bouchaud, Ke Chen, Doug Durian, Wouter Ellenbroek, Jerry Gollub, Peter Harrowell, Tom Haxton, Silke Henkes, Heinrich Jaeger, Alexandre Kabla, Randy Kamien, Steve Langer, Haiyi Liang, Tom Lubensky, L. Mahadevan, Xiaoming Mao, Kerstin Nordstrom, Corey O'Hern, David Reichman, Leo Silbert, Anton Souslov, Brian Tighe, Vincenco Vitelli, Martin van Hecke, Tom Witten, Erik Woldhuis, Ning Xu, Arjun Yodh, Zorana Zeravcic, and Zexin Zhang for valuable discussions, input, criticism and collaborations.

In addition, funding from DOE via DE-FG02-03ER46088 (SRN) and DE-FG02-05ER46199 (AJL) as well as the NSF MRSEC program via DMR-0820054 (SRN) and DMR05-20020 (AJL) is gratefully acknowledged. AJL and SRN both thank the Kavli Institute for Theoretical Physics, Santa Barbara, and AJL, SRN and MW thank the Aspen Center for Physics for their hospitality. WvS would also like to thank in particular Martin van Hecke for his help and support during the last few years, and FOM for its generous support over the years.

References

Abate, A. R. and Durian, D. J. (2006). *Phys. Rev. E*, **74**, 031308.

Agnolin, I. and Roux, J.-N. (2007). *Phys. Rev. E*, **76**, 061304.

Alexander, S. (1998). *Phys. Rep.*, **296**, 65–236.

Allen, P. B. and Feldman, J. L. (1989*a*). *Phys. Rev. Lett.*, **62**, 645.

Allen, P. B. and Feldman, J. L. (1989*b*). *Phys. Rev. B*, **48**, 12581.

Anderson, A. C. (1981). In *Amorphous Solids, Low Temperature Properties* (ed. W. A. Phillips), Springer, Berlin.

Anderson, P. W., Halperin, B. I., and Varma, C. M. (1972). *Philos. Mag.*, **25**, 1.

Angelani, L., Di Leonardo, R., Ruocco, G., Scala, A., and Sciortino, F. (2000). *Phys. Rev. Lett.*, **85**, 5356–5359.

Ashcroft, N. W. and Mermin, N. D. (1976). *Solid State Physics.* Brooks Cole, Stamford.

Berthier, L. and Witten, T. A. (2009*a*). *Europhys. Lett.*, **86**, 10001.

Berthier, L. and Witten, T. A. (2009*b*). *Phys. Rev. E*, **80**, 021502.

Biroli, G. and Bouchaud, J. P. (2009). *ArXiv e-prints*, 0912.2542.

Biroli, G., Bouchaud, J.-P., Miyazaki, K., and Reichman, D. R. (2006). *Phys. Rev. Lett.*, **97**, 195701.

Biswas, R., Bouchard, A. M., Kamitakahara, W. A., Grest, G. S., and Soukoulis, C. M. (1988). *Phys. Rev. Lett.*, **60**, 2280.

Bonneau, L., Andreotti, B., and Clément, E. (2008). *Phys. Rev. Lett.*, **101**, 118001.

Boolchand, P., Lucovsky, G., Phillips, J. C., and Thorpe, M. F. (2005). *Philos. Mag.*, **85**, 3823.

Bouchaud, J.-P. (2004). In *Les Houches Session LXXVII* (ed. J.-L. Barrat, M. Feigelman, J. Kurchan, and J. Dalibard). Springer, Heidelberg.
Brito, C., Dauchot, O., Biroli, G., and Bouchaud, J.-P. (2010). *Soft Matter*, **6**, 3013–3022.
Brito, C and Wyart, M. (2006). *Europhys. Lett.*, **99**, 149.
Brito, C and Wyart, M. (2007). *J. Stat. Mech. Theory - Exp*, L08003.
Brito, C. and Wyart, M. (2009). *J. Chem. Phys.*, **131**, 024504.
Broderix, K., Bhattacharya, K. K., Cavagna, A., Zippelius, A., and Giardina, I. (2000). *Phys. Rev. Lett.*, **85**, 5360–5363.
Buchenau, U., Galperin, Y. M., Gurevich, V. L., Parshin, D. A., Ramos, M. A., and Schober, H. R. (1992). *Phys. Rev. B*, **46**, 2798.
Buchenau, U., Pecharroman, C., Zorn, R., and Frick, B. (1996). *Phys. Rev. Lett.*, **77**, 659.
Busse, L. E. and Nagel, S. R. (1981). *Phys. Rev. Lett.*, **47**, 1848.
Cahn, R. W. (1980). *Contemp. Phys*, **21**, 43.
Candelier, R., Dauchot, O., and Biroli, G. (2009). *Phys. Rev. Lett.*, **102**, 088001.
Candelier, R., Dauchot, O., and Biroli, G. (2010). *Europhys. Lett.*, **92**, 24003.
Castellani, T. and Cavagna, A. (2005). *J. Stat. Mech.: Th. and Exp.*, **2005**, P05012.
Chaudhuri, P., Berthier, L., and Sastry, S. (2010). *Phys. Rev. Lett.*, **104**, 165701.
Chen, K., Ellenbroek, W. G., Zhang, Z., Chen, D. T. N., Yunker, P. J., Henkes, S., Brito, C., Dauchot, O., van Saarloos, W., Liu, A. J., and Yodh, A. G. (2010). *Phys. Rev. Lett.*, **105**, 025501.
Cheng, X. (2010). *Phys. Rev. E*, **81**, 031301.
Clusel, M., Corwin, E. I., Siemens, A. O. N., and Brujic, J. (2009). *Nature*, **460**, 611.
Debenedetti, P. G. and Stillinger, F. H. (2001). *Nature*, **410**, 259.
Donev, A., Cisse, I., Sachs, D., Variano, E. A., Stillinger, F. H., Connelly, R., Torquato, S., and Chaikin, P. M. (2004). *Science*, **303**, 990–993.
Donev, A., Connelly, R., Stillinger, F. H., and Torquato, S. (2007). *Phys. Rev. E*, **75**, 051304.
Drocco, J. A., Hastings, M. B., Olson Reichhardt, C. J., and Reichardt, C. (2005). *Phys. Rev. Lett.*, **95**, 088001.
Durian, D. (1995). *Phys. Rev. Lett.*, **75**, 4780.
Durian, D. (1997). *Phys. Rev. E*, **55**, 1739.
Ediger, M. D., Angell, C. A., and Nagel, S. R. (1996). *J. Phys. Chem.*, **100**, 13200.
Ellenbroek, W. G., Somfai, E., van Hecke, M., and van Saarloos, W. (2006). *Phys. Rev. Lett.*, **97**, 258001.
Ellenbroek, W. G., van Hecke, M., and van Saarloos, W. (2009*a*). *Phys. Rev. E*, **80**, 061307.
Ellenbroek, W. G., Zeravcic, Z., van Hecke, M., and van Saarloos, W. (2009*b*). *Europhys. Lett.*, **87**, 34004.
Fabian, J. and Allen, P. B. (1997). *Phys. Rev. Lett.*, **79**, 1985.
Falk, M. L. and Langer, J. S. (1998). *Phys. Rev. E*, **57**, 7192.
Ferguson, A., Fisher, B., and Chakraborty, B. (2004). *Europhys. Lett.*, **66**, 277–283.
Finney, J. L (1977). *Nature*, **266**, 309.
Forterre, Y. and Pouliquen, O. (2008). *Ann. Rev. Fluid Mech.*, **40**, 1–24.

Franz, S and Montanari, A (2007). *J. Phys. A: Math. Theor.*, **40**, F251.
Garboczi, E. J. and Thorpe, M. F. (1985). *Phys. Rev. B*, **31**, 7276.
Gardel, M. (2009). private communication.
Ghosh, A., Chikkadi, V. K., Schall, P., Kurchan, J., and Bonn, D. (2010). *Phys. Rev. Lett.*, **104**, 248305.
Goldenberg, C. and Goldhirsch, I. (2008). *Phys. Rev. E*, **77**, 041303.
Goldhirsch, I. and Goldenberg, C. (2002). *Eur. Phys. J E*, **9**, 245–251.
Goldstein, M. (1969). *J. Chem. Phys.*, **51**, 3728.
Götze, W. and Mayr, M. R. (2000). *Phys. Rev. E*, **61**, 587–606.
Grigera, T. S., Cavagna, A., Giardina, I., and Parisi, G. (2002). *Phys. Rev. Lett.*, **88**, 055502.
Grigera, T. S., Martín-Mayor, V., Parisi, G., and Verrocchio, P. (2001). *Phys. Rev. Lett.*, **87**, 085502.
Hatano, T. (2008*a*). ArXiv 0804.0477.
Hatano, T. (2008*b*). *Phys. Rev. E*, **79**, 050301(R).
Hatano, T. (2008*c*). *J. Phys. Soc. Jpn.*, **77**, 123002.
Haxton, T. *et al.* (2009). (private communication).
He, H. and Thorpe, M. F. (1985). *Phys. Rev. Lett.*, **54**, 2107–2110.
Head, D. A. (2005). *Phys. Rev. E*, **72**, 021303.
Head, D. A., Tkachenko, A. V., and Witten, T. A. (2001). *Eur. Phys. J. E*, **6**, 99–105.
Henkes, S, Brito, C., Dauchot, O., and van Saarloos, W (2010*a*). *Soft Matter*, **6**, 2939–2943.
Henkes, S., van Hecke, M., and van Saarloos, W. (2010*b*). *EPL*, **90**, 14003.
Jacob, X., Aleshin, V., Tournat, V., Leclaire, P., Lauriks, W., and Gusev, V. E. (2008). *Phys. Rev. Lett.*, **100**, 158003.
Jacobs, D. J. and Thorpe, M. F. (1995). *Phys. Rev. Lett.*, **75**, 4051.
Jacquin, H. and Berthier, L. (2010). *Soft Matter*, **6**, 2970–2974.
Jia, X., Caroli, C., and Velicky, B. (1999). *Phys. Rev. Lett.*, **82**, 1863–1866.
Johnson, K. L. (1985). *Contact Mechanics.* Cambridge University Press.
Katgert, G., Latka, A., Moebius, M. E., and van Hecke, M. (2009*a*). *Phys. Rev. E*, **79**, 066318.
Katgert, G., Moebius, M. E., and van Hecke, M. (2009*b*). *Phys. Rev. E*, **101**, 058301.
Keys, A. S., Abate, A. R., Glotzer, S. C., and Durian, Douglas J. (2007). *Nature Phys.*, **3**, 260–264.
Kittel, C. (1949). *Phys. Rev. B*, **57**, 972.
Kurchan, J. and Laloux, L. (1996). *J. Phys. A*, **29**, 1929–1948.
Langer, J. S. (2008). *Phys. Rev. E*, **57**, 021502.
Lechenault, F., Dauchot, O., Biroli, G., and Bouchaud, J.-P. (2008*a*). *Europhys. Lett.*, **83**, 46003.
Lechenault, F., Dauchot, O., Biroli, G., and Bouchaud, J.-P. (2008*b*). *Europhys. Lett.*, **83**, 46002.
Liu, A. J. and Nagel, S. R. (1998). *Nature*, **396**, 21–22.
Lubachevsky, B. D. and Stillinger, F. H. (1990). *J. Stat. Phys.*, **60**, 561.
Mailman, M., Schreck, C. F., O'Hern, C. S., and Chakraborty, B. (2009). *Phys. Rev. Lett.*, **102**, 255501.

Majmudar, T. S., Sperl, M., Luding, S., and Behringer, R. P. (2007). *Phys. Rev. Lett.*, **98**, 058001.

Majmudar, T. S. and Behringer, R. P. (2005). *Nature*, **435**, 1079–1082.

Maloney, C. E. and Lemaitre, A. (2006). *Phys. Rev. E*, **74**, 016118.

Manning, L. M. and Liu, A. J. (2009). private communication.

Mao, X., Xu, N., and Lubensky, T. C. (2010). *Phys. Rev. Lett.*, **104**, 085504.

Mari, R., Krzakala, F., and Kurchan, J. (2009). *Phys. Rev. Lett.*, **103**, 025701.

Mason, T. G., Bibette, J., and Weitz, D. A. (1995). *Phys. Rev. Lett.*, **75**, 2051–2054.

Mezard, M. and Parisi, G. (1996). *J. Phys. A*, **29**, 6515.

Monaco, G. and Mossa, S. (2009). *PNAS*, **106**, 16907.

Moukarzel, C. F. (1998). *Phys. Rev. Lett.*, **81**, 1634–1637.

O'Hern, C. S., Langer, S. A., Liu, A. J., and Nagel, Sidney R. (2002). *Phys. Rev. Lett.*, **88**, 075507.

O'Hern, C. S., Silbert, L. E., Liu, A. J., and Nagel, S. R. (2003). *Phys. Rev. E*, **68**, 011306.

Olsson, P. and Teitel, S. (2007). *Phys. Rev. Lett.*, **99**, 178001.

Olsson, P. and Teitel, S. (2009). (private communication).

Pelton, R. (2000). *Adv. Colloid Int. Sci.*, **85**, 1.

Phillips, J. C. (1979). *J. Non-Cryst. Sol.*, **34**, 153–181.

Phillips, W. A. (1972). *J. Low Temp. Phys.*, **7**, 351.

Roux, J.-N. (2000). *Phys. Rev. E*, **61**, 6802–6836.

Sacanna, S., Rossi, L., Wouterse, A., and Philipse, A. P. (2007). *J. Phys. C: Condens. Matter*, **19**, 6108.

Saint-Jalmes, A. and Durian, D. J. (1999). *J. Rheol.*, **43**, 1411.

Saunders, B. R. and Vincent, B. (1999). *Adv. Colloid Int. Sci.*, **80**, 1.

Schirmacher, W. (2006). *Europhys. Lett.*, **73**, 892.

Schirmacher, W., Ruocco, G., and Scopigno, T. (2007). *Phys. Rev. Lett.*, **98**, 025501.

Schober, H.R., Oligschleger, C., and Laird, B.B. (1993). *J. Non-Cryst. Solids*, **156-158**, 965–968.

Schober, H. R. and Oligschleger, C. (1996). *Phys. Rev. B*, **53**, 11469.

Schober, H. R. and Ruocco, G. (2004). *Philos. Mag.*, **84**, 1361.

Shundyak, K., van Hecke, M., and van Saarloos, W. (2007). *Phys. Rev. E*, **75**, 010301 (R).

Silbert, L. E., Ertaş, D., Grest, G. S., Halsey, T. C., and Levine, D. (2002). *Phys. Rev. E*, **65**, 051307.

Silbert, L. E., Liu, A. J., and Nagel, S, R. (2006). *Phys. Rev. E*, **73**, 041304.

Silbert, L. E., Liu, A. J., and Nagel, S. R. (2005). *Phys. Rev. Lett.*, **95**, 098301.

Silbert, L. E., Liu, A. J., and Nagel, S. R. (2009). *Phys. Rev. E*, **79**, 021308.

Sokolov, A. P., Buchenau, U., Steffen, W., Frick, B., and Wischnewski, A. (1995). *Phys. Rev. B*, **52**, R9815.

Somfai, E., Roux, J.-N., Snoeijer, J. H., van Hecke, M., and van Saarloos, W. (2005). *Phys. Rev. E*, **72**, 021301.

Somfai, E., van Hecke, M., Ellenbroek, W. G., Shundyak, K., and van Saarloos, W. (2007). *Phys. Rev. E*, **75**(2), 020301(R).

Song, C., Wang, P., and Makse, H. A. (2008). *Nature*, **453**, 629–632.

Soven, P. (1969). *Phys. Rev.*, **178**, 1136.

Taraskin, S. N. and Elliott, S. R. (1999). *Phys. Rev. B*, **59**, 8572.
Thorpe, M. F. (1983). *J. Non-Cryst. Sol.*, **57**, 355.
Tighe, B. P., Woldhuis, E., Remmers, J. J. C., van Saarloos, W., and van Hecke, M. (2010). *Phys. Rev. Lett.*, **105**, 088303.
Tkachenko, A. V. and Witten, T. A. (2000). *Phys. Rev. E*, **62**, 2510–2516.
Tkachenko, K., Dove, M. T., Brazhkin, V., and El'kin, F. S. (2004). *Phys. Rev. Lett.*, **93**, 135502.
Tkachenko, K. O, Dove, M. T, Harris, M. J, and Heine, V. (2000). *J. Phys.: Condens. Matter*, **12**, 8041–8064.
Torquato, S. and Stillinger, F. H. (2001). *J. Phys. Chem. B*, **105**, 11849.
Torquato, S., Truskett, T. M., and Debenedetti, P. G. (2000). *Phys. Rev. Lett.*, **84**, 111.
Vainer, Yu G., Naumov, A. V., Bauer, M., and Kador, L. (2006). *Phys. Rev. Lett.*, **97**, 185501.
Vitelli, V., Xu, N., Wyart, M., Liu, A. J., and Nagel, S. R. (2010). *Phys. Rev. E*, **81**, 021301.
Widmer-Cooper, A., Perry, H., Harrowell, P., and Reichman, D. R. (2009*a*). *Nature Phys.*, **4**, 711.
Widmer-Cooper, A., Perry, H., Harrowell, P., and Reichman, D. R. (2009*b*). *J. Chem. Phys.*, **131**, 194508.
Wouterse, A., Williams, S. R., and Philipse, A. P. (2007). *J. Phys. C: Condens. Matter*, **19**, 6215.
Wyart, M. (2005). *Ann. Phys. Fr.*, **30**((3)), 1–96.
Wyart, M. (2009*a*). *Europhys. Lett.*, **85**, 24003.
Wyart, M. (2009*b*). In *Rigidity and Boolchand Intermediate Phases in Nanomaterials* (ed. M. P. M. Micoulaut), pp. 159–177. INOE Bucarest, Bucarest.
Wyart, M. (2010). *Europhys. Lett.*, **89**, 64001.
Wyart, M., Liang, H., Kabla, A., and Mahadevan, L. (2008). *Phys. Rev. Lett.*, **101**, 215501.
Wyart, M., Nagel, S. R., and Witten, T. A. (2005*a*). *Europhys. Lett.*, **72**, 486–492.
Wyart, M., Silbert, L. E., Nagel, S. R., and Witten, T. A. (2005*b*). *Phys. Rev. E*, **72**, 051306.
Xu, N., Haxton, T. K., Liu, J., A., and Nagel, S. R. (2009). *Phys. Rev. Lett.*, **103**, 245701.
Xu, N., Vitelli, V., Liu, A. J., and Nagel, S. R. (2010). *Europhys. Lett.*, **90**, 56001.
Xu, N., Vitelli, V., Wyart, M., Liu, A. J., and Nagel, S. R. (2009). *Phys. Rev. Lett.*, **102**, 038001.
Xu, N., Wyart, M., Liu, A. J., and Nagel, S. R. (2007). *Phys. Rev. Lett.*, **98**, 175502.
Zallen, R. (1998). *Physics of Amorphous Solids.* Interscience, New York.
Zeller, R. C. and Pohl, R. O. (1971). *Phys. Rev. B*, **4**, 2029–2041.
Zeravcic, Z., Xu, N., Liu, A. J., Nagel, S. R., and van Saarloos, W. (2009). *Europhys. Lett.*, **87**, 26001.
Zhang, H. P. and Makse, H. A. (2005). *Phys. Rev. E*, **72**, 011301.
Zhang, Z., Xu, N., Chen, D. T. N., Yunker, P., Alsayed, A. M., Aptowicz, K. B., Habdas, P., Liu, A. J., Nagel, S. R., and Yodh, A. G. (2009). *Nature*, **459**, 230.

10
Kinetically constrained models

Juan P. Garrahan, Peter Sollich and Cristina Toninelli

Abstract

In this chapter we summarize recent developments in the study of kinetically constrained models (KCMs) as models for glass formers. After recalling the definition of the KCMs that we cover, we study the possible occurrence of ergodicity breaking transitions and discuss in some detail how, before any such transition occurs, relaxation time scales depend on the relevant control parameter (density or temperature). Then we turn to the main issue: the predictions of KCMs for dynamical heterogeneities. We focus in particular on multipoint correlation functions and susceptibilities, and decoupling in the transport coefficients. Finally, we discuss the recent view of KCMs as being at first order coexistence between an active and an inactive space-time phase.

10.1 Motivation

Kinetically constrained models (KCMs) are simple lattice models of glasses. They furnish a perspective on the glass-transition problem that has its origin in the work of Glarum (Glarum, 1960), Anderson (Anderson, 1979) and coworkers (Palmer *et al.*, 1984), and Andersen and coworkers (Fredrickson and Andersen, 1984). This perspective assumes that most of the interesting properties of glass-forming systems are dynamical in origin, while thermodynamics plays a very limited role. KCMs tend to have simple and uninteresting thermodynamics, typically that of a non-interacting lattice gas. In contrast, they display rich dynamical behavior as a consequence of kinetic constraints. This combination of simple thermodynamics and locally constrained dynamics is often assumed to be the result of coarse graining of a dense molecular system (Garrahan and Chandler, 2003): dense fluids are structureless at distances beyond the molecular length, but interatomic forces at high densities are highly constraining, giving rise to local restrictions in the dynamics. As such, KCMs are meant as models of glass-forming systems at high densities or low temperatures, and aim to capture their dynamical behavior for motion beyond the intermolecular distance and for long times.

KCMs generally use local constraints. Nevertheless, as we discuss below, these give rise to collective dynamics due to a form of dynamical frustration: at low temperatures/high densities there is a conflict between the scarcity of excitations/vacancies and the need for them to "facilitate" local motion, leading to hierarchical and cooperative relaxation. Note that this frustration, in contrast with, e.g., the random first-order approach (Lubchenko and Wolynes, 2007), does not arise from quenched disorder, an ingredient that at any rate is not trivial to justify in models of real liquids. A further distinction from the latter approach is that KCMs offer a "non-topographic" (Berthier and Garrahan, 2003) view of the glass-transition problem. By this we mean that it is not a change in the topographic structure of the potential-energy landscape (such as a transition between a saddle-dominated and a minima-dominated regime) that determines glassy relaxation, but a change in the degree of connectivity of the configuration space. Indeed, due to the presence of the constraints, the effective connectivity of this space depends on temperature/density, as we discuss in more detail in Section 10.3. This picture provided by the KCM approach is appealing as an effective description of the physics for example in the case of hard spheres, where all allowed configurations have the same energy. For a more detailed discussion on the non-topographic KCM approach versus the energy-landscape scenario (and additional references on the latter) we refer the reader to (Berthier and Garrahan, 2003; Whitelam and Garrahan, 2004). Further information on disorder or landscape-based approaches, along with other explanations of glassy behavior that invoke a thermodynamic transition, can be found in Chapter 1 of this book.

KCMs are simple enough to allow for detailed analysis. The current interest in KCMs originates in the fact that they exhibit explicit mechanisms for super-Arrhenius slowdown and stretched relaxation (Sollich and Evans, 1999) as a consequence of local, disorder-free interactions, and without the emergence of finite temperature singularities (Toninelli *et al.*, 2004*b*). At the same time they provide a natural explanation (Garrahan and Chandler, 2002) for the phenomenon of dynamical heterogeneity (for reviews see (Ediger, 2000; Glotzer, 2000; Andersen, 2005)). KCMs are to the constrained dynamics/facilitation perspective (Palmer *et al.*, 1984; Fredrickson and Andersen, 1984) what the random energy model and the p-spin spin glass are to the random-first-order transition approach (Lubchenko and Wolynes, 2007). Insights from the analysis of KCMs allow one to construct a comprehensive theoretical picture of the glass-transition problem (see (Chandler and Garrahan, 2010) for a review) that is quite distinct from other competing theories (Götze and Sjögren, 1992; Debenedetti and Stillinger, 2001; Lubchenko and Wolynes, 2007; Kivelson and Tarjus, 2008; Cavagna, 2009). Beyond their importance as models for describing glassy phenomenology, KCMs are of interest for the mathematical community. Indeed, even though they belong to the class of interacting particle systems with Glauber and Kawasaki dynamics, the rate at which elementary moves (birth/death or jumps of particles) occur may degenerate to zero due to the presence of the kinetic constraints. This prevents the use of the standard probabilistic tools developed for such systems. Furthermore, it gives rise to peculiar phenomena including the presence of several invariant measures, ergodicity breaking transitions (Toninelli and Biroli, 2008*a*), unusually long mixing times (Cancrini *et al.*, 2008) and aging phenomena (Faggionato *et al.*, 2010).

The most recent comprehensive review of KCMs is that of Ritort and Sollich (Ritort and Sollich, 2003) that covers the literature until roughly the end of 2001. Earlier surveys were given by Jäckle (1986), Fredrickson (1988) and Palmer (1989). The aim of this chapter is to briefly describe the developments in the last decade or so. It is organized as follows. In Section 10.2 we introduce the KCMs that we will cover, grouping them according to whether they are non-conservative (Glauber dynamics) or conservative (Kawasaki dynamics). Section 10.3 is devoted to a discussion of probably the most basic mathematical manifestation of glassiness in KCMs, i.e. whether they possess ergodicity breaking transitions. Beyond such a transition, the relaxation time to equilibrium is effectively infinite. Physically, of course, this may not be distinguishable from a finite but very long time, and so we consider in Section 10.4 how relaxation time scales in KCMs depend on the relevant control parameters (temperature/density). These times do indeed generically increase extremely fast, in particular in the so-called cooperative models. In Section 10.5, finally, we turn to the predictions of KCMs for dynamical heterogeneity in glasses. Here, the models provide an appealingly intuitive picture that is directly based on considering the dynamics in real (rather than e.g. Fourier mode) space. We cover definitions and predictions for multipoint correlation functions and susceptibilities, decoupling in the temperature or density dependence of transport coefficients, and finally the view of KCMs as being at a first-order coexistence between two different dynamical (space-time) phases. We conclude in Section 10.6 with a summary and a critical discussion of the advantages and drawbacks of KCMs. We also provide a brief comparison with alternative approaches, and finally an outlook towards future work.

10.2 The models

The majority of KCMs are defined as stochastic lattice models with binary degrees of freedom, and it is on such models that we focus in this chapter. We do not attempt to give here an exhaustive survey of the wider range of models studied previously, but refer for this to the reviews listed in the introduction (Ritort and Sollich, 2003; Jäckle, 1986; Fredrickson, 1988; Palmer, 1989). For our present purposes, then, a KCM has on each lattice site i an occupation variable $n_i \in \{0, 1\}$, and the collection of the n_i defines the overall configuration $\boldsymbol{n}$. *Conservative* KCMs are lattice gases where the n_i indicate the presence ($n_i = 1$) or absence ($n_i = 0$) of particles. The dynamics follows a continuous-time Markov process that consists of a sequence of particle jumps, subject to kinetic constraints as explained below. The total particle number $\sum_i n_i$ is conserved.

We will also discuss *non-conservative* KCMs. These are motivated by a conceptual coarse graining to a length scale several times the particle diameter of the underlying physical glass that is being modelled. Each lattice site i then represents a small region of material containing at least a few particles, and one sets $n_i = 1$ (respectively, $n_i = 0$) if the density in this region is above (respectively, below) a certain threshold where local rearrangements—of the particles inside the region—become possible. This representation no longer contains the precise values of all local densities, and consequently in non-conservative KCMs moves that change $\sum_i n_i$ are allowed. The reader should be aware that in much of the literature a reverse convention for the two states of n_i is used, with $n_i = 1$ standing for high mobility, i.e. low density, and vice versa for $n_i = 0$. (These states are then often further identified as up- and down-spins, but we will avoid this terminology.) We opt for the version discussed above as it makes for a unified discussion of conservative and non-conservative KCMs: $n_i = 0$, an empty or low-density site, is mobile in both contexts and facilitates the dynamics on neighboring sites. For brevity we use the terms "empty" and "low density" interchangeably below, and similarly for "occupied" and "high density". When we talk about the density of the system, we correspondingly mean the fraction of sites with $n_i = 1$, both in conservative and non-conservative KCMs.

The key property of all KCMs is that in order to perform a move the configuration must satisfy a local constraint, which usually corresponds to requiring a minimal number of empty sites in some appropriate neighborhood. This represents the physical intuition that, when particles rearrange in a glass, motion in any given region requires the presence of mobile regions around it (Fredrickson and Andersen, 1984; Kob and Andersen, 1993). When the density of these mobile or facilitating sites decreases, the dynamics slows down. Another property shared by all models is that the rates satisfy detailed balance with respect to (w.r.t.) a Boltzmann distribution that factorizes over sites (mathematically, a Bernoulli product measure), so that there are no static interactions, beyond the effective hard-core repulsion implemented by the restriction $n_i \in \{0, 1\}$. As explained in the introduction, this idealization means that KCMs can also be viewed as an attempt to find out how much of glass phenomenology can be explained on purely dynamical grounds, without recourse to, e.g., static phase transitions.

KCMs as described above are plainly quite simplistic compared to more realistic interacting atomic or molecular systems. As we shall explain, however, for appropriate choices of the constraints they display a behavior that is in agreement with the broad phenomenology of glass-forming liquids including super-Arrhenius slowing down of the dynamics, stretched exponential relaxation, dynamical heterogeneities, aging phenomena and ergodicity-breaking transitions.

10.2.1 Facilitated spin models: FA, East and spiral models

We next describe the non-conservative KCMs that we will consider. Because mobile ($n_i = 0$) sites facilitate motion on neighboring sites, and because binary degrees of freedom with non-conservative dynamics can be interpreted as spins, such models are also called "facilitated spin models". In these models, the rate for changing the state of site i is $f_i(\boldsymbol{n})[(1-\rho)n_i + \rho(1-n_i)]$, where f_i depends on the configuration $\boldsymbol{n}$ in a finite neighborhood of i, but *not* on n_i itself. It is then immediate to verify that detailed balance holds w.r.t. a Boltzmann distribution with energy function $-\sum_i n_i$ and inverse temperature β linked to the density by $\rho = 1/(1 + e^{-\beta})$: in equilibrium we have $n_i = 1$ with probability ρ and $n_i = 0$ with probability $1 - \rho$, independently at each site. In general, f_i is chosen to be non-zero (allowed move) when an appropriate neighborhood of i contains a minimal number of empty sites, otherwise it is zero (forbidden move).

As the temperature $1/\beta$ decreases, the density ρ of occupied sites increases and the density $q = 1 - \rho = 1/(1 + e^{\beta})$ of empty sites decreases: the dynamics must then slow down.

Non-conservative KCMs can be divided further into two classes: non-cooperative and cooperative models. For the former it is possible to construct an allowed path—a sequence of configurations linked by transitions with non-zero rates—which completely empties *any* configuration provided that it contains somewhere an appropriate finite cluster of empty sites, the *mobile defect*. For cooperative models, no such finite mobile defects exist. As we will detail in Section 10.4, non-cooperative and cooperative models display, respectively, Arrhenius and super-Arrhenius slowing down.

We now give the definitions of some specific models that cover these two different classes of behavior. The first class of non-conservative KCMs was proposed by Fredrickson and Andersen (FA) (Fredrickson and Andersen, 1984), hence the name Fredrickson–Andersen models. The simplest of these models, which we write as FA-1 ("one-spin facilitated FA"), allows a change of state at site i only if at least one of the nearest neighbors is empty: $f_i(\boldsymbol{n}) = 1$ if $\sum_{j\sim i}(1 - n_j) > 0$, $f_i(\boldsymbol{n}) = 0$ otherwise, where the sum runs over the nearest neighbors j of site i. It is easy to check that the presence of a single empty site allows one to empty the whole lattice: the model is non-cooperative. For theoretical calculations it is often easier to define f_i somewhat differently, as $f_i(\boldsymbol{n}) = \sum_{j\sim i}(1 - n_j)$. This leaves the kinetic constraints enforced by vanishing rates as they are and just changes the rates for allowed moves when more than one empty neighbor is present; as such, it makes no qualitative difference to the behavior of the model.

One can similarly define m-constrained FA models, FA-m, by setting $f_i(\boldsymbol{n}) = 1$ if at least m neighboring sites of i are empty, and $f_i(\boldsymbol{n}) = 0$ otherwise. These models are normally considered on hypercubic lattices of dimension d, with $2 \leq m \leq d$ (Fredrickson and Andersen, 1984). As can be checked directly, it is not possible in any of these models to devise a finite cluster of empty sites that always lets one empty the entire lattice. Consider for example the case $d = 2$, $m = 2$ in the infinite volume limit and focus on a configuration that contains two adjacent infinite rows of occupied sites. Even if the rest of the lattice is completely empty, none of the sites in these two rows can ever change state because each has at most one empty neighbor. Thus, it is not possible to devise a mobile defect that can unblock (empty) every configuration and the model is cooperative. The same conclusion applies on a finite lattice, e.g. with periodic boundary conditions. The restriction on m comes from the fact that, if $m > d$, it is possible to construct *finite* structures of occupied sites that are blocked, i.e. can never change state. They are therefore not suitable for describing the slow dynamics close to a glass or jamming transition because a finite fraction of the system is jammed at *any* density.

Another example of a cooperative model is the one-dimensional East model (Eisinger and Jäckle, 1991). In this case the constraint requires for an allowed move that the left nearest neighbor be empty, that is $f_i = 1 - n_{i-1}$. Note that on a finite chain the presence of a single empty site at the left boundary lets one empty the entire chain; but this is not a mobile defect because it will work only when it is found in a specific position. In a chain that is full except for a finite cluster of empty sites in a generic location, the part of the chain to the left of the cluster cannot be emptied, and so the model is cooperative. One can show that due to the directed nature of the constraint the relaxation involves the cooperative rearrangements of increasingly large regions as q becomes small. This leads to super-Arrhenius behavior of relaxation time scales (Evans and Sollich, 1999; Sollich and Evans, 2003). (The East model shows up a slight shortcoming of our definition of cooperativity: with periodic boundary conditions, we would formally label this model non-cooperative because all sites can be emptied starting from a single empty site. Physically, the model is clearly cooperative independently of the boundary conditions.) We also discuss below higher-dimensional generalizations of the East model, e.g. in $d = 3$ the North-East-front (NEF) model, where a local change of state is allowed if at least one of the neighboring sites to the North or East or front is empty.

The models described above have been closely studied since they were proposed in the 1980s (Fredrickson and Andersen, 1984) and 1990s (Eisinger and Jäckle, 1991), with some work continuing into the present decade (Ritort and Sollich, 2003), particularly regarding the non-equilibrium behavior (see Section 10.4.3) and detailed comparisons with glass phenomenology (Garrahan and Chandler, 2003; Jung *et al.*, 2004; Berthier *et al.*, 2005*a*; Toninelli *et al.*, 2005*b*; Chandler *et al.*, 2006; Elmatad *et al.*, 2009). In parallel, progress has come from the development of new KCMs (Garrahan and Chandler, 2003). We discuss below one of these, which reproduces in particular the mixed character of the glass transition, where diverging length scales characteristic of second-order transitions combine with a first-order jump in the fraction of frozen degrees of freedom. This two-dimensional spiral model (Toninelli *et al.*, 2007; Toninelli and Biroli, 2008*a*,*b*) is defined as follows. Consider a square lattice and, for each site i, divide its first and second neighbors into the North-East (NE), South-West (SW), North-West (NW) and South-East (SE) pairs, as illustrated in Fig. 10.1. Then the constraint for a move to occur at i is that "both its NE *and/or* both its SW neighbors should be empty" *and* "both its SE *and/or* both its NW neighbors should be empty" (see Fig. 10.1); in other words, $f_i(\boldsymbol{n}) = 1$ if this condition is satisfied and $= 0$ otherwise. Stated in a more geometrical form, the constraint requires that all four sites in one of four contiguous sets of neighbors ("NE and SE", or "SE and SW", or "SW and NW", or "NW and NE") have to be empty. The motivation for the somewhat involved form of this constraint will become clear when we analyze the dynamical transition in the model below. We note that before the spiral model a somewhat more involved choice

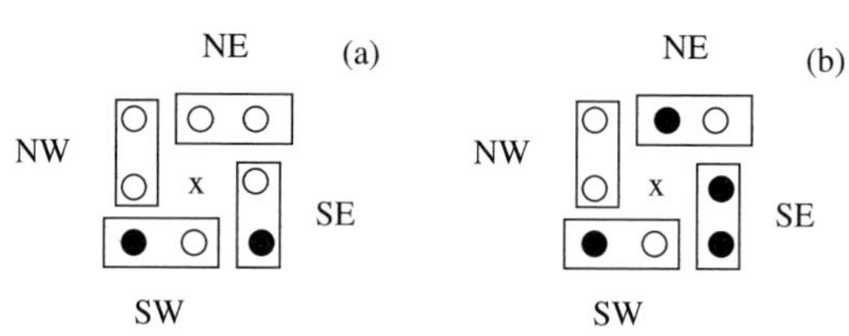

Fig. 10.1 Site i and its NE, NW, SE and SW neighbors. The constraint is (is not) verified at the central site in case (a) (in case (b)).

of the constraints (the so-called Knights model) was proposed (Toninelli *et al.*, 2006) which leads to the same mixed first/second-order transition. Other choices in the same spirit have also been investigated since, see for example (Jeng and Schwarz, 2010).

10.2.2 Kinetically constrained lattice gases: KA and TLG models

Conservative KCMs are defined in a similar spirit to their non-conservative analogs, but with dynamics that conserves total particle number. A particle at i attempts at a fixed rate ($= 1$ without loss of generality) to jump to any empty nearest-neighbor site j. The rate for this move is thus $n_i(1 - n_j)f_{ij}(\boldsymbol{n})$. The factor $f_{ij}(\boldsymbol{n})$ again implements a kinetic constraint and is chosen not to depend on the configuration at sites i and j. The resulting dynamics preserves the number of particles, and so detailed balance on finite lattices is satisfied w.r.t. the Boltzmann distribution, which is uniform on configurations with the appropriate fixed particle number. By linearly combining distributions with different particle numbers one can of course also create grand-canonical Boltzmann distributions where each site independently contains a particle with some probability ρ. Detailed balance is also satisfied w.r.t. these distributions, and they extend naturally to the limit of an infinite lattice.

Again, we can classify conservative KCMs into non-cooperative and cooperative models. For the former it is possible to construct a finite group of empty sites, the *macrovacancy*, such that for *any* configuration the macrovacancy can be moved all over the lattice and any jump of a particle to a neighboring empty site can be performed when the particle is adjacent to the macrovacancy. Clearly, the baseline lattice gas in which there are no kinetic constraints ($f_{ij}(n) = 1$), which is the symmetric simple exclusion process (SSEP), is non-cooperative and the minimal macrovacancies are just isolated empty sites.

Chronologically, the first conservative KCM is the Kob–Andersen model (KA) (Kob and Andersen, 1993), which has cooperative dynamics. Here, a particle can jump to a neighboring site only if both in the initial and final position at least m of its nearest-neighboring sites are empty. The original KA model had $m = 3$, with particles on a cubic lattice, but one can similarly define KA-m models on hypercubic lattices of dimension d and for different values of the parameter m, with $2 \leq m \leq d$. The restrictions on m arise from the fact that $m = 1$ corresponds to an unconstrained lattice gas, while a model with $m > d$ has a finite fraction of frozen particles at any density (as for FA models in the same parameter regime). Similarly to the case of FA-m models with $m \geq 2$, one can directly check that all KA-m models are cooperative. On other lattices, KA models can be non-cooperative. Consider for example the KA model with $m = 2$ on a triangular lattice. Here, one can check that a “dimer” of two neighboring empty sites forms the required macrovacancy, which can move across the lattice in a tumbling motion even when all other sites are occupied. Two other KCMs on triangular lattices were introduced in (Jäckle and Krönig, 1994) and have more recently been analyzed in (Hedges and Garrahan, 2007). For the (1)-TLG a particle can move from site i to a neighboring site j if at least one of the two mutual neighbors of i and j is empty. For the (2)-TLG a particle can move from i to j if *both* the two mutual neighbor sites are empty. It is easy to verify that the (1)-TLG is non-cooperative, with—as for KA with $m = 2$ on the same lattice—a dimer of vacancies being a macrovacancy. The (2)-TLG, on the other hand, is cooperative: a chain of particles occupying an entire row of the lattice, for example, can never be destroyed, thus it is not possible to construct a macrovacancy.

10.3 Ergodicity-breaking transitions

As explained above, the equilibrium distribution in KCMs is trivial because it factorizes over sites. In particular, then, no equilibrium phase transition can occur. On the other hand, the presence of constraints might induce transitions of purely dynamical type: detailed balance alone does not guarantee that the distribution over configurations will converge for large times to the Boltzmann equilibrium distribution. To see this, return to the example of the FA-2 model on a square lattice ($m = d = 2$), either finite with periodic boundary conditions or infinite, and a starting configuration $\boldsymbol{n}(0)$ that has two adjacent rows of occupied sites. Then, since these sites are forever blocked, for any site i within the two rows, $\lim_{t\to\infty}\langle n_i\rangle_t = 1 \neq \langle n_i\rangle_{\text{eq}} = \rho$, where $\langle\cdot\rangle_t$, is the average over the stochastic dynamics that starts from $\boldsymbol{n}(0)$ and $\langle\cdot\rangle_{\text{eq}}$ is the equilibrium average.

Motivated by this, we can ask whether convergence to the equilibrium distribution is recovered at least in the thermodynamic limit: if we sample an initial configuration $\boldsymbol{n}(0)$ from the equilibrium distribution, then is the large-time limit of the average of any function $g(\boldsymbol{n})$ equal (with probability one over random draws of an initial configuration from the equilibrium distribution) to its equilibrium average, i.e. is $\lim_{t\to\infty}\langle g\rangle_t = \langle g\rangle_{\text{eq}}$? If this is the case we will say that the system is *ergodic*. The models we consider are all ergodic at sufficiently high density $q = 1 - \rho$ of facilitating (empty) sites. If they do become non-ergodic as q is reduced, we call q_{c} the critical density of empty sites at which this transition occurs. As will be discussed below, the models defined in the previous sections are ergodic at any positive $q = 1 - \rho$, so that $q_{\text{c}} = 0$, with the exception of the spiral model and FA models on Bethe lattices that display ergodicity breaking transitions at $0 < q_{\text{c}} < 1$.

Owing to the factorized form of the equilibrium distribution, ergodicity corresponds to the fact that with probability one in the thermodynamic limit the configuration space is covered by a single irreducible set, i.e. a set of configurations that are connected to each other by allowed paths; for a proof see Cancrini *et al.*, 2008, Proposition 2.4 and Cancrini *et al.*, 2010*a*, Section 2.3. This in turn is equivalent to the requirement that for any site i (in non-conservative KCMs) or pair of nearest-neighbor sites i, j (in conservative KCMs) there is an allowed path that transforms the configuration into one where the constraint is satisfied at i (i, j). In other words, the probability that any site i belongs to a cluster of forever blocked sites must vanish in the thermodynamic limit.

For *non-cooperative* KCMs one sees easily that ergodicity holds at any density $\rho < 1$, i.e. at any $q > 0$. In the non-conservative case, the probability of finding at least one mobile defect in an equilibrium configuration goes to one in the infinite-volume limit, and then starting from this defect one can empty all sites. Thus, any configuration is connected to the "all empty" configuration with probability one. The situation is similar for non-cooperative but conservative KCMs.

The case of *cooperative* KCMs is more delicate and here an ergodicity-breaking transition can occur. The non-conservative case is again simpler, so we begin with this. In order to determine the probability that a site belongs to a blocked cluster, consider the following deterministic procedure: iteratively empty all sites for which the constraint is verified until we reach either the completely empty configuration, or one in which there is a "backbone" of mutually blocked occupied sites. Then it is easy to verify that all the sites that are occupied at the end of this deterministic procedure are blocked forever under the stochastic dynamics of the actual KCM. In other words, the problem of the existence of an ergodicity breaking transition for cooperative non-conservative KCMs can be reformulated as a percolation transition for the final configuration of the above deterministic dynamics. For FA-m models this deterministic dynamics coincides with the well-known algorithm of bootstrap percolation and the results in (Aizenman and Lebowitz, 1988; Schonmann, 1992) establish that $q_c = 0$ on hypercubic lattices in any dimension d and for any facilitating parameter $1 \leq m \leq d$ (while trivially $q_c = 1$ for $m > d$ since finite blocked structures can occur). This work disproved a long-standing conjecture from the original FA paper (Fredrickson and Andersen, 1984), namely that an ergodicity-breaking transition would occur at some $q_c > 0$. Such a transition does take place, however, when one considers FA models on a Bethe lattice as will be explained in Section 10.3.1. An example of a finite-dimensional model displaying such a transition is provided by the spiral model (see Section 10.3.2).

For cooperative conservative KCMs, the proof of ergodicity is more involved. For example, for the KA-2 model on a square lattice ($d = 2$), it has been shown that the irreducible set of configurations that has unit probability in the thermodynamic limit is the one containing all configurations that can be connected by an allowed path to a configuration that has a frame of empty sites on the last shell before the boundary (Toninelli *et al.*, 2005*a*). Establishing that this set has unit probability in the thermodynamic limit is more complicated than in the corresponding non-conservative KCMs, i.e. FA models. This is because there is no deterministic bootstrap-like procedure that allows one to establish whether or not a configuration does or does not belong to the irreducible set. Nevertheless, a formal proof can be constructed (Toninelli *et al.*, 2005*a*; Cancrini *et al.*, 2010*a*), and demonstrates that the ergodicity-breaking transition originally conjectured by Kob and Andersen (Kob and Andersen, 1993) does not exist. Analogous arguments can be constructed for the other choices of m and d (Toninelli *et al.*, 2005*a*) to rule out the occurrence of a transition.

10.3.1 FA models on Bethe lattices

As a simple example of a model that does have an ergodicity-breaking transition we consider next the FA-m model on a Bethe lattice, i.e. a random regular graph of connectivity $k+1$. Exploiting the local tree-like structure of such a graph, it is easy to write a self-consistent equation for the probability P that a given site is occupied and blocked, conditional on the fact that its ancestor is occupied and blocked:

$$P = (1-q) \sum_{i=0}^{m-1} \binom{k}{i} P^{k-i} (1-P)^i. \tag{10.1}$$

The factor $1 - q$ is the probability for the given site to be occupied; the sum gives the probability that at most $m - 1$ of the descendants of the site are empty or unblocked, so that the site in question is itself blocked. The leading term on the r.h.s. for small P is $O(P^{k-m+1})$. For $m = k$, the largest value that does not produce finite blocked structures, this is linear and so one gets a continuous transition at $(1 - q_c)m = 1$. For all smaller m, the small P increase is with a higher power of P and so a discontinuous transition results, with $P = 0$ for $q > q_c$ and $P = P_c + O(\sqrt{q_c - q})$ for $q < q_c$, where $P_c > 0$. The mechanism behind this combination of a discontinuous onset and a critical singularity has been analyzed in detail (Dorogovtsev and Mendes, 2006; Schwarz and Chayes, 2006). In particular, the singular square-root behavior is due to the extreme fragility of the infinite spanning frozen or "jammed" cluster at the transition, and to the existence of a related length scale that diverges as a power law.

10.3.2 Spiral model

In this section we will explain the mechanism behind the ergodicity-breaking transition of the spiral model (Toninelli *et al.*, 2007; Toninelli and Biroli, 2008*b*). Consider the directed lattice that is obtained from the square lattice by putting two arrows from each site towards the neighbors in the NE pair, i.e. pointing North and North-East. The resulting lattice is a tilted and squeezed version of a two-dimensional oriented square lattice (see Fig. 10.2). Therefore, if the density is larger than the critical density

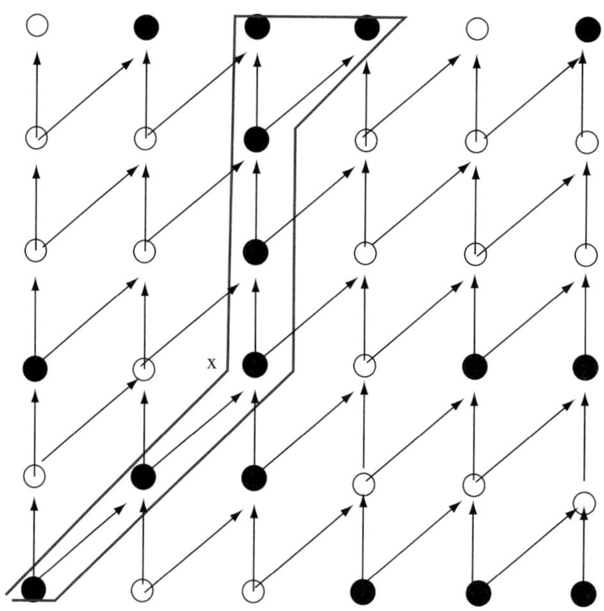

Fig. 10.2 The tilted and squeezed two-dimensional oriented lattice obtained by drawing arrows from each site to its neighbors to the North and East. The site marked by the cross belongs to an oriented occupied cluster and is therefore blocked.

of oriented site percolation (directed percolation, DP), ρ_c^{DP}, there exists a cluster of occupied sites that spans the lattice following the direction of the arrows. Consider now a site in the interior of this directed cluster (see Fig. 10.2): by definition there is at least one occupied site in both its NE and SW neighboring pairs, therefore the site is blocked with respect to the constraints of the spiral model. Thus, for $\rho > \rho_c^{DP}$ the system contains one or more blocked clusters and is therefore non-ergodic. This suggests that the ergodicity-breaking transition occurs at $q_c = 1 - \rho_c^{DP}$, but the argument so far does not exclude a transition earlier, i.e. at a larger density q of empty sites. Indeed, since blocking can occur along either the NE-SW or the NW-SE direction (or both), the presence of a blocked cluster does not imply that there is a directed path through the lattice as in the DP argument. To establish that, nevertheless, blocked clusters do not occur for $q > 1 - \rho_c^{DP}$, one shows that empty regions of linear size much larger than the parallel length of DP, $\xi_\parallel$, act as "critical defects": starting from any such empty region we can very likely empty the whole lattice. Because $\xi_\parallel$ is finite, so is the size of these defects. In the thermodynamic limit at least one such defect will occur in the system, and so for $q > 1 - \rho_c^{DP}$ any configuration can be completely emptied: the system is ergodic. This confirms that indeed $q_c = 1 - \rho_c^{DP}$.

An important feature of the ergodicity-breaking transition in the spiral model is that it is discontinuous, in the sense that the fraction of blocked sites jumps to a non-zero value at $q = q_c$. This is relevant for the connection to real glasses, because it implies that two-point correlation and persistence functions must display a plateau in their time dependence when the transition is approached ($q > q_c$), as is observed experimentally. Earlier models with a transition at a finite q_c, e.g. a two-dimensional generalization of the East model (Reiter *et al.*, 1992), develop only a fractal cluster of blocked particles at the transition, which occupies a vanishing fraction of the system. The fact that in the spiral model blocked clusters are compact at q_c follows by a direct construction of blocked structures (Toninelli *et al.*, 2007; Toninelli and Biroli, 2008*a*,*b*) and is a consequence of the presence of the two transverse blocking directions in the constraints, and of the anisotropy of DP.

10.3.3 Summary: Presence/absence of ergodicity-breaking transition

For the sake of clarity let us summarize the results of this section by listing the KCMs we have discussed according to whether or not they display an ergodicity-breaking transition at a non-trivial (different from 0 or 1) critical defect density q_c.

The models that do display such a transition are the FA and KA models on Bethe lattices, the spiral model and the North-East model. The latter was not mentioned above because it is not of direct interest for modeling the glass transition, due to the continuous character of its ergodicity-breaking transition. It is a spin-facilitated model on a square lattice with the constraint requiring both the North and East neighbor to be empty. It is easy to verify along the lines of the argument in the previous subsection that the North-East model as defined by this constraint displays a transition at the critical density of oriented percolation.

The models that do not display a transition are: the FA-m and KA-m models for any choice of m and on hypercubic lattices in any finite dimension d; the East model; and the (1)-TLG and (2)-TLG models.

10.4 Bulk dynamics of KCMs

10.4.1 Glassy time scale divergences and static length scales

For KCMs to be useful as models of physical glasses they need to satisfy the basic requirement of dynamical slowing down as the glassy regime approached. In our case this corresponds to increasing density ρ, or decreasing density of empty sites q. We will therefore now give an overview of numerical and analytical results on how relaxation time scales τ diverge in KCMs when q approaches q_c. In concrete terms, we will take τ as the typical time in the relaxation of density–density correlation and persistence functions. In cooperative KCMs, τ turns out to be connected to statically defined blocking lengths, as we also discuss.

Starting with FA-1 models, relaxation occurs via the effective diffusion of empty sites: an empty site facilitates the emptying of a neighbor site, with rate $q = 1 - \rho$, and the original site can then fill and does so with probability 1/2 before the new empty site does. Empty sites can thus diffuse freely, with an effective diffusion constant $q/2$. This suggests that τ should grow as an inverse power of q. Because $q \simeq \exp(-\beta)$ at low temperatures $1/\beta$, this corresponds to Arrhenius scaling. Indeed, by an exact mapping to a diffusion-limited aggregation model one can derive (Jack *et al.*, 2006*b*) that $\tau \sim 1/q^z$ with $z = 3$ in $d = 1$ and $z = 2$ in $d \geq 2$. The $d = 1$ result is simple to understand: empty sites are typically a distance $1/q$ apart, and relaxation requires diffusion across this distance with the effective diffusion constant $q/2$, so $\tau \sim (1/q)^2/(q/2) \sim 1/q^3$.

For FA-m with $m > 1$, an isolated empty site is unable to move on its own and has to wait for a mobile defect, i.e. an appropriate region of empty sites, to move cooperatively into its neighborhood and so facilitate its motion. As we will detail below, the typical number of moves involved in this cooperative process increases as $q \to 0$, thus a super-Arrhenius scaling of the relaxation time has to result. This has been confirmed by several numerical investigations that proposed different forms for the density dependence of τ (Butler, 1991; Fredrickson and Brawer, 1986; Graham *et al.*, 1997). In order to better understand the cooperative mechanism let us consider for example the case $m = d = 2$. An $\ell \times \ell$ square of empty sites can be expanded by one lattice spacing in both directions, provided that at least one empty site is present on two adjacent sides of the square. This is very likely to be the case when $\ell \gg 1/q$. On the other hand, the probability of being able to empty all sites in a region of size $1/q$ without using external empty sites can be shown to be proportional to $\exp(-c/q)$ with c a constant of order one (Aizenman and Lebowitz, 1988; Holroyd, 2003). This result is obtained by iterating the process outlined above, multiplying the relevant probabilities at each step: one can iteratively remove all the particles in any region starting from its interior provided there is a central 2×2 square of empty sites and one additional empty site somewhere on each of the four sides of each subsequent shell. In (Reiter, 1991) it was conjectured that relaxation occurs via the diffusion of these critical defects leading to a relaxation time diverging as τ_D/ρ_D, with $\rho_D \simeq \exp(-c/q)$ the defect density and τ_D their diffusion time. The latter, as detailed in Toninelli *et al.*, 2005*a*, Section 6, should be a subleading correction, giving $\tau \sim 1/\rho_D$ to leading order. In principle one might think that this is not the optimal relaxation mechanism and that other relaxation processes could be much faster, avoiding the super-Arrhenius scaling. However, the results in (Aizenman and Lebowitz, 1988; Holroyd, 2003) also give the typical size of the incipient blocked cluster to which any fixed site belongs and that has to be eroded via successive moves from its boundary before the site in question can be unblocked. This size, $L_{\rm c}$, diverges as $\exp(c/2q)$, thus providing (due to finite speed of propagation) a lower bound on time scales of the form $\tau \geq \exp(c/2q)$, which confirms the super-Arrhenius scaling. Note that $L_{\rm c}^2$ scales as the inverse of $\rho_{\rm D}$, which (as the probability that a region of site $1/q \times 1/q$ is a critical defect) is essentially the number density of critical defects. Indeed, clusters of linear size larger than $L_{\rm c}$ are typically unblocked because they contain at least one critical defect from which relaxation can occur. In (Schonmann, 1992) the typical size of such incipient blocked clusters was derived for generic m and d leading to τ growing at least as fast as $\exp^{\circ(m-1)}(c/q^{1/(d-m+1)})$. Here, $\exp^{\circ s}$ is the exponential iterated s times, so that the divergence of the time scale is extremely rapid for any $m \geq 3$.

For KA models the basic relaxation mechanism is very similar (Toninelli *et al.*, 2005*a*), the main difference being that the critical defects are now regions in which empty sites and particles are arranged in such a way that one can find an allowed path—a sequence of allowed moves—to perform any nearest-neighbor exchange. As shown in (Toninelli *et al.*, 2005*a*, 2004*a*), the properties as a function of q of these regions coincide with those for the FA models, leading to the same estimates for the scaling of the relaxation time.

Relaxation processes in the spiral model proceed in a broadly analogous manner. But the size of the critical defects that can expand further diverges already at $q_{\rm c}$, proportionally to the parallel length $\xi_{\parallel}$ of DP clusters. The typical size of incipient blocked clusters grows much more quickly, as $L_{\rm c} \simeq \exp[c/(q-q_{\rm c})^{\mu}]$ with $\mu = \nu_{\parallel}(1-z)$, where $\nu_{\parallel} \simeq 1.73$ is the critical exponent of $\xi_{\parallel}$ and $z \simeq 0.64$ the exponent relating the parallel and transverse lengths of DP. As for FA models, since these clusters can be unblocked only from the boundary, τ should diverge at least as $L_{\rm c}$.

Interestingly, in the other model we have considered that has an ergodicity-breaking transition at non-zero $q_{\rm c}$, the FA model on a Bethe lattice, simulation measurements of persistence and correlation functions (Sellitto *et al.*, 2005) show that relaxation times grow only as power laws on approaching the transition, $\tau \propto (q-q_{\rm c})^{-\gamma}$ with $\gamma \simeq 2.9$. Qualitatively, thus, the transition for FA models on Bethe lattices has the characteristics of a mode-coupling theory (MCT) (Götze and Sjögren, 1992) arrest transition.

Finally, we discuss the East model, where the origin of the super-Arrhenius time scale is possibly easiest to see. The basic relaxation mechanism can be understood in terms of the dynamics of domains of occupied sites separated by empty sites. In the limit of small q one can then argue (Evans and Sollich, 1999) that the typical relaxation time should scale as the minimal time required for an empty site to facilitate the motion of the first empty site to its right, which is typically at distance $1/q$. The optimal path to create an empty site at a distance d involves an energy barrier of order $\log_2(d)$ (the minimum over all paths of the maximal number of empty sites that we encounter along the path). Setting $d = 1/q$ thus leads to the relaxation time estimate

$\tau \sim q^{-\log_2(1/q)} \sim \exp[\beta^2/\ln 2]$ (Evans and Sollich, 1999). The above argument does not take into account the behavior on scales smaller than the typical distance between empty sites. Incorporating this turns out to halve the coefficient in the exponent, to $\tau \sim \exp[\beta^2/(2\ln 2)]$ (Cancrini *et al.*, 2007, 2008).

We have discussed so far only the overall time scale for the decay of correlation or persistence functions in KCMs, but not the time dependence of this decay. For the cooperative models one generically finds stretched exponential forms, as seen experimentally, while non-cooperative models can exhibit power-law tails reflecting the diffusive motion of the mobile defects. We refer to (Ritort and Sollich, 2003) for an overview and quote only the example of the East model, where one can show that the stretching becomes extremely strong at low temperatures, with correlations and persistence both decaying as scaling functions of $(t/\tau)^{1/[\beta \ln 2]}$ (Evans and Sollich, 1999; Buhot and Garrahan, 2001; Sollich and Evans, 2003).

10.4.2 Some rigorous results

In the recent years KCMs have been also analyzed in the mathematical community. We summarize here those rigorous results that help to further understand the slow relaxation of KCMs, and have in some cases corrected conjectures resulting from numerical simulations or intuitive arguments.

The analysis of the large-time behavior in the ergodic regime was started by (Aldous and Diaconis, 2002) where for the East model it was established that $[1/\ln 2 - o(1)]\ln^2(1/q) \leq \ln(1/\text{gap}(q)) \leq [1/(2\ln 2) + o(1)]\ln^2(1/q)$ with $1/\text{gap}(q)$ the inverse of the spectral gap of the Liouvillian operator generating the dynamics. The latter, which in a finite system is just the inverse of the smallest non-zero eigenvalue of the transition matrix, represents the longest relaxation time for all one-time quantities and so can be identified with the relaxation time scale τ discussed above. The bounds of (Aldous and Diaconis, 2002) quoted above then say that $\ln \tau$ scales for small q as $\ln^2(1/q) = \beta^2$, with a pre-factor between $1/(2\ln 2)$ and $1/\ln 2$, the upper bound being the naive estimate of (Evans and Sollich, 1999). These bounds were sharpened in (Cancrini *et al.*, 2008), establishing that it is in fact the lower bound that gives the correct asymptotics: written in terms of the gap, $\lim_{q\to 0}\ln(1/\text{gap})/\ln^2(1/q) = 1/(2\ln 2)$.

Positivity of the spectral gap guarantees in particular exponential convergence in the large-time limit: for any $g(\boldsymbol{n})$ one has $\langle g(\boldsymbol{n}(t))g(\boldsymbol{n}(0))\rangle - \langle g(\boldsymbol{n}(0))\rangle^2 \leq \text{const} \times \exp(-2\,\text{gap}\,t)$, where $\langle \ldots \rangle$ is the mean over the initial Boltzmann equilibrium at empty site density q and over the stochastic process governing the evolution in time. In (Cancrini *et al.*, 2008) it was shown that positivity of the spectral gap also guarantees exponential convergence of the persistence function $P(t)$, which is the probability that a site does not change its state during a time interval of length t: $P(t) \leq \exp(-q\,\text{gap}\,t) + \exp(-(1-q)\text{gap}\,t)$. In the same paper, a multiscale approach was developed that allows one to prove positivity of the spectral gap in the whole ergodic region $q > q_c$ for all the choices of constraints described in Section 10.2. With this technique one can also derive (Cancrini *et al.*, 2008) the following (sometimes optimal) bounds when $q \downarrow q_c$. For FA-1 in $d = 1$, $\tau \propto 1/q^3$; in $d = 2$, $1/q^2 < \tau \leq$

$\ln(1/q)/q^2$; and in $d \geq 3$, $1/q^{1+2/d} < \tau \leq 1/q^2$. These bounds are in agreement with the analytical results in (Jack *et al.*, 2006*b*), and confirm that the initial findings (Berthier *et al.*, 2005*b*) in $d = 2$ and $d = 3$, based on a mapping to DP, were incorrect. For the cooperative KCMs FA-2 and FA-3, results in (Cancrini *et al.*, 2008) show that $\exp(q^{-1}) \leq \tau \leq \exp(q^{-2})$ and $\exp[\exp(q^{-1})] \leq \tau \leq \exp[\exp(q^{-2})]$, respectively, thus establishing a super-Arrhenius scaling compatible with (Butler, 1991; Reiter, 1991).

For conservative KCMs with non-cooperative behavior, the diffusive scaling $1/\text{gap} = O(L^2)$ in a volume of linear size L and the positivity of the self-diffusion coefficient at any density was established in (Bertini and Toninelli, 2004). Moreover, the hydrodynamic limit was studied in (Goncalves *et al.*, 2010) for a special class of models leading to a porous-medium equation, namely a degenerate partial differential equation $\partial_t \rho = \nabla(D(\rho)\nabla\rho)$ with a diffusion coefficient $D(\rho)$ vanishing as a power law of $1 - \rho$ when $\rho \to 1$. The methods used to establish these results use as a key ingredient the existence of a path between configurations that allows any two particles to be exchanged, by exploiting the presence of mobile macrovancancies. This approach, then, cannot be extended to cooperative models. However, more recently a different technique has been devised (Cancrini *et al.*, 2010*a*) that proves also for cooperative models the diffusive scaling $1/\text{gap} = O(L^2)$ and establishes in $d = 2$ the diffusive decay in time $\sim 1/t$ of the density–density autocorrelation function. The self-diffusion coefficient for a specific cooperative model, namely the KA-m, was analyzed in (Toninelli and Biroli, 2004) where its positivity at any q was proved, modulo a conjecture on the behavior of a random walk in a random environment.

10.4.3 Non-equilibrium behavior

Due to space constraints we focus in this chapter almost exclusively on the equilibrium dynamics of KCMs. Non-equilibrium behavior results if, for example, a non-conservative KCM is prepared in a configuration with a high density q of empty sites, and then q is reduced quickly. This corresponds to a sudden lowering of the temperature $1/\beta$ and so mimicks quench experiments in real glasses. If the final q is low enough, i.e. sufficiently close to q_c, the time scale τ for relaxation to the new equilibrium will be very long, and aging will occur: the properties of the system depend on the time elapsed since the quench. Aging can be monitored via two-time response and correlation functions, and one asks, for example, whether these are linked by a non-equilibrium fluctuation–dissipation theorem with some effective temperature (Crisanti and Ritort, 2003). Surprisingly, in some simple KCMs such as the East and FA-1 models, aging at low q is in fact easier to analyze than the dynamics at equilibrium (Evans and Sollich, 1999; Mayer and Sollich, 2007). Work in this area up to late 2001 is summarized in (Ritort and Sollich, 2003), with more recent studies revealing a very rich phenomenology in the aging of KCMs (Mayer *et al.*, 2006; Léonard *et al.*, 2007; Corberi and Cugliandolo, 2009). There has also been progress in rigorous approaches, with e.g. Refs. (Faggionato *et al.*, 2010; Cancrini *et al.*, 2010*b*) establishing the non-equilibrium behavior of the East model derived in (Evans and Sollich, 1999).

10.5 Dynamical heterogeneity and its consequences

The main success of KCMs has been the ability to account, at least qualitatively and sometimes also quantitatively, for many aspects of dynamical heterogeneity (DH) (Ediger, 2000; Glotzer, 2000; Andersen, 2005) observed in glass-forming systems. DH emerges naturally in KCMs at high density or low temperature (Garrahan and Chandler, 2003; Pan *et al.*, 2005), and this is the topic of this section. We focus on the simpler KCMs, such as the FA-1, East and TLG constrained lattice gases. For these models results are easier to obtain both analytically and computationally than for, say, the spiral model or the FA-m model with $m > 1$; nevertheless, they display all the important physics. In particular, the dynamical scaling properties of the East model are close to those observed in realistic glass formers (see e.g. (Chandler and Garrahan, 2010)).

The simplicity of KCMs allows one to make detailed predictions about the growth of dynamical correlation lengths and their scaling relations to growing relaxation times (Garrahan and Chandler, 2002; Toninelli *et al.*, 2004*b*; Whitelam *et al.*, 2004; Pan *et al.*, 2005; Toninelli *et al.*, 2005*b*; Berthier and Garrahan, 2005; Jack *et al.*, 2006*b*). These dynamical length scales are an indication of spatial correlations that build up over time, so they are most easily extracted from multipoint correlation functions, which we discuss in Section 10.5.1. DH also tells us that the dynamics of glass formers is fluctuation dominated. A central consequence is transport decoupling, the breakdown in standard transport relations of liquid-state theory that are obtained under the assumption of homogeneous dynamics, such as that of Stokes–Einstein relating self-diffusion rate to viscosity (Swallen *et al.*, 2003). KCMs provide a direct explanation for decoupling based on the response of a moving molecule to the distribution of local relaxation time scales in the host fluid (Jung *et al.*, 2004). Such motion can be approximately quantified within a continuous-time random walk (CTRW) formalism (Berthier and Garrahan, 2005). Decoupling and CTRW are discussed in Section 10.5.2. Finally, the observed DH is only a mesoscopic phenomenon: dynamical length scales are always finite (and transient) at non-zero temperature or at less than maximal density, that is, for $q > q_c$ (note that all models discussed in this section have $q_c = 0$). Nevertheless, it is a precursor to a fully fledged non-equilibrium, or "space-time", phase transition (Merolle *et al.*, 2005; Jack *et al.*, 2006*a*; Garrahan *et al.*, 2007), which we discuss briefly in Section 10.5.3.

10.5.1 Multipoint-correlations and susceptibilities

Figure 10.3 illustrates DH in KCMs. A convenient observable to quantify local relaxation is the persistence field, $p_i(t) = 0, 1$, where 0 indicates that site i has changed its state at least once up to time t, and 1 otherwise. The ensemble average, $P(t) \equiv \langle p_i(t) \rangle = \lim_{N\to\infty} N^{-1} \sum_i p_i(t)$, is the persistence function discussed above. Figure 10.3(a) shows the persistence field $p_i(t_{1/2})$ in a three-dimensional version of the East model, or NEF (for North-East-Front model) (Berthier and Garrahan, 2005), where $t_{1/2}$ is the time at which $P(t_{1/2}) = 1/2$, i.e. half the system has relaxed and half has not. We see from Fig. 10.3(a) that relaxation is heterogeneous: there is a clear spatial segregation of sites that have relaxed, $p_i = 0$ (colored black),

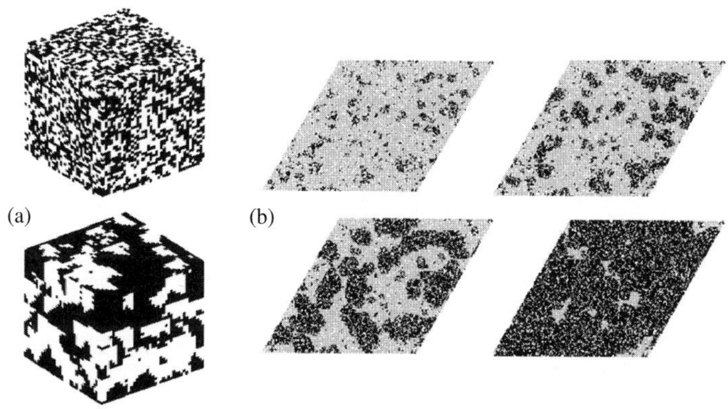

Fig. 10.3 Dynamical heterogeneity in KCMs. (a) Spatial distribution of the local persistence in the NEF model, at time $t_{1/2}$ such that $P(t_{1/2}) = 1/2$ (i.e., 50% of the sites, shown in black, have flipped by time $t_{1/2}$) at temperatures $T = 1.0$ (top) and $T = 0.15$ (bottom) for a system of size $N = 40^3$. Adapted from Ref. (Berthier and Garrahan, 2005), with permission. (b) Same, but for (2)-TLG model, at fixed density $\rho = 0.77$, and varying observation times, $t = 10^3,\ 10^4,\ 10^5,\ 10^6$. From Ref. (Pan *et al.*, 2005), with permission.

from those that have not, $p_i = 1$ (colored white). Relaxation dynamics is spatially correlated. Figure 10.3(a) shows the persistence field at two different values of q (two different temperatures). The spatial extension of dynamical correlations increases with decreasing temperature, and therefore with increasing relaxation time. Note that these dynamical correlations are unrelated to thermodynamic correlations since the equilibrium measure of the NEF is trivial at all temperatures. Figure 10.3(b) shows similar persistence field plots for a density-conserving model, the two-vacancy assisted (2)-TLG, but at a fixed high density ρ and for different observation times t. This figure shows that DH is a transient effect, with dynamical correlations becoming maximal at some intermediate time t^* (see below).

The extent of the dynamical correlations evident from Fig. 10.3 can be quantified by means of multipoint functions (Garrahan and Chandler, 2002; Lačević *et al.*, 2003; Toninelli *et al.*, 2005*b*; Chandler *et al.*, 2006). Consider for example the "four-point" structure factor,

$$S_4(\boldsymbol{k}, t) \equiv N^{-1}\mathcal{N}(t) \sum_{i,j} \left[\langle p_i(t) p_j(t)\rangle - P^2(t)\right] e^{i\boldsymbol{k}\cdot(\boldsymbol{r}_i - \boldsymbol{r}_j)}. \tag{10.2}$$

This is the Fourier transform of the spatial correlation function of the persistence field. The factor $\mathcal{N}(t)$ is a convenient normalization factor. We adopt the choice $\mathcal{N}(t) = \left[P(t) - P^2(t)\right]^{-1}$, which makes S_4 equal to unity if all the non-zero contributions come from "self" terms $i = j$. Figure 10.4(a) shows S_4 for the NEF. This is the structure factor of the DH pictures of Fig. 10.3. It has the characteristic shape of that of a system correlated over finite distances. The zero-wavevector limit of S_4 gives the "four-point" susceptibility, $\chi_4(t) = S_4(k \to 0, t)$, which provides an estimate of the

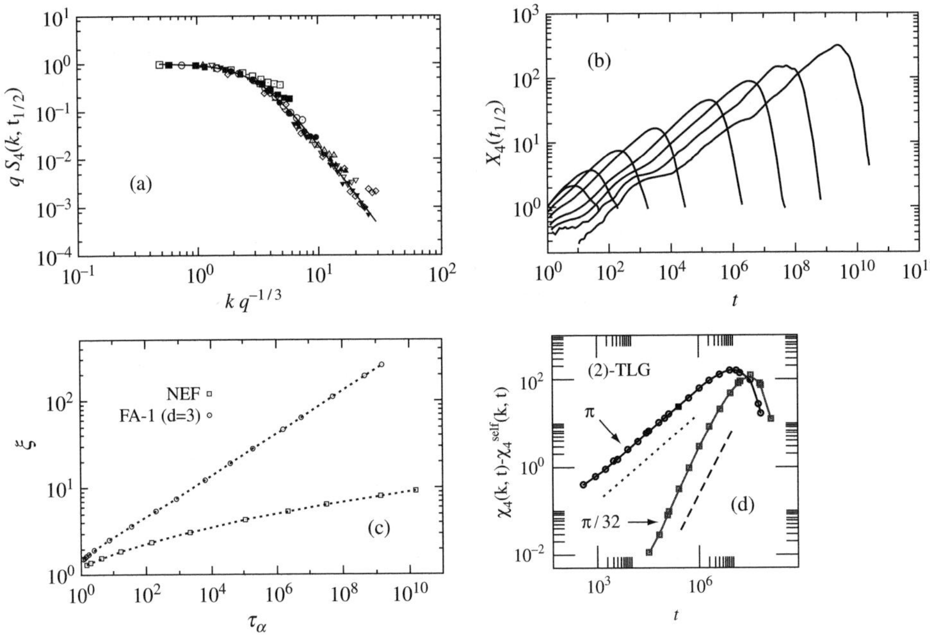

Fig. 10.4 Dynamical correlations and scaling. (a) DH structure factor, $S_4(k, t_{1/2})$ in the NEF model at various temperatures. The data collapses under the scaling $S_4 \to qS_4$ and $k \to k\xi$ with $\xi \sim q^{-1/3}$, suggesting the values $\gamma = 1$ and $\nu = 1/3$ for the scaling exponents (see text). (b) Time dependence of four-point susceptibility $\chi_4(t)$ in the NEF. Dynamical fluctuations probed by this function are maximal at time t^*. The peak susceptibility $\chi_4^* = \chi_4(t^*)$ grows with decreasing temperature (left to right). The peak time is approximately the relaxation time of the system, $t^* \approx \tau$, as extracted from the persistence function. (c) Scaling of DH correlation lengths to relaxation times: FA-1 models follow a simple scaling law (Jack *et al.*, 2006*b*) ($\tau \sim \xi^4$ for the three-dimensional FA-1 shown in the figure), while for East-like models $\tau \sim \xi^{z(q)}$, where the dynamical exponent $z(q)$ increases with decreasing q (i.e. decreasing temperature) (Sollich and Evans, 1999). Adapted from Ref. (Berthier and Garrahan, 2005), with permission. (d) The behavior of four-point susceptibilities depends on the wavelength k of the observable that is used to define them. $\chi_4(k,t)$ is plotted with its self-part, i.e. the terms with $i = j$ in its definition, subtracted. From Ref. (Chandler *et al.*, 2006), with permission.

correlation volume of DH. Figure 10.4(b) shows $\chi_4(t)$ for the NEF model as a function of observation time t, at different temperatures. Several things are apparent. χ_4 is non-monotonic in time, indicating that DH is a transient phenomenon. It peaks at around the relaxation time of the persistence function. The peak value increases with decreasing temperature: dynamical correlations increase with increasing relaxation time.

Multipoint functions reveal the scaling properties of DH. The dynamic susceptibilities defined above have their peak, $\chi_4^* = \chi_4(t^*)$, at times t^* close to the relaxation times τ of the corresponding persistence functions. Scaling is controlled by the distance to the dynamical critical point at zero concentration of excitations or vacancies, so we expect $\chi_4^* \sim q^{-\gamma}$ (Toninelli *et al.*, 2005*b*; Chandler *et al.*, 2006). Furthermore, we expect the four-point structure factor to behave as $S_4(k, t^*) \approx \chi_4^* f(k\xi)$, where f is a scaling function, and $\xi \sim q^{-\nu}$ a correlation length for DH at times $t = t^*$. These exponents—and indeed scaling forms—may vary from model to model. For the simpler KCMs they can be calculated analytically. Figure 10.4(a) shows the numerical collapse of S_4 in the NEF for $\gamma = 1$ and $\nu = 1/3$ (Berthier and Garrahan, 2005). For other models, such as the FA-1, these exponents can be calculated analytically (Toninelli *et al.*, 2005*b*). For example, in $d = 3$ one finds $\tau \sim \xi^4$. This is most easily understood from the fact that the upper critical dimension of the model is $d_c = 2$, so exponents are d independent above this (Jack *et al.*, 2006*b*). But in $d = 2$, the characteristic length scale is the distance between vacancies of density q, $\xi \sim q^{-1/2}$, and the time scale is related to this by the vacancy diffusion constant $q/2$ (see Section 10.4.1), giving $\tau \sim \xi^2/(q/2) \sim \xi^4$.

Four-point and similar dynamical susceptibilities measure the dynamical fluctuations of global observables. These observables probe relaxation over a certain length scale, so it is important to note that the scaling properties of the corresponding susceptibilities depend on such length scales (Chandler *et al.*, 2006). In the examples above the observable was the persistence function, which probes relaxation on a length scale of one lattice spacing. Let us now consider instead the following susceptibility for the case of conservative (conserved density) KCMs, $\chi_4(k, t) = \lim_{k' \to 0} N^{-1} \sum_{ij} \langle \delta F_i(\boldsymbol{k}, t) \delta F_j(-\boldsymbol{k}, t) \rangle e^{i\vec{k}' \cdot (\vec{r}_i - \vec{r}_i)}$, where i, j label the N particles in the system, the position of the ith particle at time t is $\boldsymbol{r}_i(t)$, and $\delta F_i(\boldsymbol{k}, t) \equiv e^{i\boldsymbol{k} \cdot [\boldsymbol{r}_i(t) - \boldsymbol{r}_i(0)]} - \langle e^{i\boldsymbol{k} \cdot [\boldsymbol{r}_i(t) - \boldsymbol{r}_i(0)]} \rangle$ The wavevector $\boldsymbol{k}'$ appearing here is the analog of $\boldsymbol{k}$ in $S_4(\boldsymbol{k}, t)$ above. A normalization factor $\mathcal{N}(t)$ could be included as before but this turns out to make no qualitative difference. This χ_4 measures the system to system fluctuations of the self-intermediate scattering function of wavevector $F(\boldsymbol{k}, t)$. It thus probes structural relaxation at length scales comparable to $2\pi/k$. Figure 10.4(d) shows how $\chi_4(k, t)$ changes in behavior as we go from large to small k in the (2)-TLG: as we probe larger length scales (smaller k) χ_4 peaks at later times, and the initial power-law growth changes exponent (Chandler *et al.*, 2006). This change in behavior is related to the non-Fickian to Fickian crossover of particle diffusion (Berthier *et al.*, 2005*a*), which we discuss in the next subsection.

10.5.2 Transport decoupling

A central consequence of DH is transport decoupling. A prominent example is Stokes–Einstein breakdown (Swallen *et al.*, 2003): in deeply supercooled liquids the rate for self-diffusion is orders of magnitude larger than what would be predicted from the Stokes–Einstein relation between self-diffusion constant and viscosity, $D_s \propto \eta^{-1}$. Similar transport relations of liquid-state theory also break down near the glass transition (Chang and Sillescu, 1997; Ediger, 2000). This is a consequence of the dynamical fluctuations associated with DH.

A major success of KCMs is the ability to rationalize this phenomenon. It does so in terms of the decoupling between the different fundamental time scales for local relaxation (Jung *et al.*, 2004, 2005). How this comes about is illustrated in Fig. 10.5. Panels (a) and (b) show trajectories of a probe particle embedded in a KCM. This can be thought of as a molecule in a liquid that has been labeled for tracking while coarse graininig over the rest of the system, thus describing its effective dynamics via a KCM. The motion of the probe particle is determined by the underlying fluctuations of the host KCM to which it is coupled. A natural dynamical rule is that the probe can make a diffusive jump from site i to site j only if both sites i and j are excited, $n_i = n_j = 0$ (Jung *et al.*, 2004). Figures 10.5(a) and (b) show the difference in the probe motion between high and low temperatures. At high T, Fig. 10.5(a), excitations in the KCM are plentiful and probe motion appears Brownian. At low T, Fig. 10.5(b), excitations are scarce, the dynamics of the KCM is heterogeneous, and probe motion is intermittent: the probe is immobile if immersed in an inactive space-time "bubble"; in order to move it has to wait for an excitation to come along.

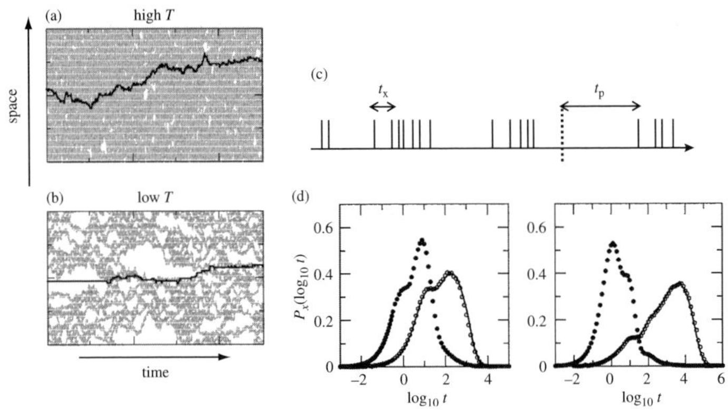

Fig. 10.5 Persistence/exchange decoupling. (a,b) Trajectory of a probe particle diffusing in a KCM. We show the case of the one-dimensional FA-1 for ease of illustration; similar behavior is observed in other KCMs. The probe particle can make a diffusive step only if allowed by the local excitations in the KCM. (Sites with $n_i = 0$ are shown as gray in the background.) At high temperatures/low densities ($T = 3$) diffusion becomes Fickian after short times and distances. Plot (b) is on the same scale as (a), showing that at lower temperatures ($T = 0.8$) diffusion is intermittent and non-Fickian over much longer time scales and length scales. (c) Timeline of displacement events. The waiting time between events is termed the local *exchange* time; it is the time measured to the next event with the knowledge of when the previous one took place. The waiting time until the next event from an arbitrarily chosen starting observation time is termed the *persistence* time. When the time series of events is non-Poissonian, as a consequence of dynamical correlation in the KCMs, typical persistence and exchange times are different: exchange times are dominated by the clustering of events, while persistence times are determined by the long quiescent periods. (d) The decoupling between the distributions of exchange (earlier curves) and persistence (later curves) times becomes more pronounced the lower the temperature. Results are shown for the East model, with $T = 1$ on the left and $T = 0.5$ on the right. Adapted from Refs. (Jung *et al.*, 2004, 2005), with permission.

There are two fundamental time scales that control this intermittent motion (Jung *et al.*, 2004, 2005). The first one is the "persistence time", $t_{\rm p}$, that is, the time the probe needs to wait to start moving *for the first time*, given an arbitrary start time for observation. The second time scale is the (local) "exchange time", $t_{\rm x}$, the time *between moves*. Crucially, DH jump events are not Poissonian, but display "bunching", and typical persistence times can become much larger than typical exchange times, see Fig. 10.5(c). This decoupling between persistence and exchange becomes more pronounced as temperatures is decreased, as illustrated in Figs. 10.5(d).

We can approximately quantify the motion of the probe particle (Berthier *et al.*, 2005*a*) by means of a continuous-time random walk (CTRW) approach (Montroll and Weiss, 1965). The probe makes random-walk steps of unit size at random times determined by the fluctuations of the host KCM. Let us assume that this random clock ticks at time intervals drawn from the exchange-time distribution, $\phi(t_{\rm x})$, but that other than that are independently distributed. The probability for the probe to be at position r at time t, or van Hove function, is $G_{\rm s}(r,t) = \sum_{m=0}^{\infty} \pi_m(t)\Gamma^{(m)}(r)$ (we assume $d=1$ for simplicity, extension to higher dimensions is straightforward). Here, $\pi_m(t)$ is the probability that the probe has made m steps by time t, and $\Gamma^{(m)}(r)$ is the probability that a random walker is at a distance r after m steps. After Laplace transforming in time and Fourier transforming in space we obtain,

$$\hat{F}_{\rm s}(k,\sigma) = \hat{P}(\sigma) + \cos{(k)}\frac{\hat{p}(\sigma)}{\sigma}\frac{1-\hat{\phi}(\sigma)}{1-\cos{(k)}\hat{\phi}(\sigma)}, \tag{10.3}$$

where $\hat{F}_{\rm s}(k,\sigma)$ is the Laplace transform of the self-intermediate scattering function $F_{\rm s}(k,t)$, which in turn is the Fourier transform of $G_{\rm s}(r,t)$. Equation (10.3) is the Montroll–Weiss equation for the motion of the probe in the CTRW approximation. $\hat{\phi}(\sigma)$, $\hat{p}(\sigma)$ and $\hat{P}(\sigma)$ are the Laplace transforms of the exchange time distribution, the persistence time distribution, and the persistence function $P(t)$, respectively. The last two functions appear in Eq. (10.3) because the first step is determined by the persistence time: its distribution $p(t_{\rm p})$ is related to the persistence function $P(t) = \pi_0(t)$ by $P(t) = \int_t^\infty p(t_{\rm p})dt_{\rm p}$. If the dynamics is stationary, then there is nothing special about time zero and $p(t_{\rm p}) = \langle \tau_x \rangle^{-1} \int_{t_{\rm p}}^\infty \phi(t')dt'$, corresponding to a uniform average over all earlier jump times; the normalization factor is the inverse of the average exchange time $\langle \tau_{\rm x} \rangle = \int_0^\infty dt \phi(t) t$. From Eq. (10.3) we can define the wavelength-dependent relaxation time: $\tau(k,T) = \lim_{\sigma\to 0} \hat{F}_{\rm s}(k,\sigma)$. We obtain (Berthier *et al.*, 2005*a*)

$$\tau(k) = \tau_{\rm p} + \frac{\cos{(k)}}{1-\cos{(k)}}\tau_{\rm x}, \tag{10.4}$$

where $\tau_{\rm x}$ and $\tau_{\rm p}$ are the average exchange and persistence times, respectively. At low temperatures we have persistence/exchange decoupling, $\tau_{\rm p} \gg \tau_{\rm x}$ (Jung *et al.*, 2005). The structural or alpha relaxation time is often defined as $\tau_\alpha = \tau(k=\pi/a)$, where a is the lattice spacing, which we set to unity. For these large wavevectors the first term of Eq. (10.4) dominates, and the structural relaxation time is set by the persistence time, $\tau_\alpha \approx \tau_{\rm p}$. For small enough wavevectors the second term in Eq. (10.4) dominates, $\tau(k) \approx \tau_{\mathrm{x}}/k^2$, and since the limit of $k \to 0$ defines the diffusion rate we find that $D \approx \tau_{\mathrm{x}}^{-1}$. This explains Stokes–Einstein breakdown at low temperatures: $D\tau_\alpha \approx \tau_{\mathrm{p}}/\tau_{\mathrm{x}} \neq \mathrm{const}$ (Jung *et al.*, 2004).

The decoupling between persistence and exchange times is an effect of dynamical fluctuations. At low temperatures the self-diffusion constant seems to scale as a fractional power of the relaxation time $D \sim \tau_{\mathrm{p}}^{-\delta}$, with $\delta < 1$ (the Stokes–Einstein relation is $\delta = 1$). This is the case for all KCMs with strong enough constraints, as shown in Fig. 10.6(a) for the FA-1 in $d = 1$, the East model and various constrained lattice gases (Jung *et al.*, 2004; Pan *et al.*, 2005; Ashton, 2009). The fractional exponent $\delta \approx 0.6 - 0.8$ is not distinct from that observed in experiments (Swallen *et al.*, 2003, 2009).

Equation (10.4) indicates that there is a spectrum of time scales that interpolate between a k-independent value τ_{p} at shorter length scales, to a diffusive time scale τ_{x}/k^2 at large length scales (Berthier *et al.*, 2005*a*). This crossover is shown in Fig. 10.6(b) for the FA-1 and East models. It is the crossover from non-Fickian diffusion at short length scales to eventual Fickian diffusion at long enough ones. The length l^* at which this crossover takes place, or "Fickian length scale", is given by $l^* \propto \sqrt{\tau_{\mathrm{p}}/\tau_{\mathrm{x}}}$, and grows with decreasing temperature/increasing density. It is the distance a particle has to move before it forgets how long it took to make the first step. Figure 10.6(b)

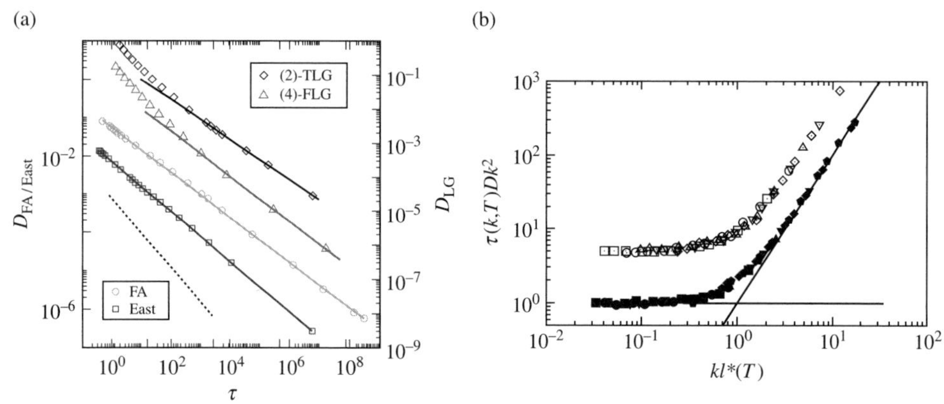

Fig. 10.6 Transport decoupling in KCMs. (a) Scaling of probe or self-diffusion constant with structural relaxation time in the one-dimensional FA-1 and East models (left scale) and two- and three- dimensional lattice gases (right scale). Solid lines indicate power-law fits $D \sim \tau^{-\delta}$; the dashed line is the Stokes–Einstein relation $D \sim \tau^{-1}$. In all cases $\delta < 1$ at low enough q, indicating a breakdown of the Stokes–Einstein relation, as a consequence of dynamical fluctuations. (For the FA-1 model $\delta = 1$ for dimensions larger than its critical dimension $d_c = 2$ (Jack *et al.*, 2006*b*).) Adapted from Refs. (Jung *et al.*, 2004; Pan *et al.*, 2005; Ashton, 2009), with permission. (b) Lengthscale-dependent relaxation time $\tau(k)$ in the FA-1 and East models. At short length scales the relaxation time is k independent and determined by the persistence time, $\tau(k) \approx \tau_{\mathrm{p}}$. At larger length scales it becomes diffusive, $\tau(k) \approx \tau_{\mathrm{x}}/k^2$. This crossover is controlled by the Fickian length scale $l^* \sim \sqrt{\tau_{\mathrm{p}}/\tau_{\mathrm{x}}}$. The figure shows that $\tau(k)$ at different q collapse under $k \to kl^*$. From Ref. (Berthier *et al.*, 2005*a*), with permission.

shows how $\tau(k)$ at different temperatures collapse under $k \to l^* k$. The non-Fickian to Fickian crossover is also responsible for the wavelength dependence of four-point functions (Chandler *et al.*, 2006), Fig. 10.4(d).

The CTRW analysis can be extended to describe the effect of driving, for example by externally forcing the probes. The competition between time scales in this case leads to interesting non-linear response behavior, such as non-monotonic differential mobility and giant diffusivity (Jack *et al.*, 2008). Furthermore, a study of Eq. (10.3) in the crossover regime between non-Fickian and Fickian explains (Chaudhuri *et al.*, 2007) the exponential tails observed (Stariolo and Fabricius, 2006) in van Hove functions at intermediate times. The waiting-time distributions used in the CTRW analysis above are the ones that are obtained from the study of KCMs. The CTRW approach can also be used by assuming a different origin for the waiting-time distributions, as for example in the analysis of metabasin transitions (Heuer, 2008) in atomistic liquids.

10.5.3 Space-time phase transitions

Thermodynamically KCMs are trivial, so all interesting behavior is dynamical. Nevertheless, DH pictures such as those of Fig. 10.3 are suggestive of phase separation between two distinct phases. The phases of Fig. 10.3 are distinguished by their dynamics: dark regions are dynamically *active*, while light ones are dynamically *inactive*. Furthermore, the phase separation in the spatial projection of low-temperature/high-density equilibrium trajectories, such as those of Fig. 10.3, is only mesoscopic: when coarse grained over large enough length scales, irrespective of time, the space-projected trajectories are homogeneous. Here, we show how these observations are directly related to a true non-equilibrium phase transition (Garrahan *et al.*, 2007), which in contrast to thermodynamic transitions, occurs in ensembles of trajectories and is driven by non-equilibrium driving fields. This phase transitions can be studied by recourse to the large-deviation method (Lecomte *et al.*, 2007; Touchette, 2009).

A convenient order parameter to discern active and inactive dynamics is the "dynamical activity" K (Garrahan *et al.*, 2007), defined as the total number of configuration changes in a trajectory. In a non-conservative KCM it would amount to the total number of local changes from empty to occupied or vice versa, and in a lattice gas (conservative KCM) to the total number of particle displacements. The activity is extensive in space-time volume, i.e. typically $K = O(Nt)$, where N is the number of lattice sites and t the time extension of the trajectory. Each trajectory $x(t)$ in the ensemble of (equilibrium) trajectories of length t has a total activity, $\hat{K}[x(t)]$. The activity is thus distributed, $P_t(K) = \left\langle \delta\left(K - \hat{K}[x(t)]\right)\right\rangle$, where the average is over the set of equilibrium trajectories, $\{x(t)\}$. At long enough times this probability acquires a large-deviation form (Lecomte *et al.*, 2007; Touchette, 2009). $P_t(K) \approx e^{-t\varphi(K/t)}$. The function $\varphi(k)$ is called a large-deviation function, and it plays in this dynamical context the same role as, for example, the entropy density in the microcanonical ensemble of equilibrium statistical mechanics. Alternatively, we can consider the generating function of K,

$$Z_t(s) \equiv \sum_K e^{-sK} P_t(K) \approx e^{t\psi(s)}, \tag{10.5}$$

which also displays a large-deviation form. The large-deviation function $\psi(s)$ is akin to a free-energy density, and is related to $\varphi(K/t)$ by a Legendre transform, $\psi(s) = -\min_k [\varphi(k) + sk]$. Just like a free energy in a thermodynamic problem, the function $\psi(s)$ carries the information of *dynamical* phase behavior. Specifically, its singularities indicate *dynamical* phase transitions (Garrahan *et al.*, 2007).

The calculation of the large-deviation function $\psi(s)$ is simplified greatly by the following observation (Lebowitz and Spohn, 1999): if $\mathbb{W}$ is the master operator that generates the stochastic dynamics, then $\psi(s)$ is the largest eigenvalue of a modified operator $\mathbb{W}_s$, where $\mathbb{W}_0 = \mathbb{W}$. This reduces the calculation from that of computing a "partition sum", Eq. (10.5), to an eigenvalue problem. For example, for a spin facilitated model, $\mathbb{W}_s = \sum_i f_i(\boldsymbol{n}) \left\{e^{-s}\left[(1-\rho)\sigma_i^+ + \rho\sigma_i^-\right] - \rho n_i - (1-\rho)\right\}$, where $f_i(\boldsymbol{n})$ is

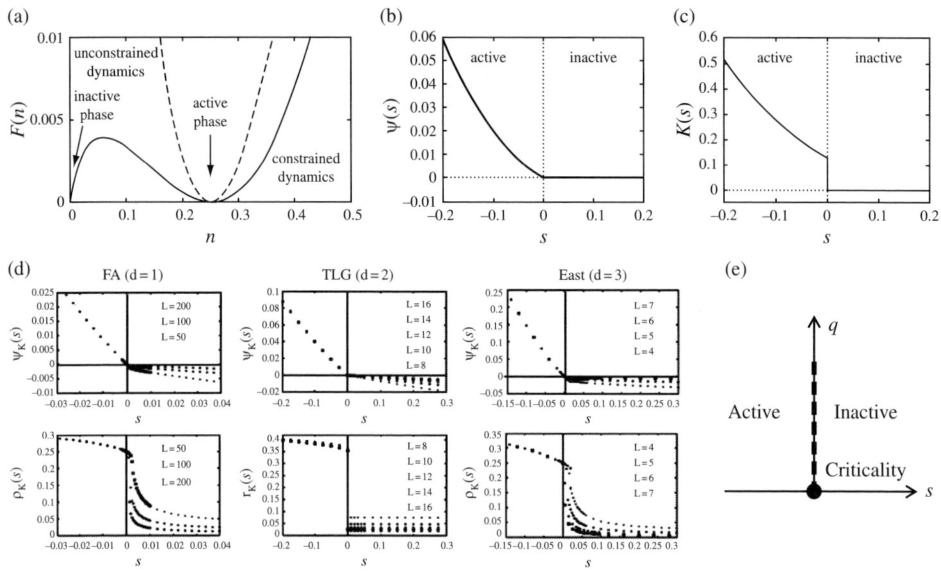

Fig. 10.7 Space-time phase transitions in KCMs. (a) Variational dynamical free energy $\mathcal{F}(n)$ in terms of mean excitation density. Kinetic constraints give rise to a bistable $\mathcal{F}(n)$ (full line). In the absence of kinetic constraints the corresponding variational function is unistable (dashed line). (b) Mean-field estimate of large-deviation function $\psi(s)$. There is a singularity at $s = 0$, indicating a dynamical phase-transition. (c) Mean activity $K(s) \equiv -\psi'(s)$ as a function of s. This dynamical order parameter shows a discontinuous jump at $s = 0$: the transition is a first-order one between an active dynamical phase and an inactive dynamical phase. (d) The dynamical first-order scenario is also present in finite dimensions, as shown from numerical computation of $\psi(s)$ and $K(s)$ for various KCMs. (e) Dynamical phase diagram. The line $s = 0$ is one of first-order coexistence between the active dynamical phase ($s < 0$) and the inactive one ($s > 0$). It extends all the way along the q-axis. The critical point at $q = 0$ controls the scaling behavior discussed in Section 10.4. Adapted from Refs. (Garrahan *et al.*, 2007, 2009*a*), with permission.

the kinetic constraint on site i, and $\sigma_i^{\pm}$ are the raising/lowering operators on site i. While it is not always possible to diagonalize such an operator analytically, bounds for its largest eigenvalue can be estimated variationally. For the FA-m this amounts to minimizing a Landau free energy $\mathcal{F}(n) = -n^m \left(2e^{-s}\sqrt{qn} - q - n\right)$. The factor n^m comes from the kinetic constraint, and makes $\mathcal{F}(n)$ non-linear enough to allow for multiple minima, Fig. 10.7(a). The corresponding $\psi(s)$ has a singular structure, the first derivative being discontinuous at $s = 0$, Figs. 10.7(b) and (c). The meaning of this is the following. Trajectories are organized into two dynamical phases, an active one with $K > 0$ and an inactive one with $K = 0$. The field s determines the bias for or against activity. For $s < 0$ the active phase is the dominant one, while for $s > 0$ the inactive phase dominates. At $s = 0$ the probability of trajectories in either phase is equal in the $t \to \infty$ limit and we have dynamical first-order phase coexistence (Garrahan *et al.*, 2007). This situation occurs in all KCMs, Fig. 10.7(d). Interestingly, similar dynamical phase structure is observed in spin-glass models (van Duijvendijk *et al.*, 2010; Jack and Garrahan, 2010) and in atomistic liquids (Hedges *et al.*, 2009).

Actual dynamics takes place at $s = 0$. The results above show that this is the condition for dynamical coexistence *in the bulk*, i.e. infinitely far from boundaries active (ergodic) or inactive (non-ergodic) trajectories are equally likely. However, just like in the case of ordinary phase transitions, boundary fields can bias the bulk into one of the coexisting phases. In the case of dynamics, initial conditions play the role of a (time) boundary. In particular, almost all possible initial configurations chosen from the equilibrium static distribution at non-zero temperature or less than maximal density will select the active dynamical phase. In this case, the inactive phase manifests via rare region effects, giving rise to DH only at the mesoscopic scale. An important open question is whether there are physical controllable fields that play the role of s in the analysis above.

10.6 Summary and outlook

In this chapter we have attempted to summarize recent developments in the study of KCMs as models of glass formers. In the long tradition of statistical mechanics KCMs provide simplified models that capture important ideas about the fundamental physics behind the phenomenology of glassy systems. Their simplicity allows for detailed study, which in turn gives rise to further physical insights into the glass-transition problem. The central message from KCMs is that the complex and cooperative dynamics of glass-forming systems can be achieved without recourse to complex thermodynamic behavior: in KCMs thermodynamics plays essentially no role, and complex dynamics emerge from rather simple local kinetic rules. These rules are local and free of disorder, but nevertheless give rise to dynamical frustration.

The irrelevance of thermodynamics for glassy dynamics that the study of KCMs suggest contrasts sharply with approaches such as that of the random first-order transition theory (Lubchenko and Wolynes, 2007; Mezard and Parisi, 2000) where thermodynamics is essential. Whether thermodynamic aspects are relevant or not to glass-transition phenomena is still a matter of debate, but to the extent that they are KCMs can say very little about them. This can either be seen as a flaw of the KCM-based approach (Biroli *et al.*, 2005) or as an indication that these aspects are described by degrees of freedom that do not contribute too much to the long-time dynamics and have therefore been coarse-grained out (Chandler and Garrahan, 2005). Furthermore, by their coarse-grained and lattice-based nature KCMs can in principle say very little about short-distance/short-time dynamics, such as beta-relaxation or anomalous vibrations. There is, however, evidence that these short-scale phenomena are coupled to longer-scale dynamic heterogeneity (see for example (Widmer-Cooper *et al.*, 2004; Brito and Wyart, 2007)) so it may be possible to capture some of these effects with generalizations of KCMs (Moreno and Colmenero, 2007; Ashton and Garrahan, 2009). In any case, KCMs provide an explicitly real-space picture of glassy dynamics. Their main success has been the rationalization of dynamical heterogeneity. While DH can be analyzed with other approaches, such as generalizations of mode-coupling theory (Biroli *et al.*, 2006), the immediacy of the results and explanations for DH-related phenomena obtained from KCMs is remarkable.

While we know a lot about KCMs we still do not have a satisfactory understanding of how they emerge as an effective description from realistic systems. (This is also the case in alternative approaches, be it the random first-order transition (Lubchenko and Wolynes, 2007), or frustration limited domains (Kivelson and Tarjus, 2008), where the idealized models that display the proposed behavior cannot be readily obtained from realistic liquid systems.) It is usually argued (Garrahan and Chandler, 2003) that KCMs ought to emerge from some form of local coarse graining of a microscopic system, but this procedure has not been shown to work just yet (except in highly simplified situations (Garrahan and Newman, 2000; Garrahan *et al.*, 2009*b*)). Proving a direct connection between atomistic liquids and KCMs is arguably the central open problem in this field.

References

Aizenman, M and Lebowitz, J L (1988). *J. Phys. A*, **21**, 3801–3813.

Aldous, D and Diaconis, P (2002). *J. Stat. Phys.*, **107**, 945–975.

Andersen, H C (2005). *Proc. Nat. Acad. Sci.*, **102**, 6686–6691.

Anderson, P W (1979). In *Ill-Condensed Matter* (ed. R. Balian, R. Maynard, and G. Toulouse), Les Houches Session XXXI. Elsevier Science.

Ashton, D. and Garrahan, J.P. (2009). *Eur. Phys. J. E*, **30**, 303–307.

Ashton, D J (2009). Ph.D. thesis, University of Nottingham.

Berthier, L, Chandler, D, and Garrahan, J P (2005*a*). *Europhys. Lett.*, **69**, 320–326.

Berthier, L and Garrahan, J P (2003). *Phys. Rev. E*, **68**, 041201.

Berthier, L and Garrahan, J P (2005). *J. Phys. Chem. B*, **109**, 3578–3585.

Berthier, L, Garrahan, J P, and Whitelam, S (2005*b*). *Phys. Rev. E*, **71**, 026128–026142.

Bertini, L and Toninelli, C (2004). *J. Stat. Phys.*, **117**, 549–580.

Biroli, G, Bouchaud, J P, Miyazaki, K, and Reichman, D R (2006). *Phys. Rev. Lett.*, **97**, 195701.

Biroli, G, Bouchaud, J P, and Tarjus, G (2005). *J. Chem. Phys.*, **123**.

Brito, C and Wyart, M (2007). *J. Stat. Mech. Theor. Exp.*, **L08003**.

Buhot, A and Garrahan, J P (2001). *Phys. Rev. E*, **64**, 021505.

Butler, S and Harrowell, P (1991). *J. Chem. Phys.*, **85**, 4466.

Cancrini, N, Martinelli, F, Roberto, C, and Toninelli, C (2007). *J. Stat. Mech. Theor. Exp.*, **L03001**.

Cancrini, N, Martinelli, F, Roberto, C, and Toninelli, C (2008). *Probab. Theory Related Fields*, **140**, 459–504.

Cancrini, N, Martinelli, F, Roberto, C, and Toninelli, C (2010*a*). *Commn. Math. Phys.*, **297**, 299–344.

Cancrini, N, Martinelli, F, Schonmann, R H, and Toninelli, C (2010*b*). *J. Stat. Phys.*, **138**, 1109–1123.

Cavagna, A (2009). *Phys. Rep.*, **476**, 51–124.

Chandler, D and Garrahan, J P (2005). *J. Chem. Phys*, **123**, 044511.

Chandler, D and Garrahan, J P (2010). *Ann. Rev. Phys. Chem.*, **61**, 191.

Chandler, D, Garrahan, J P, Jack, R L, Maibaum, L, and Pan, A C (2006). *Phys. Rev. E*, **74**, 051501.

Chang, I and Sillescu, H (1997). *J. Phys. Chem. B*, **101**, 8794–8801.

Chaudhuri, P, Berthier, L, and Kob, W (2007). *Phys. Rev. Lett.*, **99**.

Corberi, F and Cugliandolo, L F (2009). *J. Stat. Mech. Theor. Exp.*, **P09015**.

Crisanti, A and Ritort, F (2003). *J. Phys. A*, **36**, R181–R290.

Debenedetti, P G and Stillinger, F H (2001). *Nature*, **410**, 259–267.

Dorogovtsev, S N, Goltsev, A V and Mendes, J F F (2006). *Phys. Rev. Lett.*, **96**, 040601.

Ediger, M D (2000). *Ann. Rev. Phys. Chem.*, **51**, 99–128.

Eisinger, S and Jäckle, J (1991). *Z. Phys. B*, **84**, 115–124.

Elmatad, Y S, Chandler, D, and Garrahan, J P (2009). *J. Phys. Chem. B*, **113**, 5563–5567.

Evans, M R and Sollich, P (1999). *Phys. Rev. Lett*, **83**, 3238–3241.

Faggionato, A, Martinelli, F, Roberto, C, and Toninelli, C (2010). arXiv:1012.4912.

Fredrickson, G and Andersen, H (1984). *Phys. Rev. Lett.*, 1244–1247.

Fredrickson, G H (1988). *Ann. Rev. Phys. Chem.*, **39**, 149–180.

Fredrickson, G H and Brawer, S A (1986). *J. Chem. Phys*, **84**, 3351.

Garrahan, J P and Chandler, D (2002). *Phys. Rev. Lett.*, **89**, 035704.

Garrahan, J P and Chandler, D (2003). *Proc. Natl. Acad. Sci.*, **100**, 9710–9714.

Garrahan, J P, Jack, R L, Lecomte, V, Pitard, E, van Duijvendijk, K, and van Wijland, F (2007). *Phys. Rev. Lett.*, **98**, 195702.

Garrahan, J P, Jack, R L, Lecomte, V, Pitard, E, van Duijvendijk, K, and van Wijland, F (2009*a*). *J. Phys. A*, **42**, 075007.

Garrahan, J P and Newman, M E J (2000). *Phys. Rev. E*, **62**, 7670–7678.

Garrahan, J P, Stannard, A, Blunt, M O, and Beton, P H (2009*b*). *Proc. Nat. Acad. Sci.*, **106**, 15209–15213.

Glarum, S H (1960). *J. Chem. Phys.*, **33**, 1371–1375.

Glotzer, S C (2000). *J. Non-Cryst. Solids*, **274**, 342–355.

Goncalves, P, Landim, C and Toninelli, C (2009). Annales IHP - Probability and Statistics, **45**, 887–909.

Götze, W and Sjögren, L (1992). *Rep. Prog. Phys.*, **55**, 241–376.
Graham, I S, Piché, L, and Grant, M (1997). *Phys. Rev. E*, **55**, 2132–2144.
Hedges, L O and Garrahan, J P (2007). *J. Phys. Condens. Matter*, **19**, 205124.
Hedges, L O, Jack, R L, Garrahan, J P, and Chandler, D (2009). *Science*, **323**, 1309–1313.
Heuer, A (2008). *J. Phys. Condens. Matter*, **20**, 373101.
Holroyd, A E (2003). *Probab. Theory Related Fields*, **125**, 195–224.
J M Schwarz, A J Liu and Chayes, L Q (2006). *Europhys. Lett.*, **73**, 560.
Jack, R L and Garrahan, J P (2010). *Phys. Rev. E*, **81**, 011111.
Jack, R L, Garrahan, J P, and Chandler, D (2006*a*). *J. Chem. Phys.*, **125**, 184509.
Jack, R L, Kelsey, D, Garrahan, J P, and Chandler, D (2008). *Phys. Rev. E*, **78**, 011506.
Jack, R L, Mayer, P, and Sollich, P (2006*b*). *J. Stat. Mech. Theor. Exp.*, **P03006**.
Jäckle, J (1986). *Rep. Prog. Phys.*, **49**, 171–231.
Jäckle, J and Krönig, A (1994). *J. Phys. Condens. Matter*, **6**, 7633–7653.
Jeng, M and Schwarz, J M. (2010). *Phys. Rev. E*, **81**, 011134.
Jung, YJ, Garrahan, J P, and Chandler, D (2004). *Phys. Rev. E*, **69**, 061205.
Jung, Y, Garrahan, J P, and Chandler, D (2005). *J. Chem. Phys.*, **123**, 084509.
Kivelson, S A and Tarjus, G (2008). *Nature Mater.*, **7**, 831–833.
Kob, W and Andersen, H C (1993). *Phys. Rev. E*, **48**, 4359–4363.
Lačević, N, Starr, F W, Schroder, T B, and Glotzer, S C (2003). *J. Chem. Phys.*, **119**, 7372–7387.
Lebowitz, J L and Spohn, H (1999). *J. Stat. Phys.*, **95**, 333–365.
Lecomte, V, Appert-Rolland, C, and van Wijland, F (2007). *J. Stat. Phys.*, **127**, 51–106.
Léonard, S, Mayer, P, Sollich, P, Berthier, L, and Garrahan, J P (2007). *J. Stat. Mech. Theor. Exp.*, **P07017**.
Lubchenko, V and Wolynes, P G (2007). *Ann. Rev. Phys. Chem.*, **58**, 235–266.
Mayer, P, Leonard, S, Berthier, L, Garrahan, J P, and Sollich, P (2006). *Phys. Rev. Lett.*, **96**, 030602.
Mayer, P and Sollich, P (2007). *J. Phys. A*, **40**, 5823–5856.
Merolle, M, Garrahan, J P, and Chandler, D (2005). *Proc. Nat. Acad. Sci.*, **102**, 10837–10840.
Mezard, M and Parisi, G (2000). *J. Phys. Condens. Matter*, **12**, 6655–6673.
Montroll, E W and Weiss, G H (1965). *J. Math. Phys.*, **6**, 167.
Moreno, A J and Colmenero, J (2007). *J. Phys. Condens. Matter*, **19**, 205144.
Palmer, R G (1989). In *Cooperative Dynamics in Complex Physical Systems* (ed. H. Takayama), Volume 43, Springer Series in Synergetics, pp. 118–127. Springer, Heidelberg.
Palmer, R G, Stein, D L, Abrahams, E, and Anderson, P W (1984). *Phys. Rev. Lett.*, **53**, 958–961.
Pan, A C, Garrahan, J P, and Chandler, D (2005). *Phys. Rev. E*, **72**, 041106.
Reiter, J (1991). *J. Chem. Phys.*, **95**, 544–554.
Reiter, J, Mauch, F, and Jäckle, J (1992). *Physica A*, **184**, 458–476.
Ritort, F and Sollich, P (2003). *Adv. Phys.*, **52**, 219–342.

Schonmann, R H (1992). *Ann. Probab.*, **20**, 174–193.
Sellitto, M, Biroli, G, and Toninelli, C (2005). *Europhys. Lett.*, **69**, 496.
Sollich, P and Evans, M R (1999). *Phys. Rev. Lett.*, **83**, 3238–3241.
Sollich, P and Evans, M R (2003). *Phys. Rev. E*, **68**, 031504.
Stariolo, D A and Fabricius, G (2006). *J. Chem. Phys.*, **125**.
Swallen, S F, Bonvallet, P A, McMahon, R J, and Ediger, M D (2003). *Phys. Rev. Lett.*, **90**, 015901.
Swallen, S F, Traynor, K, McMahon, R J, Ediger, M D, and Mates, T E (2009). *J. Phys. Chem. B*, **113**, 4600–4608.
Toninelli, C and Biroli, G (2004). *J. Stat. Phys.*, **117**, 27–54.
Toninelli, C and Biroli, G (2008*a*). *J. Stat. Phys.*, **130**, 83–112.
Toninelli, C and Biroli, G (2008*b*). *Eur. Phys. J. B*, **130**.
Toninelli, C, Biroli, G, and Fisher, D S (2004*a*). *Phys. Rev. Lett.*, **92**, 185504.
Toninelli, C, Biroli, G, and Fisher, D S (2004*b*). *Phys. Rev. Lett.*, **92**.
Toninelli, C, Biroli, G, and Fisher, D S (2005*a*). *J. Stat. Phys.*, **120**, 167–238.
Toninelli, C, Biroli, G, and Fisher, D S (2006). *Phys. Rev. Lett.*, **96**, 035702.
Toninelli, C, Biroli, G, and Fisher, D S (2007). *Phys. Rev. Lett.*, **98**, 129602.
Toninelli, C, Wyart, M, Berthier, L, Biroli, G, and Bouchaud, J P (2005*b*). *Phys. Rev. E*, **71**, 041505.
Touchette, H (2009). *Phys. Rep.*, **478**, 1–69.
van Duijvendijk, K, Jack, R L, and van Wijland, F (2010). *Phys. Rev. E*, **81**, 011110.
Whitelam, S, Berthier, L, and Garrahan, J P (2004). *Phys. Rev. Lett.*, **92**, 185705.
Whitelam, S and Garrahan, J P (2004). *J. Phys. Chem. B*, **108**, 6611–6615.
Widmer-Cooper, A, Harrowell, P and Fynewever, H. (2004). *Phys. Rev. Lett.*, **93**, 135701.

11
Growing length scales in aging systems

Federico CORBERI, Leticia F. CUGLIANDOLO and Hajime YOSHINO

Abstract

We summarize studies of growing lengths in different aging systems. The chapter is structured as follows. We recall the definition of a number of observables, typically correlations and susceptibilities, that give access to dynamic and static correlation lengths. We use a growing length perspective to review three out-of-equilibrium cases: domain-growth phenomena; the evolution of Edwards–Wilkinson and Kardar–Parisi–Zhang manifolds and other directed elastic manifolds in random media; spin and structural glasses in relaxation and under an external drive. Finally, we briefly report on a mechanism for dynamic fluctuations in aging systems that is based on a time-reparametrization invariance scenario and may be at the origin of the dynamic growing length in glassy materials.

11.1 Introduction

For a long time, and somewhat paradoxically, the majority of theoretical physicists interested in glasses were reluctant to study these systems in the truly glassy regime, say below the glass temperature T_{g} or above the glass density ρ_{g}. One of the reasons to resist entering the glassy regime was the lack of insight on which questions to ask and, more concretely, which quantities to measure in out-of-equilibrium conditions.

The situation changed dramatically around 20 years ago when it was accepted that glassy systems do not freeze out below T_{g} but they just continue to relax at a slower and slower rate as time goes by—the aging phenomenon. Although this fact was well known experimentally and to a certain extent also numerically it was not fully assimilated theoretically. The solution to a number of simple models [namely, droplet models for disordered systems (Fisher and Huse, 1988*a*), random walks among traps in phase space (Bouchaud, 1992; Bouchaud and Dean, 1995), and mean-field spin systems with disordered interactions (Cugliandolo and Kurchan, 1993, 1994)] demonstrated the possibility of capturing aging phenomena analytically. These solutions also provided a guideline as to which are the simplest preparation protocol and the most relevant observables to focus on dynamically. More importantly, these studies also opened the way to confront glassy dynamics to the behavior of other macroscopic non-equilibrium systems considered to be simpler, such as coarsening phenomena (Bray, 1994) or the motion of manifolds in different kinds of embedding volumes [for reviews on different properties of elastic manifolds that do not, however, deal with their aging properties see (Barabási and Stanley, 1995; Halpin-Healey and Zhang, 1995)]. Experimental and numerical data on theoretically motivated measurements in various physical systems have rapidly accumulated and some of these results are summarized in other chapters in this book.

As for the measuring protocols, in short, one first chooses the way in which the system is taken into the glassy regime and defines the origin of the time axis accordingly. In the simplest setting, the "initial time" $t = 0$, is the instant when the system is suddenly prepared in an out-of-equilibrium condition. For instance, this is achieved by performing a rapid temperature quench from a high temperature above T_g down to a target value $T < T_g$ but other routes to the glassy regime, changing other control parameters, are also feasible. As will be discussed below, the physical properties of the system become functions of the waiting time t_w, the time spent after the preparation of the initial state, stationarity is lost and the relaxation time increases with t_w. The basic idea to analyze such an aging relaxation is to introduce a laboratory time scale—the waiting time—that can be controlled at will when the intrinsic time scale—the equilibrium relaxation time—goes beyond accessible times.

Following this procedure one can study, for instance, the energy relaxation as a function of t_w. However, such a macroscopic "one-time quantity" does not capture all the richness of the relaxation. Much more detailed information is contained in two (or more) time-dependent global quantities. In Fig. 11.1 we show the typical behavior of two *macroscopic* two-time quantities in many glassy systems: the spontaneous decorrelation of a chosen global observable measured at two subsequent times t and t_w (a), and the linear response of the same observable measured at t to a perturbation that couples linearly to it between t_w and t (b).

As shown in panel (a), the autocorrelation function exhibits a two-step decay. At short time scales, say $t - t_w \ll t_w$, there is a relaxation towards a plateau that we call the Edwards–Anderson (EA) order parameter q_{ea} (see its precise definition in Eq. (11.6)). The relaxation in this stage is essentially independent of the waiting time t_w and time-translational invariance (TTI) holds $C(t, t_w) = C_{st}(t - t_w)$ to a certain accuracy. The second relaxation at longer time scales, say $t - t_w \gg t_w$, exhibits t_w dependence and is not stationary. The latter reflects the strong out-of-equilibrium nature of aging. The separation of time scales is confirmed in panel (b) where we sketch the behavior of the susceptibility $\chi(t, t_w)$. The plateau occurs here at $\chi_{ea} = (1 - q_{ea})/T$ with q_{ea} the value of C at its plateau and T the temperature of the environment that is henceforth measured in units of k_B, the Boltzmann constant.

The relaxation of the correlation in Fig. 11.1(a) reminds us, in a sense, of the two-step relaxation in the supercooled liquid phase—the so-called α and β

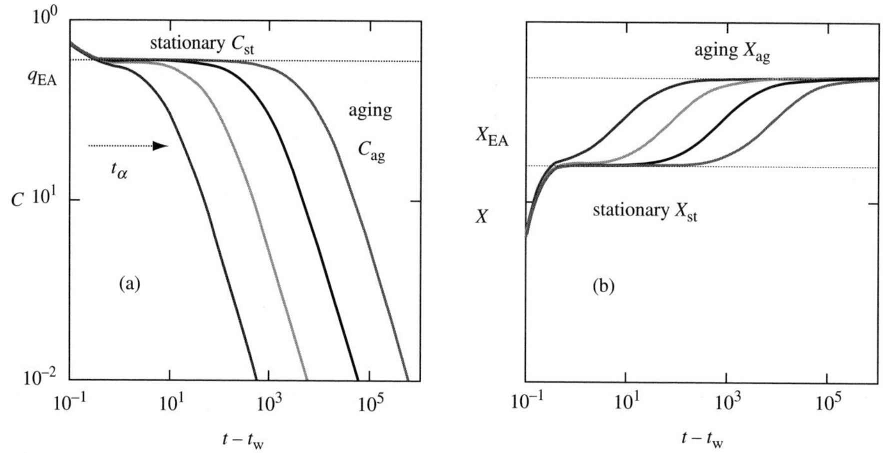

Fig. 11.1 Typical behavior of macroscopic two-time quantities during aging. (a) Autocorrelation function $C(t, t_w)$; (b) linear susceptibility $\chi(t, t_w)$. The waiting time t_w increases from the left to the right curves, typically in a logarithmic scale such that, say, $t_{w\,k+1} = 10 t_{w\,k}$. The α relaxation time, t_α, or the time difference needed to decorrelate significantly, is an increasing function of t_w and is indicated with an arrow in the left panel. The plateau in χ occurs at $\chi_{ea} = (1 - q_{ea})/T$, where q_{ea} is the height of the plateau in C.

relaxations (Donth, 2001; Binder and Kob, 2005). Even more so, it is basically indistinguishable from the one found in phase-ordering processes after a temperature quench (see Section 11.3 for a detailed description of coarsening) (Bray, 1994). In such processes domains of the different equilibrium phases progressively grow in competition. At each instant, the equilibrium order parameter is essentially uniform within each domain. When only short time differences are explored the domain walls are basically static and the correlations decay just as in equilibrium. Instead, when longer time differences are reached the walls move appreciably and the correlations decay from the plateau in a manner that depends explicitly on the waiting time. A key quantity in the description of coarsening is the *typical domain linear size*, $L(t)$, which plays the role of a dictionary between time t and length L. All properties of coarsening systems are invariant under rescaling of all lengths by $L(t)$—in a statistical sense.

The question naturally arises as to whether the dynamics in glassy systems is also of the coarsening type albeit the growing order had not been identified yet. From the analysis of the correlations in many glasses, and one example is shown in Fig. 11.2, where scaled data from an aging colloidal suspension are displayed (Chamon *et al.*, 2008), one could easily conclude this to be the case. The figure shows the two-time correlation data taken using different t_ws as a function of the ratio t/t_w. For sufficiently long t_w the scaling is quite good, suggesting that there might be a growing length $L(t) \propto t^a$, with a an undetermined power. However, the scaling of the global two-time correlation as $f(L(t)/L(t_w))$ is just a quite generic property of two-time

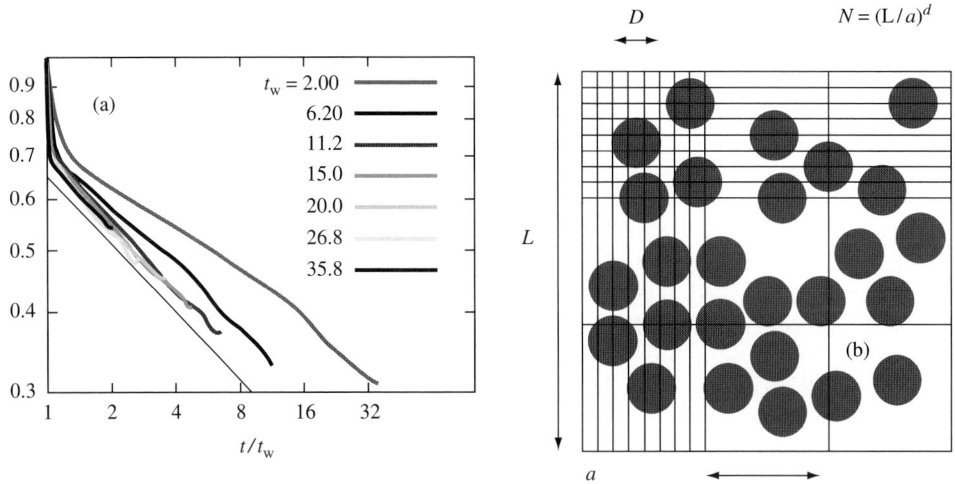

Fig. 11.2 (a) Scaling of two-time correlation data in an aging colloidal suspensions. Waiting times are given in the key in units of 10^2 s. The relatively good scaling with t/t_w suggests $L(t) \simeq t^a$. The data are taken from and analyzed along the lines explained in (Chamon *et al.*, 2008). (b) A sketch of the pixelization used to describe the particle configurations with binary variables, $\sigma_\alpha = 0, 1$; and the coarse-graining procedure (see Section 11.2.1). a is the lattice spacing, D the particle diameter, ℓ the coarse-graining length and L the system size.

correlations (Cugliandolo and Kurchan, 1994), and the space-time correlations do not signal the growth of structure, contrary to what happens in simple phase ordering kinetics in which it scales as a function of $r/L(t)$. Therefore, to reach a conclusion on the existence (or not) of a growing length one should, ideally, analyze the dynamics at *mesoscopic time and length scales* or, at least, define and explore more complex quantities giving access to the mesoscopic details of the process. Eventually, these studies might also enlighten us about the reason for the growth.

Taking the case of usual phase-ordering processes as a reference, we realize that in order to settle this issue we need to follow a number of basic steps, summarized by the following questions: (1) how can we define an order parameter for a glass? (2) how can we define domains—within which the order parameter is essentially constant—or dynamically correlated volumes? (3) can one identify a growing linear size and does it reflect the nature of the dynamics—*e.g.* smooth (such as, for instance, the curvature-driven kinetics observed in some coarsening systems) vs. activated? (4) can one understand the mechanism leading to such a growth? In the body of this chapter we develop these questions.

Having explained that the aim of this chapter is to discuss searches of growing lengths in *aging glasses*, we announce that we shall also deal with *a priori* simpler cases such as a particle's diffusion in random media or the motion of a finite-dimensional manifold, both in their aging regime. Although the glassy nature of these problems might not be complete, it turns out that some aspects of their dynamics are similar to the ones encountered in conventional glasses. Moreover, their mere definition is clearer and their analytic treatment as well as the identification of a growing length are simpler. Their discussion should then be instructive.

The organization of the chapter is the following. In Section 11.2 we give a number of definitions and we discuss how to deal with quenched disorder, if present. In Section 11.3 we review subcritical and critical coarsening phenomena in clean systems. In this section we also introduce, briefly, a relatively simpler problem with an easy to identify growing length: the motion of a free elastic manifold. Section 11.4 is devoted to the phenomenological decription of activated processes—that are specially relevant in problems with quenched disorder—starting with a short account of aging in the Sinai model of diffusion and developing afterwards the droplet model predictions for the dynamics of elastic manifolds in random media and spin-glasses. Section 11.5 recaps mean-field predictions for the growing length issued from the study of fully connected spin models and the related mode-coupling theories as well as growth phenomena in kinetically constrained models for glasses. We also discuss growing lengths in other systems with aging dynamics: the interplay between flow and internal relaxation and the effect of the former on the aging properties of coarsening and glassy systems, granular matter and quantum glasses. Finally, in Section 11.6 we briefly recall a possible mechanism for dynamic fluctuations and the associated growing correlation length in aging systems: the asymptotic development of time-reparametrization invariance, that has been extensively reviewed elsewhere (Chamon and Cugliandolo, 2007).

We do not attempt to present a comprehensive review of analytic, numeric and experimental studies of all kinds of aging systems. We rather focus on the search for growing length scales and their possible use to describe aging materials. With this in mind, the reader should not expect to find a complete list of references on aging studies.

11.2 Definitions

In this section we summarize a number of definitions relevant to the study of cooperative motion in glassy systems. These definitions do not assume equilibrium and can be used to study out-of-equilibrium and, in particular, aging samples. They depend, in general, on various times independently and no fluctuation–dissipation theorem is assumed. Although we are not going to dwell on the question of the pertinence of effective temperatures in the description of glasses, we recall its definition since it will appear in the rest of this chapter.

11.2.1 Euler and Lagrange descriptions

Two ways of studying the heterogeneous dynamics in a many-particle (or many higher-dimensional object) system are the following.

11.2.1.1 Lagrangian description

One can follow the evolution of each individual particle, labeled by an index i, and detect which are the fast- and slow-moving ones during a previously chosen time window, say $t - t_w$, around some time, say t_w, after preparation. This is the route followed in early studies of dynamic heterogeneities in supercooled liquids, see (Bennemann *et al.*, 1999; Kob *et al.*, 1997; Perera and Harrowell, 1996; Donati and et al., 1999; Onuki and Yamamoto, 1998; Doliwa and Heuer, 2002; Heuer *et al.*, 2002) and (van Blaaderen and Wiltzius, 1995; Kegel and van Blaaderen, 2000; Weeks and Weitz, 2000; Weeks *et al.*, 2000) for a few of many molecular dynamics and confocal microscopy papers that use this type of measurement. One of the main outcomes of these studies is the observation of clustering of fast particles, in the form of strings the length of which increases—and may diverge—close to the glassy arrest. In the truly glassy phase fewer studies exist. A precursor is (Miyagawa *et al.*, 1988). More recently, Vollmayr-Lee *et al.* studied the geometry and statistical properties of clusters of mobile and immobile particles with molecular dynamics (Vollmayr-Lee, 2004; Vollmayr-Lee and Zippelius, 2005; Vollmayr-Lee and Baker, 2006). Confocal microscopy has been used with similar aims (Courtland and Weeks, 2003; Weeks *et al.*, 2007).

11.2.1.2 Eulerian description

In a first step, one can define Ising spin variables as the ones that describe uniaxial magnetic systems. The choice is done for notation convenience but also because, as proposed in (Chamon *et al.*, 2008), a simple mapping between particle positions and Ising spin variables, σ_α, defined on the $N = (\mathrm{L}/a)^3$ vertices of a cubic lattice captures the dynamics of "atomic" glassy systems as well (see Fig. 11.2(b)). (L is the linear length of the box and a the length of the lattice spacing.) In a few words, one partitions space with a lattice with very fine mesh—smaller than the particle radii—and the mapping assigns a spin one to each cell occupied by a piece of particle and zero otherwise. Once this is done, all observables are written in terms of the bimodal variables, as in spin models, and the identities of the particles are ignored.

In a second step, one can coarse grain the spin variables over boxes of a chosen linear size, ℓ, with ideally $L \gg \ell \gg a$. The locality is then given by the position of the box that is labeled by an index i. The coarse graining amounts to a partial averaging, the effective spin on each box, s_i, is now a continuous variable, $0 \leq s_i \leq 1$.

In a system with quenched randomness, realized as fixed obstacles, local fields or else, the coarse-graining procedure over a sufficiently large volume averages over the peculiar local features of the fixed disorder. In this way, the "finger-print" (Castillo *et al.*, 2002, 2003) of disorder, that is to say local fluctuations determined by the particular local disorder—such as Griffiths singularities in random ferromagnets—should be washed away. One can then expect to arrive at a description of the noise-induced fluctuations present in problems with or without quenched disorder.

This method appears to be more adequate for analytic treatment through a field theory.

11.2.2 Space-time correlation

In usual coarsening systems, see Section 11.3, the averaged space-time correlation function

$$NC(r,t) = \sum_{ij/|\vec{r}_i - \vec{r}_j| = r} \langle \delta s_i(t) \delta s_j(t) \rangle, \tag{11.1}$$

with $\delta s_i(t) = s_i(t) - \langle s_i(t)\rangle$ and $\langle s_i(t)\rangle = \langle s_i\rangle_{eq}$ in all the cases we shall deal with, allows for the identification of a growing length from, for example, $L_a(t) \equiv \int d^d r r^{a+1} C(r,t) / \int d^d r r^a C(r,t)$. ($a$ is a parameter chosen to weight preferentially short or long distances; the time dependence of $L_a(t)$ should not depend on a.) Here, and in the following $\langle \ldots \rangle$ stands for an average over different realizations of thermal histories at heat-bath temperature T and/or initial conditions. In the presence of quenched disorder one adds an average over it and denotes it $[\ldots]$. The stochastic time-dependent function $N^{-1}\sum_{ij/|\vec{r}_i-\vec{r}_j|=r} s_i(t)s_j(t)$ after a quench from a random initial condition does not fluctuate in the thermodynamic limit. Therefore, the averages are not really necessary but they are usually written down. In spin-glasses and glasses this observable does not yield information on the existence of any growing length, as we shall discuss below.

11.2.3 Two-time quantities

The autocorrelation function and linear susceptibility are defined as

$$\begin{aligned} NC(t,t_w) &\equiv \sum_{i=1}^{N} \langle C_i(t,t_w)\rangle = \sum_{i=1}^{N} \langle \delta s_i(t)\delta s_i(t_w)\rangle, \\ N\chi(t,t_w) &\equiv \sum_{i=1}^{N} \langle \chi_i(t,t_w)\rangle = \sum_{i=1}^{N} \int_{t_w}^{t} dt'\ R(t,t'), \end{aligned} \tag{11.2}$$

with $R_i(t,t') = \delta\langle s_i(t)\rangle / \delta h_i(t')|_{h=0}$ and $NR = \sum_{i=1}^{N} R_i$ the local and global instantaneous linear responses, respectively.

A number of fully general relations between the linear response and a correlation computed as an average over unperturbed system trajectories have been recently derived. Quite generally they read

$$R(t,t_w) = \frac{1}{2T}\left[\frac{\partial C(t,t_w)}{\partial t_w} + N^{-1}\sum_{i=1}^{N}\langle s_i(t)B_i(t_w)\rangle\right]\theta(t-t_w). \tag{11.3}$$

The explicit form of the factor B_i in the second term in the right-hand side (r.h.s.) depends on the microscopic dynamics; it has been computed for Langevin processes (Cugliandolo *et al.*, 1994) and Markov processes for discrete variables (Lippiello *et al.*, 2005). In all cases B_i is the deterministic drift in the sense that $\langle \dot{s}_i\rangle = \langle B_i\rangle$.

In equilibrium, the second term in the r.h.s. of Eq. (11.3) is equal to the first one, and R (or χ) and C are related by the fluctuation–dissipation theorem

$$R(t,t_w) = \frac{1}{T}\frac{\partial C(t,t_w)}{\partial t_w}\theta(t-t_w), \tag{11.4}$$

$$\chi(t,t_w) = \frac{1}{T}[C(t,t) - C(t,t_w)]\theta(t-t_w), \tag{11.5}$$

and all two-time quantities depend on $t - t_w$ only.

11.2.4 Order parameter

The very concept of a glass order parameter is not obvious. Although some kind of static amorphous order may develop with time in a glassy regime, searches have given negative results so far (not surprisingly since one does not really know what is being looked for. See, however, the recent proposal by (Kurchan and Levine, Kurchan and Levine)). The simplest possibility, $\langle s_i(t) \rangle$, is void of information and the space-time spin–spin correlation $C(r,t)$ (the Fourier transform of the structure factor) is, to a first approximation, time independent and very similar to the one in the supercooled liquid (Chamon *et al.*, 2008).

11.2.4.1 Edwards–Anderson order parameter

A *dynamic* order parameter, named after Edwards and Anderson (EA) who introduced it for spin-glasses, is defined as

$$q_{\rm ea} = \lim_{t\to\infty} \lim_{t_{\rm w}\to\infty} C(t, t_{\rm w}), \tag{11.6}$$

where $C(t, t_{\rm w})$ is the autocorrelation function. The order of the two long time limits is crucial since the weak long-term memory (Bouchaud, 1992; Bouchaud and Dean, 1995; Cugliandolo and Kurchan, 1993) ensures that $\lim_{t_{\rm w}\to\infty} \lim_{t\to\infty} C(t, t_{\rm w}) = 0$ (see Fig. 11.1(a)). The EA order parameter detects ergodicity breaking: in the liquid (paramagnetic) phase $q_{\rm ea} = 0$ because the system loses memory at finite time scales but in the glass phase such memory remains. The idea is quite generic and it can be applied to glassy systems made of constituents of any kind.

11.2.4.2 Replica overlap—a static counterpart

The theory of spin-glasses (Mézard *et al.*, 1987) suggests the definition of a *static* order parameter as the *overlap* between two replicas (two systems with the same quenched randomness) a and b subjected to a fictitious attractive coupling of strength ϵ that forces them to be in the same thermodynamic state:

$$q = \lim_{\epsilon\to 0} \lim_{N\to\infty} \frac{1}{N} \sum_{i=1}^{N} \langle s_i^a s_i^b \rangle_\epsilon . \tag{11.7}$$

The *spin-glass susceptibility* is its linear response to an infinitesimal variation of the coupling:

$$\chi_{\rm sg} = \left. \frac{\partial q}{\partial \epsilon} \right|_{\epsilon=0} = \frac{\beta}{N} \sum_{i,j} \left(\langle s_i^a s_j^a s_i^b s_j^b \rangle - \langle s_i^a s_j^a \rangle \langle s_i^b s_j^b \rangle \right). \tag{11.8}$$

The averages $\langle \ldots \rangle_\epsilon$ and $\langle \ldots \rangle$ are taken here with the corresponding Gibbs–Boltzmann distributions. A crucial problem with these definitions is that below $T_{\rm g}$ aging persists at any finite (with respect to some function of N) time scale. Nonetheless, $\chi_{\rm sg}$ motivates the definition of a dynamic analog that could detect the growth of static order during aging, see Section 11.2.6.

11.2.5 Dynamically correlated volume

A more natural proposal is to consider the spatial variation of the *speed of relaxation* in different regions of space. This point of view is very close in spirit to the idea of dynamical heterogeneity in the supercooled-liquid regime as characterized by local relaxation times. As discussed later, this point of view also emerges naturally in the theoretical analysis of soft modes associated with the time-reparametrization invariance in the effective dynamical equations of motion at long time scales (Chamon and Cugliandolo, 2007). We present below different ways to access a correlation related to the local dynamics.

11.2.5.1 Spatial correlation of local dynamics

Let us regard the local two-time dependent autocorrelation function $C_i(t, t_{\rm w})$ as the *local order parameter.* Its spatial variation can be easily characterized by the spatial correlation function (Castillo *et al.*, 2002, 2003; Parisi, 1999; Jaubert *et al.*, 2007)

$$G_4(i, j; t, t_{\rm w}) = \langle C_i(t, t_{\rm w}) C_j(t, t_{\rm w})\rangle - [C(t, t_{\rm w})]^2. \tag{11.9}$$

The subscript 4 is to remind one that it is a 4-point function since the autocorrelation function is already a two-point correlation function (in time).

The key question is whether there is a characteristic dynamical length scale, $\xi_4(t, t_{\rm w})$, beyond which G_4 decorrelates, e.g.

$$-\ln G_4(i, j; t, t_{\rm w}) \simeq \frac{r}{\xi_4(t, t_{\rm w})} + b \ln r, \tag{11.10}$$

with $r = |\vec{r}_i - \vec{r}_j|$. [This expression assumes the usual decay $G_4 \sim r^{-b} \exp(-r/\xi_4)$ but more general forms of the type $G_4 \simeq r^{-b} f(r/\xi_4)$ have also been considered.]

By integrating Eq. (11.10) over space one obtains

$$N\chi_4(t, t_{\rm w}) = \sum_{i,j} G_4(i, j; t, t_{\rm w}) = \langle [\sum_i C_i(t, t_{\rm w})]^2\rangle - \langle \sum_i C_i(t, t_{\rm w})\rangle^2, \tag{11.11}$$

which measures the dynamically correlated volume. Note that in spite of being usually called a "susceptibility", strictly speaking χ_4 is not one (Semerjian *et al.*, 2004; Lippiello *et al.*, 2008*a*,*b*). The same proviso applies to $\chi_{\rm sg}$ in Eq. (11.13).

The 4-point correlation function in Eq. (11.9) and the integrated one in Eq. (11.11) are extensions of similar expressions defined for the supercooled (stationary) liquid state. In the glassy regime one needs to keep the $t_{\rm w}$ dependence in $\xi_4(t, t_{\rm w})$. Still, it can be measured in numerical simulations and in some experiments from real-time real-space images. One has to keep in mind, though, that this quantity detects how similar motion in different regions is but not necessarily whether these are ordering in the same state.

11.2.5.2 Higher-order susceptibilities

Although G_4 has been studied in numerical simulations (Jaubert *et al.*, 2007; Belletti, 2008, 2009) its direct experimental investigation remains a challenge as, in general, multipoint correlators. (Lucky exceptions are colloidal suspensions in which confocal microscopy allows one to store the full particle configuration (Chamon *et al.*, 2008).) A natural way out would be to measure responses to external perturbations, actual susceptibilities, as suggested in (Huse, 1988, 1991; Bouchaud and Biroli, 2005) and done experimentally in (Berthier *et al.*, 2005). The basic idea is that G_4 should be related to a non-linear susceptibility by some sort of generalization of the FDT, much in the same way as the ordinary correlation function is linked to the linear susceptibility, see Eq. (11.5). One could then measure the latter to extract information on the former. In order to fulfill this program it is necessary to establish which are the non-linear susceptibilities associated to multipoint correlators and which is the generalization of the FDT, holding possibly out of equilibrium. Exact general relations between multipoint correlators and non-linear response functions (Semerjian *et al.*, 2004; Lippiello *et al.*, 2008*a*,*b*; Bouchaud and Biroli, 2005) derived for systems subjected to a Markovian dynamics show that beyond linear order the susceptibilities are related not only to multispin correlations (such as G_4) but also to more complicated correlators. For instance, for the second-order response of two spins $R^{(2)}_{ij}(t,t_1,t_2) = \delta^2\langle s_i(t)s_j(t)\rangle/\delta h_i(t_1)h_j(t_2)|_{h=0}$ in equilibrium one has

$$R^{(2)}_{ij}(t,t_1,t_2) = \frac{1}{2T}\left[\frac{\partial}{\partial t_1}\frac{\partial}{\partial t_2}\langle s_i(t)s_j(t)s_i(t_1)s_j(t_2)\rangle - \frac{\partial}{\partial t_2}\langle s_i(t)s_j(t)B_i(t_1)s_j(t_2)\rangle\right], \tag{11.12}$$

for $t_1 \neq t_2$. This feature poses the problem of choosing the best suited non-linear susceptibility to detect cooperative effects. In a series of papers both third-(Bouchaud and Biroli, 2005) and second-(Lippiello *et al.*, 2008*a*,*b*) order susceptibilities have been considered. Analytical and numerical studies show that the non-linear susceptibilities and G_4 obey analogous scaling forms from which one can extract a cooperative length. Experimental studies are on the way (Ladieu *et al.*, 2007; Thibierge *et al.*, 2008).

The second-order susceptibility $\chi^{(2)}(t,t_w) = \int_{t_w}^t dt_1 \int_{t_w}^t dt_2\, R^{(2)}_{ij}(t,t_1,t_2)$ is strictly related to the fluctuations of $\chi_i(t,t_w)$ (Lippiello *et al.*, 2008*a*,*b*; Corberi *et al.*, 2010).

11.2.5.3 Distributions of coarse-grained two-time quantities

Another way of extracting a growing length, ξ, alternative to that expressed by Eq. (11.10), is to study the full distribution of local two-time functions (Castillo *et al.*, 2002, 2003; Chamon *et al.*, 2002; Castillo, 2008; Chamon *et al.*, 2004, 2006). Indeed, for $\ell \gg \xi$ the coarse-graining boxes naturally become independent and one should recover a Gaussian distribution. The crossover from non-trivial to trivial dependence can then be used to estimate ξ that should, presumably, behave as ξ_4. The same argument can be applied to the pdf of the local linear susceptibility χ_is.

11.2.6 Growth of underlying static glass order?

An intriguing problem is whether any static glass order develops during aging. The spin-glass susceptibility, Eq. (11.8), suggests to define a 4-point correlation function (Belletti, 2008, 2009; Huse, 1988, 1991; Rieger, 1993; Marinari *et al.*, 2000; Komori *et al.*, 1999*a*,*b*, 2000; Yoshino *et al.*, 2002),

$$N\chi_{\rm sg}(t) = \sum_{i,j} G_{\rm sg}(i,j;t) = \sum_{i,j} \langle s_i^a(t)s_j^a(t)s_i^b(t)s_j^b(t)\rangle \,. \tag{11.13}$$

Time t is measured after the temperature quench at which the two replicas, labeled a and b, are prepared in independent random initial configurations. One is interested in finding a dynamic length scale, $\xi_{\rm sg}(t)$, beyond that $G_{\rm sg}$ decorrelates,

$$-\ln G_{\rm sg}(i,j;t) \sim \frac{r}{\xi_{\rm sg}(t)} + c\ln r. \tag{11.14}$$

$\xi_{\rm sg}(t)$ is simpler than $\xi_4(t,t_{\rm w})$ in that it is a one-time quantity like the domain size $L(t)$ in usual phase ordering processes.

In the context of spin-glasses the value of the parameter c is used to distinguish between a disguised ferromagnet picture, as proposed in the droplet model (Fisher and Huse, 1988*a*), and a more complex equilibrium structure, such as the one predicted by mean-field models (Mézard *et al.*, 1987). In the former $\lim_{r\to\infty} G_{\rm sg}(r) \to$ const and $c = 0$ while in the latter $\lim_{r\to\infty} G_{\rm sg}(r) \to 0$ and $c > 0$.

It is interesting to compare $\chi_{\rm sg}(t)$ and $\chi_4(t,t_{\rm w})$. Since there are no interactions between the two replicas a and b, purely dynamic correlation cannot exist between them. Thus, the dynamical SG susceptibility $\chi_{\rm sg}(t)$ can only detect growth of (if any) static order, much as domains in usual phase ordering. On the other hand, $\chi_4(t,t_{\rm w})$ can detect both static and dynamic order.

11.2.7 Effective temperature

Although we shall not develop the $T_{\rm eff}$ ideas here we include a short paragraph recalling its definition; the concept will appear in a number of places later in the chapter.

The deviation from the fluctuation–dissipation theorem (FDT) found in mean-field glassy models (Cugliandolo and Kurchan, 1993, 1994) and in a number of finite-dimensional systems (Crisanti and Ritort, 2003) can be rationalized in terms of the generation of an effective temperature (Cugliandolo *et al.*, 1997*b*) in the system. The identification of a temperature out of equilibrium makes sense in the asymptotic limit of small entropy production (either long times or very weak applied drive, see Section 11.5.4) in which the system evolves slowly. $T_{\rm eff}$ is defined as minus the inverse slope of the asymptotic parametric plot $\chi(C)$, where χ is the linear integrated response and C is the two-time correlation. The thermodynamic character of $T_{\rm eff}$ has been checked in mean-field models and mode-coupling theories as well as in simulations of Lennard-Jones mixtures, models of silica and many others. A basic condition is that $T_{\rm eff}$ obtained from different (interacting) observables should be equal whenever measured in the same dynamic regime. A different scenario is found at the lower critical dimension and in critical dynamics, and in trap models with unbounded trap depths.

11.3 Phase ordering

Phase-ordering kinetics is an important problem for material science but also for our generic understanding of pattern formation in non-equilibrium systems.

Let us consider a physical macroscopic system in contact with an external reservoir in equilibrium. Imagine now that one changes a parameter instantaneously in such a way that the system is taken from a disordered phase to an ordered one in its (equilibrium) phase diagram. Two paradigmatic examples are spinodal decomposition, *i.e.* the process whereby a mixture of two or more substances separate into distinct regions with different concentrations, and magnetic domain growth in ferromagnetic materials quenched below the Curie temperature (Bray, 1994).

Closely related to the above is the process whereby a critical state is approached *via* a quench from the disorder state right to the temperature at which the phase transition occurs.

In both subcritical and critical coarsening the dynamical process starts from an equilibrium high-temperature disordered state and progressively evolves building a new phase, stable at a lower temperature, either ordered or critical.

11.3.1 Growing length, aging and scaling

The evolution of an initial condition that is not correlated with the final equilibrium state (and with no bias fields) does not reach equilibrium in finite times. More explicitly, domains of all the phases of the equilibrium state at the final temperature T keep on growing, until their typical size $L(t)$ becomes of the order of the system size L. For any shorter time the system is out of equilibrium and, in particular, the non-equilibrium evolution does not come to an end whenever the thermodynamic limit $\mathrm{L} \to \infty$ is taken at the outset.

The very existence of a growing length $L(t)$ is at the heart of the aging behavior observed in these systems, as can be easily understood. Since $L(t_{\mathrm{w}})$ increases with t_{w} the system needs larger rearrangements to decorrelate from older configurations. This simple fact is the origin of the strong t_{w} dependence of the autocorrelation and linear response described in Section 11.1 and depicted in Fig. 11.1.

In the asymptotic time domain, when $L(t)$ has grown much larger than any microscopic length in the system, a *dynamic scaling* symmetry sets in, similarly to the usual scaling symmetry observed in equilibrium critical phenomena. According to this hypothesis, the growth of $L(t)$ is the only relevant process and the whole time dependence enters only through $L(t)$. Observables such as correlation and response functions take precise scaling forms that will be discussed in Sections 11.3.2 and 11.3.3. Exceptional cases where dynamic scaling is not observed will be briefly discussed in Section 11.3.3.4.

11.3.2 Aging at a critical point

The scaling behavior of binary systems quenched to the critical point is quite well understood since this issue can be addressed via scaling arguments (Godrèche and

Luck, 2000, 2002) and renormalization group approaches (Janssen *et al.*, 1989; Calabrese and Gambassi, 2002*a*,*b*, 2005) which give explicit expressions for many of the quantities of interest up to two loops order. Numerical simulations (Chatelain, 2004; Pleimling and Gambassi, 2005; Mayer *et al.*, 2005; Lippiello *et al.*, 2006; Henkel *et al.*, 2006; Corberi *et al.*, 2008) confirm the analytic results and probe exponents and scaling functions beyond the available perturbative orders. In this case the system builds correlated critical clusters with fractal dimension $D = (d + 2 - \eta)/2$, where η is the usual static critical exponent, in regions growing algebraically as $L(t) \sim t^{1/z_{\rm eq}}$, $z_{\rm eq}$ being the dynamic equilibrium critical exponent relating times and lengths.

In the asymptotic time domain the correlation function (11.1) has the scaling form

$$C(r, t) = L(t)^{-2(d-D)} f\left(\frac{r}{L(t)}\right). \tag{11.15}$$

The pre-factor $L(t)^{-2(d-D)}$ takes into account that the growing domains have a fractal nature (hence their *density* decreases as their size grows) and the dependence on $r/L(t)$ in $f(x)$ expresses the similarity of configurations at different times once lengths are measured in units of $L(t)$.

For two-time quantities, when $t_{\rm w}$ is sufficiently large one has

$$C(t, t_{\rm w}) = C_{\rm st}(\tau)\ f_{\rm c}\left(\frac{L(t)}{L(t_{\rm w})}\right), \qquad R(t, t_{\rm w}) = R_{\rm st}(\tau)\ f_{\rm r}\left(\frac{L(t)}{L(t_{\rm w})}\right). \tag{11.16}$$

Here, $C_{\rm st}(\tau) \simeq L(\tau)^{-2(d-D)}$ and $R_{\rm st}(\tau) \simeq \beta L(\tau)^{-2(d-D)-z_{\rm eq}}$, where $\tau = t - t_{\rm w}$. The scaling functions $f_{\rm c}$ and $f_{\rm r}$ describe the non-equilibrium behavior and take the limiting values $f_{\rm c}(0) = f_{\rm r}(0) = 1$ and $f_{\rm c}(\infty) = f_{\rm r}(\infty) = 0$. The correlation and response function of the equilibrium state at $T_{\rm c}$ obey FDT, $R_{\rm st}(\tau) = \beta \dot{C}_{\rm st}(\tau)$. In the scaling forms the equilibrium and non-equilibrium contributions enter in a *multiplicative* structure. Non-equilibrium effects are taken into account by taking ratios between the sizes of the correlated domains at the observation times $t_{\rm w}$ and t in the scaling functions. Of a certain interest is the limiting fluctuation–dissipation ratio $T/T_{\rm eff} = X_\infty = \lim_{t\to\infty} \lim_{t_{\rm w}\to\infty} = TR(t, t_{\rm w})/[\partial C(t, t_{\rm w})/\partial t_{\rm w}]$, due to its universal character (Godrèche and Luck, 2000, 2002).

Experiments on the non-equilibrium kinetics near a critical point are reported in (Joubaud *et al.*, 2009), where the orientation fluctuations of the director of a liquid crystal are measured after a sudden change of the control parameter (in this case an AC voltage) from a value in the ordered phase to one near the critical point where the Fréedericksz second-order transition occurs. In this *quenching* procedure the initial state is ordered. Experimental data show a behavior of two-time quantities in substantial agreement with the scaling pattern described in Eqs. (11.16), as expected since the system can be described in terms of a time-dependent Ginzburg–Landau equation, similarly to ordinary magnetic systems. Interestingly enough, even the limiting fluctuation–dissipation ratio X_∞ turns out to be in good agreement with the value found in the two-dimensional Ising model.

11.3.3 Aging in the low-temperature phase

The late stage of phase ordering in binary systems is characterized by a patchwork of large domains the interior of which is basically thermalized in one of the two equilibrium phases, while their boundaries are slowly moving producing the power law $L(t) \sim t^{1/z}$. This picture suggests the splitting of the degrees of freedom (spins) into two categories, providing statistically independent contributions to observables such as correlation or response functions. More precisely, a quasi-equilibrium stationary contribution arises as due to bulk spins, while boundaries account for the non-equilibrium part. Then (Bouchaud *et al.*, 1997; Franz and Virasoro, 2000), asymptotically one has

$$C(r,t) \simeq C_{\mathrm{st}}(r) + C_{\mathrm{ag}}(r,t). \tag{11.17}$$

The first term describes the equilibrium fluctuations in the low-temperature broken-symmetry pure states

$$C_{\mathrm{st}}(r) = \left(1 - \langle s_i \rangle_{\mathrm{eq}}^2\right) \, g\left(\frac{r}{\xi_{\mathrm{eq}}}\right), \tag{11.18}$$

where $\langle s_i \rangle_{\mathrm{eq}}$ is the equilibrium expectation value of the local spin, and $g(x)$ is a function with the limiting values $g(0) = 1$, $\lim_{x\to\infty} g(x) = 0$. The second term takes into account the motion of the domain walls through

$$C_{\mathrm{ag}}(r,t) = \langle s_i \rangle_{\mathrm{eq}}^2 \, f\left(\frac{r}{L(t)}\right), \tag{11.19}$$

with $f(1) = 1$ and $\lim_{x\to\infty} f(x) = 0$. Both C_{st} and C_{ag} obey (separately) scaling forms with respect to the equilibrium and the non-equilibrium lengths ξ, $L(t)$. In particular, Eq. (11.19) expresses the fact that system configurations at different times are statistically similar provided that lengths are measured in units of $L(t)$, namely the very essence of dynamical scaling.

An analogous *additive* separation holds for two time quantities

$$C(t,t_{\mathrm{w}}) \simeq C_{\mathrm{st}}(\tau) + C_{\mathrm{ag}}(t,t_{\mathrm{w}}), \tag{11.20}$$

and similarly for the response function. The equilibrium character of the first term implies

$$C_{\mathrm{st}}(\tau) = \left(1 - \langle s_i \rangle_{\mathrm{eq}}^2\right) \, g_{\mathrm{c}}\left(\frac{L(\tau)}{\xi_{\mathrm{eq}}}\right), \tag{11.21}$$

where $g_{\mathrm{c}}(x)$ is a function with the limiting values $g_{\mathrm{c}}(0) = 1$, $\lim_{x\to\infty} g_{\mathrm{c}}(x) = 0$. The non-equilibrium term obeys

$$C_{\mathrm{ag}}(t,t_{\mathrm{w}}) = \langle s_i \rangle_{\mathrm{eq}}^2 \, f_{\mathrm{c}}\left(\frac{L(t)}{L(t_{\mathrm{w}})}\right), \tag{11.22}$$

with $f_{\mathrm{c}}(1) = 1$ and $\lim_{x\to\infty} f_{\mathrm{c}}(x) = 0$. In the long t_{w} limit the two terms in Eq. (11.20) vary in completely different two-time scales. The first one changes when the second one is fixed to $q_{\text{ea}} \equiv \langle s_i \rangle_{\text{eq}}^2$, see Eq. (11.6), at times such that $L(t)/L(t_{\text{w}}) \simeq 1$. In this regime C decays to a plateau at $q_{\text{ea}} = \langle s_i \rangle^2$, see Fig. 11.1. The second one varies when the first one has already decayed to zero. The mere existence of the second term is the essence of the aging phenomenon below T_{c} with older systems (longer t_{w}) having a slower relaxation than younger ones (shorter t_{w}). Such a sharp separation of time scales is the hallmark of quenches in the ordered phase, at variance with critical quenches where equilibrium and non-equilibrium contributions are entangled in the multiplicative form (11.16). Although we have used the terminology of binary systems, the scaling structure discussed sofar is believed to hold quite generally (Bray, 1994) (some exceptions will be discussed below), including systems with more than two low-temperature equilibrium phases (as described, e.g. by Potts or clock models) or with a continuous symmetry (i.e. vector $O(N)$ models). Contrarily to the case of critical quenches, however, a systematic expansion method in subcritical quenches is much more difficult (Mazenko, 2004) and general results comparable to those at T_{c} are not yet available. The scaling structure discussed above has been proven analytically only in special cases, such as the one-dimensional Ising chain with Glauber dynamics (Godrèche and Luck, 2000, 2002; Lippiello and Zannetti, 2000) or the Langevin dynamics of the d-dimensional $O(N)$ model with non-conserved order parameter in the large N limit (Corberi *et al.*, 2002*c*). It is supported by semianalytical arguments in two dimensions (Arenzon *et al.*, 2007; Sicilia *et al.*, 2007, 2008*a*; Henkel and Pleimling, 2008*a*), numerical simulations (Corberi *et al.*, 2003*b*, 2006; Aron *et al.*, 2008) and approximate theories (Berthier *et al.*, 1999; Corberi *et al.*, 2001*a*,*b*). It has also been confirmed in experiments (Chou and Goldburg, 1981; Katano and Iizumi, 1984; Komura *et al.*, 1985; Gaulin *et al.*, 1987; Sicilia *et al.*, 2008*b*).

11.3.3.1 The growing length

The growing length depends on a few characteristics of the ordering process—the dimension of the order parameter, whether there are conservation laws, the presence of quenched disorder—and may serve to classify systems in classes akin to universality ones (Bray, 1994). Although the following results are hard to obtain with rigorous arguments, they are by now well established. Basically, one distinguishes two important cases: in clean models $L(t) \simeq t^{1/z}$; in problems with quenched randomness, where topological defects are pinned, the growing length is expected to slow down from a power law to a logarithmic dependence on time (due to thermal activation above barriers with a power-law distribution) see Section 11.4. The dynamic exponent z depends on the type of microscopic dynamics and dimension of the order parameter. For example, in curvature-driven growth with scalar order parameter $z = 1/2$ and in phase separation with scalar order parameter (and no hydrodynamics) $z = 1/3$. A crossover at a static length explains how the activated scaling can be confused with a temperature-dependent power law in dirty cases (Iguain *et al.*, 2009) as explained in Section 11.4.

11.3.3.2 The scaling functions

The full theoretical description of a coarsening process necessitates the determination of the scaling functions. This is a hard task and there is no powerful and systematic method to attack this problem yet.

Still, Fisher and Huse proposed that the scaling functions in the aging regime, e.g. f_c, should be robust (Fisher and Huse, 1988*a*). Changes in the model definition that do not modify the nature of the equilibrium initial nor target ordered state—such as weak quenched disorder not leading to frustration—should not alter the scaling functions. This is the so-called *superuniversality hypothesis*. In this way, the scaling functions in spin models with random ferromagnetic bonds or random fields should be identical to those found in the pure limit. Numerical tests in $d > 1$ systems point in the direction of validating this hypothesis (Sicilia *et al.*, 2008*a*; Henkel and Pleimling, 2008*a*,*b*; Aron *et al.*, 2008) while very recent studies of the scaling properties of the linear response in the $d = 1$ random bond ferromagnet tend to falsify it (Lippiello, Mukherjee, Puri and Zannetti, 2010).

11.3.3.3 The four-point correlation G_4

The behavior of G_4, and the way $L(t)$ is encoded in it, can be easily understood in coarsening systems. In Fig. 11.3 a configuration of the system with two interfaces (denoted 1 and 2, continuous lines) at time t_w is sketched. The dashed interface denoted as 3 is the location of interface 2 at the later time t (such that $t - t_w < t_w$). For $t - t_w \ll t_w$ one has $C_i(t, t_w) = 1$—for concreteness we think in terms of Ising spins—everywhere except in the regions spanned by an interface in the interval (t_w, t)—the region between 2 and 3 in Fig. 11.3—where $C_i(t, t_w) = -1$. Hence, G_4 decays on a volume of order $L^d(t) - L^d(t_w)$ that grows in time, and ξ_4 increases. For longer time differences ($t - t_w \gtrsim t_w$), however, another interface present at t_w (1 in the sketch) may supercede at time t the position of interface 2 at the previous time t_w. From this time on, χ_4 stops growing and saturates to a value of order $L^d(t_w)$, namely the typical volume contained between two interfaces at t_w (i.e. between 1 and 2). The saturation can also be explained by observing that for long time differences, $t - t_w \gg t_w$, χ_4 factorizes as $(1/N) \sum_{i,j} \langle s_i(t) s_j(t) \rangle \langle s_i(t_w) s_j(t_w) \rangle$. Enforcing scaling, $\langle s_i(t) s_j(t) \rangle \simeq f[r/L(t)]$, one recovers $\lim_{t-t_w \to \infty} \chi_4(t, t_w) \propto L^d(t_w)$. In this way, $L(t_w)$

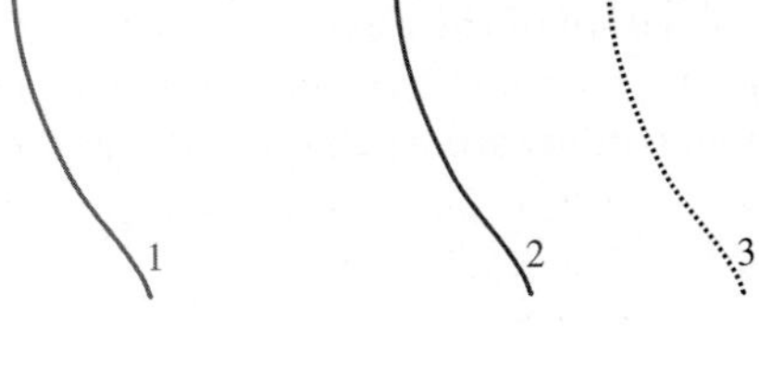

Fig. 11.3 Schematic representation of the mechanism of growth and saturation of $\chi_4(t, t_w)$ in coarsening systems.

can be extracted from the long $t - t_w$ behavior of χ_4 but, of course, this is not really necessary in such coarsening problems. Importantly enough, $L(t)$ is a measure of growth of static order in these cases.

11.3.3.4 Exceptional cases: breakdown of dynamic scaling

There are also a number of coarsening systems in which dynamic scaling breaks down (Bray, 1994). Among these are the XY model [$O(2)$ symmetry] in one (Rutenberg and Bray, 1995) and two (Bray and Jervis, 2000) dimensions, the clock model with $p > 4$ states (Andrenacci *et al.*, 2006), the one-dimensional Heisenberg model [$O(3)$ symmetry] (Burioni *et al.*, 2009) and the large-N model with conserved dynamics (Coniglio and Zannetti, 1989; Coniglio *et al.*, 1994). Interestingly, in most cases this is due to the presence of more than one length growing macroscopically during phase ordering. Let us consider the XY model in $d = 1$ as a simple paradigm. The order parameter in this case is a planar vector. For long times after a quench from a high to zero temperature, when the excess energy is greatly reduced, the continuous nature of the order parameter allows only soft (Goldstone) modes, namely smooth rotations of the order parameter, called textures. A winding length L_w can be defined as the typical distance over which a 2π rotation of the spin is detected. In addition, due to the symmetry of the disordered initial state and of the Hamiltonian between clockwise (textures) and counterclockwise (anti-textures) rotations, both topological defects are formed in the evolution, separated by winding inversions. Denoting with L the typical distance between winding inversions, it is clear that the reduction of energy implies the growth of both L_w and L as time elapses. It was shown (Rutenberg and Bray, 1995) that these lengths grow with different exponents and this phenomenon prevents dynamic scaling setting in.

11.3.4 Elastic manifolds

The dynamics of a directed d-dimensional elastic manifold embedded in an N-dimensional transverse space is simply modelized by a Gaussian scalar field theory that goes under the name of Edwards–Wilkinson (EW) model. Non-linear effects can be accounted for by including a Kardar–Parisi–Zhang (KPZ) term. It has been known for some time that these systems age and that the aging dynamics is governed by a growing correlation length (Bray, 1994; Cugliandolo *et al.*, 1994; Bustingorry *et al.*, 2007; Bustingorry, 2007).

The aging regime, before saturation at a time difference that depends on the length of the line, L, is characterized by a multiplicative scaling of two-time quantities. For instance, a two-time generalization of the roughness, $\langle w^2\rangle(t, t_w) \equiv L^d \int d^dx \langle[\delta h(\vec{x}, t) - \delta h(\vec{x}, t_w)]^2\rangle$, with $\delta h(\vec{x}, t) = h(\vec{x}, t) - \langle h(\vec{x}, t)\rangle$, h the height of the manifold and $\vec{x}$ the position on the d-dimensional substrate, scales as

$$\langle w^2\rangle(t, t_w) \simeq L^{2\zeta}(t_w)\; f_{w^2}\left(\frac{L(t)}{L(t_w)}\right). \tag{11.23}$$

Assuming $\lim_{x\to\infty} f_{w^2}(x) = x^{2\zeta}$, in the limit $\tau \gg t_w$ this observable reaches a stationary regime in which $\langle w^2\rangle(t, t_w) \simeq L^{2\zeta}(\tau)$. For even longer time delays such that

$L(t) \to L$ one finds saturation at $L^{2\zeta}$. The roughness exponent ζ is due to thermal fluctuations and it is simply equal to $(2-d)/2$ in the EW manifold. The crossover to saturation is then described by an extension of the Family–Vicsek scaling that takes into account the out-of-equilibrium relaxation and aging effects. Correlation and response functions are related by a modified fluctuation-dissipation relation after removing the diffusive factor and a well-defined effective temperature (Cugliandolo *et al.*, 1997*b*) exists and depends on the initial (T_0) and final (T) values of the temperature before and after the quench with $T_{\rm eff} > T$ if $T_0 > T$ and $T_{\rm eff} < T$ if $T_0 < T$ (Bustingorry *et al.*, 2007; Bustingorry, 2007).

11.4 Role of activation: the droplet theory

The dynamics of glassy systems at low enough temperatures should be dominated by thermal activation. Although it is very difficult to study activated processes from first principles, several phenomenological proposals for models with quenched disorder exist. In particular, a droplet picture has been put forward for spin-glasses (Fisher and Huse, 1988*a*; Bray and Moore, 1987; Fisher and Huse, 1988*b*) and related systems including elastic manifolds in random media (Fisher and Huse, 1991) and vortex glasses (Fisher *et al.*, 1991). This model assumes that a static low-temperature phase, associated with a zero-temperature glassy fixed point in a renormalization-group sense, exists in these systems. Droplet-like low-energy excitations of various sizes L on top of the ground state render the dynamics strongly heterogeneous both in space and time. The typical free-energy gap of a droplet with respect to the ground state and the free-energy barrier to nucleate a droplet are assumed to scale as L^θ and L^ψ, respectively, with θ and ψ two non-trivial exponents. Static order is assumed to grow as in standard coarsening systems. Dynamical observables such as the two-time autocorrelation function should then follow universal scaling laws in terms of a growing length $L(t)$ originated in Arrhenius activation over barriers growing as a power of the length

$$t \sim \tau_0 \, e^{\frac{L^\psi}{T}} \qquad \Rightarrow \qquad L(t) \simeq [T \ln(t/\tau_0)]^{1/\psi}. \tag{11.24}$$

The strong-disorder renormalization approach in configuration space (Monthus and Garel, 2008*a*,*b*) yields an explicit construction in favor of the droplet logarithmic scaling.

In the case of spin-glasses the very existence of static glassy states is accepted but their detailed nature is a much debated issue. Moreover, it is far from obvious whether coarsening of only two competing states occurs and whether the growth is determined by thermal activation over such barriers. In the present section we discuss a recipe to examine the droplet picture quantitatively.

11.4.1 Efficient strategy for data analysis

In practice, the asymptotic dynamic scaling features associated with the putative zero-temperature glassy fixed point are difficult to access in numerical simulations and experiments. The following strategy helps avoiding the control of pre-asymptotic effects that last for very long (Yoshino, 2001):

I) Measure the dynamical length $L(t)$ and analyze it over the full time duration by taking care of the crossover from the initial non-activated dynamics, that could be diffusive, critical or else, to the asymptotic activated regime.
II) Reparametrize the time-dependent quantity of interest, say $A(t)$, using $L(t)$ obtained in I) as a time-length dictionary.

In this way, pre-asymptotic corrections are dealt with separately: those due to L and those due to the scaling functions. Once the "dictionary" $L(t)$ is determined, step II) is very much straightforward: no uncontrolled fits are needed since the essential exponents (such as the energy exponent θ) are provided by independent studies of static properties. We prove the efficiency of this strategy by analyzing numerical results for Sinai diffusion (Sinai, 1982) and we recall the study of the random manifold and Edwards–Anderson spin-glass along these lines.

11.4.1.1 Sinai model: a test case

The Sinai model is a random walker hopping on a one-dimensional lattice $i = 1, 2, \dots N$ over which a quenched random potential U_i is defined. The statistics of the random potential are such that $[(U_i - U_j)^2] = r_{ij}$, where r_{ij} is the distance between sites i and j and $[\cdots]$ stands for the average over different realizations of the random potential. The statistical analysis of disorder yields $\theta = \psi = 1/2$.

The walker starts from a randomly chosen initial point at time $t = 0$. The mean-squared displacement is $\mathrm{B}(t, t_w) = [\langle (x(t) - x(t_w))^2 \rangle]$ with $x(t)$ the position at time t and $\langle \cdots \rangle$ an average over thermal histories and initial conditions. The rigorous analysis of $\sqrt{\mathrm{B}(t,0)} = L(t)$ gives the diffusion law $L(t) \propto [T \ln(t/\tau_0)]^2$ at temperature T (Sinai, 1982). Also of interest is the linear susceptibility, $\chi(t, t_{\mathrm{w}})$, of the averaged position with respect to a small bias field acting on the walker after the waiting time t_{w}. In equilibrium, the FDT relates the two quantities as $\chi(t, t_{\mathrm{w}}) = \mathrm{B}(t, t_{\mathrm{w}})/2T$.

The asymptotic scaling properties have been almost fully uncovered by a real-space renormalization group (RSRG) approach (Fisher *et al.*, 1998; Le Doussal *et al.*, 1998) that is believed to become exact asymptotically. For illustrative purposes, we reproduce some of these results with droplet-scaling arguments.

In the quasi-equilibrium regime $L(\tau = t - t_{\mathrm{w}}) < L(t_{\mathrm{w}})$ the RSRG predicts $\mathrm{B}(\tau + t_{\mathrm{w}}, t_{\mathrm{w}}) = 2T\chi(\tau + t_{\mathrm{w}}, t_{\mathrm{w}}) \sim TL^{3/2}(\tau) g_{\mathrm{b}}(L(\tau)/L(t_{\mathrm{w}}))$ where g_{b} is a scaling function. The exponent $3/2$ can be explained as follow. In first approximation, at time t_{w} the particle sits in the lowest energy minimum within a length scale $L(t_{\mathrm{w}})$ around the initial point. Since $L(\tau) \ll L(t_{\mathrm{w}})$ the particle does not have time to explore regions that are far from this minimum and diffusion typically vanishes in the quasi-equilibrium regime. However, with a small probability $\delta U_{\mathrm{H}}/L^{\theta}(\tau) \sim T/L^{\theta}(\tau)$ with $\theta = 1/2$ there is a secondary energy minimum within the length scale $L(\tau)$ with the energy barrier between them being lower than the thermal energy T. In such a rare event the particle can hop to the secondary minimum yielding the disorder-averaged behavior $\mathrm{B}(\tau) \simeq T/L^{1/2}(\tau) L^2(\tau) \simeq TL^{3/2}(\tau)$.

In the aging regime $L(t) > L(t_{\rm w})$ the particle diffuses over longer and longer lengths looking for lower and lower energy minima. Within a given time scale t the particle moves to the right or to the left by an amount $L(t)$. However, the direction of motion is almost deterministically given by the direction with the lowest energy barrier since the difference between energy barriers (not minima) is of order $L^{1/2}(t)$, a diverging quantity in the long-times limit. One then has $\mathrm{B}(t, t_{\rm w}) \sim L^2(t) f_{\rm b}(L(t)/L(t_{\rm w}))$. This also implies that the linear susceptibility typically vanishes in the aging regime since a change of the potential of order $\delta U_{\rm H} = hL(t)$ (an analog of the Zeeman energy in magnets) induced by an *infinitesimal* field h does not affect the difference in energy barriers of order $L^{1/2}(t)$. However, there are rare events such that the energy barriers to the left and right are almost degenerate with only a small difference in height $\delta U_{\rm H}$. The probability to find such degenerate barriers scales as $\delta U_{\rm H}/L^{\psi}$ with $\psi = 1/2$. The infinitesimal h then induces a change in the direction of diffusion resulting in a disorder-averaged displacement of order $(\delta U_{\rm H}/L^{\psi})L$ and thus $\chi(t, t_{\rm w}) \sim L^{3/2}(t) f_{\chi}(L(t)/L(t_{\rm w}))$.

We performed Monte Carlo (MC) simulations to examine the anticipated asymptotic scaling (Yoshino, 2001; Corberi *et al.*, 2002*a*). In Fig. 11.4 (a) we show the mean-squared displacement $B(t, 0)$ of the Sinai walker at different temperatures, decreasing from top to bottom. It defines unambiguously a dynamical length $L(t) = \sqrt{B(t, 0)}$ that we shall use below to reparametrize time-dependent quantities. The data demonstrate that extremely long times are needed to reach the $L(t) \simeq \ln^2 t/\tau_0$ asymptotic result known to be exact analytically. At finite T the short-time diffusion is normal $B(t, 0) = Dt$ with a diffusion constant D, see the inset in Fig. 11.5. The crossover to the activated regime takes place at a "thermal length scale" $L_0(T)$ such that

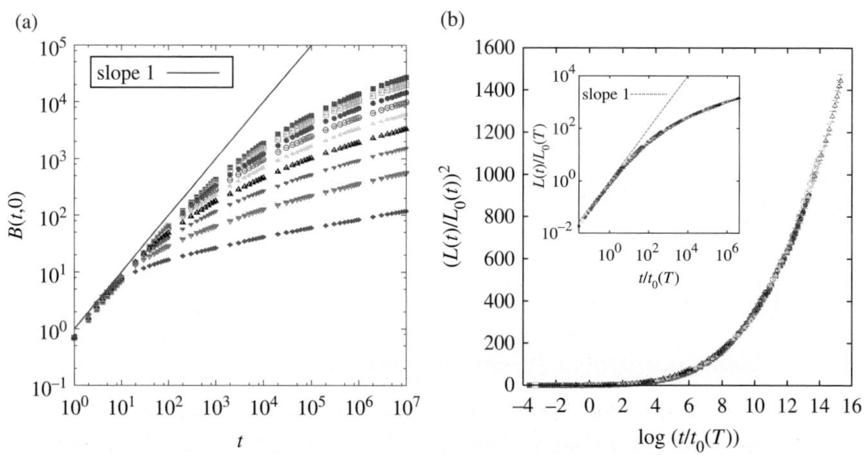

Fig. 11.4 Diffusion in the Sinai model: the mean-squared displacement from MC simulations. (a) Raw data at $T = 1.8, 1.6, \ldots, 0.4, 0.2$ from top to bottom. The unit of time t is one MCs. The average is taken over 10^5 realizations of the random potential. (b) Scaling plot of the dynamic length $L^2(t) = B(t, 0)$ showing the crossover behavior from normal diffusion at short time t—see the inset—to the expected $L^2 \simeq \ln^4(t/\tau_0)$ asymptotic law.

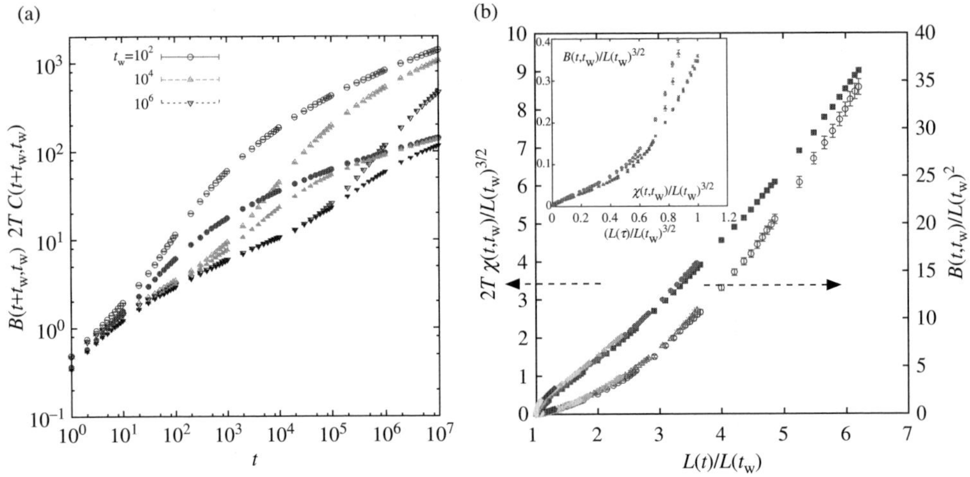

Fig. 11.5 Diffusion in the Sinai model: two-time mean-squared displacement and linear susceptibility from MC simulations at $T = 0.6$. Note the absence of a developing plateau. (a) Raw data of the mean-squared displacement $B(t, t_w)$ (open symbols) and the linear susceptibility $\chi(t, t_w)$ (filled symbols). (b) Scaling plots in terms of the dynamic length. Inset: zoom over the quasi-equilibrium regime.

$\delta U(L_0(T)) = L_0^{1/\theta}(T) \sim T$ and the corresponding time scale $\tau_0(T)$ is fixed using $L_0^2(T) = D\tau_0(T)$. As shown in Fig. 11.4 (b) the slow crossover is well described by proposing that the scaled length $L(t)/L_0(T)$ is a universal function of the scaled time $t/\tau_0(T)$. The asymptotic Sinai's diffusion law can be parametrized precisely with $L(t) \sim L_0(T)[\ln(t/\tau_0(T))]^2$.

Next we proceed to step II. In Fig. 11.5 (a) we show the aging effects observed in the mean-squared displacement and the corresponding linear-susceptibility. In Fig. 11.5 (b), the numerical data are reparamerized by the dynamical length scale obtained in the step I and follow the expected scaling laws. This means that the correction to the asymptotic behavior is not large in terms of length scales. The absence of a plateau in B and χ in their raw and scaled forms demonstrates the fact that there is no additive separation of time scales—as in Eq. (11.20)—in this problem.

11.4.1.2 Elastic manifolds in random media

The dynamics of a directed d-dimensional elastic manifold embedded in an N-dimensional transverse space under the effect of a quenched random potential play an important role in a variety of physical systems ranging from coarsening in dirty systems to fracture. An application is the one in which the directed lines are vortices in superconductors aligned in the direction of the magnetic field and simultaneously pinned by impurities.

These systems are intimately related to the problem of Sinai's diffusion. The simplest case is a (1+1)-dimensional directed polymer in random media (DPRM).

Each configuration is described by the transverse displacement $x(z,t)$ at position z along the directed polymer at time t. Two natural quantities used to characterize aging are the mean-squared displacement $B(z,t,t_{\rm w}) = [\langle(x(z,t) - x(0,t_{\rm w}))^2\rangle]$ and the linear susceptibility $\chi(z,t,t_{\rm w})$. The FDT $\chi(z,t,t_{\rm w}) = {\rm B}(z,t,t_{\rm w})/2T$ is satisfied in equilibrium.

Scaling arguments of droplet type naturally apply to the present case (Yoshino, 2001). To this end one just needs to keep in mind the scaling relation between the longitudinal length L and transverse length $L_\perp \propto L^\zeta$ with ζ the roughness exponent. Depending on the explored length scale, the latter takes a thermal, $\zeta_{\rm T} = 1/2$ or a disorder, $\zeta_{\rm D} = 2/3$, value with the crossover given at a static temperature-dependent length $L_0(T)$. The disorder-dominated roughness exponent is related to the energy exponent θ by an exact scaling relation $\theta = 2\zeta_{\rm D} - 1$ implying $\theta = 1/3$. It is also conjectured that $\psi = \theta$. The time evolution of a local segment $x(z,t)$ may be viewed as diffusion of a particle that feels an effective (renormalized) potential created by the rest of the system, the variation of which scales with the transverse length as $\Delta U \simeq L^\theta \simeq L_\perp^{\theta/\zeta_{\rm D}} \simeq L_\perp^{2-1/\zeta_{\rm D}}$ in the disorder-dominated regime (Le Doussal and Vinokur, 1995).

Using the analogy with the Sinai model one derives the scaling behavior of different observables. At a time t after the quench, the system is equilibrated up to longitudinal length $L(t)$ over which the energy is higher than the equilibrium one by an amount of order L^θ. Thus, the energy density per unit longitudinal length, $e(t) \equiv U(t)/L$, is expected to decay as $e(t) = e(\infty) + ct/L(t)^{1-\theta}$. In the quasi-equilibrium regime the consideration of the degeneracy of minima suggests ${\rm B}(z,\tau+t_{\rm w},t_{\rm w}) = 2T\chi(z,\tau+t_{\rm w},t_{\rm w}) = TL(\tau)g_{\rm b}(z/L(t_{\rm w}),L(\tau)/L(t_{\rm w}))$ [the pre-factor $TL(\tau)$ is due to $T/L^\theta(\tau)L^{2\zeta}(\tau) = TL(\tau)$ since $\theta = 2\zeta - 1$.] In the aging regime we find ${\rm B}(z,t,t_{\rm w}) = L(t)^{2\zeta} f_{\rm B}(z/L(t_{\rm w}),L(t)/L(t_{\rm w}))$, while $\chi(z,t,t_{\rm w}) = L(t)f_\chi(z/L(t_{\rm w}),L(t)/L(t_{\rm w}))$ due to the degeneracy of barriers.

In order to test the above, the simplest protocol is to choose a flat initial condition, i.e. $x(z,0) = 0$ for all z. During isothermal aging the roughness of the system develops progressively from short to long wavelengths. The dynamical length $L(t)$ can be extracted from the growth of the static roughness $B(0,t,t) = [\langle(x(z,t) - x(z,t)^2\rangle] = B_{\rm eq}(z = L(t))$. Here $B_{\rm eq}(z) = \lim_{t\to\infty} B(z,t,t)$ is the equilibrium roughness, which can be computed numerically with a transfer matrix method in the 1+1 case (Yoshino, 2001). Another way to determine $L(t)$ is given in (Iguain *et al.*, 2009).

By performing Monte Carlo simulations of the 1+1 DPRM the analysis of steps I) and II) can be done precisely (Yoshino, 2001; Iguain *et al.*, 2009) helped by the knowledge of the exact values of the exponents $\zeta_{\rm T} = 1/2$, $\zeta_{\rm D} = 2/3$ and $\theta = 1/3$. Quite interestingly the variation of the energy scales as $\Delta U \simeq L_\perp^{1/2}$ just as in the Sinai model discussed before.

Concomitantly with the change in roughness exponent from thermal to disorder-dominated values, $L(t)$ exhibits a gradual crossover from pure diffusion with $L(t) \simeq t^{1/z}$ (and $z = 2$) to an activated regime consistent with an algebraic growth of barriers (Fisher and Huse, 1991). This is similar to what is found in the Sinai model (see Fig. 11.4) but with $\psi = 1/3$. The crossover occurs at a static temperature-dependent correlation length $L_0(T)$. In the analysis of numerical data the two regimes tend to be confused into a single one with an effective temperature-dependent power law, $L(t) \sim t^{1/\overline{z}(T)}$ (Barrat, 1997; Yoshino, 1996, 1998; Bustingorry *et al.*, 2006), the T dependence of which is inherited from the one in $L_0(T)$ (Iguain *et al.*, 2009), but this is just an approximation. Further support to the asymptotic logarithmic scaling is given by the renormalization group study in (Monthus and Garel, 2008*a*,*b*).

As regards the scaling properties of the two-time quantities $\mathrm{B}(z, t, t_\mathrm{w})$ and $\chi(z, t, t_\mathrm{w})$ in the quasi-equilibrium and aging regimes one also faces the difficulty of going beyond the thermal regime and reaching, for sufficiently long time scales, the disorder dominated one. In (Iguain *et al.*, 2009) it was shown that in the early effective power-law regime, approximated by $L(t) \simeq t^{1/\overline{z}(T)}$, and using the thermal roughness exponent, two-time linear response and correlation functions conform to the scalings discussed above and a finite effective temperature exists [basically because $2\zeta_\mathrm{T} = 1$ and the diffusive prefactors in B and χ are both equal to $L(t)$]. Simulations entering the truly activated regime suggest that the dynamic scaling above with the disorder roughness exponent ζ_D sets in (and the effective temperature progressively vanishes) (Yoshino, 2001).

11.4.1.3 Spin-glasses

The isothermal aging of the Edwards–Anderson (EA) model was analyzed along the same steps. The model is supposed to have a finite-temperature phase transition and thus an additive separation of time scales of the form (11.20)—the precise nature of the aging term depending on a two-state droplet picture (Fisher and Huse, 1988*a*; Bray and Moore, 1987; Fisher and Huse, 1988*b*) or a more complicated dynamics of the Sherrington–Kirkpatrick (SK) type (Cugliandolo and Kurchan, 1994).

The growing length is extracted from the real-replica overlap, see Eq. (11.14) (Huse, 1988, 1991). It is common to use the fit $\xi_\mathrm{sg}(t) \sim t^{1/\overline{z}(T)}$ with $\overline{z}(T)$ an effective temperature-dependent exponent (Belletti, 2008, 2009; Huse, 1988, 1991; Rieger, 1993; Marinari *et al.*, 2000; Komori *et al.*, 1999*a*,*b*, 2000). The best currently available data are from the JANUS collaboration suggesting $\overline{z}(T) \simeq \overline{z}(T_\mathrm{c})T_\mathrm{c}/T$ (Belletti, 2008, 2009). However, $\xi_\mathrm{sg}(t)$ should exhibit a gradual crossover from critical dynamics at short times to activated dynamics asymptotically (Yoshino *et al.*, 2002; Berthier and Bouchaud, 2002). The crossover is expected at $L_0(T) \sim \xi$ and $\tau_0(T) \sim \xi^{z_\mathrm{eq}}$ with the equilibrium correlation length $\xi = |T - T_\mathrm{c}|^{-\nu}$ and z_eq and ν the usual critical exponents at the critical temperature T_c. The values of T_c, ν, z_eq and θ are fixed with equilibrium studies. In the rest of the discussion we call $L(t)$ the growing length [$\xi_\mathrm{sg}(t) \to L(t)$].

Similar arguments to the ones exposed for Sinai diffusion and the directed manifold imply that the energy density per spin should decay as $e(t) = e(\infty) + \mathrm{c}t/L(t)^{d-\theta}$ and provide scaling forms for the self-correlation and linear susceptibility of very similar type to the ones in previous subsections (but with different values of the exponents).

Numerical tests of the droplet picture predictions were done by a number of groups. The energy density decay in the $d = 3$ and $d = 4$ EA models were checked in (Komori *et al.*, 1999*a*,*b*, 2000) and (Yoshino *et al.*, 2002), respectively. The numerical spin autocorrelation function $C(t, t_\mathrm{w})$ and the linear susceptibility $\chi(t, t_\mathrm{w})$ were analyzed along the proposed scaling forms. In the quasi-equilibrium regime where the FDT $C(\tau + t_w, t_w) = T\chi(\tau + t_w, t_w)$ holds, the expected scaling $C(\tau + t_w, t_w) = q_{ea} + ct\ T/L(\tau)^\theta g_c(L(\tau)/L(t_w))$ with $g_c(\lambda) \propto 1 - \text{const}\ \lambda^{d-\theta} + \ldots$ was verified in $d = 3$ (Komori *et al.*, 1999*a*,*b*, 2000) and $d = 4$ (Yoshino *et al.*, 2002) (the fact that $L(t)$ is far from a logarithm was attributed to pre-asymptotics and the analysis was perfomed using the method sketched above.). Evidence for the validity of droplet aging scaling for the correlation function $C_{ag}(t, t_w) = f_c(L(t)/L(t_w))$ and the field-cooled susceptibility in which the sample is cooled in a field that it switched off at t_w, $\chi_{fc}(t, t_w) = \chi(t, 0) - \chi(t, t_w) = L(t)^{-\theta} f_\chi(L(t)/L(t_w))$ with $\chi(t, 0) = \chi_D - \text{const}\ T/L(t)^\theta$ (Fisher and Huse, 1988*a*) in $d = 4$ was given in (Yoshino *et al.*, 2002) (note that $\psi \simeq \theta$ is assumed here). These results were used as support to the standard droplet theory. However, the asymptotic value of the dynamical susceptibility χ_D is significantly higher than $\chi_{ea} = \beta(1 - q_{ea})$ (Yoshino *et al.*, 2002; Marinari *et al.*, 2000). This is clearly at odds with the standard droplet theory (Fisher and Huse, 1988*a*) and suggests the existence of excessive contributions from some unknown soft modes other than the droplets in equilibrium. Moreover, it was found that the FDT holds up to significantly longer time differences than time-translational invariance (Yoshino *et al.*, 2002). These observations suggest that, even within a droplet perspective, modifications of the conventional approach are needed.

A different school of thought (Marinari *et al.*, 2000; Belletti, 2008, 2009) pushes for a picture *à la* mean-field (namely, the dynamics of the SK model (Cugliandolo and Kurchan, 1994)) that should be accompanied by a complex relaxation of correlations and susceptibilities in multiple time scales, presumably linked to multiple length scales. So far, this *ultrametric* organization of time scales has not been found numerically (Jaubert *et al.*, 2007; Belletti, 2008, 2009; Berthier *et al.*, 2001*a*).

11.4.2 Other random systems

Let us finalize this section by briefly recalling studies of growing lengths in two other systems with quenched randomness: the random ferromagnet and the sine-Gordon model with random phases.

On the basis of a droplet theory with barriers increasing as a power law of distance, the random ferromagnet should coarsen with a logarithmically increasing typical length (Huse and Henley, 1985). Early Monte Carlo simulations (Puri *et al.*, 1991; Bray and Humayun, 1991) showed agreement with this prediction but more recent numerical results from Rieger's group were interpreted, instead, as evidence for a power law with variable exponent (Paul *et al.*, 2004, 2005; Rieger *et al.*, 2005). This result would imply a logarithmic dependence of barriers on the domain size. It was argued in (Iguain *et al.*, 2009) that the effective power law may be the effect of a very long crossover from curvature driven to activation with conventional power-law growth of barriers with size. Recent simulations support this claim (Park and Pleimling, 2010)—although they were performed in a dilute ferromagnet.

The Cardy–Ostlund or random sine-Gordon model describes the relaxation dynamics of 2D periodic elastic manifolds under the effect of a quenched random potential. The peculiarity of the model is that disorder-induced barriers grow only logarithmically with size, $\simeq \ln L$. A "superrough" or "marginal" glassy phase exists below a critical temperature $T_{\rm g}$. The aging dynamics were studied with Coulomb gas and renormalization methods close to $T_{\rm g}$, and with the functional renormalization group (FRG) (Le Doussal, 2010) and numerical simulations in the full low-T phase (Schehr and Le Doussal, 2003, 2004; Schehr and Rieger, 2005). A dynamic length $L(t) \sim t^{1/z(T)}$ with $z(T) \simeq 2 + c/T$ develops in time, consistently with an Arrhenius argument of the kind described in Section 11.4. Intriguingly, $X_\infty = z$ was also found analytically and numerically.

11.5 Growing length scales in aging glasses

In this section we summarize predictions from mean-field theory as well as the outcome of measurements of two-time lengths in several glassy systems.

11.5.1 Mean-field models

It has been argued that the Langevin dynamics of mean-field disordered models are equivalent to mode-coupling theories of glasses. More precisely, the Sherrington–Kirkpatrick (SK) spin-glass has a dynamic transition similar to the one found in so-called type-A mode coupling, while p-spin models with $p > 3$ have a random first-order transition (RFOT) similar to the one found in type-B mode-coupling theories (Bouchaud *et al.*, 1996). The asymptotic out-of-equilibrium relaxation of these models, when the thermodynamic limit has been taken at the outset, approaches a region of phase space that is not the one where equilibrium configurations lie. In models of the RFOT class this region was named the *threshold* since its energy density is at $O(1)$ distance from the equilibrium one (Cugliandolo and Kurchan, 1993). In models of the SK or type-A class, although the configurations visited dynamically are not the ones of equilibrium their energy density is the same (Cugliandolo and Kurchan, 1994). In both cases the region reached dynamically is *flat*, in the sense that a dynamic free-energy landscape can be defined and its geometric properties studied, and one finds that the eigenvalues of its Hessian mostly vanish (Cugliandolo and Kurchan, 1993). The flat directions provide, on the one hand, channels of aging relaxation and, on the other, their associated zero modes give rise to diverging susceptibilities. In the case of mean-field models the latter cannot be directly associated to diverging length scales since these models do not contain any notion of distance. Having said this, if one proposes that the same mechanism, namely the relaxation to a flat region of phase space, is at the root of aging phenomena in finite-dimensional glassy systems, the diverging susceptibility should be linked to a diverging length scale. In the supercooled liquid regime this argument was used by Franz and Parisi to argue for the existence of a diverging length scale within the RFOT scenario from the analysis of the p-spin model (Franz and Parisi, 2000), see the chapter by Franz and Semerjian. These ideas were extended to the actual MCT in (Biroli and Bouchaud, 2004). In the trully glassy phase a very similar mechanism is at work and gives support to the symmetry argument (Chamon and Cugliandolo, 2007) that we discuss in Section 11.6 and provides a scenario for dynamic fluctuations in aging glassy systems.

11.5.2 Measurements

In Section 11.2.1 we already mentioned simulations (Vollmayr-Lee, 2004; Vollmayr-Lee and Zippelius, 2005; Vollmayr-Lee and Baker, 2006) and experiments (Courtland and Weeks, 2003; Weeks *et al.*, 2007) in which the tagged motion of particles was followed with the intention of characterizing clusters of more or less-mobile ones in aging glasses.

A two-time-dependent dynamic growing length, as defined in Eq. (11.10), was measured with numerical simulations of aging soft spheres and Lennard-Jones mixtures (Parisi, 1999; Parsaeian and Castillo, 2009; Castillo and Parsaeian, 2007), spin-glasses (Castillo *et al.*, 2002, 2003; Jaubert *et al.*, 2007; Belletti, 2008, 2009) and with confocal microscopy data of colloidal suspensions (Chamon *et al.*, 2008; Cianci *et al.*, 2006; Yunker *et al.*, 2009). An example is shown in Fig. 11.6 where we display the outcome of the data analysis performed in (Chamon *et al.*, 2008).

In all cases, ξ_4 is described by

$$\xi_4(t, t_w) \simeq \begin{cases} \xi_{st}(t - t_w) & C > q_{ea} \\ t_w^a \ \kappa(C) & C < q_{ea}, \end{cases}$$

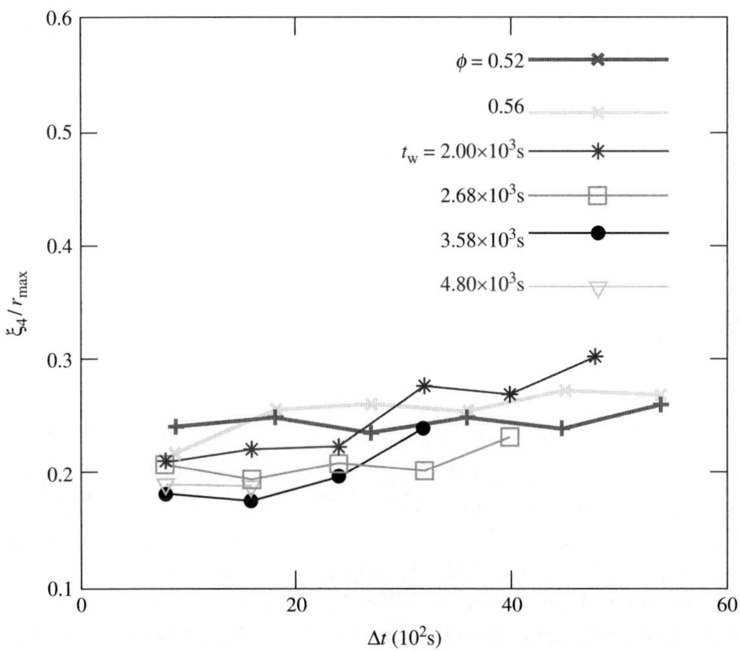

Fig. 11.6 The two-time dependent growing length in a colloidal suspension. The data is taken from (Chamon *et al.*, 2008). ϕ is the packing fraction and $r_{\max}$ is a cutoff used to compute the integrals (see (Chamon *et al.*, 2008) for more details).

with $\kappa(C)$ a monotonically decreasing function of C and the exponent a taking a very small value. It is important to stress that a clear identification of a finite Edwards–Anderson parameter in some of these systems is hard and the separation of time scales might not be as clear-cut as expressed in the formula above.

11.5.3 Growth processes in kinetically constrained models

Kinetically constrained models were originally introduced in the 1980s by Fredrickson and Andersen (Fredrickson and Andersen, 1984, 1985) as toy models for understanding glassy dynamics. Generally, these are stochastic models with Markovian dynamics obeying detailed balance with respect to a (usually) trivial energy function. Some constraints prevent particular local transitions between configurations (Jäckle, 2002; Sollich and Ritort, 2003; Leonard *et al.*, 2007). Despite their trivial equilibrium measure, they capture many features of real glass-forming systems summarized in the chapter by Garrahan, Sollich and Toninelli.

The so-called spin-facilitated Ising models, are described by the Hamiltonian $H = -\sum_i n_i$, where $n_i = 0, 1$ are spin variables on a d-dimensional lattice that can be regarded as a coarse-grained density of moving particles. The basic idea is that the rearrangement in a certain region around i will be facilitated if it is surrounded by a relatively low-density neighborhood. In some of these systems, the existence of a growing length in the non-equilibium evolution after a quench can be easily recognized. Some time after a sudden change in temperature towards $T \simeq 0$ all sites are in a dense state with $n_i = 1$ and large frozen regions, the size of which increases as the number of sites with $n_i = 0$ decreases, are present (Crisanti *et al.*, 2000; Sollich and Evans, 1999; Garrahan and Newman, 2000; Majumdar *et al.*, 2001).

A particularly interesting kinetically constrained system is the $2d$ spiral model. It has the peculiar feature of having a glass-jamming transition at the critical density $\langle n_i \rangle = n_c$ of directed percolation (Biroli and Toninelli, 2008; Toninelli and Biroli, 2008). In equilibrium above n_c a finite fraction of sites are frozen by the kinetic constraint. This means that the configuration space is partitioned into mutually inaccessible regions; namely, the dynamics is reducible. Then, all configurations without blocked regions are dynamically connected to the empty state with $n_i = 0\ \forall i$, usually denoted as the high-temperature partition, and they are disconnected from any partition containing frozen regions. Therefore, quenching the system from (say) the empty state to the jammed phase, the dynamics do not freeze and remain confined in the high-temperature partition by means of a coarsening process with a growing correlation length that resembles in some sense the one observed in binary systems without kinetic constraints (Corberi and Cugliandolo, 2009).

11.5.4 Interplay between aging and drive

The classical and quantum dynamics of large systems under an external, non-conservative, drive is a subject of active research. An example in the first family is the analysis of the rheological properties of soft glassy materials, that are a result of a competition between the response to the shearing forces and their intrinsic slow dynamics. An example in the second class is the study of quantum magnets under a current generated to the coupling to leads at different chemical potential. We shall recap here and in Section 11.5.6 a few issues in this field that are related to the growth of a length and how this is affected by the drive.

11.5.4.1 Flow in coarsening systems

The effect of an external drive on coarsening systems is not completely understood and in many cases a general consensus is lacking. This is at odds with the wide technological interest of these and similar systems in various application areas (Larson, 1999). In particular, the case of binary systems in a plane shear flow (Couette flow) has been thoroughly studied recently. In this case, domains grow elongated and stretched along the flow direction. This causes ruptures of the network (Otha *et al.*, 1990, 1991) that may render the segregation incomplete. The characteristic lengths in the parallel or perpendicular directions to the flow could keep on growing indefinitely as without shear or, due to domain break-up, the system may eventually enter a stationary state (similarly to what happens in the mean-field models discussed above) characterized by domains with a finite thickness. In general, there is no agreement on this point neither from experimental (Chan *et al.*, 1991; Takebe *et al.*, 1993; Hashimoto *et al.*, 1995; Läuger *et al.*, 1995; Yamamoto and Zeng, 1997; Shou and Chakrabarti, 2000; Liu and Chakrabarti, 2000) nor theoretical points of view. On the theoretical side, large-N calculations (Corberi *et al.*, 1998, 2000*a*; Rapapa and Bray, 1999) show the existence of an asymptotic regime with ever-growing lengths in all directions. Stabilization of the domain size by rupture is excluded also by different approaches based on approximate theories for scalar systems complemented with RG analysis (Bray and Cavagna, 2000; Bray *et al.*, 2001, 2002), although with the unexpected feature of domains growing in the flow direction while shrinking in the perpendicular one in $d = 2$. Numerical simulations (Padilla and Toxvaerd, 1997; Cates *et al.*, 1999; Corberi *et al.*, 1999, 2000*b*) still cannot establish clear-cut evidence due to finite-size and discretization effects. It was also argued (Stansell *et al.*, 2006) that hydrodynamic effects are crucial in stabilizing a stationary state, while in the diffusive regime domains might keep on growing.

Another difference with driven mean-field models is provided by the fluctuation–dissipation relation $\chi(C)$, the exact computation of which in the large-N model (Corberi *et al.*, 2002*b*, 2003*a*) excludes a simple scenario with a single effective temperature. Much more details are given in the contribution by Barrat and Lemaître.

11.5.4.2 Flow in glassy systems

The Langevin dynamics of mean-field disordered models under non-potential forces has been studied in a series of papers (Cugliandolo *et al.*, 1997*a*; Berthier *et al.*, 1999; Corberi *et al.*, 2001*a*,*b*). In all cases (apart from the spherical model with two-body interactions at zero temperature) aging is suppressed and a stationary state is reached. Two-time observables decay in two steps and the relaxation time decreases with the strength of the drive as in shear-thinning systems. In the limit of weak drive the effective temperature takes two values, the one of the bath in the first relaxation step and a higher value in the second slower stage. These results have been verified in molecular dynamic simulations of Lennard-Jones mixtures (Barrat and Berthier, 2002*a*,*b*).

The fact that a weak non-potential perturbation can change the relaxation of a classical system so dramatically, by rendering the dynamics stationary, can be understood in different ways. First, the type of non-potential force used in (Cugliandolo *et al.*, 1997*a*) can be loosely associated to an infinite temperature noise. Secondly, and more importantly, the mechanism for fluctuations based on time-reparametrization invariance gives a reason for this fact, as discussed in Section 11.6.

Experimentally, the interplay between aging and flow in an aqueous suspension of Laponite has been studied by diffusive-wave spectroscopy (Bonn *et al.*, 2002; Viasnoff and Lequeux, 2002) and light-scattering echo experiments (Petekis *et al.*, 2002) (multiple scattering) and dynamic light scattering in the single-scattering regime (Di Leonardo *et al.*, 2005). Using the latter technique Di Leonardo *et al.* measured the density auto-correlation function and found that as long as the characteristic relaxation time is smaller than the inverse shear rate aging in unaffected by the perturbation but, when the relaxation time reaches the inverse shear rate, aging is strongly reduced.

A recent study of microscopic structural relaxation of colloidal suspensions under shear together with relevant references appeared in (Chen *et al.*, 2010). Numerical studies of aging and plastic deformations have been performed by several groups, see the reviews in (Rottler and Warren, 2008; Barrat *et al.*, 2004). Nevertherless, as far as we know, an analysis of growing dynamic lengths, as defined from G_4 or χ_4 has not been performed in driven systems yet.

11.5.4.3 Aging at the depinning transition of elastic manifolds

A series of analytic and numerical papers have recently showed that a non-steady critical regime, limited only by the steady correlation length or the system size, exists at the zero-temperature depinning transition of an elastic manifold in a disordered medium (Schehr and Le Doussal, 2005; Kolton *et al.*, 2006, 2009). The FRG analysis (Le Doussal) as well as molecular dynamics studies prove that there is a driven transient aging regime that displays universal features in which a growing length $L(t) \sim t^{1/z}$ participates in the critical dynamic scaling description of the dynamics. At the depinning transition the effective height of the barriers vanishes justifying the power-law scaling the effect of disorder being a reduction of z to a value smaller than 2 (at the existent one-loop FRG calculation).

11.5.5 Granular matter

The gently driven dynamics of dense granular matter—athermal systems—shows many points in common with the ones of glass-forming liquids (Dauchot, 2007). Aging in a water-saturated granular pile submitted to discrete taps has been reported in (Kabla and Debregeas, 2004). The data were obtained using multispeckle diffusive-wave spectroscopy to measure particle displacements. Evidence for dynamic heterogeneities and a growing correlation length in a horizontally vibrated amorphous assembly of hard disks close to the jamming transition was given in (Dauchot *et al.*, 2005;

Lechenault *et al.*, 2008). Presumably, a two-time growing length would also exist in such athermal systems if studied in the aging regime. On the theoretical side, the aging properties of a periodically perturbed mean-field model of the RFOT type and, more precisely, how these depend on the amplitude and frequency of the drive were studied in (Berthier *et al.*, 2001*b*).

11.5.6 Quantum fluctuations

The study of the dynamics of quantum systems has been recently boosted by the development of cold-atom experiments and advanced measurement techniques. Quantum magnets are expected to undergo coarsening dynamics in the ferromagnetic phase. The study of a mean-field model (freely relaxing and driven by an external current) has revealed an extension of the superuniversality hypothesis to the quantum realm once small scales are well separated from large ones, the former feeling the quantum fluctuations and the latter describing the motion of large objects—the walls—being basically identical to the classical ones. More precisely, in $O(N)$ field theory in the large-N limit or models of rotors in interaction the length over which static order is established grows as $L(t) \sim t^{1/z}$ with $z = 2$ as in the classical limit. This type of coarsening survives the application of a voltage difference by two leads at different chemical potential (Aron *et al.*, 2009). This prediction is open to experimental tests.

11.6 A mechanism for dynamic fluctuations

In a series of papers Chamon *et al.* proposed that the symmetry that captures the universal aging dynamics of glassy systems is the invariance of an effective dynamical action under *uniform reparametrizations of the time scales* (Castillo *et al.*, 2002, 2003; Jaubert *et al.*, 2007; Chamon *et al.*, 2002; Castillo, 2008; Chamon *et al.*, 2004, 2006), see also (Parisi, 1999). This approach is reviewed in (Chamon and Cugliandolo, 2007). Such type of invariance was first encountered in the *mean-field equilibrium dynamics* of spin-glasses (Sompolinsky, 1981; Ginzburg; Ioffe, 1988) and it was later found in the better formulated *out-of-equilibrium* dynamic of the same models (Cugliandolo and Kurchan, 1993, 1994). The invariance means that in the asymptotic regime of very long times the solution can be found up to a time-reparametrization transformation. Within such a scenario, extended to systems in finite dimensions, the remaining symmetry is responsible of the dynamic fluctuations. Support to this claim was given by the proof of global time-reparametrization invariance in the long-times *action* of *short-range spin-glasses* (Chamon *et al.*, 2002; Castillo, 2008; Castillo *et al.*, 2002, 2003). In the proof one assumes that the dynamics is causal and that there is an additive separation between a time regime with a fast relaxation and another in which it is slow [just as in Fig. 11.1 and in Eq. (11.20)]. The invariance can then be used to describe dynamic fluctuations in spin-glasses and, as conjectured, in other glassy systems as well. The fact that the dynamic action is symmetric under uniform, i.e. spatially independent, reparametrizations of time variables $[t \to h(t)]$ suggests that the dynamic fluctuations that cost little action are those associated to space-dependent, long-wavelength, reparametrizations of the form $t \to h(\vec{r}, t)$. These should be the Goldstone modes associated with breaking time-reparametrization invariance symmetry.

In the slow and aging regime the two-time correlation and linear response depend on both times and time-translation invariance is lost. As we argued in Section 11.1 the two-time dependent correlation acts as an order parameter. Dynamic fluctuations are such that *ages* can fluctuate from point to point in the sample with younger and older pieces (lower and higher values of the correlation) coexisting at the same values of the two laboratory times: this has been named *heterogeneous aging*. By looking now at spatially heterogeneous reparametrizations, we can predict the behavior of local correlations and linear susceptibilities and the relations between them. Different sites can be retarded or advanced with respect to the global behavior but they should all have the same overall type of decay. Similarly, the relation between local susceptibilities and their associated correlations should be identical all over the sample (Castillo *et al.*, 2002, 2003) leading to a uniform effective temperature (Cugliandolo *et al.*, 1997*b*). The reason why the scaling functions should not fluctuate much is that these are massive modes. Within the time-reparametrization invariance scenario the growing correlation length is due to the approach to the long-time regime in which the symmetry is fully developed, and the long-wavelength modes eventually become massless.

This hypothesis was put to the test in glassy systems with Monte Carlo simulations of the Edwards–Anderson model (Castillo *et al.*, 2002, 2003; Jaubert *et al.*, 2007), molecular dynamics of Lennard-Jones mixtures (Parsaeian and Castillo, 2009; Castillo and Parsaeian, 2007), numerical studies of kinetically constrained models (Chamon *et al.*, 2004), and in the study of solvable ferromagnetic models with the analysis of the $O(N)$ model in the large-N limit (Chamon *et al.*, 2006) and the spherical ferromagnet (Annibale and Sollich, 2009). A series of stringent tests some of them performed in these papers were summarized in (Chamon and Cugliandolo, 2007). The outcome of these studies is that while truly glassy systems conform to the consequences of the hypothesis, simple coarsening as developed in the $O(N)$ and spherical ferromagnet does not, with time-reparametrization invariance being reduced to time rescaling at the heart of the difference. This result is very suggestive since it implies that the invariance properties and the fluctuations associated to them are intimately related to the behavior of the global effective temperature (finite against infinite) in the aging regime.

In short, this idea proposes a *reason* for the development of large dynamic fluctuations in glassy dynamics.

11.7 Closing remarks

In aging systems, the relaxation of typical features in a sample of a certain age t_w takes a time that increases with t_w. Although the reason for this fact is not necessarily the growth of those features, the widespread observation of growing lengths in physical systems, as reviewed in this chapter, promotes this one as the basic mechanism at the heart of most aging phenomena. (See (Montanari and Semerjian, 2006) and the chapter by Franz and Semerjian for rigorous relations between growing lengths and growing times.) Related to this fact, the occurrence of dynamic scaling provides a parametrization of the kinetics in terms of typical lengths, giving access to universal properties.

Besides these unifying concepts, the very nature of the growing length, the growth mechanisms, the relation with equilibrium properties, etc. are issues that, in most cases, deserve clarification. In particular, the interplay between the growth of static order—as, in coarsening systems—and purely dynamical correlations—as in most kinetically constrained models—could help in understanding the behavior of glasses. In the case of spin-glasses, moreover, the comprehension of the kind of order established within the correlated volume is fundamental to distinguish between competing scenarios, namely droplet-like and *à la* mean-field pictures.

References

Andrenacci, N., Corberi, F., and Lippiello, E. (2006). *Phys. Rev. E*, **74**, 031111.

Annibale, A. and Sollich, P. (2009). *J. Stat. Mech.*, P02064.

Arenzon, J. J., Bray, A. J., Cugliandolo, L. F., and Sicilia, A. (2007). *Phys. Rev. Lett.*, **98**, 145701.

Aron, C., Biroli, G., and Cugliandolo, L. F. (2009). *Phys. Rev. Lett.*, **102**, 050404.

Aron, C., Chamon, C., Cugliandolo, L. F., and Picco, M. (2008). *J. Stat. Mech.*, P05016.

Barabási, L. and Stanley, H. E. (1995). In *Fractal Concepts in Surface Growth*. Cambridge University Press, Cambridge.

Barrat, A. (1997). *Phys. Rev. E*, **55**, 5651.

Barrat, J.-L., Baschnagel, J., and Lyulin, A. (2004) *arXiv:1002.2065*. *Soft Matter*, **6**, 3430.

Barrat, J.-L. and Berthier, L. (2002*a*). *Phys. Rev. Lett.*, **89**, 095702.

Barrat, J.-L. and Berthier, L. (2002*b*). *J. Chem. Phys.*, **116**, 6228.

Belletti, F. *et al* (2008). *Phys. Rev. Lett.*, **101**, 157201.

Belletti, F. *et al* (2009). *J. Stat. Phys.*, **135**, 1121.

Bennemann, C., Donati, C., Baschnagel, J., and Glotzer, S. C. (1999). *Nature*, **399**, 246.

Berthier, L., Barrat, J.-L., and Kurchan, J. (1999). *Eur. Phys. J. B*, **11**, 635.

Berthier, L., Barrat, J.-L., and Kurchan, J. (2001*a*). *Phys. Rev. E*, **63**, 016105.

Berthier, L., Biroli, G., Bouchaud, J.-P., Cipelletti, L., El Masri, D., L'Hôte, D., Ladieu, F., and Pierno, M. (2005). *Science*, **310**, 1797.

Berthier, L. and Bouchaud, J.-P. (2002). *Phys. Rev. B*, **66**, 054404.

Berthier, L., Cugliandolo, L. F., and Iguain, J. L. (2001*b*). *Phys. Rev. E*, **63**, 051302.

Binder, K. and Kob, W. (2005). In *Glassy Materials, and Disordered Solids: An Introduction to their Statistical Mechanics.* World Scientific, Singapore.

Biroli, G. and Bouchaud, J.-P. (2004). *Europhys. Lett.*, **67**, 21.

Biroli, G. and Toninelli, C. (2008). *Eur. Phys. J. B*, **64**, 567.

Bonn, D., Tanase, S., Abou, B., Tanaka, H., and Meunier, J. (2002). *Phys. Rev. Lett.*, **89**, 015701.

Bouchaud, J.-P. (1992). *J. Phys. I*, **2**, 1705.

Bouchaud, J.-P. and Biroli, G. (2005). *Phys. Rev. B*, **72**, 064204.

Bouchaud, J.-P., Cugliandolo, L. F., Kurchan, J., and Mézard, M. (1996). *Physica A*, **226**, 243.

Bouchaud, J.-P., Cugliandolo, L. F., Kurchan, J., and Mézard, M. (1997). In *Spin Glasses, and Random Fields* (ed. Y. A. P.). World Scientific, Singapore.

Bouchaud, J.-P. and Dean, D. S. (1995). *J. Phys. (France) I*, **5**, 265–286.

Bray, A. J. (1994). *Adv. Phys.*, **43**, 357.

Bray, A. J. and Cavagna, A. (2000). *J. Phys. A*, **33**, L305.

Bray, A. J., Cavagna, A., and Travasso, Rui D. M. (2001). *Phys. Rev. E*, **64**, 012102.

Bray, A. J., Cavagna, A., and Travasso, Rui D. M. (2002). *Phys. Rev. E*, **65**, 016104.

Bray, A. J. and Humayun, K. (1991). *J. Phys. A*, **24**, L1185.

Bray, A. J. and Jervis, D. K. (2000). *Phys. Rev. Lett*, **84**, 1503.

Bray, A. J. and Moore, M. A. (1987). *Phys. Rev. Lett.*, **58**, 57.

Burioni, R., Corberi, F., and Vezzani, A. (2009). *Phys. Rev. E*, **79**, 041119.

Bustingorry, S. (2007). *J. Stat. Mech.*, P10002.

Bustingorry, S., Iguain, J. L., Chamon, C., Cugliandolo, L., and Domínguez, D. (2006). *Europhys. Lett.*, **76**, 856.

Bustingorry, S., Iguain, J. L., and Cugliandolo, L. F. (2007). *J. Stat. Mech.*, P09008.

Calabrese, P. and Gambassi, A. (2002*a*). *Phys. Rev. E*, **65**, 066120.

Calabrese, P. and Gambassi, A. (2002*b*). *Phys. Rev. E*, **67**, 36111.

Calabrese, P. and Gambassi, A. (2005). *J. Phys. A: Math. Gen.*, **38**, R133.

Castillo, E., Chamon, C., Cugliandolo, L. F., Iguain, J. L., and Kennett, M. P. (2003). *Phys. Rev. B*, **68**, 134442.

Castillo, H. E. (2008). *Phys. Rev. B*, **78**, 214430.

Castillo, H. E., Chamon, C., Cugliandolo, L. F., and Kennett, M. P. (2002). *Phys. Rev. Lett.*, **88**, 237201.

Castillo, H. E. and Parsaeian, A. (2007). *Nature Phys.*, **3**, 26.

Cates, M. E., Kendon, V. M., Bladon, P., and Despat, J.-C. (1999). *Faraday Discuss.*, **112**, 1.

Chamon, C., Charbonneau, P., Cugliandolo, L. F., Reichman, D. R., and Sellitto, M. (2004). *J. Chem. Phys.*, **121**, 10120.

Chamon, C. and Cugliandolo, L. F. (2007). *J. Stat. Mech.*, P07022.

Chamon, C., Cugliandolo, L. F., Fabricius, G., Iguain, J. L., and Weels, E. R. (2008). *PNAS*, **105**, 15263.

Chamon, C., Cugliandolo, L. F., and Yoshino, H. (2006). *J. Stat. Mech*, P01006.

Chamon, C., Kennett, M. P., Castillo, H. E., and Cugliandolo, L. F. (2002). *Phys. Rev. Lett.*, **89**, 217201.

Chan, C. K., Perrot, F., and Beysens, D. (1991). *Phys. Rev. A*, **43**, 1826.

Chatelain, C. (2004). *J. Stat. Mech.*, P06006.

Chen, D., Semwogerere, D., Sato, J., Brredveld, V., and Weeks, E. R. (2010) *arXiv:0908.4226*. *Phys. Rev. E*, **81**, 011403.

Chou, Y. C. and Goldburg, W. I. (1981). *Phys. Rev. A*, **23**, 858.

Cianci, G. C., Courtland, R. E., and Weeks, E. R. (2006). *Solid State Commun.*, **139**, 599.

Coniglio, A., Ruggiero, P., and Zannetti, M. (1994). *Phys. Rev. E*, **50**, 1046.

Coniglio, A. and Zannetti, M. (1989). *Europhys. Lett.*, **10**, 575.

Corberi, F. and Cugliandolo, L. F. (2009). *J. Stat. Mech.*, P09015.
Corberi, F., de Candia, A., Lippiello, E., and Zannetti, M. (2002*a*). *Phys. Rev. E*, **65**, 046114.
Corberi, F., Gambassi, A., Lippiello, E., and Zannetti, M. (2008). *J. Stat. Mech.*, P02013.
Corberi, F., Gonnella, G., and Lamura, A. (1998). *Phys. Rev. Lett.*, **81**, 3852.
Corberi, F., Gonnella, G., and Lamura, A. (1999). *Phys. Rev. Lett.*, **83**, 4057.
Corberi, F., Gonnella, G., and Lamura, A. (2000*a*). *Phys. Rev. E*, **61**, 6621.
Corberi, F., Gonnella, G., and Lamura, A. (2000*b*). *Phys. Rev. E*, **62**, 8064.
Corberi, F., Gonnella, G., Lippiello, E., and Zannetti, M. (2002*b*). *Europhys. Lett.*, **60**, 425.
Corberi, F., Gonnella, G., Lippiello, E., and Zannetti, M. (2003*a*). *J. Phys. A*, **36**, 4729.
Corberi, F., Lippiello, E., Sarracino, A., and Zannetti, M. (2010). *J. Stat. Mech.*, P04003.
Corberi, F., Lippiello, E., and Zannetti, M. (2001*a*). *Phys. Rev. E*, **63**, 061506.
Corberi, F., Lippiello, E., and Zannetti, M. (2001*b*). *Eur. Phys. J. B*, **24**, 359.
Corberi, F., Lippiello, E., and Zannetti, M. (2002*c*). *Phys. Rev. E*, **65**, 046136.
Corberi, F., Lippiello, E., and Zannetti, M. (2003*b*). *Phys. Rev. E*, **68**, 046131.
Corberi, F., Lippiello, E., and Zannetti, M. (2006). *Phys. Rev. E*, **74**, 041106.
Courtland, R. and Weeks, E. R. (2003). *J. Phys.: Condens. Matter*, **15**, S359.
Crisanti, A. and Ritort, F. (2003). *J. Phys. A*, **36**, R181.
Crisanti, A., Ritort, F., Rocco, A., and Sellitto, M. (2000). *J. Chem. Phys.*, **113**, 10615.
Cugliandolo, L. F. and Kurchan, J. (1993). *Phys. Rev. Lett.*, **71**, 173.
Cugliandolo, L. F. and Kurchan, J. (1994). *J. Phys. A*, **27**, 5749.
Cugliandolo, L. F., Kurchan, J., Le Doussal, P., and Peliti, L. (1997*a*). *Phys. Rev. Lett.*, **78**, 350.
Cugliandolo, L. F., Kurchan, J., and Parisi, G. (1994). *J. Phys. I (France)*, **4**, 1641.
Cugliandolo, L. F., Kurchan, J., and Peliti, L. (1997*b*). *Phys. Rev. E*, **55**, 3898.
Dauchot, O. (2007). *Lecture Notes in Physics*, **716**, 161.
Dauchot, O., Marty, G., and Biroli, G. (2005). *Phys. Rev. Lett.*, **95**, 265701.
Di Leonardo, R., Ianni, F., and , G. Ruocco. (2005). *Phys. Rev. E*, **71**, 011505.
Doliwa, B. and Heuer, A. (2002). *J. Non-Cryst. Solids*, **307**, 32.
Donati, C. *et al.* (1999). *Phys. Rev. E*, **60**, 3107.
Donth, E. (2001). In *The Glass Transition: Relaxation Dynamics in Liquids and Disordered Materials*. Springer Berlin.
Fisher, D. S., Fisher, M. P. A., and Huse, D. A. (1991). *Phys. Rev. B*, **43**, 130.
Fisher, D. S. and Huse, D. A. (1988*a*). *Phys. Rev. B*, **38**, 373.
Fisher, D. S. and Huse, D. A. (1988*b*). *Phys. Rev. B*, **38**, 386.
Fisher, D. S. and Huse, D. A. (1991). *Phys. Rev. B*, **43**, 10728.
Fisher, D. S., Le Doussal, P., and Monthus, C. (1998). *Phys. Rev. Lett.*, **80**, 3539.
Franz, S. and Parisi, G. (2000). *J. Phys. C*, **12**, 6335.
Franz, S. and Virasoro, M. A. (2000). *J. Phys. A: Math. Gen.*, **33**, 891.
Fredrickson, G. H. and Andersen, H. C. (1984). *Phys. Rev. Lett.*, **53**, 1244.

Fredrickson, G. H. and Andersen, H. C. (1985). *J. Chem. Phys.*, **83**, 5822.
Garrahan, J. P. and Newman, M. E. J. (2000). *Phys. Rev. E*, **62**, 7670.
Gaulin, B. D., Spooner, S., and Mori, Y. (1987). *Phys. Rev. Lett.*, **59**, 668.
Ginzburg, S. L. *Zh. Eksp. Teor. Fiz.*, **90**.
Godrèche, C. and Luck, J. M. (2000). *J. Phys. A*, **33**, 9141.
Godrèche, C. and Luck, J. M. (2002). *J. Phys.: Condens. Matter*, **14**, 1589.
Halpin-Healey, T. and Zhang, Y.-C. (1995). *Phys. Rep.*, **254**, 215.
Hashimoto, T., Matsuzaka, K., Moses, E., and Onuki, A. (1995). *Phys. Rev. Lett.*, **74**, 126.
Henkel, M., Enss, T., and Pleimling, M. (2006). *J. Phys. A*, **39**, L589.
Henkel, M. and Pleimling, M. (2008*a*). *Phys. Rev. B*, **78**, 224419.
Henkel, M. and Pleimling, M. (2008*b*). *Phys. Rev. B*, **78**, 224419.
Heuer, A., Kunow, M., Vogel, M., and Banhatti, R.D. (2002). *Phys. Rev. B*, **66**, 224201.
Huse, D. A. (1988). *J. Appl. Phys.*, **64**, 5776.
Huse, D. A. (1991). *Phys. Rev. B*, **43**, 8673.
Huse, D. A. and Henley, C. L. (1985). *Phys. Rev. Lett.*, **54**, 2708.
Iguain, J. L., Bustingorry, S., Kolton, A. B., and Cugliandolo, L. F. (2009). *Phys. Rev. B*, **80**, 094201.
Ioffe, L. B. (1988). *Phys. Rev. B*, **38**, 5181.
Jäckle, J. (2002). *J. Phys. Condens. Matter*, **14**, 1423.
Janssen, H. K., Schaub, B., and Schmittman, B. (1989). *Z. Phys. B Condens. Matter*, **73**, 539.
Jaubert, L. D. C., Chamon, C., Cugliandolo, L. F., and Picco, M. (2007). *J. Stat. Mech.*, P05001.
Joubaud, S., Percier, B., Petrosyan, A., and Ciliberto, S. (2009). *Phys. Rev. Lett.*, **102**, 130601.
Kabla, A. and Debregeas, G. (2004). *Phys. Rev. Lett.*, **92**, 035501.
Katano, S. and Iizumi, M. (1984). *Phys. Rev. Lett.*, **52**, 835.
Kegel, W. K. and van Blaaderen, A. (2000). *Science*, **287**, 290.
Kob, W., Donati, C., Plimpton, S. J., Poole, P. H., and Glotzer, S. C. (1997). *Phys. Rev. Lett.*, **79**, 2827.
Kolton, A. B., Rosso, A., Albano, E. V., and Giamarchi, T. (2006). *Phys. Rev. B*, **74**, 140201(R).
Kolton, A. B., Schehr, G., and Le Doussal, P. (2009). *Phys. Rev. Lett.*, **103**, 160602.
Komori, T., Takayama, H., and Yoshino, H. (1999*a*). *J. Phys. Soc. Jpn.*, **68**, 3387.
Komori, T., Takayama, H., and Yoshino, H. (1999*b*). *J. Phys. Soc. Jpn.*, **69**, 1192.
Komori, T., Takayama, H., and Yoshino, H. (2000). *J. Phys. Soc. Jpn. 69 (Suppl. A)*, **69**, 228.
Komura, S., Osamura, K., Fujii, H., and Takeda, T. (1985). *Phys. Rev.B*, **31**, 1278.
Kurchan, J. and Levine, D. *arXiv:0904.4850*.
Ladieu, F., Thibierge, C., and L'Hôte, D. (2007). *J. Phys.: Condens. Matter*, **19**, 20.
Larson, R. G. (1999). In *The Structure and Rheology of Complex Fluids*. Oxford University Press, New York.
Läuger, J., Laubner, C., and Gronsky, W. (1995). *Phys. Rev. Lett.*, **75**, 3576.

Le Doussal, P. (2010) *arXiv:0809.1192*. *Ann. Phys.*, **325**, 49.
Le Doussal, P., Monthus, C., and Fisher, D. S. (1998). *Phys. Rev. E*, **59**, 4975.
Le Doussal, P. and Vinokur, V.M. (1995). *Physica C: Superconductivity*, **254**, 63–68.
Lechenault, F., Dauchot, O., Biroli, G., and Bouchaud, J.-P. (2008). *Europhys. Lett.*, **83**, 46002.
Leonard, S., Mayer, P., Sollich, P., Berthier, L., and Garrahan, J. P. (2007). *J. Stat. Mech.*, P07017.
Lippiello, E., Corberi, F., Sarracino, A., and Zannetti, M. (2008*a*). *Phys. Rev. B*, **77**, 212201.
Lippiello, E., Corberi, F., Sarracino, A., and Zannetti, M. (2008*b*). *Phys. Rev. E*, **78**, 041120.
Lippiello, E., Corberi, F., and Zannetti, M. (2005). *Phys. Rev. E*, **71**, 036104.
Lippiello, E., Corberi, F., and Zannetti, M. (2006). *Phys. Rev. E*, **74**, 041113.
Lippiello, E., Mukherjee, A., Puri, S., and Zannetti, M. (2010). *Europhys. Lett.*, **90**, 46006.
Lippiello, E. and Zannetti, M. (2000). *Phys. Rev. E*, **61**, 3369.
Liu, H. and Chakrabarti, A. (2000). *J. Chem. Phys.*, **112**, 10582.
Majumdar, S. N., Dean, D. S., and Grassberger, P. (2001). *Phys. Rev. Lett.*, **86**, 2301.
Marinari, E., Parisi, G., Ricci-Tersenghi, F., and Ruiz-Lorenzo, J. J. (2000). *J. Phys. A*, **33**, 2373.
Mayer, P., Berthier, L., Garrahan, J. P., and Sollich, P. (2005). *Phys. Rev.E*, **68**, 016116.
Mazenko, G. (2004). *Phys. Rev. E*, **69**, 0116114.
Mézard, M., Parisi, G., and Virasoro, M. A. (1987). In *Spin Glasses and Beyond*. World Scientific, Singapore.
Miyagawa, H., Hiwatari, Y., Bernu, B., and Hansen, J.-P. (1988). *J. Chem. Phys.*, **88**, 3879.
Montanari, A. and Semerjian, G. (2006). *J. Stat. Phys.*, **125**, 23.
Monthus, C. and Garel, T. (2008*a*). *J. Stat. Mech.*, P07002.
Monthus, C. and Garel, T. (2008*b*). *J. Phys. A*, **41**, 499801.
Onuki, A. and Yamamoto, Y. (1998). *J. Non-Cryst. Solids*, **235-237**, 19.
Otha, T., Nozaki, N., and Doi, M. (1990). *Phys. Lett. A*, **145**, 304.
Otha, T., Nozaki, N., and Doi, M. (1991). *J. Chem. Phys.*, **93**, 2664.
Padilla, P. and Toxvaerd, S. (1997). *J. Chem. Phys.*, **106**, 2342.
Parisi, G. (1999). *J. Chem. Phys. B*, **103**, 4128.
Park, H. and Pleimling, M. (2010) *arXiv:1009.1677*. *Phys. Rev. B*, **82**, 144406.
Parsaeian, A. and Castillo, H. E. (2009). *Phys. Rev. Lett.*, **102**, 055704.
Paul, R., Puri, S., and Rieger, H. (2004). *Europhys. Lett.*, **68**, 881.
Paul, R., Puri, S., and Rieger, H. (2005). *Phys. Rev. E*, **71**, 061109.
Perera, D. and Harrowell, P. (1996). *Phys. Rev. E*, **54**, 1652.
Petekis, G., Moussald, A., and Pusey, P. N. (2002). *Phys. Rev. E*, **66**, 051402.
Pleimling, M. and Gambassi, A. (2005). *Phys. Rev. B*, **71**, 180401.
Puri, S., Chowdhury, D., and Parekh, N. (1991). *J. Phys. A*, **24**, L1087.
Rapapa, N. P. and Bray, A. J. (1999). *Phys. Rev. Lett.*, **83**, 3856.
Rieger, H. (1993). *J. Phys. A*, **26**, L615.

Rieger, H., Schehr, G., and Paul, R. (2005). *Prog. Thoer. Phys. Suppl.*, **157**, 111.
Rottler, J. and Warren, M. (2008). *Eur. Phys. J. Sp. Topics*, **161**, 55.
Rutenberg, A. D. and Bray, A. J. (1995). *Phys. Rev. Lett.*, **74**, 3836.
Schehr, G. and Le Doussal, P. (2003). *Phys. Rev. E*, **68**, 046101.
Schehr, G. and Le Doussal, P. (2004). *Phys. Rev. Lett.*, **93**, 217201.
Schehr, G. and Le Doussal, P. (2005). *Europhys. Lett.*, **71**, 290.
Schehr, G. and Rieger, H. (2005). *Phys. Rev. B*, **71**, 184202.
Semerjian, G., Cugliandolo, L. F., and Montanari, A. (2004). *J. Stat. Phys.*, **115**, 493.
Shou, Z. and Chakrabarti, A. (2000). *Phys. Rev. E*, **61**, R2200.
Sicilia, A., Arenzon, J. J., Bray, A. J., and Cugliandolo, L. F. (2007). *Phys. Rev. E*, **76**, 061116.
Sicilia, A., Arenzon, J. J., Bray, A. J., and Cugliandolo, L. F. (2008*a*). *Europhys. Lett.*, **82**, 10001.
Sicilia, A., Arenzon, J. J., Dierking, I., Bray, A. J., Cugliandolo, L. F., Martínez-Perdiguero, J., Alonso, I., and Pintre, I. C. (2008*b*). *Phys. Rev. Lett.*, **101**, 197801.
Sinai, Y. G. (1982). *Theor. Probab. Appl.*, **27**, 247.
Sollich, P. and Evans, M. R. (1999). *Phys. Rev. Lett.*, **83**, 3238.
Sollich, P. and Ritort, F. (2003). *Adv. Phys.*, **52**, 219.
Sompolinsky, H. (1981). *Phys. Rev. Lett.*, **47**, 935.
Stansell, P., Stratford, K., Despat, J.-C., Adhikari, R., and Cates, M. E. (2006). *Phys. Rev. Lett.*, **96**, 085701.
Takebe, T., Fujioka, F., Sawaoka, R., and Hashimoto, T. (1993). *J. Chem. Phys.*, **98**, 717.
Thibierge, C., L'Hôte, D., Ladieu, F., and Tourbot, R. (2008). *Rev. Sci. Instrum.*, **79**, 103905.
Toninelli, C. and Biroli, G. (2008). *J. Stat. Phys.*, **130**, 83.
van Blaaderen, A. and Wiltzius, P. (1995). *Science*, **270**, 1177.
Viasnoff, B. and Lequeux, F. (2002). *Phys. Rev. Lett.*, **89**, 065701.
Vollmayr-Lee, K. (2004). *J. Chem. Phys.*, **121**, 4781.
Vollmayr-Lee, K. and Baker, E. A. (2006). *Europhys. Lett.*, **76**, 1130.
Vollmayr-Lee, K. and Zippelius, A. (2005). *Phys. Rev. E*, **72**, 041507.
Weeks, E. R., Crocker, J. C., Levitt, A. C., Schofield, A., and Weitz, D. A. (2000). *Science*, **287**, 627.
Weeks, E. R., Crocker, J. C., and Weitz, D. A. (2007). *J. Phys.: Condens. Matter*, **19**, 205131.
Weeks, E. R. and Weitz, D. A. (2000). *Phys. Rev. Lett.*, **89**, 095704.
Yamamoto, R. and Zeng, X. C. (1997). *Phys. Rev. E*, **59**, 3223.
Yoshino, H. (1996). *J. Phys. A*, **29**, 1421.
Yoshino, H. (1998). *Phys. Rev. Lett.*, **81**, 1493.
Yoshino, H. (2001). *unpublished.*
Yoshino, H., Hukushima, K., and Takayama, H. (2002). *Phys. Rev. B*, **66**, 064431.
Yunker, P., Zhang, Z. X., Aptowicz, K. B., and Yodh, A.G. (2009). *Phys. Rev. Lett.*, **103**, 115701.

12
Analytical approaches to time- and length scales in models of glasses

Silvio FRANZ and Guilhem SEMERJIAN

Abstract

The goal of this chapter is to review recent analytical results about the growth of a (static) correlation length in glassy systems, and the connection that can be made between this length scale and the equilibrium correlation time of its dynamics. The definition of such a length scale is first given in a generic setting, including finite-dimensional models, along with rigorous bounds linking it to the correlation time. We then present some particular cases (finite connectivity mean-field models, and Kac limit of finite-dimensional systems) where this length can be actually computed.

12.1 Introduction

The "random first-order theory" (RFOT) is one of the most widely discussed theories of glass formation and glassy phenomena in fragile systems. Its origin is based upon the observation by Kirkpatrick, Wolynes and Thirumalai (Kirkpatrick and Wolynes, 1987; Kirkpatrick *et al.*, 1989) that a family of abstract long-range spin-glass models with "one-step replica symmetry breaking" (1RSB) (Mézard *et al.*, 1987) presents freezing phenomena that in several aspects resemble the observed phenomenology in freezing of fragile liquids and other glassy systems. In this perspective, long-range spin-glasses provide a unifying description of glassy phenomena, including dynamical aspects (MCT dynamical singularity and aging phenomena) and thermodynamical ones (metastability and Kauzmann-like entropy crisis), which have given rise to predictions verified in simulations and experiments (Sciortino and Tartaglia, 2001; Berthier *et al.*, 2005). Despite the appeal of the resulting picture, it immediately became clear that two problems had to be overcome to be able to apply convincingly the theory to supercooled liquids: (1) The disordered interactions of spin-glass models is not a realistic microscopic description of liquid systems (2) The 1RSB picture is strongly dependent on the long-range character of the interactions and presents some typical pathologies of mean-field theories. Both problems have attracted a lot of attention.

The first problem has been basically circumvented by the fact that the most sophisticated mean-field theories based on realistic liquid models of particles in interaction give back the 1RSB scenario (Mézard and Parisi, 1999*a*,*b*; Zamponi and Parisi, 2009). This suggest a high level of universality of glassy phenomena that goes from spin models with random interactions to supercooled liquids, one can thus expect to understand general properties of the latter from the study of the former, which are simpler microscopically.

The second problem, that is how to effectively take into account the finite-range character of the interactions, remains, in our opinion, the main obstacle against an accomplished theory of the glass transition, despite important contributions. The RFOT, based on scaling and phenomenological arguments, proposes an intriguing scenario about which features of mean-field theory survive in short-range systems and what aspects of the picture should be modified. However, a theory based on a first-principles analysis of microscopic models is unfortunately lacking. Understanding the role of finite-range interactions includes both important questions of principle, but not directly related to physical observables such as the possibility of an ideal glass transition in systems with short-range interactions, and questions of direct practical interest such as finding a theory for activated dynamics and the crossover between mode-coupling behavior and activation.

This chapter reviews some of the theoretical efforts to understand the relation between the physics of finite-dimensional glassy systems and their mean-field description and to include finite-range interactions in microscopic models of glassy phenomena. A crucial point that an accomplished theory of the glass transition should address is the growth of correlations that accompanies the increase of relaxation time as temperature is decreased. The relaxation time of a supercooled liquid increases dramatically upon lowering its temperature, until reaching experimentally accessible time scales at the laboratory glass-transition temperature T_{g}. In this range of temperature the usual static spatial correlations (for instance the structure factor measured in scattering experiments) remain essentially the same as those of a high-temperature liquid. These two facts seem contradictory: the physical intuition relates a large correlation time to cooperative relaxation mechanisms involving a large volume of the sample (which incidentally is an argument in favor of the universality of glassy phenomena, the microscopic details being "averaged out" in this case), hence the very strong increase of the relaxation time around T_{g} should have a trace in spatial correlations. One way to address this puzzle is to define dynamical correlation lengths. Convincing experiments (Berthier *et al.*, 2005), following numerical simulations (Donati *et al.*, 1998) and dynamical theories (Donati *et al.*, 2002; Franz and Parisi, 2000; Biroli and Bouchaud, 2004*a*) have, after a long search, demonstrated for the first time growing dynamical correlations (Donati *et al.*, 2002; Franz and Parisi, 2000; Biroli and Bouchaud, 2004*a*; Berthier *et al.*, 2005). Another way, which we shall review in this contribution, is to define a static correlation length through a point-to-set procedure slightly more involved than the two-point function underlying the definition of the structure factor.

A large part of this chapter will be devoted to the discussion of this length and its relation with the relaxation time. In Section 12.2 we explain in detail the definition of the point-to-set correlation length and we show how it allows us to prove bounds between this correlation length and the equilibrium correlation time that agrees with the intuition sketched above. We then discuss two classes of disordered spin models of the 1RSB type where spatial aspects can be addressed through analytic techniques, namely models on diluted random graphs and finite-dimensional models in the Kac limit. In the first class of models, considered in Section 12.3, each spin interacts with a finite number of other spins chosen at random. These models can be solved exactly through the cavity method; in agreement with the general bounds, at the point of dynamical transition both the relaxation time and the correlation length are divergent. The second class, studied in Section 12.4, consists in genuine finite-dimensional models with a tunable interaction range r_0. In the limit of large r_0 one can compute the point-to-set correlation function and associated correlation lengths. This leads to a detailed picture of glassy phenomena, with dynamical and static correlation lengths that do not necessarily coincide.

We hope the style of presentation adopted in this chapter will provide the reader with a global view and some mathematical and theoretical tools that should make easier the reading of the original, more formal, literature on the subjects we address.

12.2 Definition of the point-to-set correlation function and its relation to correlation time

12.2.1 Heuristic discussion

In this section we want to introduce the notion of point-to-set correlation functions and to show that the correlation length derived from it is relevant for glassy systems, as upper and lower bounds on the correlation time can be inferred from this length. The presentation will be first done in an informal way, following the thought experiment first discussed in (Biroli and Bouchaud, 2004*b*). We shall then revisit it with more mathematical definitions and sketch the results and the methods of proof of (Montanari and Semerjian, 2006*a*).

Let us consider the thought experiment of (Biroli and Bouchaud, 2004*b*), schematized in Fig. 12.1. First, one takes a snapshot of an equilibrium configuration of the system under study, i.e. for instance the values of the spins, or the positions of particles depending on the model investigated. Let us call $\underline{\sigma}$ this first configuration, depicted on the first panel of Fig. 12.1. Suppose now that the configuration of the system is frozen to the value it has in $\underline{\sigma}$ outside a given volume $\bar{B}$ around an arbitrary point, for instance the center of the system (see middle panel in Fig. 12.1), and that the interior is thermalized in the presence of this boundary condition. One thus obtains another equilibrium configuration $\underline{\sigma}'$, which is forced to coincide with $\underline{\sigma}$ outside $\bar{B}$. Consider now the following question: how similar are $\underline{\sigma}$ and $\underline{\sigma}'$ around the center of the system? A precise notion of similarity shall be given below, in any case it is natural to expect that the larger the volume $\bar{B}$, the less similar should $\underline{\sigma}$ and $\underline{\sigma}'$ be at its center. Indeed, the influence of the boundary conditions, which force $\underline{\sigma}'$ to be very close to $\underline{\sigma}$ when $\bar{B}$ is small, becomes less and less efficient when the boundary is pushed

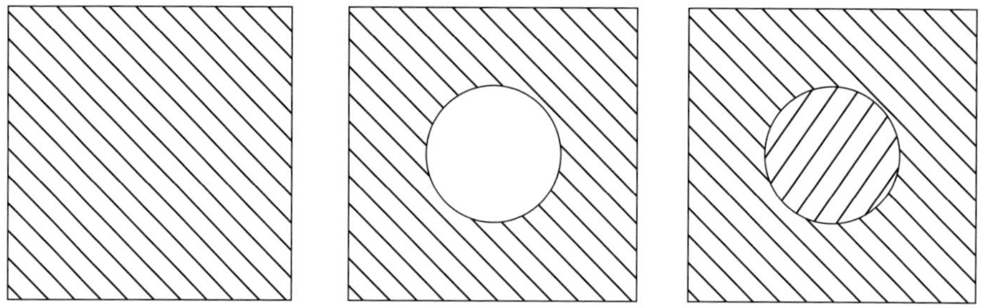

Fig. 12.1 Scheme of the thought experiment (Biroli and Bouchaud, 2004*b*) underlying the definition of the point-to-set correlation length.

away. This procedure thus allows to define a correlation function between a point (the center of the system) and a set of points (the boundary of $\bar{B}$), hence the name already mentioned. It is understood that in the correlation function the similarity measure should be averaged with respect to the configurations $\underline{\sigma}$ and $\underline{\sigma}'$. From this function one can further define a correlation length. Taking for simplicity $\bar{B}$ to be a spherical ball of radius ℓ, we shall indeed define the correlation length $\ell_{\rm c}$ as the minimal radius that brings the point-to-set correlation function (i.e. the average measure of similarity of the center of $\underline{\sigma}$ and $\underline{\sigma}'$) below a small threshold fixed beforehand.

The equilibrium correlation time $\tau_{\rm c}$ of the system can be defined in a similar fashion, as the minimal time necessary for the autocorrelation function (average similarity measure at the same point, between one equilibrium configuration and the outcome of its evolution during a certain amount of time) to drop below a given threshold. It turns out that the intuition discussed in the introduction, namely that large correlation times and large correlation lengths are two intertwined phenomena, can be given a precise content with these two definitions of $\tau_{\rm c}$ and $\ell_{\rm c}$. Indeed, the rigorous proof of (Montanari and Semerjian, 2006*a*) we shall sketch below implies that $\ell_{\rm c} \leq \tau_{\rm c} \leq \exp\left\{\ell_{\rm c}^d\right\}$, where we have hidden for simplicity several constants. The interpretation of these two inequalities might sound disappointingly simple. As $\ell_{\rm c}$ measures the radius of a correlated region of the system, and as for the center of the system to decorrelate it must receive some information from the boundary of the correlated region, the lower bound merely states that this information cannot propagate faster than ballistically. On the other hand, the upper bound follows from the fact that the dynamics of the center of the system is weakly sensitive to the outside of the correlated zone, hence it should closely resemble the dynamics of the ball of radius $\ell_{\rm c}$ without its surrounding environment. The latter case being the dynamics of a finite system of volume $\ell_{\rm c}^d$ (where d denotes the dimension of the system), its relaxation cannot be slower than exponential in its volume.

The fact that these two inequalities have natural interpretations does not mean they have a trivial content, but rather demonstrates the relevance of the point-to-set definition of the correlation length $\ell_{\rm c}$. Recall indeed that one of the puzzles of the glass phenomenology is the drastic growth of the relaxation time without significant traces in the structure factor, hence in the 2-point correlation length. The bounds between τ_c and ℓ_c show that despite this fact the growth of the relaxation time must be accompanied by a growth of a (static) correlation length if the latter is appropriately defined.

12.2.2 More precise definition of the correlation function

We want now to provide the reader with more formal definitions of the quantities discussed informally above, before restating with more details the bounds between the correlation length and the correlation time. Note that if point-to-set correlations are relatively new in physics they are quite common in the mathematical literature, in particular in the context of the so-called tree reconstruction problem, see for instance (Martinelli *et al.*, 2004; Mossel, 2004).

For the sake of concreteness we shall consider a model of N Ising spins $\sigma_i = \pm 1$, whose global configuration will be denoted $\underline{\sigma} = (\sigma_1, \ldots, \sigma_N)$. For a subset S of the variable indexes $\{1, \ldots, N\}$ we will call $\underline{\sigma}_S$ the configuration of the variables in S. The energy function (Hamiltonian) is decomposed as $E(\underline{\sigma}) = \sum_{a=1}^{M} E_a(\underline{\sigma}_{\partial a})$, that is a sum of M terms E_a, with the ath term involving a subset denoted ∂a of the variables. For instance, in a two-dimensional square lattice model with nearest-neighbor interactions there would be an interaction a for each edge of the lattice. The definition of E encompasses, however, more general cases, in particular multispin interactions involving more than a pair of spins. It can be convenient in such a case to represent the network of interactions as a so-called factor graph (Kschischang *et al.*, 2001), see Fig. 12.2 for an illustration. Each variable $\sigma_1, \ldots, \sigma_N$ is associated to a circle vertex, while the interactions $E_1, \ldots, E_M$ are symbolized with square vertices. An edge is drawn between a variable i and an interaction a if and only if E_a depends on σ_i. On the right panel of Fig. 12.2 is drawn a portion of the factor graph corresponding to a square lattice model.

We will use in the following the notation $d(i, j)$ for the distance between two variables. This will be taken as the graphical distance on the factor graph, that is the number of interactions that have to be crossed along a shortest path linking i and j. This notion of distance has the virtue of being well defined for any topology of the interaction network, not only for finite-dimensional models. In the latter case the graph distance is equivalent to the Euclidean one.

The Gibbs–Boltzmann probability measure is defined on the space of configurations by $\mu(\underline{\sigma}) = \exp[-\beta E(\underline{\sigma})]/Z$, with the partition function Z ensuring its normalization. Angular brackets $\langle \cdot \rangle$ will be used to denote averages with respect to this law. In order to quantify the strength of the correlations induced by the Gibbs–Boltzmann probability between one variable, say σ_i, and a set B of other variables, one can consider a function $F(\underline{\sigma}_B)$ that depends only on the values of the variables in B, and compute the correlation between the two as $\langle \sigma_i F(\underline{\sigma}_B) \rangle - \langle \sigma_i \rangle \langle F(\underline{\sigma}_B) \rangle$. If for instance σ_i were completely independent from the status of the variables in B this quantity would vanish. To make contact with the thought experiment explained above one can choose as a particular function F the magnetization of the ith spin *conditioned* on $\underline{\sigma}_B$, denoted $\langle \sigma_i' \rangle_{\underline{\sigma}_B}$. This conditional average is defined for an arbitrary function f as

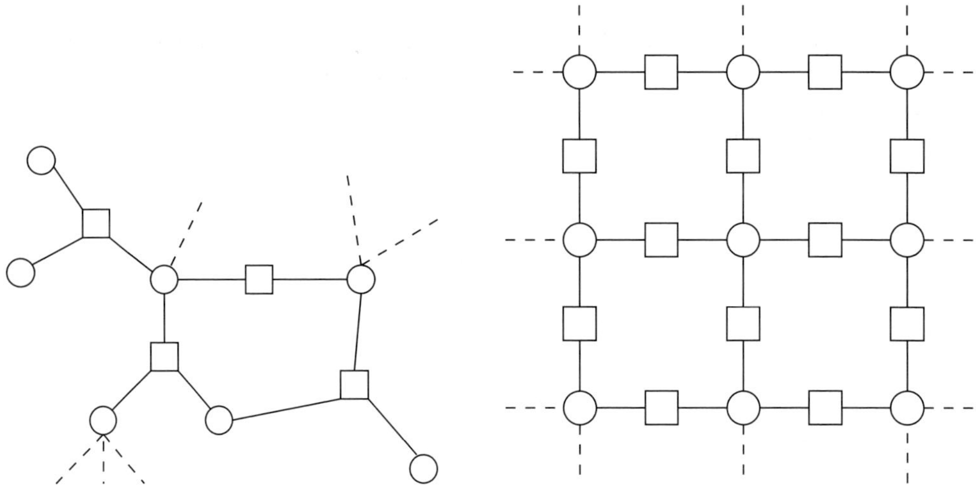

Fig. 12.2 Left: general example of a factor graph representing an energy $E(\underline{\sigma}) = \sum_{a=1}^{M} E_a(\underline{\sigma}_{\partial a})$. Each circle represents a variable σ_i, each square an interaction term E_a. An edge is drawn between E_a and σ_i whenever the ath interaction depends on the ith variable. Right: the case of a square lattice with nearest-neighbor interactions.

$$\langle f(\underline{\sigma}')\rangle_{\underline{\sigma}_B} = \sum_{\underline{\sigma}'} f(\underline{\sigma}')\mu(\underline{\sigma}'|\underline{\sigma}_B), \tag{12.1}$$

where the conditional probability is restricted to configurations that coincide with $\underline{\sigma}_B$ on the variables in B,

$$\mu(\underline{\sigma}'|\underline{\sigma}_B) = \begin{cases} \frac{1}{Z(\underline{\sigma}_B)} e^{-\beta E(\underline{\sigma}')} & \text{if } \underline{\sigma}'_B = \underline{\sigma}_B \\ 0 & \text{otherwise} \end{cases}, \tag{12.2}$$

$Z(\underline{\sigma}_B)$ ensuring its normalization. With this choice for F we have thus obtained the correlation function between a point i and a set B as

$$G(i,B) = \langle \sigma_i \langle \sigma'_i \rangle_{\underline{\sigma}_B} \rangle - \langle \sigma_i \rangle^2 = \sum_{\underline{\sigma}} \mu(\underline{\sigma}) \sum_{\underline{\sigma}'} \mu(\underline{\sigma}'|\underline{\sigma}_B)\sigma_i\sigma'_i - \langle \sigma_i \rangle^2, \tag{12.3}$$

where the last expression enlightens the connection with the heuristic discussion above: the configuration $\underline{\sigma}'$ corresponds to the second equilibrium configuration drawn conditioned on the value in B of the first configuration $\underline{\sigma}$.

We would like to emphasize that the configuration $\underline{\sigma}'$ constructed in the second place is as representative as $\underline{\sigma}$ of the Gibbs–Boltzmann equilibrium measure, in other words for any function f of the configuration one has (Zarinelli and Franz, 2010)

$$\langle \langle f(\underline{\sigma}')\rangle_{\underline{\sigma}_B} \rangle = \langle f(\underline{\sigma})\rangle. \tag{12.4}$$

This property follows simply from the properties of conditional probability and its proof will be omitted. In the following sections we will consider correlations between replicas of the system with the same constraint (i.e. $\underline{\sigma}'$ and $\underline{\sigma}''$ are generated independently from $\mu(\cdot|\underline{\sigma}_B)$ for fixed $\underline{\sigma}_B$ and quenched disorder in the energy function E). Given any two functions $f(\underline{\sigma})$ and $g(\underline{\sigma})$ one can define the correlation

$$C(f,g) = \langle \langle f(\underline{\sigma}')\rangle_{\underline{\sigma}_B} \langle g(\underline{\sigma}'')\rangle_{\underline{\sigma}_B}\rangle. \tag{12.5}$$

Similarly to Eq. (12.4) one can show that this coincides with the correlation between $\underline{\sigma}'$ and $\underline{\sigma}$ according to

$$C(f,g) = \langle f(\underline{\sigma}) \langle g(\underline{\sigma}')\rangle_{\underline{\sigma}_B}\rangle. \tag{12.6}$$

This identity will play an important role in the analysis of the point-to-set function in the Kac limit discussed in Section 12.4, and we shall call it the conditional equilibrium condition.

We come back now to the definition of the point-to-set correlation function, and consider more specifically the case where B is the outside of a ball of radius ℓ around i, i.e. the set of variables at distance larger than or equal to ℓ from variable i, $B(i,\ell) = \{j|d(i,j) \geq \ell\}$. We shall call $G(i,\ell) = G(i,B(i,\ell))$ the correlation function for this geometry. It is intuitively clear that $G(i,\ell)$ decreases when the radius ℓ of the ball increases, as farther away sites are less correlated with i. One can thus set a small threshold ε and define the correlation length for site i as the minimal distance ℓ necessary to make the correlation $G(i,\ell)$ drops below the threshold ε. In formula, $\ell_i(\varepsilon) = \min\{\ell|G(i,\ell) \leq \varepsilon\}$.

We turn finally to the definition of the correlation time. We shall consider a single spin-flip dynamics in continuous time, defined through transition rates $W(\underline{\sigma} \to \underline{\sigma}')$. The single spin-flip assumption means that these rates vanish whenever $\underline{\sigma}$ and $\underline{\sigma}'$ differ in more than one variable. We assume the rates to verify the detailed balance condition

$$\mu(\underline{\sigma})W(\underline{\sigma} \to \underline{\sigma}') = \mu(\underline{\sigma}')W(\underline{\sigma}' \to \underline{\sigma}), \tag{12.7}$$

which ensures that the Gibbs–Boltzmann probability is stationary under this dynamics. Moreover, the rate from a configuration $\underline{\sigma}$ to the configuration where a single variable i has been flipped is assumed to depend on $\underline{\sigma}$ only through the configuration of σ_i and of the variables at unit distance from i. This last condition is obviously fulfilled by the usual Monte Carlo dynamics like the Metropolis or heat-bath (Glauber) rules. The equilibrium dynamics of the model is defined by these rates and the initial condition $\underline{\sigma}(t=0)$ that is drawn from the equilibrium law μ. The average over the initial condition and the subsequent evolution shall be denoted again by angular brackets $\langle \ldots \rangle$. The autocorrelation function of variable i is hence defined as $C_i(t) = \langle \sigma_i(0)\sigma_i(t)\rangle - \langle \sigma_i \rangle^2$, and we can assign a correlation time τ_i to this variable as $\tau_i(\varepsilon) = \min\{t|C_i(t) \leq \varepsilon\}$, i.e. as the minimal time for the autocorrelation function to drop below a given threshold ε. To avoid confusion in the following let us emphasize that in the definition of the correlation time we consider the dynamics of the *whole* system; the constraints on the finite ball B only appear in the definition of the correlation length.

12.2.3 Relation between the correlation length and the equilibrium correlation time

Now that the definitions of the correlation time τ_i and correlation length ℓ_i have been given more precisely we can restate in a more accurate way the bounds derived rigorously in (Montanari and Semerjian, 2006*a*):

$$C_1\, \ell_i(\varepsilon') \leq \tau_i(\varepsilon) \leq 1 + \exp\left\{C_2|\bar{B}(i, \ell_i(\varepsilon''))|\right\}, \tag{12.8}$$

where $|\bar{B}|$ denotes the number of sites in the ball $\bar{B}$. In a finite-dimensional setting one has $|\bar{B}(\ell)| \sim \ell^d$, but Eq. (12.8) is valid for any topology of the interaction graph. The small thresholds ε' and ε'' are functions of ε that go to zero when ε vanishes, and $C_{1,2}$ are numerical constants that depend on the microscopic details of the Hamiltonian and of the dynamics.

Let us make a series of remarks on this result:

- For simplicity we have explained this relationship between length- and time scales for a discrete system of Ising spins, with arbitrary interactions. Its extension to more general discrete degrees of freedom is simple and was the case considered in (Montanari and Semerjian, 2006*a*). It is natural to expect that a similar result will hold for particle systems evolving in the continuum, for instance by means of a coarse-grained occupation number field, see e.g. (Cavagna *et al.*, 2007).
- The relationship between ℓ_i and τ_i holds site by site. This is particularly important for inhomogeneous systems with quenched disorder, where the correlation lengths and times can vary wildly from site to site.
- In the limit of zero temperature the constant C_2 will diverge: the presence of trivial "energetic barriers" can lead to very large correlation times without a growing correlation length.
- The lower and upper bounds in Eq. (12.8) are widely separated when ℓ grows, which could suggest that the bounds are very far from optimal. However, it can be argued that with the very weak hypotheses made for its derivation, this result can only be marginally enhanced (i.e. at the level of the numerical constants and with a dynamical exponent $z \geq 2$ in the lower bound). Indeed, both bounds can be saturated in the low/high-temperature regimes of some models that enter in the range of validity of the result (see below for a discussion of diluted mean-field models).
- We would like to emphasize the *static* character of the point-to-set definition of the correlation length. Indeed, the expression of the correlation function stated in Eq. (12.3) involves only equilibrium averages, and does not make any reference to the dynamical evolution of the system. Of course, if one wants in practice to evaluate this function for a system or a model that does not admit an analytical solution, Monte Carlo simulations will probably have to be used to generate

thermalized configurations of the full and constrained systems, yet this dynamics is here only a computational tool and not a part of the definition.

- The dependence of the correlation length and time on the arbitrary threshold ε may look rather unsatisfactory at first sight. Fortunately, this is not a real issue for glassy systems: one expects for them a discontinuous behavior upon approaching the glass transition. The spatial and temporal correlation function decay to zero in a two-step fashion, with the appearance of a growing plateau. As long as ε is smaller than the height of the plateau (called Edwards–Anderson or non-ergodicity parameter) the asymptotic behavior of $\tau(\varepsilon)$ and $\ell(\varepsilon)$ is essentially independent of the choice of this threshold. This is, however, a concern for more conventional critical phenomena where the order parameter grows continuously at the transition.
- It should be acknowledged that these bounds do not apply directly for kinetically constrained models (KCM) (Fredrickson and Andersen, 1984; Kob and Andersen, 1993). As a matter of fact the equilibrium measure of these models is a trivial product measure factorized over the sites, hence the correlation length as defined above is always equal to 1. However, the dynamical rules defining the KCM violates a technical assumption of permissivity that is necessary for the derivation of these bounds.

12.2.4 Determination of the correlation length in finite-dimensional systems

In the following subsection we shall give some explanations as to how these bounds can be proven, then the focus of the rest of the chapter will be put on two families of models where analytical computations can be pushed further. Before that we want, however, to mention some papers, mostly numerical, where finite-dimensional models have been investigated with a perspective somewhat related to the point-to-set correlation function.

To the best of our knowledge the first papers where glassy systems in confined geometry with boundary conditions self-consistently generated according to the prescription described in the previous subsections were in Refs. (Scheidler *et al.*, 2000, 2002, 2004). These works investigated the effect of the presence of a wall or confining geometries on the dynamics of Lennard-Jones binary mixtures.

More recent literature has focused on the thought experiment of (Biroli and Bouchaud, 2004*b*). In (Jack and Garrahan, 2005) a plaquette spin model of a glass was studied analytically and numerically. The effect of boundary conditions on finite-size subsystems could be analyzed and led to a determination of a static correlation length along the lines of the thought experiment of (Biroli and Bouchaud, 2004*b*).

A binary mixture of soft-sphere particles (i.e. a fragile glass-former liquid modeled microscopically with particles in the continuum) was considered in (Cavagna *et al.*, 2007; Biroli *et al.*, 2008; Cammarota *et al.*, 2009). The procedure described above of freezing an equilibrium configuration outside a cavity of a given radius and thermalizing its interior was thus implemented with Monte Carlo simulations. This led

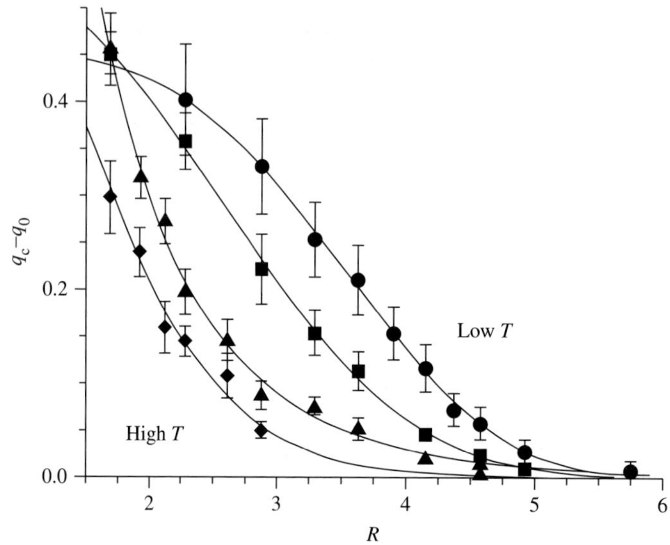

Fig. 12.3 Point-to-set correlation function for a binary mixture of soft-sphere particles, from (Biroli *et al.*, 2008). When the temperature is lowered the influence of the boundary of a spherical cavity persists for larger radius R.

to a demonstration of the growth of the static correlation length upon lowering the temperature of the liquid. In Fig. 12.3 we reproduce a result of (Biroli *et al.*, 2008), which shows the point-to-set correlation as a function of the radius of the cavity, for various temperatures. The reader is also referred to (Cavagna *et al.*, 2007; Biroli *et al.*, 2008; Cammarota *et al.*, 2009) for the interpretation of these results along the lines of RFOT and for a critical discussion of the scaling exponents of RFOT.

12.2.5 Tools for the proof

12.2.5.1 Couplings

In this subsection we shall sketch in an informal way the method of proof of Eq. (12.8), the reader being referred to (Montanari and Semerjian, 2006*a*) for the details. We want in particular to introduce a very useful probabilistic concept on which the proof of both the upper and the lower bounds of (12.8) relies, namely the construction of a coupling between stochastic processes (Lindvall, 2002). Let us first explain what is a coupling on the simplest case of two random variables, for instance two biased coins. The first coin $X^{(1)}$ takes value Head with probability p_1, Tail otherwise, while the probability of Head of the coin $X^{(2)}$ is p_2. A coupling of these two biased coins is a random variable $(X^{(1)}, X^{(2)})$ that can take four values $\{(H,H),(H,T),(T,H),(T,T)\}$, such that if one observes only the first (respectively, second) element of the couple $(X^{(1)}, X^{(2)})$, one sees Head occurring with probability p_1 (respectively, p_2). Obviously one trivial way to construct such a coupling is to take $X^{(1)}$ and $X^{(2)}$ as independent copies of the original biased coins. The power of the notion of coupling relies in the possibility to introduce a dependency between the two elements of the couple, without spoiling the partial (marginal) frequency of observations of head in the first or second position. It is indeed easy to realize that there exists an infinity of couplings of these two biased coins parametrized by a real number. Let us give one explicit example. Suppose without loss of generality that $p_1 \geq p_2$, and set the value of the coupling to be

$$(X^{(1)}, X^{(2)}) = \begin{cases} (H, H) & \text{with probability } \; p_2, \\ (H, T) & \text{with probability } \; p_1 - p_2, \\ (T, T) & \text{with probability } \; 1 - p_1. \end{cases} \tag{12.9}$$

One can easily check that this is indeed a coupling (the marginal probabilities for the *H*ead of the two coins are, respectively, p_1 and p_2), and that, for given values of p_1 and p_2, it minimizes the probability that $X^{(1)} \neq X^{(2)}$. For this reason it is termed the *greedy coupling* of the two random variables. In particular, if the two coins are identical, $p_1 = p_2$, then $X^{(1)}$ is always equal to $X^{(2)}$ in this coupling.

12.2.5.2 Lower bound

This notion of coupling extends naturally to random variables more complicated than biased coins, and also to stochastic processes, that is sequences of random variables indexed by a time parameter. The proof of the lower bound in Eq. (12.8) relies indeed on such a construction that we shall now explain.

Let us choose a variable i and a positive integer ℓ, and denote as above B (respectively, $\bar{B}$) the outside (respectively, inside) of the ball of radius ℓ around i. Consider now the stochastic process $(\underline{\sigma}^{(1)}(t), \underline{\sigma}^{(2)}(t))$, where two configurations of the same system evolves simultaneously, defined by the following rules:

- at the initial time $t = 0$, the two configurations coincide, $\underline{\sigma}^{(1)}(0) = \underline{\sigma}^{(2)}(0) = \underline{\sigma}$, with $\underline{\sigma}$ drawn from the equilibrium Gibbs–Boltzmann measure;
- at any time $t > 0$ where a variable j attempts an update of its value, both configurations are modified together according to:
  - * if $j \in B$, the configuration $\underline{\sigma}^{(2)}(t)$ is kept unchanged, while the spin $\sigma_j^{(1)}(t)$ is flipped or not according to the transition rates $W(\underline{\sigma}^{(1)}(t) \to \underline{\sigma}')$;
  - * if $j \in \bar{B}$, one determines the probability of $\sigma_j^{(1)}(t)$ (respectively, $\sigma_j^{(2)}(t)$) right after the update according to $W(\underline{\sigma}^{(1)}(t) \to \underline{\sigma}')$ (respectively, $W(\underline{\sigma}^{(2)}(t) \to \underline{\sigma}')$). Then the new values of $\left(\sigma_j^{(1)}(t), \sigma_j^{(2)}(t)\right)$ are drawn according to the greedy coupling (as defined above) of these two Ising spin random variables.

Observed separately each element of this coupling corresponds:

- for $\underline{\sigma}^{(1)}(t)$ to the original dynamics of the whole system;
- for $\underline{\sigma}^{(2)}(t)$ to the equilibrium dynamics for the inside of the ball $\bar{B}$, submitted to a time-independent boundary condition in B, with $\underline{\sigma}_B^{(2)}(t)$ fixed forever to the value $\underline{\sigma}_B$ it has at time $t = 0$.

It turns out that the original temporal and point-to-set correlation functions can be computed from appropriate averages over this coupling, that we shall still denote $\langle \dots \rangle$ with a slight abuse of notation. Assuming that the equilibrium magnetization $\langle \sigma_i \rangle$ vanishes to simplify the discussion, one realizes that $C_i(t) = \langle \sigma_i^{(1)}(0)\sigma_i^{(1)}(t) \rangle$, while $G(i,B)$ is the limit of $\langle \sigma_i^{(2)}(0)\sigma_i^{(2)}(t) \rangle$ as t goes to infinity. To obtain the lower bound of Eq. (12.8) one has to relate in some way the behavior of the spatial and temporal correlation functions. From the above observation this translates into a comparison of the value of σ_i in the two processes $\underline{\sigma}^{(1)}(t)$ and $\underline{\sigma}^{(2)}(t)$. The idea of the proof is then to exploit the properties of the above-defined coupling. In fact for "small" times, that is of order smaller than ℓ, most probably $\sigma_i^{(1)}(t) = \sigma_i^{(2)}(t)$. Indeed, at initial times the two configurations coincide everywhere in the system. Moreover, when $j \in \bar{B}$ tries to update, if the two configurations coincide on j and its immediate neighborhood, then necessarily $\sigma_j^{(1)} = \sigma_j^{(2)}$ right after the update, because of the use of the greedy coupling of the two update probabilities. In other words, disagreement between the two configurations can only initiate in the outside B of the ball of radius ℓ, and must propagate from the surface of the ball to its center (van den Berg, 1993; Hayes *et al.*, 2007).

Let us rephrase this reasoning with more explicit formulas. We start from a simple inequality on the point-to-set correlation function,

$$G(i,B) = \lim_{t\to\infty} \left\langle \sigma_i^{(2)}(0)\sigma_i^{(2)}(t) \right\rangle \leq \left\langle \sigma_i^{(2)}(0)\sigma_i^{(2)}(t) \right\rangle, \tag{12.10}$$

which holds for any value of $t \geq 0$. Indeed, the equilibrium autocorrelation functions of reversible Markov processes are decreasing in time.[1] To relate the two coupled processes let us define the indicator function $X(t) = \delta_{\sigma_i^{(1)}(t),\sigma_i^{(2)}(t)}$. We can thus upper-bound the spatial correlation function as

$$\begin{aligned} G(i,B) &\leq \left\langle \sigma_i^{(1)}(0)\sigma_i^{(1)}(t) \right\rangle - \left\langle \sigma_i^{(1)}(0)\sigma_i^{(1)}(t)(1-X(t)) \right\rangle + \left\langle \sigma_i^{(2)}(0)\sigma_i^{(2)}(t)(1-X(t)) \right\rangle \\ &\leq \left\langle \sigma_i^{(1)}(0)\sigma_i^{(1)}(t) \right\rangle + 2\langle (1-X(t)) \rangle \\ &= C_i(t) + 2\, \mathcal{P}_{\text{dis}}(t), \end{aligned} \tag{12.11}$$

where $\mathcal{P}_{\text{dis}}(t)$ is the probability that the two parts of the coupling disagree on the value of σ_i at time t. As explained above disagreement between the two configurations of the coupling has to travel from the boundary of B towards i along a a sequence of ℓ adjacent spins on a path from B to i. An upper bound on $\mathcal{P}_{\text{dis}}(t)$ can thus be obtained by multiplying the number of such paths with the probability that in the interval of time $[0,t]$ all ℓ sites of a given path attempts to update their configurations in the right order (from the boundary inwards). The first factor is obviously smaller than the maximum connectivity of a spin raised to the power ℓ. Because the times where the spin attempts an update form a Poisson process the second factor is the

[1] This is a simple property that can be proven using the spectral decomposition of the evolution operator $W(\underline{\sigma} \to \underline{\sigma}')$.

probability that a Poisson random variable of average t is greater than ℓ, and this last probability is smaller than $(e\,t/\ell)^\ell$. Putting these two factors together one obtains that $\mathcal{P}_{\rm dis}(t) \le (Ct/\ell)^\ell$, where C is a constant that depends on the connectivity of the interaction graph. Now, if one sets $t = \tau_i(\varepsilon)$, the r.h.s. of Eq. (12.11) can be made smaller than 2ε by taking ℓ larger than some constant multiplied by $\tau_i(\varepsilon)$, hence the lower bound in Eq. (12.8).

12.2.5.3 Upper bound

The upper bound of Eq. (12.8) is obtained by showing that the autocorrelation function $C_i(t)$ of the ith spin is weakly sensitive to the configuration of the system out of the ball of radius $\ell_i(\varepsilon)$. One can thus approximate $C_i(t)$ with the value it would have in the system made only of the interior $\bar{B}$ of the ball of radius $\ell_i(\varepsilon)$. It is natural that the autocorrelation time of the latter cannot grow faster than exponentially in its volume. Consider indeed an arbitrary system with n variables, and the following coupling between two copies of its dynamics. It is initialized with two arbitrary configurations $(\underline{\sigma}^{(1)}(0), \underline{\sigma}^{(2)}(0))$, and one performs at later times the updates of $\left(\sigma_j^{(1)}(t), \sigma_j^{(2)}(t)\right)$ according to the greedy coupling of the two transition rates. This implies that as soon as the two copies coincide, they remain the same for all subsequent times. Moreover, we assume the dynamics to be "permissive", that is there is a constant $\kappa > 0$ such that when j attempts an update, irrespectively of the neighborhood of j in $(\underline{\sigma}^{(1)}, \underline{\sigma}^{(2)})$, the probability that $\sigma_j^{(1)} = \sigma_j^{(2)}$ after the update is larger than κ. This condition of permissivity is satisfied at any strictly positive temperature if there are no hard constraints in the model (no infinite energy configurations), with possibly κ vanishing when $T \to 0$. We can upper-bound the relaxation time by the coalescence time of the coupling: if for two arbitrary initial configurations (as different as they can be) the coupled dynamics has coalesced at a given time t_0, hence "forgotten" its initial conditions, then the equilibrium dynamics of a single configuration would have as well. What remains to be proven is that, with a large probability, this coalescence time is not larger than exponential in the number of variables n. This follows from considering a particular sequence of update events that brings the two evolving copies of the system to coincide, namely that all variables have attempted to update at least once, and that their last update brought the variables to the same value in the two parts of the coupling. The probability of the last condition is by definition larger than κ^n, hence on time intervals larger than κ^{-n} coalescence is very probable.

12.3 Computation of the correlation function in mean-field (random graph) models

In this and the next section we shall discuss simplified models, of mean-field nature (in a sense that shall be specified), for which analytical computations of the point-to-set correlation function are possible.

12.3.1 A reminder on mean-field glassy models

Many researchers agree that important insights on the physics of glassy systems have been gained by the study of apparently remote models, namely mean-field spin-glasses with multispin interactions. The paradigmatic example is the so-called p-spin model, with $p \geq 3$, defined by the Hamiltonian

$$E(\underline{\sigma}) = - \sum_{1 \leq i_1 < \cdots < i_p \leq N} J_{i_1 \dots i_p} \sigma_{i_1} \dots \sigma_{i_p} \; , \tag{12.12}$$

where the $J_{i_1 \dots i_p}$ are independent Gaussian random variables of zero mean and variance $\mathbb{E}\left[J^2_{i_1 \dots i_p}\right] = \frac{p!}{2N^{p-1}}$. In the Ising version of the model the degrees of freedom are $\sigma_i = \pm 1$, while the σ_i are reals in the spherical case, with the additional constraint $\sum_i \sigma_i^2 = N$. The properties of this model have been extensively studied, we refer the reader to the lecture notes (Parisi, 2002; Cugliandolo, 2002; Castellani and Cavagna, 2005) and references therein for the references to the original works. We shall content ourselves here with a very brief reminder of its salient features. From a thermodynamical point of view, the free-energy of this model exhibits a non-analytic behavior at the critical temperature $T_{\rm c}$ (also called Kauzmann temperature). However, the dynamical properties of the model changes at the dynamic transition temperature $T_{\rm d} > T_{\rm c}$: the Gibbs measure gets split onto an exponential number of pure states, the equilibration time of the system diverges in the thermodynamic limit for $T < T_{\rm d}$, hence the dynamics exhibit aging (with a non-trivial effective temperature) if the system is initialized in a random configuration. The absence of any thermodynamic singularity at $T_{\rm d}$ is due to a compensation phenomenon between the exponential degeneracy of the pure states (complexity, or configurational entropy) and the internal free energy of the pure states, such that the total free energy of the system in the intermediate phase between $T_{\rm d}$ and $T_{\rm c}$ coincides with the analytical continuation of the high-temperature (liquid) phase. In technical words one has one step of replica symmetry-breaking (1RSB) behavior with Parisi breaking parameter $m = 1$. On the contrary, at $T_{\rm c}$ the complexity vanishes, and for $T < T_{\rm c}$ the system enters a true 1RSB phase with $m < 1$.

Of course the main criticism that can be made against the relevance of such a model for the study of structural glasses is its completely different microscopic nature, in particular because of the quenched disorder present in its definition. The two variations of the p-spin model presented in this and in the next section shall not escape this criticism, yet they partially cure some of the pathologies of the model and present an even richer phenomenology, which motivates the theoretical interest for them.

12.3.2 Definition of diluted mean-field models

The model defined by the Hamiltonian (12.12) together with the above-stated variance of the couplings constants $J_{i_1 \dots i_p}$ will be termed the fully connected p-spin model. Indeed all of the $\binom{N}{p}$ possible interactions between p-uplets of spins are present in the system, even if each of them is individually weak in order to have an extensive scaling of the Hamiltonian in the thermodynamic limit. In this case each spin interacts with all the others, in other words the distance $d(i, j)$ between any pair of spins defined in Section 12.2.2 is equal to one. This peculiarity, very different from the finite-dimensional models with short-range interactions where any degree of freedom interacts only with a finite number of neighbors, can be cured with the following definition of the coupling constants:

$$J_{i_1 \dots i_p} = \begin{cases} \pm 1 & \text{with probability } \alpha \frac{p!}{2N^p} \\ 0 & \text{with probability } 1 - \alpha \frac{p!}{N^p} \end{cases} . \tag{12.13}$$

This case will be called the diluted p-spin model. At variance with the fully connected case, only an extensive number (concentrated around its average αN) of interactions are now present, but each of them is strong (of order 1). Moreover, a short computation reveals that the degree of one spin, that is the number of interactions it belongs to, converges in the thermodynamic limit to a Poisson random variable of average αp, hence finite with respect to the system size N.

This kind of diluted model, pioneered in the case $p = 2$ by Viana and Bray (Viana and Bray, 1985), can be viewed as intermediary between fully connected and finite-dimensional ones. They share with the latter the finiteness of the connectivity of each degree of freedom, but are mean-field as the former ones. Indeed, there is no *a priori* underlying Euclidean space in their definition. This allows for detailed analytical computation of their properties, and provides a family of models where the Bethe–Peierls approximation is actually exact in the thermodynamic limit. Moreover, there is one more control parameter besides the temperature, namely the "density" parameter α that controls the number of interactions present in the system.

There has been an important research effort in the last decades in the community of statistical mechanics of disordered systems to develop analytical tools for tackling such diluted spin-glass models. An important motivation has been the intimate connection between these models and random combinatorial optimization problems, such as the k-satisfiability and the coloring of random graphs. An extensive account of this line of research can be found in the recent book (Mézard and Montanari, 2009).

Before turning to some explanations on the computations in diluted models, in particular the determination of the point-to-set correlation function, let us plot in Fig. 12.4 the shape of the phase diagram in the (density α, temperature T) plane. The two transition temperatures $T_{\rm d}$ and $T_{\rm c}$ of the fully connected model (which corresponds to the $\alpha \to \infty$ limit up to a rescaling of the energy) become two lines of transitions, which ends up at zero temperature for finite values of the connectivity α, respectively $\alpha_{\rm d}$ and $\alpha_{\rm c}$. The zero-temperature properties of this model have been largely studied in the context of optimization problems (where it is known as XORSAT), the thresholds $\alpha_{\rm d}$ and $\alpha_{\rm c}$ corresponding there to the clustering and satisfiability transitions (Mézard *et al.*, 2003; Cocco *et al.*, 2003).

12.3.3 Computation of the correlation function

We shall now explain the main steps in the computation of the point-to-set correlation function in mean-field diluted systems. Let us first discuss the properties of the (random) factor graph associated with the diluted p-spin model defined by Eqs. (12.12) and (12.13). To make contact with the general definitions we rewrite it as

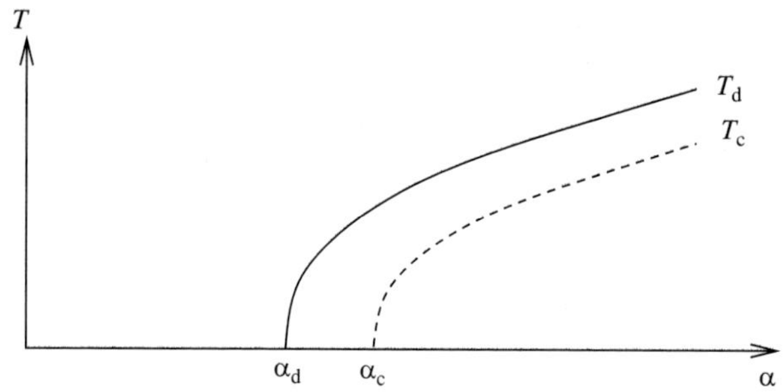

Fig. 12.4 Shape of the phase diagram for the diluted p-spin model.

$$E(\underline{\sigma}) = \sum_a E_a(\underline{\sigma}_{\partial a}), \qquad E_a(\underline{\sigma}_{\partial a}) = -J_a \prod_{i\in\partial a} \sigma_i, \tag{12.14}$$

where we have only retained the p-uplets with a non-vanishing coupling constant and renumbered them in terms of the interaction index a. Each interaction a involves p variables, while each variable appears in a random number of interactions, which is easily found to be a Poisson random variable of average αp. What is slightly less obvious is the fact that these random factor graphs have a local tree structure, that is short loops are relatively rare, and only loops of length of order $\log N$ begin to proliferate in the thermodynamic limit ($N \to \infty$ with α finite). It is this locally tree-like property that allows for analytical computations in these models; note that this property is shared by other definitions of the random factor graph model, in particular with degree distributions other than Poissonian, for instance deterministic.

Let us forget for a moment the existence of (long) loops in the factor graph and see how the problem can be solved easily on a finite tree. The key point is the natural recursive structure of a tree: removing a vertex from a tree leads to a set of smaller subtrees, which, by definition, are disconnected one from the others. So, if one can solve the considered problem on trees of some given size, and glue together the solutions to compute the solution on the concatenated tree, then by recurrence the solution can be found for any tree. Consider for instance the computation of the local magnetization $\langle\sigma_i\rangle$ of a spin in a tree factor graph. According to the recursive strategy just explained, this can be expressed in terms of the magnetization in the subtrees $F_{a\to i}$ where only one interaction a around i is retained, in formula one defines the effective magnetic field H_i acting on i,

$$\langle\sigma_i\rangle = \tanh(\beta H_i), \tag{12.15}$$

and deduces the latter from the effective fields in the subtrees $F_{a\to i}$,

$$H_i = \sum_{a\in\partial i} u_{a\to i}. \tag{12.16}$$

This relation is represented schematically on the left panel of Fig. 12.5. These in turn can be computed from the effective fields in the subtrees of the form $F_{j\to a}$, in which all interactions around the variable j except a are retained,

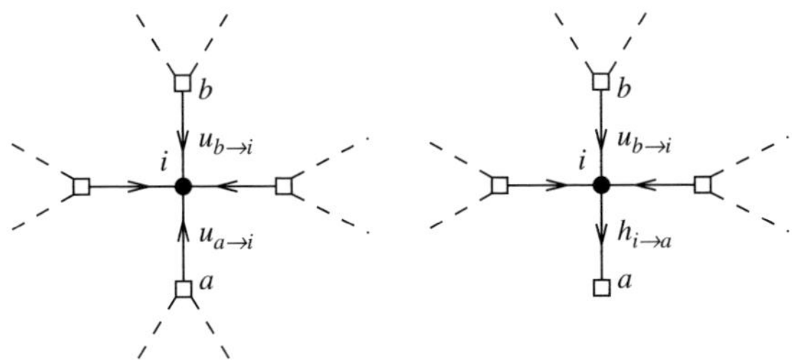

Fig. 12.5 Pictorial representation of the recursive computation on a tree. Left: the effective magnetic field H_i is obtained from the effective fields in the subtrees $F_{a\to i}$, see Eq. (12.16). Right: the construction of the field for $F_{i\to a}$ from the $F_{b\to i}$, as shown in the second part of Eq. (12.17).

$$u_{a\to i} = f_{a\to i}(\{h_{j\to a}\}_{j\in\partial a\setminus i}), \qquad h_{i\to a} = \sum_{b\in\partial i\setminus a} u_{b\to i}, \tag{12.17}$$

see the right panel of Fig. 12.5 for an illustration. The function $f_{a\to i}$ depends on the precise form of the interaction energy E_a; in the case of the p-spin interaction defined in Eq. (12.14) one finds

$$u_{a\to i} = \frac{1}{\beta}\text{arctanh}\left[\tanh(\beta J_a)\prod_{j\in\partial a\setminus i}\tanh(\beta h_{j\to a})\right]; \tag{12.18}$$

the present discussion applies to any kind of interaction E_a, provided one takes the function $f_{a\to i}$ corresponding to the model under study.

To summarize, the equilibrium properties of a tree factor graph model can be obtained by solving the recursion equations (12.17) for the effective fields $\{h_{i\to a}, u_{a\to i}\}$ living on the directed edges of the factor graph. These equations admit a unique solution that can be easily obtained by starting from the leaves of the tree (the variables with a unique neighboring interaction), where the boundary condition is $h_{i\to a} = 0$.

We can now proceed with the computations of the conditional averages $\langle\ldots\rangle_{\underline{\sigma}_B}$, where the configurations of the variables in a subset B of the variables is fixed to $\underline{\sigma}_B$. A moment of thought reveals that this is a simple modification of the previous computation. In fact, one has to look for a set of effective fields $\{h^{\underline{\sigma}_B}_{i\to a}, u^{\underline{\sigma}_B}_{a\to i}\}$, which depends on the imposed configuration $\underline{\sigma}_B$. They obey the same recursive equations (12.17), but with an additional boundary condition: $h^{\underline{\sigma}_B}_{i\to a} = \sigma_i \times \infty$ if $i \in B$. Indeed, a fixed variable feels an infinite effective field (positive or negative according to the value it is fixed to). Once the fields $\{h^{\underline{\sigma}_B}_{i\to a}, u^{\underline{\sigma}_B}_{a\to i}\}$ have been determined on all edges of the factor graph one obtains in a simple way the conditional magnetizations, for instance

$$\langle\sigma_i'\rangle_{\underline{\sigma}_B} = \tanh\left(\beta H_i^{\underline{\sigma}_B}\right), \qquad H_i^{\underline{\sigma}_B} = \sum_{a\in\partial i} u^{\underline{\sigma}_B}_{a\to i}. \tag{12.19}$$

To complete the computation of the correlation function defined in Eq. (12.3), one has to average the correlation $\sigma_i \langle \sigma'_i \rangle_{\underline{\sigma}_B}$ with respect to σ_i and $\underline{\sigma}_B$, which are both parts of a reference equilibrium configuration $\underline{\sigma}$. The generation of an equilibrium configuration on a tree factor graph is most naturally done in a recursive, broadcasting like, way. This means that one has first to choose the value of a reference variable, say i, according to its marginal probability law. The latter is known on a tree, since the probability of σ_i to be $+1$ in an equilibrium configuration is by definition $e^{\beta H_i}/(2\cosh(\beta H_i)) = (1 + \langle \sigma_i \rangle)/2$. Once the value of σ_i is fixed in this way, the problem of the generation of the other variables decouples between the various branches emanating from i, thanks again to the tree structure of the factor graph. One can treat independently each interaction a around i, and generate the value of the other variables appearing in a, $\underline{\sigma}_{\partial a \setminus i}$, conditional on the value of σ_i. The correct probability law for this generation reads

$$\mu(\underline{\sigma}_{\partial a\setminus i}|\sigma_i) = \frac{1}{z} \exp\left[-\beta E_a(\sigma_i, \underline{\sigma}_{\partial a \setminus i}) + \beta \sum_{j\in\partial a\setminus i} h_{j\to a}\sigma_j \right], \tag{12.20}$$

where z stands for a normalization constant. Once all the variables at distance 1 from i have been fixed, one can proceed in the same way with variables at distance 2, and so on and so forth until the complete configuration $\underline{\sigma}$ has been generated.

We can now wrap together the recursive computation of the conditional magnetizations and the recursive broadcast generation of the reference equilibrium configuration. We shall indeed define $P^{\sigma_i}_{i\to a}(h'_{i\to a})$ as the (density of) probability that $h^{\underline{\sigma}_B}_{i\to a}$ is equal to $h'_{i\to a}$ when $\underline{\sigma}_B$ is part of an equilibrium configuration $\underline{\sigma}$ drawn conditional on the value of σ_i. A similar definition holds for $Q^{\sigma_i}_{a\to i}(u'_{a\to i})$; one can derive recursion relations between these probability distributions,

$$P^{\sigma_i}_{i\to a}(h'_{i\to a}) = \int \prod_{b\in\partial i\setminus a} \mathrm{d}Q^{\sigma_i}_{b\to i}(u'_{b\to i})\ \delta\left(h'_{i\to a} - \sum_{b\in\partial i\setminus a} u'_{b\to i} \right), \tag{12.21}$$

$$Q^{\sigma_i}_{a\to i}(u'_{a\to i}) = \sum_{\underline{\sigma}_{\partial a\setminus i}} \mu(\underline{\sigma}_{\partial a\setminus i}|\sigma_i) \int \prod_{j\in\partial a\setminus i} \mathrm{d}P^{\sigma_j}_{j\to a}(h'_{j\to a}) \\ \delta\left(u'_{a\to i} - f_{a\to i}\left(\{h'_{j\to a}\}_{j\in\partial a\setminus i} \right) \right), \tag{12.22}$$

complemented with the boundary condition $P^{\sigma_i}_{i\to a}(h'_{i\to a}) = \delta(h'_{i\to a} - \sigma_i\infty)$ when $i \in B$. From the solution of these recursion equations one finally obtains

$$G(i,B) = \sum_{\sigma_i} \sigma_i \frac{e^{\beta\sigma_i H_i}}{2\cosh(\beta H_i)} \int \prod_{a\in\partial i} \mathrm{d}Q^{\sigma_i}_{a\to i}(u'_{a\to i}) \tanh\left(\beta \sum_{a\in\partial i} u'_{a\to i} \right) \\ - \tanh^2(\beta H_i). \tag{12.23}$$

The computation of the point-to-set correlation function we just presented is exact for any model whose factor graph is a finite tree. This is not precisely the case of problems on random graphs, such as for instance the diluted p-spin model defined by Eqs. (12.12) and (12.13). In such models the factor graph is only locally tree-like, loops do exist but most of them have a length that diverge in the thermodynamic limit. In consequence, the properties of random graph models can be inferred from such computations on finite trees, provided one takes into account the presence of the loops as self-consistent boundary conditions. This kind of reasoning has been extensively developed in recent years under the name of the cavity method (Mézard and Parisi, 2001; Mézard and Montanari, 2009), which is equivalent to the replica method (Monasson, 1998) yet more practical for diluted systems. Following this methodology one can perform the average of the point-to-set correlation function (12.23) with respect to the quenched disorder of the model. Without entering the details, this leads to functional recursion equations of the form $P_{\ell+1} = F[P_\ell]$, where P is a probability distribution over real numbers (the h' in Eq. (12.21)), and the recursion is over the radius ℓ of the ball on which the correlation is computed. Numerical algorithms to deal with such equations between probability distributions have been extensively used to solve the equations of the cavity method. The idea, known in this context as population dynamics (Mézard and Parisi, 2001), is to approximate a probability distribution by a sample of elements, with an empirical distribution as close as possible to the true distribution.

We show in the left panel of Fig. 12.6 the results of such a computation for the average point-to-set correlation, as a function of the radius of the ball, in the case of the $p = 3$ diluted model. The density parameter $\alpha = 1$ is larger than the zero-temperature limit $\alpha_{\rm d} \approx 0.818$ of the dynamic transition line (recall the sketch of Fig. 12.4). Indeed one clearly sees a plateau developing a diverging length upon decreasing the temperature towards $T_{\rm d}(\alpha) > 0$. As discussed above the presence of this plateau makes the correlation length $\ell(\varepsilon)$ asymptotically independent of the precise choice of ε, as long as it is smaller than the plateau value. This length is plotted in the right panel of Fig. 12.6 as a function of temperature. As the dashed line shows the divergence of the length occurs with a critical exponent 1/2, i.e. $\ell(\varepsilon) \sim (T - T_{\rm d}(\alpha))^{-1/2}$. One can expect this exponent to be generic for almost all random graph models with such a discontinuous transition. It is indeed a natural consequence of the recursive structure of the equation underlying the computation, namely $P_{\ell+1} = F[P_\ell]$. In the liquid phase the only fixed point of the equation $P = F[P]$, reached when $\ell \to \infty$, corresponds to a vanishing correlation. When the parameters (α, T) are taken to the glassy phase a bifurcation in this fixed point equation occurs, a non-trivial solution with a strictly positive correlation appears discontinuously. To simplify the discussion let us consider a toy model where the recursion is upon a real x instead of a function P, $x_{\ell+1} = f(x_\ell)$ (this corresponds actually to the zero temperature limit of the diluted p-spin model, see (Montanari and Semerjian, 2006*b*) for more details). The behavior of the function f upon approaching such a discontinuous bifurcation is shown in Fig. 12.7. It is then a simple exercise to expand the function f when its parameter is slightly in the liquid phase, and its argument x is around the value of the close-by bifurcation, and to show in such a way that the length of the plateau in x_ℓ does indeed diverge with an exponent

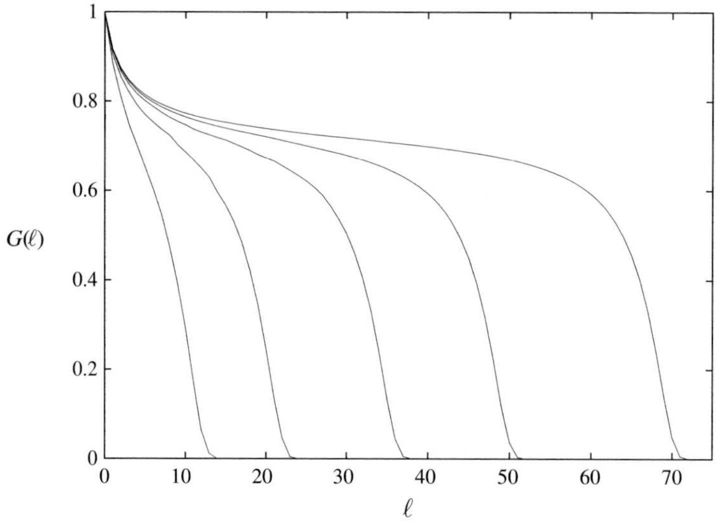

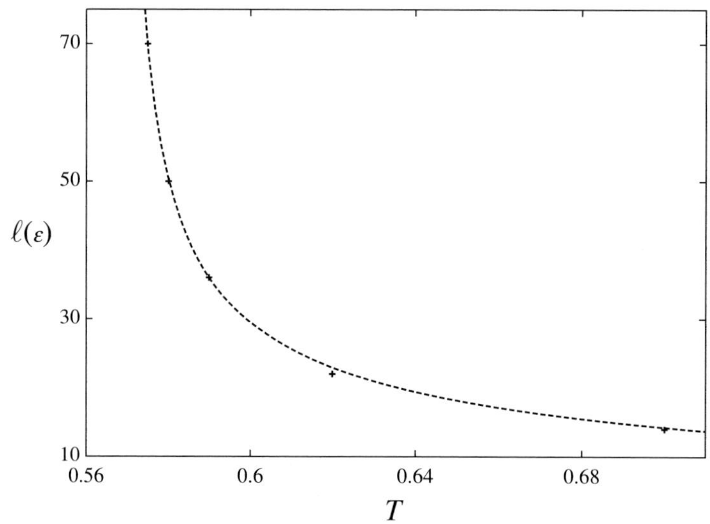

Fig. 12.6 Left: Point-to-set correlation function for the $p = 3$ diluted mean-field model with $\alpha = 1$, from left to right, $T = 0.7$, 0.62, 0.59, 0.58 and 0.575. Right: the point-to-set correlation length with $\varepsilon = 0.1$, the dashed line is a fit to this data according to the asymptotic expression $\ell_*(\varepsilon) = C\,(T - T_{\rm d}(\alpha))^{-1/2}$, with $T_{\rm d}(\alpha = 1) = 0.51695$.

1/2. The functional nature of the recursion in the general computation should not spoil this argument, one can thus expect to find generically such an exponent for the divergence of the point-to-set correlation length in mean-field diluted models.

An important point should be emphasized here: for any kind of transition studied on a Bethe lattice where a correlation length (shortest path on the tree) diverges as

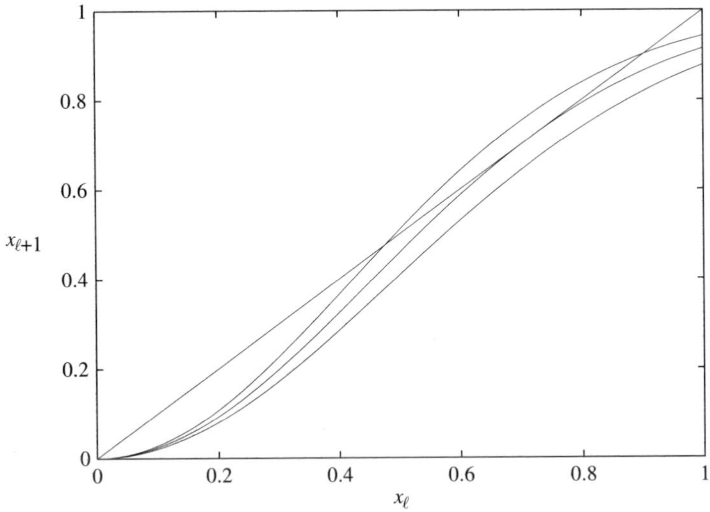

Fig. 12.7 A representation of the bifurcation for the equation $x_{\ell+1} = f(x_\ell)$. From bottom to top the parameters in f are in the liquid phase, precisely at the transition, in the glass phase.

$\epsilon^{-\nu_{\rm tree}}$ ($\epsilon \to 0$ controlling the approach to the critical point), the mean-field exponent of the transition should be deduced as $\nu_{\rm mf} = \frac{\nu_{\rm tree}}{2}$ and not $\nu_{\rm tree}$. Indeed, the finite-connectivity Bethe lattice should be pictured as embedded in a hypercubic lattice of very large dimension d. In this limit, a path of length ℓ along the tree corresponds to a random walk (almost independently of the other branches of the tree) on the embedding lattice, with an end-to-end separation of order $\sqrt{\ell}$ in Euclidean distance.[2] The mean-field value of ν for the transition discussed here is hence 1/4, we shall come back on this point in Section 12.4.1.

12.3.4 Numerical simulations of diluted mean-field models

The mean-field diluted models provide an ideal playground to assess the relevance of the bounds (12.8) between correlation lengths and correlation times: we have seen above how to compute the correlation length in the liquid phase and its divergence at the dynamical transition temperature $T_{\rm d}$. In addition, the dynamics of these models can be easily investigated with Monte Carlo simulations (for out-of-equilibrium dynamics see for instance (Barrat and Zecchina, 1999; Montanari and Ricci-Tersenghi, 2004)). The equilibrium correlation times have been thus determined in (Montanari and Semerjian, 2006*a*), for a variation of the diluted p-spin model with constant degrees for the spins, and allowed to justify the claim of almost optimality of Eq. (12.8) without additional hypotheses. Indeed for $T > T_{\rm d}$ one finds that the equilibrium correlation time remains finite in the thermodynamic limit, and diverges algebraically when

[2]The validity of this argument, which is found for instance in (Vannimenus, 1981; Straley, 1982), can be checked on the simple example of a particle freely diffusing on a Bethe lattice.

$T \to T_{\rm d}^+$. Hence, the lower bound of Eq. (12.8) could only been improved to include a dynamical exponent z and be promoted to $\tau \geq \ell^z$, with probably $z \geq 2$ for a large class of models. On the contrary, in the low-temperature phase, $T < T_{\rm d}$, it is the upper bound of Eq. (12.8) that is saturated (up to an improvement of the constant C_2): the equilibrium correlation time diverges exponentially with the size N of the system, and the correlation volume is itself proportional to N.

Coming back to the diluted p-spin model with Poissonian degrees, let us mention a particular feature that is apparent in the sketch of Fig. 12.4. The dynamic transition line $T_{\rm d}(\alpha)$ reaches the zero-temperature axis at a finite value $\alpha_{\rm d}$. The neighborhood of the point $(\alpha = \alpha_{\rm d}, T = 0)$ thus exhibits a rich crossover phenomenon between low-temperature activated dynamics controlled by energy barriers and positive temperature transitions of the schematic mode-coupling theory kind (which is an exact description of the $\alpha \to \infty$ limit (Bouchaud *et al.*, 1996)). We refer the reader to (Montanari and Semerjian, 2006*b*) for more details.

12.3.5 Connection with the replica symmetry-breaking formalism

We shall close this section on mean-field diluted systems with a discussion of the relationship between the computation of the point-to-set correlation function and the usual cavity formalism at the level of one step of replica symmetry breaking (1RSB), first unveiled in (Mézard and Montanari, 2006). The existence of such a connection should be expected by the reader accustomed to the fully connected p-spin model: we argued that in the phase diagram sketched in Fig. 12.4 the transition line $T_{\rm d}(\alpha)$, defined as the point where the point-to-set correlation length diverges, can be meaningfully called the dynamic transition. Indeed, the lower bound in Eq. (12.8) implies a divergence of the equilibrium correlation time on this line. However, the large α limit of the diluted p-spin model gives back the fully connected one. For the latter the dynamic transition has been studied for a long time, and its appearance is known to be related to the static properties of the model studied by various methods, in particular the replica and the TAP ones. One of these characterizations is the existence of a non-trivial solution of the 1RSB equations with the Parisi breaking parameter $m = 1$.

One can indeed recover this equivalent definition of the dynamic transition in generic mean-field diluted systems. Let us return to the computation on a finite tree presented in Section 12.3.3. We introduced there the distribution $P^{\sigma_i}_{i\to a}(h'_{i\to a})$ as the probability of $h^{\underline{\sigma}_B}_{i\to a}$ with respect to the choice of the boundary configuration $\underline{\sigma}_B$ conditioned on σ_i. One can instead define the unconditional version of this distribution, $P_{i\to a}(h'_{i\to a})$, which is related to the conditional one by

$$P_{i\to a}(h'_{i\to a}) = \sum_{\sigma_i} \frac{e^{\beta h_{i\to a}\sigma_i}}{2\cosh(\beta h_{i\to a})} P^{\sigma_i}_{i\to a}(h'_{i\to a}). \tag{12.24}$$

A similar definition holds for $Q_{a\to i}(u'_{a\to i})$. A computation whose details can be found in (Mézard and Montanari, 2006) shows that the conditional equation (12.21) translates into

$$P_{i\to a}\left(h'_{i\to a}\right) = \frac{1}{\mathcal{Z}_{i\to a}} \int \prod_{b\in\partial i\setminus a} \mathrm{d}Q_{b\to i}\left(u'_{b\to i}\right)\, \delta\left(h'_{i\to a} - \sum_{b\in\partial i\setminus a} u'_{b\to i}\right)$$
$$\times \left(\frac{\cosh\left(\beta h'_{i\to a}\right)}{\prod_{b\in\partial i\setminus a} \cosh\left(\beta u'_{b\to i}\right)}\right). \qquad (12.25)$$

This is simply the usual 1RSB equation with $m = 1$ (Mézard and Parisi, 2001) (in general the last term in parenthesis, called the reweighting factor or free-energy shift, is raised to the power m). A similar observation holds for the equation on $Q_{a\to i}$ obtained from Eq. (12.23). The 1RSB equations are usually solved on a random factor graph without any explicit boundary B. In consequence, the existence of a non-trivial solution of the usual 1RSB equations with $m = 1$ is equivalent to the presence of long-range point-to-set correlations (with a fictitious boundary sent to infinity).

This computation also provides an alternative interpretation of the 1RSB equations for mean-field diluted systems. $P_{i\to a}$ is usually considered as the distribution of an effective field with respect to the choice of a *pure state* of the system. This can be intuitively defined for a random graph of large but finite size as a portion of the configuration space sufficiently disconnected from the other pure states, yet a clear-cut mathematical definition is not easy to handle. One can alternatively interpret $P_{i\to a}$ as a distribution over the boundary conditions of a tree of very large depth. This is precisely the computation done for the point-to-set correlation, i.e. the 1RSB equations with $m = 1$, where the boundary condition is drawn itself from the Gibbs–Boltzmann distribution. One can, moreover, generalize this interpretation to arbitrary values of m, biasing accordingly the probability distribution of the boundary conditions (Montanari *et al.*, 2008).

12.4 Kac models

A different class of systems in which the computation of PS correlation is possible is provided by Kac models. Kac models are classical tools of statistical physics to study the relationship between finite-dimensional systems and their mean-field counterparts (Kac, 1959; Lebowitz and Penrose, 1966, 1971). In these models one considers variables interacting through a potential of growing range r_0 and intensity decreasing in such a way as to keep finite the total interaction strength of one particle with the surrounding environment. For any finite values of r_0 the general properties of finite-dimensional system such as convexity of free-energy and absence of infinite life metastable states hold. There properties break down in mean-field theory, which holds exactly in the limit of weak long-range forces in which r_0 scales as the linear system size L and both lengths tend to infinity together. The relation between the two regimes can be understood by studying the so-called Kac limit (Kac, 1959), where one considers a large interaction range $r_0 \to \infty$ but only after having taken the thermodynamic limit $L \to \infty$. Remarkably, one finds that while the mean-field predictions hold in single-phase regions, some of the typical pathologies of mean-field-theory, such as the non-convex free energies are removed and the Maxwell construction emerges naturally. The resulting theory is an improved mean-field theory that incorporarates the spatial dimension, with the possibility of describing inhomogeneous configurations and interfaces. For some systems (see, for instance, Refs. (Bovier, 1998; Lebowitz *et al.*, 1998)) it has been possible to devise asymptotic expansions around $r_0 = \infty$ that have led to the analysis of decay of metastable states or proofs of the existence of phase transitions. It is natural in the context of glassy systems to address the study of correlation lengths in the Kac limit. This provides a solvable limit case to confront with phenomenological theories based on mean-field ideas (Kirkpatrick and Wolynes, 1987; Kirkpatrick *et al.*, 1989; Biroli and Bouchaud, 2004*b*) and gives rise to a detailed picture with two relevant length scales.

12.4.1 Definition

Disordered models with Kac interactions were first described in (Froehlich and Zegarlinski, 1987), the case of spin-glass models with pair interactions. Following the general philosophy of this chapter, here we are interested in more general models with p-body interactions providing a minimal finite dimension modification of the fully connected p-spin model (12.12).

We consider then spins σ_i in the d-dimensional hypercube $i \in \Lambda = \{1, \dots, L\}^d$ with periodic boundary condition. The variables σ_i can be of various nature, for example they can be Ising variables $\sigma_i = \pm 1$ or real variables subject to some local spherical constraint, as we will specify below. The Kac p-spin Hamiltonian is defined by (Franz and Toninelli, 2004*a*)

$$E(\underline{\sigma}) = - \sum_{i_1, i_2, \dots i_p \in \Lambda} J_{i_1, \dots, i_p} \sigma_{i_1} \dots \sigma_{i_p}, \tag{12.26}$$

where, as for the mean-field model, the coupling constants $J_{i_1,\dots,i_p}$ are independent centered Gaussian variables, but now differently from the mean-field model, the variance depends on the mutual distances between the indexes according to:

$$\mathbb{E}\left[J^2_{i_1,\dots,i_p}\right] = \frac{1}{r_0^{pd}} \sum_{k \in \Lambda} \psi\left(\frac{|k - i_1|}{r_0}\right) \dots \psi\left(\frac{|k - i_p|}{r_0}\right). \tag{12.27}$$

The function $\psi(x)$ can be chosen somewhat arbitrarily, provided it fulfills the following constraints:

- it is positive, $\psi(x) \geq 0$;
- it becomes negligible when its argument is much larger than 1;
- it is normalized such that $\int d^d x \; \psi(x) = 1$.

These conditions guarantee that only spins within distances of order r_0 have an effective direct interaction. The function ψ can also be used to define a locally spherical model that without changing the physics from the Ising case, allows important simplifications in analytic studies. This is done considering $\sigma_i \in \mathbb{R}$ subject to the conditions $\frac{1}{r_0^d}\sum_{j\in\Lambda}\psi\left(\frac{|i-j|}{r_0}\right)\sigma_i^2 = 1$ in all points of space i. Note that the disorder distribution implies that the Hamiltonian itself has a Gaussian distribution with covariance specified by

$$\mathbb{E}[E(\underline{\sigma}^{(1)})E(\underline{\sigma}^{(2)})] = \sum_i Q_i(\underline{\sigma}^{(1)},\underline{\sigma}^{(2)})^p\ , \tag{12.28}$$

where $Q_i(\underline{\sigma}^{(1)},\underline{\sigma}^{(2)}) = \frac{1}{r_0^d}\sum_j \psi\left(\frac{|i-j|}{r_0}\right)\sigma_j^{(1)}\sigma_j^{(2)}$ measures the similarity of the two configurations $\underline{\sigma}^{(1)}$ and $\underline{\sigma}^{(2)}$ on a ball of scale r_0 around i. Models that compound different values of p can be defined considering gaussian Hamiltonian with correlations $\mathbb{E}[E(\underline{\sigma}^{(1)})E(\underline{\sigma}^{(2)})] = \sum_i \phi(Q_i(\underline{\sigma}^{(1)},\underline{\sigma}^{(2)}))$. The function $\phi(q)$ should satisfy suitable conditions to define a non-negative covariance matrix.

Following the original suggestion by Kac (Kac, 1959), the scaling of the interaction with r_0 is chosen in such a way to insure the existence of the limit of the free energy for $r_0 \to \infty$. Different ways of taking this limit allows study of the relation between systems with finite-range interactions and their mean-field counterpart. If r_0 is chosen to scale with the system size $r_0 = L \gg 1$ we have back the mean-field limit. All groups of p spins interact and the lattice structure and the space dimension become completely inessential. But one can consider a different procedure, called the Kac limit, where $r_0 \to \infty$ only after the thermodynamic limit $L \to \infty$. A requirement for mean field to have a qualitative relevance in the description of finite-range systems is a smooth r_0 crossover from $r_0 \ll L$ to $r_0 \sim L$ $(L \gg 1)$. Rigorous results connecting the Kac limit to the mean-field limit have been proved in (Franz and Toninelli, 2004*b*,*a*), under the technical hypothesis of parity of the function ϕ, $\phi(q) = \phi(-q)$. These findings can be informally and qualitatively stated as follows:

- The free energy of the Kac model tends for all temperatures to the free energy of the corresponding mean-field model in the Kac limit.
- In the same limit, the structure of the correlation functions on length scales of order of r_0 approaches the structure of mean-field correlations.

The first point is a continuity property, it shows that the mean-field limit is a good starting point in the evaluation of the free energy for large r_0. Of course this does not necessarily imply that the phase transitions found in mean-field are also present for some finite r_0. For example, it is clear that in 1D the free energy is an analytic function of temperature for all finite r_0, despite the fact that the large r_0 limiting function exhibits a phase transition.

The second point clarifies that the mean-field approximation is at least capable of capturing the local properties of the system. This is particularly interesting at low temperature where it implies that the system behaves as an ideal glass, at least locally on length scales of order r_0. The possibility that this description extends to scales much larger than r_0 in some finite space dimension remains an open question. To go beyond these general rigorous results and study quantitatively the behavior of correlation lengths, tools of theoretical physics and non-rigorous techniques are needed, as will be explained in the next sections.

12.4.2 Effective potential: a Landau free-energy functional for glasses

In this section we would like to review the basic tool of the *glassy effective potential* (Franz and Parisi, 1995, 1997, 1998), which has been instrumental in the computation of point-to-set correlation lengths in the Kac model (Franz and Montanari, 2007).

The effective potential is a Landau kind of free energy as a function of an order parameter explicitly devised to be sensitive to stable or metastable glassy states. While this construction is in principle very general, for the sake of simplicity we first expose it in the case of fully connected models, and later explain how to extend the analysis to Kac models. The basic idea is that metastable states can be studied through restrictions of the Boltzmann–Gibbs measure to the surroundings of a representative configuration. One then observes that in glassy systems where equilibrium is composed of an extensive number of metastable states, any equilibrium configuration can be taken as a representative configuration. Let us then choose a random configuration $\underline{\sigma}^0$ according to the equilibrium measure, and consider a second system $\underline{\sigma}$ that is in a constrained equilibrium where the overlap with the reference configuration $q(\underline{\sigma}^{(0)}, \underline{\sigma}) = \frac{1}{N}\sum_i \sigma_i \sigma_i^0$ is fixed to some pre-assigned values p.[3]

The free energy of this second system, for fixed $\underline{\sigma}^{(0)}$ is

$$F[p, \underline{\sigma}^{(0)}] = -\frac{T}{N} \log Z[p, \underline{\sigma}^{(0)}] \tag{12.29}$$

$$Z[p, \underline{\sigma}^{(0)}] = \sum_{\underline{\sigma}} e^{-\beta E(\underline{\sigma})} \prod \delta(q(\underline{\sigma}^{(0)}, \underline{\sigma}) - p). \tag{12.30}$$

In the thermodynamic limit this is self-averaging with respect to the quenched disorder in the interactions and the choice of the equilibrium configuration $\underline{\sigma}^0$,

$$F[p] = \mathbb{E}\langle F[p, \underline{\sigma}^{(0)}]\rangle, \tag{12.31}$$

where by $\langle \cdot \rangle$ we denoted the average with respect to the equilibrium distribution for $\underline{\sigma}^{(0)}$ and by $\mathbb{E}$ the average with respect to the quenched disorder. The function $F[p]$ can be used as a Landau free energy as a function of the order parameter to detect and study glassy states. It is customary to define the effective potential $V[p]$ subtracting the unconstrained value of the free energy, $V[p] = F[p] - F$. This is the large deviation function allowing evaluation of the probability $\mathbb{P}[p]$ of the overlap with a reference configuration

$$\mathbb{P}[p] = \exp\left(-\beta N V[p]\right). \tag{12.32}$$

The analysis of the effective potential, which requires the computation of the double average in Eq. (12.31) can be performed in mean-field theory through the replica method, or alternatively the cavity method. For the family of models we are interested in, the most general ansatz needed to describe $V[p]$ is of a 1RSB form that accounts

[3] Note that we used already the letter p to define the p-spin Hamiltonian. It will be clear from the context to which quantity the notation refers.

for the possibility of having multiple metastable states. In fully connected models this takes a simpler form with respect to the diluted models described in Section 12.3.5. This is just parametrized by the Parisi breaking parameter m and two real variables q_1 and q_0, which represent, respectively, the typical value of the overlap between configuration in the same metastable state and in different ones (Mézard *et al.*, 1987). The parameter m can be tuned to select different families of metastable states with different free energies. If the equilibrium phase is composed of an extensive number of metastable states, corresponding to a non-zero configurational entropy m takes the value $m = 1$.

In the spherical p-spin model, defined by the Hamiltonian (12.12), the effective potential can be computed in a closed form as a saddle point over q_1, q_0:

$$V[p] = \max_{q_1,q_0} \mathcal{V}(p, q_1, q_0, m) \tag{12.33}$$

$$\mathcal{V}(p, q_1, q_0, m) = -\frac{1}{2}\beta[2\phi(p) - m\phi(q_0) - (1-m)\phi(q_1)] + \frac{1}{2\beta}\left(\frac{p^2 - q_0}{1-(1-m)q_1 - mq_0}\right.$$
$$\left. + \frac{1-m}{m}\log[1-q_1] - \frac{1}{m}\log[1-(1-m)q_1 - mq_0]\right),$$

where we denoted $\phi(q) = \frac{1}{2}q^p$. The shape of the function $V[p]$, which depends on temperature, is sensitive to the existence of metastable states. Indeed, as can be seen in Fig. 12.8, one can identify three model-dependent temperatures of interest where qualitative changes take place. Two of them are the mean-field dynamical transition temperature T_d and the static transition temperature T_c. In addition, there is a third temperature T^*, with $T^* > T_\mathrm{d} > T_\mathrm{c}$, first identified in (Franz and Parisi, 1995, 1997, 1998). Above T^*, the function V is a convex function with a single minimum at $p = 0$. At T^* and inflection point appears, and below that temperature the potential is non-convex. For temperature between T_d and T^*, the function continues to have a single minimum for $p = 0$. At T_d a local minimum at a value $p = q_{EA} > 0$ develops. In the interval $[T_\mathrm{c}, T_\mathrm{d}]$, the point $p = 0$ is still the absolute minimum of V. The two-minima structure below T_d reflects the partition of the equilibrium measure in disjoint metastable states. The value q_EA is the typical overlap between configurations belonging to the same metastable state. For $p = q_\mathrm{EA}$ $\underline{\sigma}$ is in the state specified by $\underline{\sigma}^{(0)}$. Different metastable states have zero mutual overlap. For $p = 0$ all but the metastable state specified by $\underline{\sigma}^{(0)}$ contribute to the free-energy and $V(0) = 0$. Correspondingly, the difference in free energy between the two minima equals the system's configurational entropy $\Sigma_\infty(T)$ multiplied by temperature. The configurational entropy vanishes linearly on approaching T_c, $\Sigma_\infty(T) \sim T - T_\mathrm{c}$ and the two minima become degenerate. Below that temperature the mean-field model is in an ideal glassy state and the two minima remain degenerate.

We now generalize the previous construction to Kac models. As before, we use the overlap with an equilibrium reference configuration $\underline{\sigma}^{(0)}$ as an order parameter. In this case, however, we are interested in considering the free energy as a functional of an order-parameter profile that is space dependent. In view of considering large values

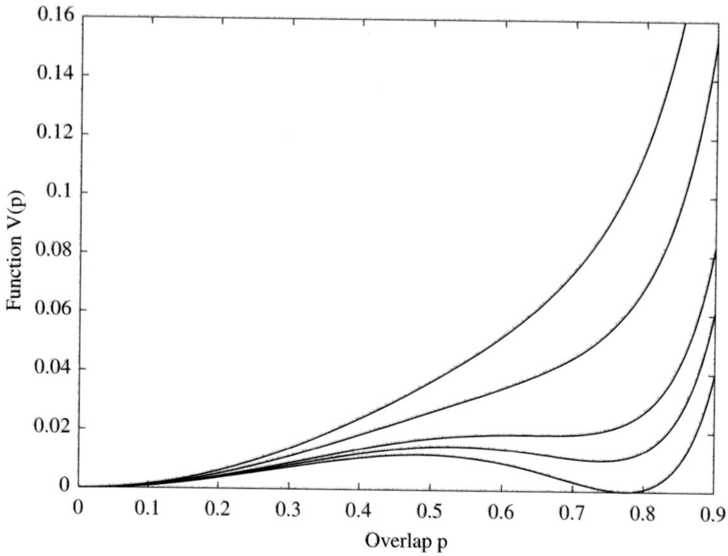

Fig. 12.8 The function $V(p)$ at different temperatures. For comparison with the case of the Kac model we consider a Hamiltonian with two-body and 4-body interactions with $\phi(p) = 1/2(0.1 \times p^2 + p^4)$. From top to bottom $T = 0.703486 > T^*$, $T = T^* = 0.633137$, $T = T_{\rm d} = 0.57525$, $T = 0.558049$, $T = T_{\rm c} = 0.541847$. The function is convex for $T > T^*$. It has an inflection point with positive slope for $T_{\rm d} < T < T^*$. In the interval $T_{\rm c} < T < T_{\rm d}$, $V(q)$ has a local minimum for a temperature-dependent value $p = q_{\rm EA}$. The difference $V(q_{\rm EA}) - V(0)$ is (T times) the bulk configurational entropy $V(q_{\rm EA}) - V(0) = T\Sigma_\infty(T)$.

of the interaction range r_0 it is natural to coarse grain the order parameter on scales δ such that $\delta \simeq dx\, r_0 \gg 1$ with $dx \ll 1$ (Alberti *et al.*, 1996; Franz and Toninelli, 2005). This allows to rescale all lengths $x = i/r_0$ that are then measured in units of the interaction range r_0. Let us denote $q_x(\underline{\sigma}^{(0)}, \underline{\sigma}) = \frac{1}{\delta^d} \sum_{i \in B_x} \sigma_i^{(0)} \sigma_i$ the local overlap in a square cell B_x of linear size δ labeled by x between two configurations $\underline{\sigma}^{(0)}$ and $\underline{\sigma}$. In complete analogy with the long-range case, one can define an effective potential functional $W[p(x)] = F[p(x)] - F$ as the average free-energy excess for imposing an overlap profile $p(x)$.

$$F[p(x)] = -T\mathbb{E} \log Z[p(x), \underline{\sigma}^{(0)}] \tag{12.34}$$

$$Z[p(x), \underline{\sigma}^{(0)}] = \sum_{\underline{\sigma}} e^{-\beta E(\underline{\sigma})} \prod_x \delta(q_x(\underline{\sigma}^{(0)}, \underline{\sigma}) - p(x)). \tag{12.35}$$

We can now relate the point-to-set correlation function to the effective potential. Consider the correlation function between point 0 and the set B considering profiles such that $p(x) = 1$ is fixed for $x \in B$. Observing that the probability of a profile can be written as $\mathbb{P}[p(x)] = \exp(-W(p(x)))$ we find

$$\langle\langle p(0)\rangle_{\underline{\sigma}_B^{(0)}}\rangle = \frac{\int_{\{p(x)=1\ ;\ x\in B\}} \mathcal{D}p(x)\ e^{-\beta W[p(x)]}p(0)}{\int_{\{p(x)=1\ ;\ x\in B\}} \mathcal{D}p(x)\ e^{-\beta W[p(x)]}} . \tag{12.36}$$

For general values of r_0 the evaluation of the large deviation functional $W[p(x)]$ is a formidable task. Fortunately, in the Kac limit $r_0 \to \infty$ the computation becomes simpler, it can be stated as a variational problem that generalize to spatially varying functions the saddle-point evaluation of Eq. (12.34). The region of temperatures $T > T_c$ can be analyzed through the 1RSB theory that we have discussed in several contexts in this chapter. This generalizes the description of the fully connected case in the fact that the overlap variational parameters become position-dependent functions $q_1(x)$ and $q_0(x)$ (the parameter m remains space independent). We give the explicit expression of the effective potential for the p-spin model (12.26) with local spherical constraint, where the dependence on the various parameters can be written in a closed form

$$W[p(x)] = \max_{q_1(x),q_0(x)} r_0^d \int d^dx \left(-\frac{1}{2}\beta \{2[\phi(p^\star(x)) - \phi(p(x))] - m [\phi(q_0^\star(x)) - \phi(q_0(x))] \right.$$
$$\left. -(1-m)[\phi(q_1^\star(x)) - \phi(q_1(x))]\} + \mathcal{V}(p(x), q_1(x), q_0(x), m)\right) . \tag{12.37}$$

We have introduced the notation $f^\star(x) = \int d^dy\ \psi(|x-y|)f(y)$ for the convolution with ψ and the function $\mathcal{V}(p, q_1, q_0, m)$ is given by the mean-field expression (12.34). Mutating the terminology from mechanics we will call action the argument of the max in Eq. (12.37). We notice that for profiles that are inhomogeneous on limited spatial extension W is proportional to the interaction volume r_0^d. The expression (12.37) can be simplified using a (physically harmless) lowest-order gradient expansion for the convolution $\phi(f^\star(x)) = \phi(f(x)) + c\phi'(f(x))\nabla^2 f(x)$ with $c = \frac{1}{2d}\int d^dz\ \psi(|z|)z^2$. This approximation does not affect the physics of the problem while the action takes the familiar field-theoretical form of the integral of a Lagrangian density, dependent on the various overlap parameters and their spatial derivatives. Notice, however, that for the expansion to be well behaved one needs $\phi''(q) > 0$ for all q. This is not verified for pure p-spin interactions if $p \geq 3$, which has $\phi''(0) = 0$. For this reason, the numerical computations based on this truncation have used functions ϕ of the kind $\phi(q) = \frac{1}{2}(q^4 + aq^2)$, which correspond to a mixed p-spin model with two-body and four-body interactions. For small enough values of a the model has the same phenomenology as the pure p-spin. Figures are presented for the value $a = 0.1$.

Thanks to the proportionality of the action to the interaction volume r_0^d, the functional integral in the expression (12.36) of the PS correlations can be evaluated by saddle points. We will then be interested in profiles $p(x)$ that are stationary points of the action $\frac{\delta W}{\delta p(x)} = 0$ for $x \in \bar{B}$ with boundary condition $p(x) = 1$ for $x \in B$. These include the absolute minima that represent equilibrium profiles and directly determine the PS correlation function, but also relative minima, saddle points and maxima that we will associate to metastable states and barriers.

The condition of equilibrium (12.6) discussed in Section 12.2.2 has interesting implications on the mutual relation between the overlap functions $q_1(x)$, $q_0(x)$ and

$p(x)$. Consider the local overlap $q_x(\underline{\sigma}', \underline{\sigma}'')$, between two replicas with the same constraint $\underline{\sigma}^{(0)}$ on B, and its kth moment $\mathbb{E}\langle q_x(\underline{\sigma}', \underline{\sigma}'')^k\rangle$. It is possible to show (Mézard *et al.*, 1987) that within the 1RSB formalism

$$\lim_{r_0 \to \infty} \mathbb{E}\langle q_x(\underline{\sigma}', \underline{\sigma}'')^k\rangle = (1-m)q_1^k(x) + mq_0^k(x), \tag{12.38}$$

where $q_1(x)$ and $q_0(x)$ are the solution of the maximization problem (12.37). On the other hand, according to Eq. (12.6), this expectation value should be simply equal to $\lim_{r_0 \to \infty} \mathbb{E}\langle q_x(\underline{\sigma}^{(0)}, \underline{\sigma}')^k\rangle = p(x)^k$, thus leading for all k and x to the identity $(1-m)q_1^k(x) + mq_0^k(x) = p(x)^k$. Two different ways of satisfying this identity are of physical relevance: (1) a simple solution describing equilibrium in a single metastable state, such that for all x

$$q_1(x) = q_0(x) = p(x). \tag{12.39}$$

(2) A solution with $m \to 1$, for which

$$\begin{aligned} q_0(x) &= p(x), \\ q_1(x) &> p(x). \end{aligned} \tag{12.40}$$

This solution (see below) is relevant when ergodicity is broken and the constrained equilibrium is composed by multiple metastable states with a non-zero configurational entropy Σ. According to general principles (Monasson, 1995) the configurational entropy can be computed from a solution with the form (12.40) as $\Sigma = -\beta \frac{1}{m^2}\frac{\partial W}{\partial m}|_{m=1}$.

Non-equilibrium stationary profiles $p(x)$, associated to transition states and barriers, will be relevant in our analysis. For these non-equilibrium solutions, there are no restrictions of principle on the functions $q_1(x)$ and $q_0(x)$. Of course, the condition $q_0(x) = q_1(x) = p(x)$ can still be used as an approximation. Despite the fact that in some regions this gives rise to clear physical inconsistencies we believe that it predicts the correct scaling of the free energy of metastable and barrier states with temperature and interaction range. More complex solutions, studied in (Dzero *et al.*, 2005) confirm this point.

Let us stick to the forms (12.39) and (12.40). In both mentioned cases, one gets great simplifications in the variational problem one needs to solve. For example, for the ergodic equilibrium profiles (12.39) the effective potential, using the mentioned gradient expansion for the convolution reads:

$$\begin{aligned} W[p(x)] &= r_0^d \int d^d x \left(-\frac{c\beta}{2}(\phi'(p(x))\nabla^2 p(x) + V(p(x)) \right) \\ V(p) &= -\frac{\beta}{2}\phi(p) - \frac{T}{2}\,[p + \log(1-p)]. \end{aligned} \tag{12.41}$$

In the 1RSB solution (12.40) an additional q_1-dependent term proportional to $1-m$ appears that allows amputation of the configurational entropy. It is natural to look for solutions respecting the symmetries imposed by the set B. We will discuss mainly the spherical geometry, where the set B is just the exterior of a spherical cavity,

$B = \{x: |x| > \ell\}$ and solutions should only depend on $|x|$. Another interesting choice is that of planar geometry, where $B = \{x: |x_1| > \ell\}$, with x_1 the first component of the d-dimensional vector x, solutions should only depend on x_1. In both cases it is interesting to study the behavior of the solutions as a function of the cavity size ℓ, that, remember, we measure in units of r_0 as all the lengths. Physical saddle points should be such that in both cases the profile is flat in the center of the cavity $\nabla p(0) = 0$. In both geometries the minimum equations for W can be easily numerically integrated, yielding solutions qualitatively independent from the physical dimension in the spherical case, and strictly independent of the physical dimension in the planar case. The main qualitative features can be obtained in the planar case where the overlap profile can be obtained by quadrature. In fact, this case is formally identical to the problem of a 1D Newtonian particle with coordinate p evolving in time in a potential equal to $-V(p)$. Unless otherwise stated the results we will discuss, and in particular the behavior of the different lengths that will be identified, are independent of the geometry and the dimension. We will denote $p(x;\ell)$ the solution $p(x)$ of the variational problem for the boundary conditions fixed at distance ℓ.

12.4.3 The physical picture

The nature of the solutions of the stationary point of Eq. (12.41) and the behavior of the PS correlations as a function of the cavity size ℓ reflect the shape of the mean-field potential $V(p)$ that we have discussed in the previous section.

In the "very high" temperature region $T > T^*$ for each value of ℓ, both in the spherical and planar geometries, there is a unique solution of the field equations with the imposed boundary conditions. As a function of ℓ the PS function $p(0;\ell)$ decreases continually from 1 to 0.

At $T = T^*$ an instability develops. Below that temperature one finds a temperature-dependent interval of lengths $I = [\ell_0, \ell_d]$ where the field equations admit three distinct solutions. For temperatures in the interval (T_d, T^*) both lengths are finite, while $\ell_d = \infty$ for $T < T_d$. The behavior of $p(0;\ell)$ for different temperatures is depicted in Fig. 12.9. The figure also shows the complete profile $p(x;\ell)$ as a function of x for specific values of ℓ and the temperature in the three solutions. According to the value of $p(0;\ell)$, we call the three profiles low-overlap solution (LOS), medium-overlap solution (MOS) and high-overlap solution (HOS). These have different free energies. The HOS and the LOS one correspond to free-energy absolute or relative minima, the MOS one is a maximum with respect to $p(0;\ell)$.

The structure of the solutions allows to identify three relevant lengths that grow for decreasing temperatures: two metastability lengths ℓ_0 and ℓ_d defining the interval in which three solutions exist, and a thermodynamic length ℓ_c, with $\ell_0 < \ell_c < \ell_d$, for which the HOS and the LOS have the same free energy. The presence of two free-energy minima separated by a barrier naturally interpreted a metastability phenomenon. The typical relaxation time depends on the size of the cavity ℓ and diverges exponentially for $r_0 \to \infty$ if $\ell_0 < \ell < \ell_d$.

We divide the remainder of this section in the discussion of the equilibrium solutions, the metastable solutions and the unstable solutions for $T < T^*$.

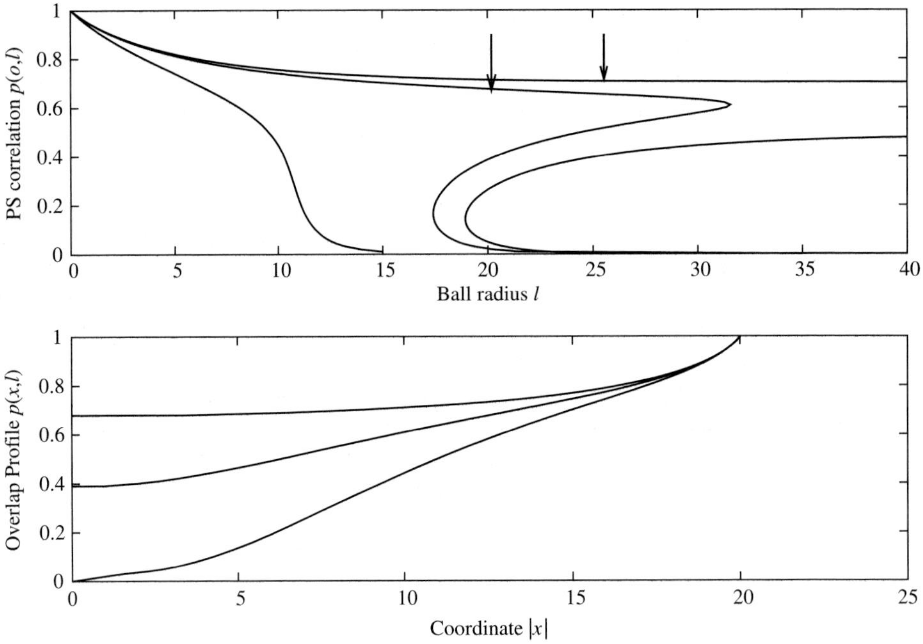

Fig. 12.9 Upper panel: values of the PS function $p(0, \ell)$ for three values of the temperature in the different solutions: $T = 0.676 > T^* = 0.633$, $T = 0.610$, in the interval $[T_{\mathrm{d}}, T^*] = [0.575, 0.633]$ and $T = 0.569$, in the interval $[T_{\mathrm{c}}, T_{\mathrm{d}}] = [0.541, 0.575]$. In the first case there is a unique solution for all values of ℓ. In the second case, three solutions are present for ℓ in the interval $[\ell_0, \ell_{\mathrm{d}}] = [17.31, 31.57]$. In the third case, three solutions are present for $\ell > \ell_0 = 19.47$. The arrows mark the points $\ell = \ell_{\mathrm{c}}$. Below ℓ_{c} the HOS is the thermodynamically favored state and the LOS is metastable, above that length the roles are interchanged. For the $T = 0.610$, $\ell_{\mathrm{c}} = 20.4$, for $T = 0.569$, $\ell_{\mathrm{c}} = 26.7$. Lower panel: The overlap profiles $p(x; \ell)$ corresponding to the three solutions for $\ell = 20$ and $T = 0.610$.

12.4.3.1 Equilibrium and metastable solutions

Let us consider the behavior of the system in spherical geometry as a function of the ball radius ℓ. Below the length ℓ_0 the HOS is the only minimizer of the free-energy functional, there are no metastable states poorly correlated with the boundary conditions. This absence has consequences in dynamics: any configuration in B chosen as initial condition for the dynamics, evolves in an r_0-independent time to the HOS. Above this length, in the interval $\ell \in (\ell_0, \ell_{\mathrm{c}})$ the LOS makes its appearance as a free-energy local minimum. Though the HOS is still thermodynamically dominant, metastable states exist such that if they are chosen as initial conditions in relaxation dynamics, they will take a time exponentially large in the interaction volume r_0^d to relax to equilibrium.

The difference in free energy between the HOS and the LOS decreases upon increasing ℓ, until, for $\ell = \ell_{\mathrm{c}}$ the two free-energy minima are degenerate. As we will discuss later, the length ℓ_c is growing for decreasing temperatures, and diverges for $T \to T_c$. Above this length the LOS becomes the dominating thermodynamic state. There is a phase transition in ℓ separating a phase strongly correlated with the boundary below ℓ_c, from an essentially uncorrelated phase above. The PS correlation function that should be identified with the value of $p(0;\ell)$ in the thermodynamically dominant solution exhibits a first-order jump in $\ell = \ell_c$. Below ℓ_c the system is thermodynamically in an ideal glassy amorphous state where, despite no sign of intrinsic order, the boundary conditions determine the bulk configuration. Conversely, above ℓ_c, the system is in a liquid situation where the boundary has little influence on the bulk. The presence of the transition is reminiscent of the analysis of PS function by Biroli and Bouchaud (Biroli and Bouchaud, 2004*b*), with ℓ_c thus identified with the mosaic length of phenomenological RFOT (Kirkpatrick and Wolynes, 1987; Kirkpatrick *et al.*, 1989). This theory postulates a competition between interfacial energy and configurational entropy, the contributions compensating exactly at the transition. However, as we have discussed for generic observables in Section 12.2.2 (see formula (12.4)), the average energy density for the constrained system is equal in all points of space to its unconstrained equilibrium value. This implies in particular that at the coexistence point both phases should have the same energy density, and therefore entropy, in each point of space. The transition should therefore depend on a different mechanism. Further analysis of the HOS and the LOS presented below reveals that it has the character of a configurational entropy crisis similar to the one that happens at T_c for the infinite system.

In the region $\ell \in (\ell_c, \ell_d)$, the HOS appears to be a local minimum of the free-energy functional. The length ℓ_d marks the limit of metastability of the HOS and was called the dynamical length in (Franz and Montanari, 2007). For $r_0 \to \infty$ and $\ell \in (\ell_c, \ell_d)$, a system prepared in the configuration $\underline{\sigma}^{(0)}$ at an initial time $t = 0$ would not be able to dynamically relax towards the thermodynamic equilibrium profile. For large but finite r_0 the relaxation time (for the dynamics in a ball of radius ℓ) to the LOS profile can be expected to behave exponentially, $\tau(r_0, \ell, T) \sim \exp\left(r_0^d \mathcal{B}(\ell, T)\right)$. In other words, for $r_0 \to \infty$, while the relaxation time is divergent, the barrier (counted in units of the interaction volume) $\lim_{r_0 \to \infty} \frac{1}{r_0^d} \log \tau(r_0, \ell, T) = \mathcal{B}(\ell, T)$ remains strictly positive. Conversely, for ℓ larger than ℓ_d the relaxation time would remain finite for $r_0 \to \infty$. ℓ_d appears as the minimal length for which relaxation is possible without barrier jumping. As Fig. 12.10 shows, the dynamic length increases monotonically and diverges as T decreases from T^* on approaching T_d. Generically, in the class of models we are interested in, it behaves as

$$\ell_d \sim (T - T_d)^{-\frac{1}{4}}, \tag{12.42}$$

at least below dimension 6. Interestingly, this scaling, first found in (Kirkpatrick and Wolynes, 1987; Kirkpatrick *et al.*, 1989), coincides with the one obtained for the length ℓ_4 quantifying the typical extension of collective motion in MCT (Donati *et al.*, 2002; Franz and Parisi, 2000; Biroli and Bouchaud, 2004*a*) relaxation processes. This seems natural since, in the $r_0 \to \infty$ limit a set of ideal inhomogeneous MCT equations can be shown to describe the dynamics of the Kac model (Franz, 2007). A simple scaling

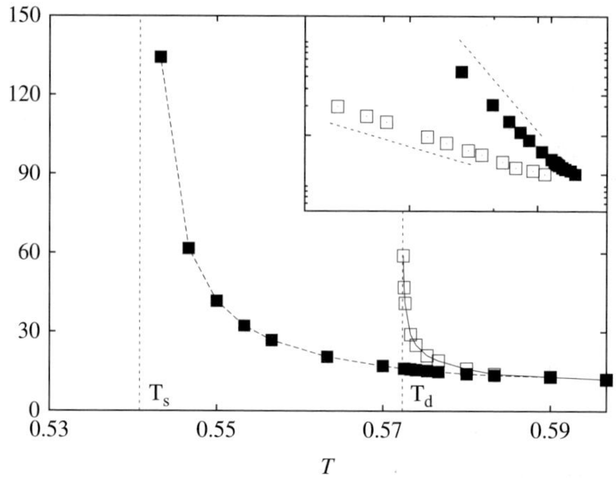

Fig. 12.10 Thermodynamic (filled squares) and dynamic (empty squares) lengths ℓ_c and ℓ_d. The vertical lines correspond to the dynamic and thermodynamic glass transitions $T_c = 0.541847$ and $T_d = 0.57525$. The lengths behave, respectively, as $\ell_c \sim (T - T_c)^{-1}$ close to T_c and as $\ell_d \sim (T - T_d)^{-1/4}$ close to T_d. In the inset, the lengths as functions of $(T - T_d)$ (for ℓ_d) and $(T - T_c)$ (for ℓ_c). The dotted lines have slope (respectively) 1/4 and 1.

argument accounts for the above behavior in the present context. For temperatures close to T_d and dimension smaller than $d_c = 6$, the properties of the solutions only depend on V in the vicinity of the inflection point (Muratov and Vanden-Eijnden, 2004). The potential can then be expanded to the cubic order around p_d, its inflection point at T_d, as $V(p,T) = V(p_d,T) + \epsilon a(p - p_d) + b(p - p_d)^3$, where generically the coefficient of the linear term is $\epsilon = T - T_d$, while a and b are positive model-dependent parameters with a weaker temperature dependence. Writing $p - p_d = \varphi$ this gives rise to a cubic field theory

$$W[\varphi] = r_0^d \int d^d x \left(\frac{c}{2}(\nabla \varphi^2(x)) + a\epsilon\varphi + b\varphi^3 \right), \tag{12.43}$$

whose critical properties dominate the behavior of observable quantities for $\epsilon \to 0$. In particular, equating the order of magnitude of the different terms in the Lagrangian $\epsilon\varphi \sim \varphi^3 \sim (\nabla\varphi)^2$ immediately leads to a length $\ell_d \sim \epsilon^{-1/4}$. It has been noticed in the context of spinodal theories that the expansion (12.43) cannot be used above dimension 6, and the critical point has a non-universal character (Muratov and Vanden-Eijnden, 2004).

The nature of the LOS is clarified considering the 1RSB solution for $m = 1$ (12.40) we mentioned in the previous section. In the region $\ell > \ell_d$ the LOS is the unique free-energy minimizer and it represents an ergodic state. Conversely, for $\ell_c < \ell < \ell_d$ this solution describes a non-ergodic state composed by a collection of $\mathcal{N} \sim \exp\left(r_0^d \ell^d \Sigma(\ell, T)\right)$ metastable states. The free-energy difference between the HOS and the LOS, which, as previously observed, should be purely entropic, in fact coincides with (T times) the configurational entropy in the constrained system, denoted

$\ell^d\Sigma(\ell, T)$. In other words there is an exponentially large number $\mathcal{N}$ of metastable states compatible with the boundary condition. The HOS just represents one of them, on the same basis as all the others. The behavior of $\Sigma(\ell, T)$ is depicted in the lower panel of Fig. 12.11, which shows curves growing with ℓ, crossing the value 0 for $\ell = \ell_c$. If $T > T_d$ the curves stop at the dynamical length ℓ_d beyond which the system recovers ergodicity. For $T < T_d$ the curves tend to the bulk value of the mean-field configurational entropy $\Sigma_\infty(T)$. The dependence of $\Sigma(\ell, T)$ upon ℓ can in both regions be described in a good approximation as $\Sigma(T, \ell) = A(T)(1 - \ell_c/\ell)$, where for $T < T_d$, $A(T) = \Sigma_\infty(T)$. Upon decreasing the values of ℓ towards ℓ_c, the transition to the confined state is driven by the reduction to zero of the configurational entropy. This is decreased by a term that scales as ℓ^{d-1} relative to the total configurational entropy $\ell^d\Sigma(\ell, T)$.

Let us mention without entering in a detailed discussion that in addition to the solutions we discussed, low-overlap RSB solutions with $m \neq 1$ (not respecting the conditional equilibrium conditions) were also found in (Franz and Montanari, 2007). These represent families of metastable states in the system with free-energy higher than the

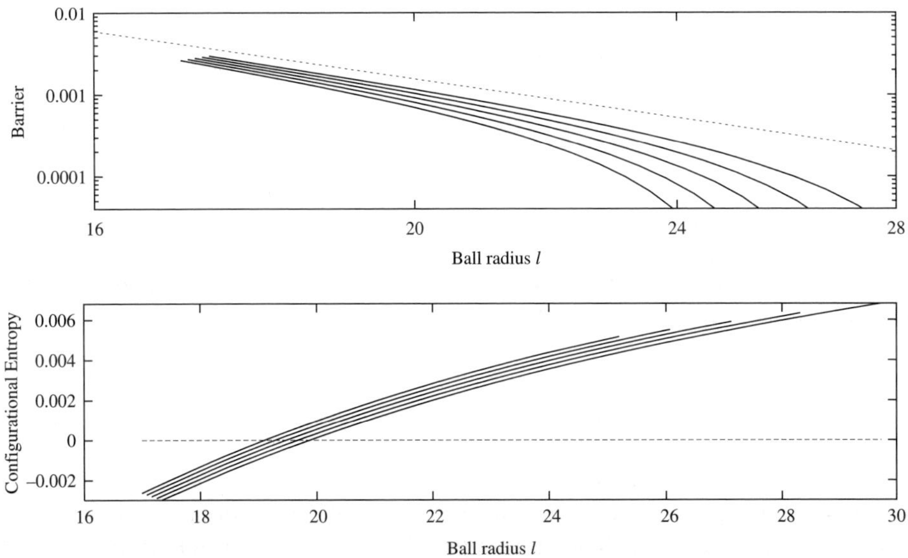

Fig. 12.11 Upper panel: The barrier density as a function of the ball radius ℓ for different temperatures. From bottom to top $T = 0.5819, 0.5816, 0.5812, 0.5809, 0.5805$, all larger than $T_d = 0.57525$. Lower temperatures correspond to higher barriers. The barrier goes to zero for $\ell \to \ell_d$. The dotted line allows the comparison with the power law ℓ^{-6} expected from scaling close to T_d. Lower panel: Configurational entropy density as a function of ℓ for the same temperatures as in the upper panel. The higher curves correspond to the higher temperatures. The configurational entropy curves touch zero at the mosaic length ℓ_c and terminate at the dynamical length ℓ_d. Both lengths increase for decreasing temperatures. Below the length ℓ_d the correct solution should include RSB.

equilibrium one. For $\ell < \ell_c$ the solution with $m = 1$ gives a negative configurational entropy and is therefore inconsistent. The dominating metastable low-overlap state is one of the RSB $m \neq 1$ solutions just mentioned. It has zero configurational entropy and positive energy difference with the HOS.

The scenario for the various glass transitions as a function of ℓ in the LOS resembles the one of the mean-field or diluted p-spin models as a function of temperature. We have an ergodic phase for large $\ell > \ell_d$, an intermediate phase with a finite configurational entropy for $\ell_c < \ell < \ell_d$, and a zero configurational entropy 1RSB phase at low $\ell < \ell_c$. We notice that the point at ℓ_d has the character of a typical dynamical transition point of 1RSB systems. This has specifc consequences on the dynamical properties (for the constrained dynamics in a ball of radius ℓ) when ℓ gets close to ℓ_d, which should be similar to the one of the mean-field model close to T_d. In particular, for $r_0 \to \infty$, the relaxation time grows and diverges if the system size goes to ℓ_d from above. On the same basis one can draw consequences on the relaxation time within the metastable states dominating below ℓ_d. One can argue that the stability of these states decreases upon increasing ℓ until they become marginal exactly at ℓ_d (Barrat, Barrat). Accordingly, the relaxation time within a metastable state diverges if ℓ tends to ℓ_d from below.

As already mentioned the dynamics of the model in the Kac limit is exactly described by inhomogeneous MCT equations (Franz, 2007), so that the bulk relaxation time behaves as $\tau(T) \sim \epsilon^{-\gamma}$ close to T_d, with a model-dependent exponent γ (Reichman and Charbonneau, 2005; Das, 2004). For $\ell > \ell_d \sim \epsilon^{-1/4}$ one can expect the scaling

$$\tau(\ell, T) = \frac{1}{\epsilon^{\gamma}} \hat{\tau}(\ell/\ell_d), \tag{12.44}$$

where the scaling function $\hat{\tau}(u)$ tends to a constant at large arguments, and has a power-law behavior $\hat{\tau}(u) \sim (u-1)^{-\delta}$ for $u \gtrsim 1$. The exponent δ should be in principle derived from the dynamical MCT-like equations (Franz, 2007) and as γ should depend on the choice of the model. An analogous scaling should be expected below ℓ_d with $\tau(\ell, T)$ interpreted as the relaxation time within a state, for the dynamics inside a ball of radius ℓ.

The thermodynamic mosaic length ℓ_c, whose temperature behavior is shown in Fig. 12.10, displays no singularity at T_d. This is coherent with the fact that for any finite r_0 the dynamical transition is rounded-off and the relaxation time remains finite, and is in agreement with the inequalities (12.8) of Section 12.2.3. Its behavior becomes singular near the ideal glass-transition point $T \gtrsim T_c$. Here, the two minima of V are almost degenerate, and ℓ_s is large. The solutions of the field equations in spherical geometry for $d > 1$ can be obtained through the "thin-wall approximation" of nucleation theory (Langer, 1967, 1969). This self-consistently neglects the thickness of the interface region, where the solution passes form values of p close to q_{EA} to values close to zero, this thickness being small with respect to ℓ_c. Let us consider $\ell \gtrsim \ell_c$. The transition is driven by a bulk contribution to the configurational entropy $\Sigma_\infty(T)\ell^d$ and an almost temperature-independent surface-entropy reduction $-Y\ell^{d-1}$. The value of Y can be computed in analogy with standard nucleation theory as the action of the "instantonic solution" to the field equation that connects the two degenerate minima of $V(p)$ at $T = T_c$:

$$Y = S_d \int_0^{q_{EA}} dp \frac{2V(p)}{\sqrt{\frac{c}{2}\beta\phi''(p)V(p)}}\Bigg|_{T=T_c}, \tag{12.45}$$

where S_d is the surface of the unitary sphere in dimension d, $S_d = 2\pi^{d/2}\Gamma(d/2)$. The length ℓ_c is then found from the cancelation of the total configuration entropy,

$$\ell^d \Sigma(\ell, T) = \Sigma_\infty(T)\ell^d - Y\ell^{d-1}, \tag{12.46}$$

namely $\ell_c = Y/\Sigma_\infty(T)$. The length ℓ_c diverges at T_c, with the scaling

$$\ell_c \sim \frac{Y}{\Sigma_\infty(T)} \sim \frac{Y}{T - T_c}. \tag{12.47}$$

We will see in the next subsection that this is also the typical spatial extension of barrier states close to T_c. Notice that the negative contribution to Eq. (12.46) is proportional to the ball surface. As we discuss in section 12.4.3.4 phenomenological theories (Kirkpatrick and Wolynes, 1987; Kirkpatrick *et al.*, 1989; Biroli and Bouchaud, 2004*b*) suggest that interface reduction terms could scale as ℓ^θ with $\theta < d - 1$. As for all qualitative results we present the value $\theta = d - 1$ can be checked to be independent on the choice of the model in the 1RSB class or the various approximations like the thin-wall approximation or the gradient expansion of the interaction that we have used to simplify the analytic treatment. Our derivation, however, depends crucially on the Kac limit, that in our analysis preceeds the limit $T \to T_c$. It is certainly possible that for finite r_0 a non-trivial crossover could change this exponent.

12.4.3.2 The barrier state

Let us complete our analysis with a discussion of the medium overlap solution (MOS). In Fig. 12.9 the value of $p(0; \ell)$ plotted in the MOS has been obtained supposing conditional equilibrium as an approximation. Unfortunately, it is difficult to go beyond this solution using numerical integration. Since by definition the MOS is an unstable saddle of W, iterative methods that work well for minima, simply do not find the solution. Experience with the replica method shows, however, that such replica symmetric solutions often give good approximations for the free energy and other quantities even in regions where the exact description should include RSB.

We analyze below the regimes of temperatures close to T_d and T_c where scaling can be expected and analytic arguments allow us to go beyond numerical integration of the field equations. We believe that our approximate solution, while possibly failing in numeric pre-factors, gives back the correct scaling behavior describing the dependence of the MOS on system size and temperature for large r_0.

The MOS is interpreted physically as a dynamical barrier state to be overcome during relaxation (Franz, 2006). The equilibrium activation barrier $\mathcal{B}(\ell, T)$ can be estimated as the free-energy difference between the MOS and the HOS

$$r_0^d \mathcal{B}(\ell, T) = W_{\mathrm{MOS}} - W_{\mathrm{HOS}}. \tag{12.48}$$

This is, for all temperatures, a decreasing function of the system size ℓ. For temperatures in the range $[T_{\mathrm{d}}, T^*]$ the barrier vanishes at ℓ_{d}, coherent activation is not needed for relaxation in large systems. For temperatures below T_{d} activation is required at all scales. The barrier for bulk relaxation can be obtained as the large ℓ limit of $\mathcal{B}(\ell, T)$.

Close to T_{d} the behavior of the MOS and the HOS becomes independent of the details of the model. Since we do not expect RSB to affect the scaling, we can obtain the behavior of the various quantities through dimensional analysis of the cubic field theory of Eq. (12.43). The main results of this analysis are that

- For $T \gtrsim T_{\mathrm{d}}$ the barrier is non-zero for $\ell < \ell_{\mathrm{d}}$ and admits the scaling form

$$\mathcal{B}(\ell, T) = \ell^{d-6} b_+(\ell/\ell_{\mathrm{d}}), \tag{12.49}$$

  where the scaling function is such that the barrier vanishes linearly as $\ell \to \ell_{\mathrm{d}}$: $b_+(x) \approx \text{Const.} \times (1 - x)$ for $x \to 1$ and $b_+(x) = 0$ for $x > 1$. The behavior of the barrier in this region is shown in the upper panel of Fig. 12.11.
- For $T \lesssim T_{\mathrm{d}}$ the barrier admits the scaling form

$$\mathcal{B}(\ell, T) = |\epsilon|^{\frac{6-d}{4}} b_-(\ell|\epsilon|^{1/4}), \tag{12.50}$$

  where the function $b_-(x)$ is a decreasing function of x that admits a positive limit for $x \to \infty$. As already noticed, from Eq. (12.50) we can obtain the barrier for relaxation in the bulk as the limit of large ℓ of $\mathcal{B}(\ell, T)$. This coincides with a direct estimate of the barrier in (Franz, 2005; Dzero *et al.*, 2005) where boundary conditions are imposed at infinity and scales as $\mathcal{B}_\infty(T) \sim |\epsilon|^{\frac{6-d}{4}}$. The scaling (12.50) shows that the asymptotic value is reached on lengths of order $\ell_{\mathrm{b}} \sim |\epsilon|^{-1/4}$. This is then the typical spacial extension of dominating barrier modes governing the relaxation in the bulk, which becomes large on approaching T_{d} from below.

Let us now turn to temperatures close to the static transition temperature T_{c}. Analogously to the case of the equilibrium solutions of the previous subsection, for small $T - T_{\mathrm{c}}$ nucleation arguments can be used to study the barrier. For large ℓ, the barrier solution $p(x; \ell)$ has the shape of a bubble of small overlap values in a sea where $q(x; \ell) \approx q_{\mathrm{EA}}$. This is surrounded by a crown around the boundary of $\bar{B}$, where p decays from 1 to $\approx q_{EA}$. For $d > 1$ the typical bubble radius ℓ_{b}, and the barrier value can be just estimated from the point of maximum of expression (12.46).[4] This gives, as announced, a common scaling for ℓ_{b} and ℓ_{c},

$$\ell_{\mathrm{b}} \sim \frac{Y}{\Sigma_\infty(T)} \sim (T - T_{\mathrm{c}})^{-1}, \tag{12.51}$$

[4]The 1D case requires a special analysis and one finds that $\ell_{\mathrm{b}} \sim -\log(T - T_{\mathrm{c}})$.

and for the barrier the behavior

$$\mathcal{B}(\ell, T) \sim Y^d(\Sigma_\infty(T))^{-(d-1)} \sim (T - T_c)^{-(d-1)}, \tag{12.52}$$

which is divergent for $d > 1$ when $T \to T_c$. Equation (12.52) has the form of a modified Adams–Gibbs inverse proportionality relation between the relaxation barrier and the configurational entropy (Adams and Gibbs, 1965). In turn, using $\Sigma_\infty(T) \sim T - T_c$, one finds a modified Vogel–Fulcher law[5] with an exponent that depends on dimensionality, with associated lower critical dimension $d_{\text{LCD}} = 1$. We note the coincidence of this result in the Kac limit with the naive scaling analysis of (Kirkpatrick and Wolynes, 1987; Kirkpatrick *et al.*, 1989). The comment we made at the end of Section 12.4.3.1 on the θ exponent can be verbatim applied here to the modified Vogel–Fulcher exponent.

12.4.3.3 Correlations in wall geometry

Before concluding our discussion on the Kac limit, let's have a closer look at the three solutions for large $\ell \gg \ell_c$ and $T \gtrsim T_c$. The large-ℓ limit is equivalent to imposing the boundary condition on a single planar wall. This condition has been the object of recent numerical experiments (Scheidler *et al.*, 2000, 2002, 2004) and deserves a short comment.

The question here is up to which distance the boundary has an effect on the configuration of the system, i.e. what is the typical distance from the boundary over which $p(x; \ell)$ decays to zero and if this distance coincides with ℓ_c. The behavior of the three solutions in this regime can be easily worked out from the field equation. The metastable HOS never loses correlations: after a rapid decay from 1 on the wall, which is common with the MOS and the LOS, stabilizes to $p(x; \ell) \simeq q_{\text{EA}}$. The unstable MOS persists to values $p(x; \ell) \approx q_{\text{EA}}$ up to distances of order ℓ from the wall. It decays to zero at a distance ℓ_c from the center and is similar to a ball of radius ℓ_c of small overlap in a sea of large overlap values. More interesting is the behavior of the LOS that gives the equilibrium profile. In this case, the overlap decays to low values over distances of the order of a "wetting" length $\ell_w \sim -\log(T - T_c)$ from the boundary before decaying to values close to zero. We conclude that the semi-infinite wall geometry fails to identify the correlations ℓ_c, while it identifies a much smaller length ℓ_w.

12.4.3.4 Beyond the Kac limit

The results presented so far are strictly valid in the Kac limit. We briefly discuss here some possible scenarios for r_0 large but finite.

We would like first to discuss the rounding off of the dynamical (MCT) transition. The analysis can be performed in analogy with ferromagnetic models with Kac interactions below the lower critical dimension where "finite-range scaling" (FRS) holds (Pelissetto and Vicari, 2002; Luijten and Binder, 1998). The basic hypothesis of FRS is that the interaction range acts as a cutoff to critical behavior in a similar way as finite size does for usual critical phenomena. This idea has been employed in a 1D Kac spin glass with a continuous transition in the Kac limit, to predict the rate of

[5] The usual Vogel–Fulcher form posits a barrier proportional to $(T - T_c)^{-1}$.

growth of the correlation length with r_0 at the mean-field critical temperature of the model (Franz and Parisi, 2006). The $r_0 \to \infty$ dynamical critical point has a mean-field character, so that one can expect mean-field scaling. An observable O that depends on T, r_0 and ℓ, should be described by scaling as:

$$O(T, r_0, \ell) = \epsilon^{-y_O} \hat{O}\left(r_0^a \epsilon^{1/4}, \ell \epsilon^{1/4}\right). \tag{12.53}$$

The scaling function $\hat{O}$ is such as to cutoff both the singularity that appear for $r_0 \to \infty$ at $\ell = \ell_{\rm d}$ for $\epsilon > 0$ and the one at $\ell \to \infty$ and $\epsilon = 0$. The value of the exponent a could be guessed if one used the potential functional as an ordinary field theory beyond the Kac limit for which it has been derived. Simple rescaling of the cubic action (12.43) leads then to $a = \frac{d}{6-d}$, implying that the crossover from mean-field MCT behavior to non-critical behavior happens on length scales of the order of $\ell_1 = r_0^{\frac{d}{6-d}}$. The same conclusion could be reached by analyzing the behavior of the barrier for $T \lesssim T_{\rm d}$ and $\ell \lesssim \ell_{\rm d}$, which scales as $r_0^d \mathcal{B}(\ell, T) \sim r_0^d \ell^{d-6}(1 - \ell/\ell_{\rm d})$. This analysis, however, relies on a continuum approximation of the potential $W[p(x)]$, which for finite r_0 neglects spatial fluctuations of the quenched disorder and heterogeneities in the reference configuration $\underline{\sigma}^{(0)}$. It has been recently observed that finite-size scaling around mean-field critical points in disordered systems can be non-trivial due to effective fluctuations of the critical temperature (Sarlat *et al.*, 2009; Franz *et al.*, 2010). Analogously here we can expect that effective local fluctuations of the critical temperatures can affect FRS. A criterion for the validity of naive FRS can be obtained in analogy with the analysis of FSS in disordered systems that in turn is an application of the Harris criterion (Harris, 1974) for relevance of disorder in phase transitions.

Let us rewrite the naive FRS of the barrier for temperatures slightly smaller than $T_{\rm d}$, which for $\ell \to \infty$ we write as

$$r_0^d \mathcal{B}_\infty(T) \sim r_0^d |\epsilon|^{\nu_{\rm MF}(d_{\rm c}-d)}. \tag{12.54}$$

In our case, $\nu_{\rm MF} = 1/4$ and $d_{\rm c} = 6$; the form (12.54) offers the possibility of studying more general cases. The scaling (12.54) gives a characteristic crossover length $\ell_1 \sim r_0^{\frac{d}{d_{\rm c}-d}}$ beyond which deviations from mean-field theory are to be expected (remember that lengths are measured in units of r_0). One can expect violations of this naive FRS when, on scale ℓ_1 the typical fluctuations of the critical temperature are larger then the average deviation $|\epsilon|$. The critical temperature fluctuations, related to local disorder fluctuations, can be expected to be of the order $(r_0 \ell_1)^{-d/2}$. Applying the above criterion, we find that naive FRS is violated if $r_0^{-\left(\frac{d}{(d_{\rm c}-d)\nu_{\rm MF}}\right)} \ll r_0^{-\frac{d}{2}\frac{d_{\rm c}}{(d_{\rm c}-d)}}$ i.e. if $d_{\rm c}\nu_{\rm MF} < 2$. It is notable that this relation coincides with the criterion for the validity of naive FSS for disordered systems above the upper critical dimension. In our case $d_{\rm c}\nu_{\rm MF} = 3/2$ and so that naive scaling cannot not be expected to hold. In order to understand the nature of the crossover for finite r_0 a deep analysis of the effect of critical temperature fluctuations is needed (Franz *et al.*, 2010). This goes beyond the scope of the present review and won't be attempted here.

The second point we would like to discuss is the behavior of the barrier close to $T_{\rm c}$. The Kac-limit prediction is that the configurational entropy reduction in a finite volume of linear size $\ell > \ell_c$ is proportional to the surface, $\delta\Sigma\ell^d = -Y\ell^{d-1}$, which, as observed, leads to a modified Vogel–Fulcher law for the relaxation time $\tau \sim \exp\left(r_0^d(T-T_c)^{-(d-1)}\right)$. Of course this does not necessarily imply an ideal glass transition if r_0 is finite. It just tells us that for growing r_0 and fixed temperature $T \leq T_c$ the relaxation time grows faster than $\exp\left(r_0^d\mathcal{B}\right)$ for any $\mathcal{B} > 0$. The result is compatible both with the absence of a singularity at finite temperature or with a singularity that could be weaker than the one found in the Kac limit. In phenomenological RFOT it is supposed that the interface exponent is renormalized by the effect of fluctuations, yielding $\delta\Sigma\ell^d = -Y\ell^\theta$ with $\theta \leq d-1$. In Ref. (Kirkpatrick and Wolynes, 1987; Kirkpatrick *et al.*, 1989), scaling arguments in favor of the value $\theta = d/2$ have been put forward, which leads to the canonical Vogel–Fulcher form in all dimensions $d \geq 2$. Correspondingly, the static length ℓ_c would behave as $(T-T_c)^{-d/2}$ rather than $(T-T_c)^{-1}$ as found for $r_0 \to \infty$. If these suggestions are correct, for large but finite r_0 there should be a crossover between the two regimes. A Ginzburg criterion involving r_0 as well as $T-T_c$ should govern this crossover.[6] Differently from the crossover close to T_d the formalism we have used does not seem to suggest a mechanism for this change of behaviour.

12.5 Conclusions

This chapter concerns the inclusion of spatial aspects in glassy theory. In particular, the emphasis has been put on recently proposed measures of correlations—the point-to-set correlations—that involve an infinite number of variables.

In the first part of the chapter we reviewed exact general bounds that relate the growth of a relaxation time to the growth of a correlation length. We first introduced with a progressive mathematical rigor the idea of point-to-set correlation functions and the associated correlation lengths, and explained to that extent this definition allows us to reconcile the phenomenology of glasses with the intuitive association between growing time- and length scales. These bounds do not spoil the interest of studying PS correlations in glassy systems. The lower and upper bounds have different forms, which correspond to different physical mechanisms of relaxation. The lower bound is, up to a non-trivial exponent, of the critical dynamics type, while the upper bound has the form of activated dynamics.

In the second part of this chapter we discussed this correlation length in the context of the mean-field picture of glassy phenomena provided by 1RSB disordered models. We have thus reviewed some analytical works dedicated to growing length scales in simplified models of glasses and their puzzling relationship with growing time scales. We discussed diluted random graph models and Kac like models that add new elements to the already rich behavior of fully connected models of the p-spin family. For random graph models we found that the divergence of the point-to-set length accompanies the one of the relaxation time at the dynamical transition.

[6] Note that if we ask here that the fluctuations of the critical temperature on a correlated volume are smaller than $T-T_c$ we get the condition $(T-T_c) \gg \frac{1}{(r_0\ell_c)^{d/2}}$, which is always verified for $\ell_c \sim (T-T_c)^{-1}$.

The study of the Kac model reveals a rich scenario with two relevant correlation lengths, the mosaic length below which a system behaves thermodynamically as an ideal glass, and a dynamical length quantifying the typical extension of relaxation modes that do not require activation. At the dynamical transition the former remain finite, coherently with the fact that for any finite interaction range the transition is rounded-off, while the latter diverges.

Let us close this discussion with a few open questions.

The upper bound in Eq. (12.8) implies that the divergence of the equilibrium correlation time is necessarily accompanied by a divergence of the point-to-set correlation length, even though with a possibly much slower form of the divergence. Both in diluted random graph models, and in spin models in the Kac limit, the divergence of the point-to-set correlation length at $T_{\rm d}$ cannot be detected by the study of static n-point correlation function, for any finite n. Hence, in these models there is a true separation between the dynamic transition and the thermodynamic one, the free energy having a singularity only at a lower temperature $T_{\rm c}$. An important open question is to determine whether this phenomenon of growth of the point-to-set correlation function without a trace in the two-point function can persist in finite-dimensional models or if this is an artifact of mean-field models. In the latter case it would mean that bounds of the form (12.8) holds with ℓ being replaced by a standard two-point correlation length (and different constants $C_{1,2}$). This should be important from a theoretical point of view, yet would not contradict the experimental and numerical situation. On the range of correlation time where measurements are possible the growth of the static correlations implied by this hypothetical extension of Eq. (12.8) could still be very weak.

Other questions relate to the study of finite-range disordered models beyond the Kac limit. An accomplished theory of finite-range scaling close to the dynamical temperature $T_{\rm d}$ should enable us to theoretically describe the cross-over between mode-coupling regime to activated dynamics. A related question is the connection between the purely static correlation length discussed here and the dynamical one extracted from four-points (two-time, two-location) dynamical susceptibilities. As we have discussed, the dynamical length found in MCT above $T_{\rm d}$ can be detected in the Kac limit by analyzing metastable solutions of the field equations in finite geometry. In this case, dynamic and static correlation length do not coincide. It would be important to better understand the quantitative relationship between these two characterizations of the correlations in a system.

Acknowledgments

We warmly thank Andrea Montanari, Giorgio Parisi and Fabio-Lucio Toninelli for fruitful collaborations on several works presented here.

References

Adams, G. and Gibbs, J.-H. (1965). *J. Chem. Phys.*, 43.

Alberti, G., Bellettini, G., Cassandro, M., and Presutti, E. (1996). *J. Stat. Phys.*, **82**, 743.

Barrat, A. (1997) *arXiv:cond-mat/970103*, unpublished.

Barrat, A. and Zecchina, R. (1999). *Phys. Rev. E*, **59**, R1299.

Berthier, L., Biroli, G., Bouchaud, J.-P., Cipelletti, L., El Masri, D., L'Hôte, D., Ladieu, F., and Pierno, M. (2005). *Science*, **310**, 1797.

Biroli, G. and Bouchaud, J.-P. (2004*a*). *Europhys. Lett.*, **67**, 21.

Biroli, G. and Bouchaud, J.-P. (2004*b*). *J. Chem. Phys.*, **121**, 7347.

Biroli, G., Bouchaud, J.-P., Cavagna, A., Grigera, T. S., and Verrocchio, P. (2008). *Nature Phys.*, **4**, 771.

Bouchaud, J.-P., Cugliandolo, L.F., Kurchan, J., and Mézard, M. (1996). *Physica A*, **226**, 243.

Bovier, A. (1998). *J. Stat. Phys.*, **91**, 459.

Cammarota, C., Cavagna, A., Gradenigo, G., Grigera, T. S., and Verrocchio, P. (2009). *preprint: arXiv:0906.3868*.

Castellani, T. and Cavagna, A. (2005). *J. Stat. Mech.*, P05012.

Cavagna, A., Grigera, T. S., and Verrocchio, P. (2007). *Phys. Rev. Lett.*, **98**, 187801.

Cocco, S., Dubois, O., Mandler, J., and Monasson, R. (2003). *Phys. Rev. Lett.*, **90**, 047205.

Cugliandolo, L. F. (2002). *Dynamics of Glassy Systems, Les Houches Summer School LXXVII.* Springer, Berlin.

Das, S. (2004). *Rev. Mod. Phys.*, **76**, 785.

Donati, C., Douglas, J. F., Kob, W., Plimpton, S. J., Poole, P. H., and Glotzer, S. C. (1998). *Phys. Rev. Lett.*, **80**, 2338.

Donati, C., Franz, S., Glotzer, S. C., and Parisi, G. (2002). *J. Non-Cryst. Solids*, **307**, 215.

Dzero, M., Schmalian, J., and Wolynes, P. G. (2005). *Phys. Rev. B*, **72**, 100201.

Franz, S. (2005). *J. Stat. Mech.*, P04001.

Franz, S. (2006). *Europhys. Lett.*, **73**, 492.

Franz, S. (2007). *J. Stat. Phys.*, **126**, 765.

Franz, S. and Montanari, A. (2007). *J. Phys. A*, **40**, F251.

Franz, S. and Parisi, G. (1995). *J. Phys. I (France)*, **5**, 1401.

Franz, S. and Parisi, G. (1997). *Phys. Rev. Lett.*, **79**, 2486.

Franz, S. and Parisi, G. (1998). *Physica A*, **261**, 317.

Franz, S. and Parisi, G. (2000). *J. Phys.: Condens. Matter*, **12**, 6335.

Franz, S. and Parisi, G. (2006). *Europhys. Lett.*, **75**, 385.

Franz, S., Parisi, G., Ricci-Tersenghi, F., and Rizzo, T. (2010) *preprint: arXiv:1001.1746*.

Franz, S. and Toninelli, F. L. (2004*a*). *J. Phys. A*, **37**, 7433.

Franz, S. and Toninelli, F. L. (2004*b*). *Phys. Rev. Lett.*, **92**, 030602.

Franz, S. and Toninelli, F. L. (2005). *J. Stat. Mech.*, P01008.

Fredrickson, G. H. and Andersen, H. C. (1984). *Phys. Rev. Lett.*, **53**, 1244.

Froehlich, J. and Zegarlinski, B. (1987). *Commun. Math. Phys.*, **112**, 553.

Harris, A. B. (1974). *J. Phys. C*, **7**, 1671.

Hayes, T. P., Sinclair, and A. (2007). *Ann. Appl. Probab.*, **17**, 931.

Jack, R. L. and Garrahan, J.P. (2005). *J. Chem. Phys.*, **123**, 164508.

Kac, M. (1959). *Phys. Fluids*, **2**, 8.

Kirkpatrick, T. R. and Wolynes, P. G. (1987). *Phys. Rev. B*, **36**, 8552.
Kirkpatrick, T. R., Thirumalai, D., and Wolynes, P. G. (1989). *Phys. Rev. A*, **40**, 1045.
Kob, W. and Andersen, H. C. (1993). *Phys. Rev. E*, **48**, 4364.
Kschischang, F., Frey, B. J., and Loeliger, H.-A. (2001). *IEEE Trans. Informat. Theory*, **47**, 498.
Langer, J. S. (1967). *Ann. Phys.*, **41**, 108.
Langer, J. S. (1969). *Ann. Phys.*, **54**, 258.
Lebowitz, J. L., Mazel, A. E., and Presutti, E. (1998). *Phys. Rev. Lett.*, **80**, 4701.
Lebowitz, J. L. and Penrose, O. (1966). *J. Math. Phys.*, **7**, 98.
Lebowitz, J. L. and Penrose, O. (1971). *J. Stat. Phys.*, **3**, 211.
Lindvall, T. (2002). *Lectures on the Coupling.* Dover Publications, New York.
Luijten, E. and Binder, K. (1998). *Phys. Rev. E*, **58**, 4060.
Martinelli, F., Sinclair, A., and Weitz, D. (2004). *Commun. Math. Phys.*, **250**, 301.
Mézard, M. and Montanari, A. (2006). *J. Stat. Phys*, **124**, 1317.
Mézard, M. and Montanari, A. (2009). *Information, Physics and Computation.* Oxford University Press, Oxford.
Mézard, M. and Parisi, G. (1999*a*). *Phys. Rev. Lett.*, **82**, 747.
Mézard, M. and Parisi, G. (1999*b*). *J. Phys. Condens. Matter A*, **11**, 157.
Mézard, M. and Parisi, G. (2001). *Eur. Phys. J. B*, **20**, 217.
Mézard, M., Parisi, G., and Virasoro, M. A. (1987). World Scientific, Singapore.
Mézard, M., Ricci-Tersenghi, F., and Zecchina, R. (2003). *J. Stat. Phys.*, **111**, 505.
Monasson, R. (1995). *Phys. Rev. Lett.*, **75**, 2847.
Monasson, R. (1998). *J. Phys. A*, **31**, 515.
Montanari, A. and Ricci-Tersenghi, F. (2004). *Phys. Rev. B*, **70**, 134406.
Montanari, A., Ricci-Tersenghi, F., and Semerjian, G. (2008). *J. Stat. Mech.*, P04004.
Montanari, A. and Semerjian, G. (2006*a*). *J. Stat. Phys.*, **125**, 23.
Montanari, A. and Semerjian, G. (2006*b*). *J. Stat. Phys.*, **124**, 103.
Mossel, E. (2004). *Survey: Information flow on trees, in Graphs, Morphisms and Statistical Physics, DIMACS series in discrete mathematics and theoretical computer science*, J. Nestril and P. Winkler eds. *155*, `arXiv:math/0406446`.
Muratov, B. and Vanden-Eijnden, E. (2004). *J. Stat. Phys.*, **114**, 605.
Parisi, G. (2002). *Glasses, Replicas and all that, Les Houches Summer School LXXVII.* Springer, Berlin.
Pelissetto, A. and Vicari, E. (2002). *Phys. Rep.*, **368**, 549.
Reichman, D. R. and Charbonneau, P. (2005). *J. Stat. Mech.*, P05013.
Sarlat, T., Billoire, A., Biroli, G., and Bouchaud, J.-P. (2009). *J. Stat. Mech.*, P08014.
Scheidler, P., Kob, W., and Binder, K. (2000). *Europhys. Lett.*, **52**, 277.
Scheidler, P., Kob, W., and Binder, K. (2004). *J. Phys. Chem. B*, **108**, 6673.
Scheidler, P., Kob, W., Binder, K., and Parisi, G. (2002). *Phil. Mag. B*, **82**, 283.
Sciortino, F. and Tartaglia, P. (2001). *Phys. Rev. Lett.*, **86**, 107.
Straley, J. P. (1982). *J. Phys. C: Solid State Phys.*, **15**, 2333.
van den Berg, J. (1993). *Commun. Math. Phys.*, **152**, 161.
Vannimenus, J. (1981). *Z. Phys. B*, **43**, 141.
Viana, L. and Bray, A. J. (1985). *J. Phys. C: Solid State Phys.*, **18**, 3037.
Zamponi, F. and Parisi, G. (2009). *J. Stat. Mech.*, P03026.
Zarinelli, E. and Franz, S. (2010). *J. Stat. Mech.*, P04008.